AF411169

Recent Advances in Cereals, Legumes and Oilseeds Grain Products Rheology and Quality

Recent Advances in Cereals, Legumes and Oilseeds Grain Products Rheology and Quality

Editor

Georgiana Gabriela Codină

MDPI • Basel • Beijing • Wuhan • Barcelona • Belgrade • Manchester • Tokyo • Cluj • Tianjin

Editor
Georgiana Gabriela Codină
Ștefan cel Mare University
Romania

Editorial Office
MDPI
St. Alban-Anlage 66
4052 Basel, Switzerland

This is a reprint of articles from the Special Issue published online in the open access journal *Applied Sciences* (ISSN 2076-3417) (available at: https://www.mdpi.com/journal/applsci/special_issues/Cereals_Legume).

For citation purposes, cite each article independently as indicated on the article page online and as indicated below:

LastName, A.A.; LastName, B.B.; LastName, C.C. Article Title. *Journal Name* **Year**, *Volume Number*, Page Range.

ISBN 978-3-0365-3146-5 (Hbk)
ISBN 978-3-0365-3147-2 (PDF)

Contents

About the Editor

Georgiana Gabriela Codină (Professor Habilitate, PhD Eng.) joined the Ștefan cel Mare University, Faculty of Food Engineering (Suceava, Romania) in 2005, where she teaches different courses in the food science and technology field. She obtained her PhD in Industrial engineering in 2009 and became a PhD supervisor in food engineering in 2017, when she sustained her habilitation thesis. Dr. Codină has expertise in food rheology, food sensory analysis, bread making, beer industry, food quality analysis through different rheological, textural, and sensory methods, as well as in the design of experiments and data analysis. Her research activities focus on improving the technology of food products and the quality of different foods, such as baked goods, beer, dairy products, etc. She has been involved in more than 10 interdisciplinary research projects, published more than 130 scientific papers, and authored of more than 35 patents under evaluation, 5 of which have been published.

Editorial

Recent Advances in Cereals, Legumes and Oilseeds Grain Products Rheology and Quality

Georgiana Gabriela Codină

Faculty of Food Engineering, "Ştefan cel Mare" University of Suceava, 13 Universitatii Street, 720229 Suceava, Romania; codina@fia.usv.ro

Citation: Codină, G.G. Recent Advances in Cereals, Legumes and Oilseeds Grain Products Rheology and Quality. *Appl. Sci.* **2022**, *12*, 1035. https://doi.org/10.3390/app12031035

Received: 31 December 2021
Accepted: 18 January 2022
Published: 20 January 2022

Publisher's Note: MDPI stays neutral with regard to jurisdictional claims in published maps and institutional affiliations.

Grains and the products obtained from them have a central importance in human nutrition, representing the main source of food for humans. The term grain encompasses multiple crop products such as cereals, legumes, oilseeds, pseudocereals and others. From these grains, cereals are the most used in the food industry in order to obtain different products such as bakery ones, beer, ethanol, starch, gluten, dextrose, e.g., In this Special Issue, the use of maize and sorghum, some of the most important cereal grains in the world as raw materials in the brewing process, whether as simple adjuvants or via the brewing of beers made from 100% sorghum or maize malt, are reviewed by Dabija et al. [1] who discussed the advantages and disadvantages of using them in brewing. In addition to the use of sorghum as raw material in brewing, it may also be used as a valuable ingredient in bakery industry, as was discussed in this Special Issue in the article by Apostol et al. [2], who demonstrated that the addition of sorghum seed flour in various percentages to wheat flour improves the nutritional value of bread by increasing its mineral, fiber and fat (monounsaturated fatty acids + polyunsaturated fatty acids) content. Moreover, from the technological point of view, the rheological behavior of composite dough obtained from sorghum and wheat flour was adequate for bakery products, which presented good quality characteristics. However, the use of a starter culture such as *L. plantarum* in bread made with different sorghum seed flour addition, further improves the bread quality especially from the nutritional and sensory point of view, compared to the bread samples in which only sorghum flour was added to wheat flour. Among maize and sorghum, the wheat and barley are also two of the most cultivated cereal grains in the world. Barley is the most used cereal for brewing [1], whereas wheat is the most used one for bakery products due to its unique viscoelastic properties we underlined in their article by Jańczak-Pieniążek et al. [3]. According to them, the wheat variety influences in a significant way the bread quality from the technological and quality point of view. Grains of different hybrid wheat cultivars grown in various climatic and agronomic conditions may be also used for the production of bread of a good quality, which leads to the conclusion that it can be considered an alternative in agricultural production for population cultivars.

Many research groups have focused on improving the bread quality especially from the nutritional point of view, by adding different ingredients of a high nutritional value or using different processing techniques, such as germination and fermentation. Bread reflects the nutritional value of the flour from which it is obtained, the ingredients used and of the technology applied to obtain it. It is an important source of proteins, B vitamins and mineral salts. However, through wheat refining, the nutritional value of bread decreased. Bread is the main source of vegetable protein, covering about 1/5–1/3 of the total protein requirement for the human body, but is deficient in essential amino acids, especially lysine, but also tryptophan and threonine. In order to correct this deficit, different ingredients may be used, such as legume protein concentrate as Belc et al. [4] have reported in this Special Issue. Through refined wheat flour substitution with pea and soy protein concentrate up to 15% addition, the bread nutritional quality was improved. Comparatively, bread obtained with soy protein concentrate presented higher quality characteristics from the

1

technological point of view (higher volume, softer crumb, stronger dough structure, higher sensory characteristics) than the bread obtained with pea protein concentrate. The protein digestibility does not dependent only on the protein source, but also on the food processing technique. The review written by Atudorei and Codină [5] underlined the impact of germination on the nutritional quality and sensory characteristics of different legume types such as bean, lentil, chickpea, lupine, soy and the effect of their addition in a germinated form in bread making. Germination induces changes in protein and starch digestibility (amino acid content, available sugars), an increase in the mineral bioavailability (calcium, copper, manganese, zinc), vitamin concentrations (riboflavin, niacin and ascorbic acid), polyphenols content and antioxidant activity and activates some hydrolytic enzymes such as amylases, proteases, lipases which may improve bread quality not only from the nutritional point of view but also from the technological point of view. The effect of germination combined with a fermentation process on dough rheological properties from different wheat and triticale varieties has also been reported in this Special Issue by Banu et al. [6]. According to them, flour obtained from germinated grains presented lower dough stability, higher protein weakening and a significant improvement of dough pasting properties when sourdough was added in wheat flour. Also, a narrower viscoelastic domain was recorded for samples subjected to the fermentation and germination process. The nutritional aspects related to the sprouts of soybean resulted from germination process and treated with different levels of chitooligosaccharide with various molecular weights have been approached in this Special Issue by Tang et al. [7]. According to them, the use of chitooligosaccharide during the germination process led to a significant increase in the bioactive compounds and antioxidant activity of the soybean sprouts. The use of a buckwheat sprout flour in a bakery product, namely a wheat bun, in combination with buckwheat, has been reported by Sturza et al. [8]. According to them, by adding the buckwheat and buckwheat sprout flours in wheat flour, the nutritional quality of buns has been improved in terms of starch digestibility, phenolic content, total flavonoid content and antioxidant activity. The use of ingredients such as pseudocereals in bread making has also become a concern in the bakery industry. One of the reasons for their use in bread is their nutritional value—these grains being richer in fibers, lipids, minerals and vitamins, compared to the wheat flour. Also, these grains present a high protein quality, thus completing the wheat proteins deficient in lysine, threonine and methionine. However, their use in bread making may be difficult, since these cereals do not contain gluten, fact that may affect bread quality. An extensive study related to the effect of the quinoa seed flour addition in wheat flour on dough rheology has been reported in this Special Issue by Coțovanu et al. [9]. According to them, the substitution of wheat flour with different milling fractions of quinoa seeds flour may affect in a different way the dough rheological properties and therefore the bread quality. The Falling Number values, water absorption, protein weakening decreased with the increasing particle size of quinoa seeds, whereas the Mixolab starch gelatinization rate and starch retrogradation speed increased. These authors predicted that the optimum bread may be obtained according to the rheological data for a medium quinoa particle size addition of 8.98% in wheat flour.

Nowadays, the bakery producers are trying to diversify the assortment range of bakery products in order to satisfy the consumer demands. Some of the consumers want special bakery products, healthier ones to meet their needs. For example, some of the consumers want bakery products to be gluten-free, others without sodium or with a low sodium content, others want a clean label, etc. These aspects were approached in this Special Issue by some research groups. Almost 1% of the world's population is affected by celiac disease and therefore they demand gluten-free bakery products. Prolamines are gluten components that are responsible for the immediate immune response. Celiac disease is not the only disease associated with gluten ingestion. In fact, gluten also causes other pathologies grouped under the term "gluten-related disorders". The only therapy to counteract gluten-related disorders, which are on the rise nowadays, is a gluten-free diet. This requires the production of bakery products from gluten-free raw materials, in which wheat flour is replaced by gluten-free flours, as Culetu et al. [10] reported in this Special Issue. In their article, they presented an overall view of different gluten-free flours: rice, brown rice, millet, maize, amaranth, teff, buckwheat, chickpea, quinoa, gram, plantain, tiger nut, e.g., which they compared from the functional properties and nutritional point of view. These may be valid alternatives to wheat flour in order to obtain gluten-free bakery products. However, there are challenges to producing gluten-free bakery products, due to the fact that gluten presents unique viscoelastic properties which lead to the good development of bakery products. A development of gluten-free bakery products is discussed in the article by Chiș et al. [11] published in this Special Issue. In their study they used quinoa flour as gluten-free flour which fermented with Lactobacillus plantarum ATCC 8014 (Lp) in order to obtain gluten-free muffins. Another challenge to obtain special bakery products for consumers who demand a low sodium diet was discussed in the article written by Voinea et al. [12], in which potassium chloride was proposed to partial substitute the sodium chloride from bread recipe. This was an interesting approach, since sodium chloride has a huge impact on dough rheology, yeast activity, bread quality especially from the sensory point of view. The article focuses only on the technological influence that sodium chloride substitution with potassium chloride may present on dough rheology. According to this study, the optimum replacement of sodium chloride with potassium chloride is of 22% in order to obtain the best dough rheological properties.

The quality of various bakery types have been reported in different ways in this Special Issue through baking loss, bread yield, bread volume, dallmann porosity, index of crumb, crumb moisture content [3], compositional analysis, loaf volume, crumb porosity, elasticity, colour parameters, texture properties, sensory ones through nose system, descriptive method [4], preferential method [2], free sugar content, total flavonoid content, total phenolic content, radical scavenging activity [8], carbohydrates, organic acids, folic acid, minerals [11], e.g., These methods are common ones used in the international literature to analyze bakery products quality from the nutritional and technological point of view. An innovative approach by using multivariate analysis as a statistical tool to evaluate bread quality in order to classified as a specific type of food product according to a standard database was reported in this Special Issue by Popescu et al. [13]. According to them, Karl Fischer titration can be a rapid and useful tool for a simple differentiation between various types of breads.

Last but not least, different research groups have focused in this Special Issue on the capitalization of by-products from the food industry in order to minimize losses resulting from the processing of agro-food raw materials. The review written by Chetrariu and Dabija [14] discussed the possibility of using brewer's spent grains, which were globally produced in 38.8 million tons in 2018 in food, as substrate for the cultivation of microorganisms and the production of enzymes, in different fermentation processes, for obtaining building materials, as adsorbent, as source of phenolic compounds, for biogas production, for food/composite packaging, for proteins, protein hydrolysates, bioactive peptides, as source of fiber, polymers, e.g., The review written by Petraru and Amariei [15] discussed the possibility of using the by-products obtained from the oil industry for extraction of bioactive compounds (protein isolate, concentrate and hydrolysate, antioxidants, dietary fiber), as substrate for functional ingredient production (enzymes, mushrooms, antibiotics, biosurfactants), animal feedstuff, in food products (bread, biscuits, snacks, desserts, dairy products), biopolymer packaging, e.g., The article written by Ursache and Gutt [16] presented the possibility to obtain bioethanol from wheat straw through different technological methods.

The works published in this Special Issue presented recent advances in cereals, legumes and oilseeds grain products rheology and quality in different directions of research, such as obtaining special food products such as gluten-free ones [10,11], with a lower sodium content [12], novel ones by partial replacement of wheat flour with other flour types [2,8,9], or by combining different ingredients [1,10] which are high nutritional quality by improving the quality of raw materials to be used [3,7] or of the bakery products, by using ingredients with a high nutritional value [4,5,9] and using different technological methods [6], etc. In addition, novel methods to evaluate bread quality [13] and by-products valorization through circular economy approaches [14–16] have been developed in an extensive way in this Special Issue.

Funding: This research received no external funding.

Institutional Review Board Statement: Not applicable.

Informed Consent Statement: Not applicable.

Data Availability Statement: Not applicable.

Acknowledgments: I want to thank all the authors and peer reviewers for their valuable contributions to this special issue. In addition, thanks to the MDPI management of *Applied Sciences* and their editorial support.

Conflicts of Interest: The author declares no conflict of interest.

References

1. Dabija, A.; Ciocan, M.E.; Chetrariu, A.; Codină, G.G. Maize and Sorghum as Raw Materials for Brewing, a Review. *Appl. Sci.* **2021**, *11*, 3139. [CrossRef]
2. Apostol, L.; Belc, N.; Gaceu, L.; Oprea, O.B.; Popa, M.E. Sorghum Flour: A Valuable Ingredient for Bakery Industry? *Appl. Sci.* **2020**, *10*, 8597. [CrossRef]
3. Jańczak-Pieniążek, M.; Buczek, J.; Kaszuba, J.; Szpunar-Krok, E.; Bobrecka-Jamro, D.; Jaworska, G. A Comparative Assessment of the Baking Quality of Hybrid and Population Wheat Cultivars. *Appl. Sci.* **2020**, *10*, 7104. [CrossRef]
4. Belc, N.; Duta, D.E.; Culetu, A.; Stamatie, G.D. Type and Amount of Legume Protein Concentrate Influencing the Technological, Nutritional, and Sensorial Properties of Wheat Bread. *Appl. Sci.* **2021**, *11*, 436. [CrossRef]
5. Atudorei, D.; Codină, G.G. Perspectives on the Use of Germinated Legumes in the Bread Making Process, A Review. *Appl. Sci.* **2020**, *10*, 6244. [CrossRef]
6. Banu, I.; Patraşcu, L.; Vasilean, I.; Horincar, G.; Aprodu, I. Impact of Germination and Fermentation on Rheological and Thermo-Mechanical Properties of Wheat and Triticale Flours. *Appl. Sci.* **2020**, *10*, 7635. [CrossRef]
7. Tang, W.; Lei, X.; Liu, X.; Yang, F. Nutritional Improvement of Bean Sprouts by Using Chitooligosaccharide as an Elicitor in Germination of Soybean (*Glycine max* L.). *Appl. Sci.* **2021**, *11*, 7695. [CrossRef]
8. Sturza, A.; Păucean, A.; Chiş, M.S.; Mureşan, V.; Vodnar, D.C.; Man, S.M.; Urcan, A.C.; Rusu, I.E.; Fostoc, G.; Muste, S. Influence of Buckwheat and Buckwheat Sprouts Flours on the Nutritional and Textural Parameters of Wheat Buns. *Appl. Sci.* **2020**, *10*, 7969. [CrossRef]
9. Coţovanu, I.; Batariuc, A.; Mironeasa, S. Characterization of Quinoa Seeds Milling Fractions and Their Effect on the Rheological Properties of Wheat Flour Dough. *Appl. Sci.* **2020**, *10*, 7225. [CrossRef]
10. Culetu, A.; Susman, I.E.; Duta, D.E.; Belc, N. Nutritional and Functional Properties of Gluten-Free Flours. *Appl. Sci.* **2021**, *11*, 6283. [CrossRef]
11. Chiş, M.S.; Păucean, A.; Man, S.M.; Vodnar, D.C.; Teleky, B.-E.; Pop, C.R.; Stan, L.; Borsai, O.; Kadar, C.B.; Urcan, A.C.; et al. Quinoa Sourdough Fermented with *Lactobacillus plantarum* ATCC 8014 Designed for Gluten-Free Muffins—A Powerful Tool to Enhance Bioactive Compounds. *Appl. Sci.* **2020**, *10*, 7140. [CrossRef]
12. Voinea, A.; Stroe, S.-G.; Codină, G.G. Use of Response Surface Methodology to Investigate the Effects of Sodium Chloride Substitution with Potassium Chloride on Dough's Rheological Properties. *Appl. Sci.* **2020**, *10*, 4039. [CrossRef]
13. Popescu, G.; Radulov, I.; Iordănescu, O.A.; Orboi, M.D.; Rădulescu, L.; Drugă, M.; Bujancă, G.S.; David, I.; Hădărugă, D.I.; Lucan, C.A.; et al. Karl Fischer Water Titration—Principal Component Analysis Approach on Bread Products. *Appl. Sci.* **2020**, *10*, 6518. [CrossRef]
14. Chetrariu, A.; Dabija, A. Brewer's Spent Grains: Possibilities of Valorization, a Review. *Appl. Sci.* **2020**, *10*, 5619. [CrossRef]
15. Ancuţa, P.; Sonia, A. Oil Press-Cakes and Meals Valorization through Circular Economy Approaches: A Review. *Appl. Sci.* **2020**, *10*, 7432. [CrossRef]
16. Ursachi, V.-F.; Gutt, G. Production of Cellulosic Ethanol from Enzymatically Hydrolysed Wheat Straws. *Appl. Sci.* **2020**, *10*, 7638. [CrossRef]

applied
sciences

Review

Maize and Sorghum as Raw Materials for Brewing, a Review

Adriana Dabija, Marius Eduard Ciocan, Ancuta Chetrariu and Georgiana Gabriela Codină *

Faculty of Food Engineering, Stefan cel Mare University of Suceava, 720229 Suceava, Romania;
adriana.dabija@fia.usv.ro (A.D.); marius_ec@yahoo.com (M.E.C.); ancuta.chetrariu@fia.usv.ro (A.C.)
* Correspondence: codina@fia.usv.ro; Tel.: +40-745-460-727

Abstract: Brewing is among the oldest biotechnological processes, in which barley malt and—to a lesser extent—wheat malt are used as conventional raw materials. Worldwide, 85–90% of beer production is now produced with adjuvants, with wide variations on different continents. This review proposes the use of two other cereals as raw materials in the manufacture of beer, corn and sorghum, highlighting the advantages it recommends in this regard and the disadvantages, so that they are removed in technological practice. The use of these cereals as adjuvants in brewing has been known for a long time. Recently, research has intensified regarding the use of these cereals (including in the malted form) to obtain new assortments of beer from 100% corn malt or 100% sorghum malt. There is also great interest in obtaining gluten-free beer assortments, new nonalcoholic or low-alcohol beer assortments, and beers with an increased shelf life, by complying with current food safety regulations, under which maize and sorghum can be used in manufacturing recipes.

Keywords: craft beer; gluten-free beer; functional beer; adjuvants; malted cereals

Citation: Dabija, A.; Ciocan, M.E.; Chetrariu, A.; Codină, G.G. Maize and Sorghum as Raw Materials for Brewing, a Review. *Appl. Sci.* **2021**, *11*, 3139. https://doi.org/10.3390/app11073139

Academic Editor: Wojciech Kolanowski

Received: 13 March 2021
Accepted: 29 March 2021
Published: 1 April 2021

Publisher's Note: MDPI stays neutral with regard to jurisdictional claims in published maps and institutional affiliations.

1. Introduction Cross-Reference

Brewing is a food process that began in the Middle East 10,000 years ago [1]. Today, at almost 200 billion liters a year, beer is one of the most commonly consumed low-alcohol beverages in the world—and, in terms of volume, after water and tea, the third most prevalent beverage in general [2–6]. The largest manufacturer of beer in the world is China, followed by the USA and Brazil. In Europe, Germany has the highest rates of beer consumption and production [7].

Dating back to antiquity, people have exploited the grains in their vicinity to develop fermented drinks [8]. Over the past hundred years, advancements in brewing have led manufacturers to use other types of cereals, in addition to barley malt or wheat malt. In any industry, innovation is the key to remaining competitive, and the beer industry is no exception. Consumers are always searching for new products on the marketplace—a novel brand, an original taste, eye-catching packaging, innovative technology, health benefits, quality improvements, etc. [9–11]. Barley is the most used cereal for brewing; however, unconventional malted grains have been used successfully. For instance: rice is used in Asia, maize is used in America, and millet and sorghum are used in Africa [12–14]. This process of replacing barley malt in beer production is increasing, and several factors shown in Figure 1 have contributed to this.

Because barley is used in such large quantities by all brewers, no matter their size, manufacturers often adopt various cost-cutting strategies. These can include the partial replacement of barley malt with adjuvants, considered supplementary sources of carbohydrates. The first adjuvants used in brewing were used to reduce the costs of obtaining the finished product [15]. For this reason, local raw materials—which may be available at lower costs and in higher quantities—can be used, provided they are in compliance with applicable legislation. Barley and maize are the most commonly used adjuvants in Europe as partial substitutes for malt [15–17].

Adjuvants are widely used in the beer industry (in variable proportions ranging from 10–50%) to provide additional sources of fermentable yeast carbohydrates, to improve foam stability, to change the color of beer, or to adjust the flavor of the finished product [18]. The literature states that 85–90% of world beer production is now produced with adjuvants, with wide variations on different continents. Between 10% and 30% of malt is substituted by unmalted materials in European countries; 40–50% or more is substituted in the United States and Australia, and in Africa the substitution is between 50 and 75% [17].

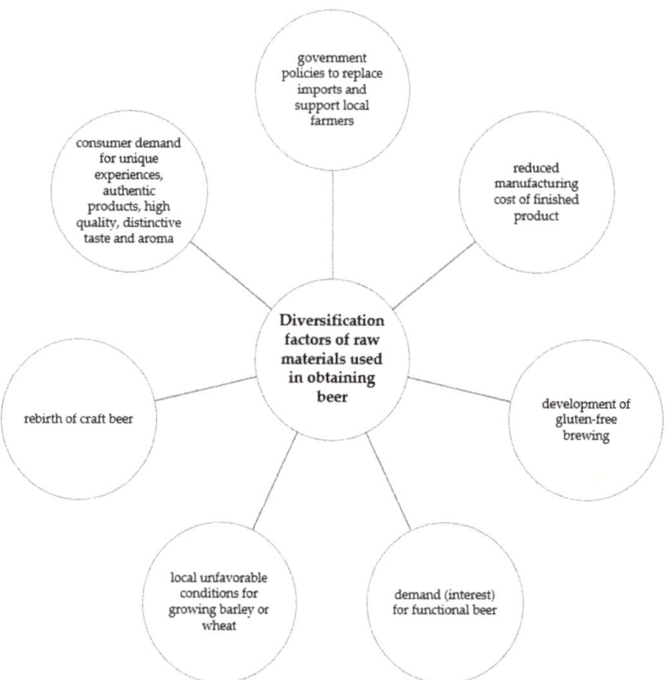

Figure 1. Diversification factors of raw materials for obtaining beer.

Beer adjuvants are ingredients—not including malted barley—that bring supplementary sources of protein and carbohydrates into the wort. Various cereals—e.g., wheat, barley, sorghum, corn, rice, and oats—are frequently used in various forms (whole grains, semolina, flakes, boiled, extruded, roasted/micronized, flour/starch, syrup, etc.) to obtain different beers [19]. For example, the use of maize as an adjuvant in the manufacture of a 30% beer assortment has led to an 8% reduction in total production costs [20]. In the Netherlands, Belgium, Luxembourg, and France, the production of beer with up to 40% unmalted cereals is permitted by law and has been put into practice. Contrastingly, beer production using higher concentrations of adjuvants (>40%) is often prohibited or impracticable. Replacing barley malt with adjuvants in beer manufacturing can potentially reduce the total cost of raw materials and obtain a distinctive beer taste or aroma [21]. Currently, the most common adjuvant of beer is rice, but as the demand for this cereal has risen severely, the quest for more adjuvants to substitute rice is ongoing [22].

Some adjuvants could feature compounds that may have beneficial effects on people's health—e.g., vitamins, minerals, and phenolic compounds such as phytoestrogens [23].

On the other hand, in recent decades, the beer industry has faced a phenomenon: the rebirth of craft beer, whose origins date back to the 1970s in the United States. The growing popularity of this beer submarket has benefited from innovation, creativity and authenticity. Craft beers are characterized as giving pleasure, joy, a feeling of identity, social recognition, self-satisfaction and sustainability. When brewing craft beer, the potential for improvement in unconventional starch ingredients is exploited to obtain a note of uniqueness, often by adding local fruits, herbs, spices and vegetables. These tactics are at the heart of local gastronomic traditions. Last but not least, the craft beer industry adds to the body of technical brewing knowledge by reinterpreting traditional styles [24–26].

Great creative innovations are possible when formulating beers. The options are plentiful; brewers only need to meet customer requirements and build a market founded on their needs. While craft beer producers have experimented to obtain new assortments, functional beer producers could similarly obtain high value products. An original product must be described in clear terms, including: its health benefits, the concentration of functional ingredients added, as well as its organoleptic profile, quality, safety and shelf life [4].

Another increasing production segment in the brewing industry is the manufacture of gluten-free beer. As stated by Codex Alimentarius and EU Regulation 41/2009 for gluten-free foods, gluten-free or non-gluten beer can be defined as beer with less than 20 mg/kg of gluten [27,28]. The manufacturing and marketing of gluten-free beers (<20 mg/L) and beers with extremely low gluten content (<100 mg/L) is still in its infancy, and the European market value is already estimated at several billion Euros per year. The majority of gluten-free beers need to use at least a fraction of malt obtained from pseudocereals and cereals that do not contain gluten or its precursors, including quinoa, amaranth, sorghum, buckwheat, maize, and rice [29]. Currently, research on the production of non-gluten beer from buckwheat malt, sorghum malt, quinoa malt and amaranth malt is the most developed. These types of malt contain high starch and moderate protein levels. Techniques for producing cereal beers other than barley and wheat are not yet well developed [30].

Beer producers have considered that the removal of gluten from beer malt is an expensive and complex process, which raises the price of the end product. On the other hand, in the USA, the Food and Drug Administration (FDA) does not allow beers made from gluten-containing raw materials to be considered non-gluten products [29].

The primary driver for research is the desire to expand efficient means of producing gluten-free malt beer. A crucial aspect in non-gluten beers is their taste; beers made from pseudocereals differ significantly from conventional beers in flavor profile. Techniques for the manufacture of non-gluten beers from other ingredients are therefore constantly being improved so that they can be marketed as beer products that comply with the regulations in force [31].

Unfavorable growing conditions for barley or wheat in some parts of the world (including those attributable to climate change) have led beer producers to the use of more versatile local raw materials to replace barley and wheat in conventional recipes. Extreme weather events can lead to substantial declines in global barley production, with a potential loss of 17% in the most severe conditions. This declining global barley supply could lead to higher beer prices, with a potential prince increase of 193% by 2099 [32]. This has led to the use of indigenous raw materials—often supported by government import substitution policies—and provided support to local farmers [33].

Consumers have become more and more aware that beer is a multifaceted product that has the potential to suit several settings and conditions. This may be why beer drinkers are especially inclined to experiment with authentic, high-quality products with distinctive tastes and aromas [25,33].

Maize and sorghum are among the most explored crops on earth, with multiple uses. These crops are grown in all agro-ecological areas of the world [34]. In this review, we address aspects of these two grains and how they might serve as possible substitutes for barley malt in brewing. Although obtaining an assortment of beer from cereals other than barley is simple, managing the efficiency of the economic process and creating a product acceptable to consumers presents a long-term research challenge.

2. Maize and Sorghum: Raw Materials for Brewing

In the process of obtaining beer, barley is the most common malted raw material in current use [35]. The advantages of using this cereal for brewing are well known. Barley malt production is among the oldest and most complex examples of applied biotechnology [36]. This review details two other cereals (maize and sorghum) which are currently used in brewing, emphasizing the characteristics that recommend them in this regard.

2.1. Overview

Maize is widely grown in over 166 countries, including all agro-ecological areas (arid, semi-arid, temperate and tropical) of the world, e.g., the tropics of Central America and Mexico, from whence it originates [34,37,38]. Main maize-producing countries include: United States, China, Brazil, Argentina, India, France, and Indonesia [39].

Maize (*Zea mays* L.) is a significant annual cereal crop belonging to the *Poaceae* family. *Zea* comes from ancient Greece (meaning "life support") and *mays* is a Latin word that means "giver of life". Maize is among the top three crops worldwide, after rice and wheat [40]. It is also known as the miracle crop and the queen of cereals, due to its high production potential [38]. Internationally, maize is grown on 184 million hectares, with 1016 million tons produced annually [41].

In addition to its use as human food and animal feed, maize is a source for a huge number of industrial products. Corn is grown in a multitude of varieties (e.g., white grain, the more common yellow grain, sweet corn, baby corn, waxy corn, popcorn, high amylase corn, flour corn, flint corn, high oil corn, dent corn, amylomaize, and quality protein maize) which appropriately meet nutritional demands. Corn varieties may differ in color, including yellow, white, red, and purple. Interest in the research and use of pigmented maize—which is full of anthocyanins, carotenoids and phenolic substances with antioxidant and bioactive characteristics—has increased due to its benefits on human health [38,39]. Maize has not been found in the wild in any part of the world. It survives only through human care [34].

Sorghum (*Sorghum vulgare*) originates from central Africa, and has reached Asia (including India). It belongs, like barley and maize, to the *Poaceae* family. Sorghum is closely related to corn in genomic organization, plant shape, developmental physiology, and even in its applications. Sorghum is the fifth most important cereal harvest in the world after corn, rice, wheat and barley, and serves as a major food grain for over 750 million people residing in the semi-arid tropical regions of Africa (Burkina Faso, Nigeria, Ethiopia, Sudan), Asia (China, India), and areas of Central and South America [21,30,42]. According to the FAOSTAT's (Statistics Division of Food Agriculture Organization of the United Nation) 2017 statistical data, Africa is the largest contributor to worldwide sorghum production, with approximately 29.7 million tons out of a total of 57 million tons [43,44].

Compared to other cereal crops, sorghum is more drought-tolerant; it is also called the camel plant. Therefore, this cereal is a vital staple food in many semi-arid areas of the developing world. In western countries, it is mainly consumed as animal feed [21]. Sorghum requires less fertilizer than corn. Sorghum is highly efficient in its use of heat from solar radiation, and uses water more efficiently than corn [36]. Sorghum is among the most versatile food crops in Africa, which has a rate of human consumption comprising about 40% of total world production [30]. During drought, sorghum rolls its leaves to reduce water loss due to perspiration. If the drought continues, it becomes latent rather than dying. The leaves are protected by waxy cuticles to reduce evaporation [42]. With continuing increases in the world's population, declining water supplies and the consequences of climate change, sorghum could be vital for human use and will be a significant crop in the future due to its drought resistance [43].

Usually, the color of sorghum grains varies from red, black and brown to coffee-colored, yellow and white. The variety of colors or pigments is due to the different quantity of polyphenols in the grain, which are usually located in the pericarp. To date, there are more than 10,000 types of sorghum, with more being generated through incessant plant growth research to choose desired properties. The majority of sorghum types vary in size, structure, pigmentation, texture, hardness and biochemical properties [45].

Sorghum also grows wild, especially on its continent of origin (Africa) [46].

2.2. Chemical Structure and Composition

Barley, maize and sorghum are part of the *Poaceae* family, all of which are monocotyledonous herbaceous plants with fasciculate roots and stems—formed by nodes and internodes—with spike-shaped inflorescence (Figure 2) [47].

Figure 2. Plants of maize and sorghum.

The grains are caryopses; barley grains have a straw-like, sticky coating, which protects the germs during the malting process, while maize and sorghum are bare grains, devoid of this coating. The average dimensions of barley grains are generally between 8–12 mm long, 3–5 mm wide and 2–4.5 mm thick. Depending on the variety and cultivation conditions, the weight of 1000 grains can vary between 37–45 g [48]. Barley grains are composed of endosperms (80–90%), embryos (2–5%), and shells (8–15%).

The maize grain is 2.5–22 mm long and 3–8 mm wide. Depending on the variety and cultivation conditions, the weight of 1000 grains vary greatly (between 30 and 1200 g). Maize grains are constituted of endosperms (82–83%), germs (10–11%), pericarps (5–6%), and peaks (0.8–1.0%) [49].

Sorghum grains are rounded and sharp with a diameter of 4–8 mm. They vary in size, shape and color depending on the variety of sorghum. The weight of 1000 grains varies between 20 and 60 g. These grains are composed of endosperms (80–84.6%), embryos (7.8–12.1%), and shells (7.3–9.3%) [50].

For the beer industry, the chemical composition of raw materials is particularly important. Table 1 summarizes the physicochemical characteristics of the maize, sorghum, and barley.

The composition of cereals (shown in Table 1) varies according to seed maturity, date of harvest, variety/species, soil and climatic conditions, crop management, storage, and drying conditions applied after harvest.

Starch is the main component of cereals, and is the main substance that is later converted to fermentable carbohydrates in beer wort. Therefore, one can determine the brewing qualities of a raw material by examining the starch content. This is important for the alcohol industry. The highest amount of starch is found in maize (62–80%), followed by sorghum (55.60–76.20%), and then barley (52.10–69.08%). The protein content varies between 8–15.25% for barley, 5.8–13.7% for maize, and 4.4–14.86% for sorghum. The fat content is 1.09–3% for barley and 2.2–5.91% for maize. Sorghum varies greatly, with a lipid content between 1.38–10.54%.

Table 1. Physicochemical characteristics of maize, sorghum and barley.

Grain	Characteristic, [% DM *]					Moisture [%]	References
	Starch	**Proteins**	**Lipid**	**Fiber**	**Ash**		
Barley	60	8–13	-	2–10	-	-	[51]
	65–68	10–17	2–3	11–24	1.5–2.5	-	[52]
	63–65	8–11	2–3	-	2	14–15	[19]
	62–64	11.09–14.68	2.01–2.35	18.7–19.5	-	-	[53]
	66.97–69.08	10.35–12.38	1.58–1.71	3.57–5.12	1.94–2.39	-	[54]
	59.50–60.98	14.53–15.25	1.82–1.87	2.85–3.25	2.42–2.52	-	[55]
	65.45–69.08	10.37–11.93	1.09–2.00	3.07–5.10	1.94–2.40	-	[56]
	52.1–64.4	8.7–13.1	2.2–3.5	13.6–23.8	2.0–2.6		[57]
Maize	71.88	8.84	4.57	2.15	2.33	10.23	[40]
	74.4–76.8	8.05–11.03	5.91	-	-	15	[39]
	76–80	9–12	4–5	-	3.87	10–14	[58]
	-	8.92–10	-	1.3–6.26	1.20–2.38	-	[41]
	70.99	9.21	5.10	2.21	1.05	11.44	[59]
	62–78	10	4.4	-	-	-	[33]
	71.7	9.5	4.3	2.6	1.4	-	[60]
	72–73	5.8–13.7	2.2–5.7	0.8–2.9	1.2–2.9	9.5–12.2	[61]
Sorghum	-	9.4	2.8	-	2.1	-	[13]
	61.0–74.8	9.0–13.5	2.8–4.8	-	1.2–1.8	9–12	[21]
	55.6–75.2	4.4–21.1	2.1–7.6	1–3.4	1.3–3.3	-	[42]
	65.15–75.2	6.23–14.86	1.38–10.54	1.65–7.94	0.90–4.20	1.39–19.02	[43]
	70.65–76.20	8.90–11.02	2.30–2.80	1.40–2.70	0.92–1.75	8.10–9.99	[62]
	-	12.5	3.30	1.7	1.9	9.8	[63]
	71.95	11.36	4.70	2.76	3.17	6.07	[64]
	64.3–73.8	8.19–14.02	2.28–4.98	1.41–2.55	1.46–2.32		[65]

* DM—dry matter.

According to the chemical composition of maize grains, carbohydrates and proteins are its key components play a significant role in the process of obtaining beer. The major chemical component of the maize kernel is starch, concentrated in the endosperm. The composition of maize starch is genetically controlled. Amylopectin and amylose are the main components of starch. Normal corn usually contains 25–30% amylose and 70–75% amylopectin. Nevertheless, maize types with amylose contents as high as 85% (amylomaize) and as low as 1% or even less (waxy maize) have been reported. In maize grains, starch can be found in granular forms which vary in size and shape depending on the type of maize. Normal maize starch has a bimodal particle size distribution, with granules <10 μm and >10 μm in relative percentages of 10.2% and 89.8%, respectively. A cell's walls contain proteins, phenolic acids and nonstarch polysaccharides such as β-glucan and arabinoxylan [39].

A second component of maize kernels is protein. Maize varieties have different levels of protein content. Waxy corn grains have higher protein content (11.03%) than normal corn (8.05–8.62%) and flint corn (8.5–8.7%). Zein represents about 60% of the total protein in corn. Zein contains very low lysine and tryptophan percentages. Zein is comprised of 21.4% glutamine, 19.3% leucine, 9.0% proline, 8.3% alanine, 6.8% phenylalanine, 6.2% isoleucine, 5.7% serine, and 5.1% tyrosine.

Depending on the type of maize, the grain can contain lipids up to 5.91%. Even though the lipid percentage can be high, the fat content of the endosperm can be relatively low (approximately 1%). Saturated fatty acids are more abundant in the lipids from the endosperm, compared to the lipids from the germ.

Pigmented maize types are distinct due to their higher quantity of phenolic compounds (flavonoids) especially anthocyanins. The color of these substances is contingent upon the quantity and location of the substituents in the molecule. A prevalence of methoxyls usually creates a red tint, whereas any growth in the amount of hydroxyl groups creates tones of blue pigment. The anthocyanins present in blue maize originate from cyanidin and malvidin, while those in red maize originate from malvidin, pelargonidin, and cyanidin [39]. In the structure of cereals, anthocyanins can be found in both the pericarp and the aleuronic layer [8]. Carotenoids with oxygen-containing molecules (xanthophylls) give maize its yellow color. Numerous researchers have emphasized the profile of phenolic acids and anthocyanins and the antioxidant activity of different types of maize; they have indicated antioxidant, anticarcinogenic and antimutagenic capacities [8]. The antioxidant activity of phytochemicals related to maize grains is stated to be 157.68 μmol/g compared to 68.74, 43.60, and 39.76 μmol/g in wheat, oats, and rice grains, respectively [39].

Maize contains vitamins C, E, K, B1, B2, B3, B5, folic acid, selenium, Np-coumaril tryptamine, N-ferulyl tryptamine, and several minerals—the most important being potassium. However, corn may contain antinutritional compounds such as α-amylase, trypsin and phytates inhibitors. These compounds can inhibit digestibility, absorption and utilization of nutrients. Microbiological activity during alcoholic fermentation in brewing can help reduce these antinutritional factors and favorably impact the nutritional properties of the finished product [40].

The composition of sorghum grains and their constituent parts is generally similar to maize, except for the lipid content (which is lower). The major component of sorghum is starch; dietary fiber present in the cell wall makes up 75% of grains. The content of starch-free polysaccharides (NSPs) in sorghum grains suggests their possible ability to lower cholesterol levels and improve bowel function in humans [43]. The total soluble sugar content of sorghum cereals ranges from 0.7 to 4.2%, while reducing sugars range from 0.05 to 0.53% [30].

As many studies have indicated, sorghum grains are a good source of energy, as well as vitamins, minerals, carbohydrates, polyunsaturated fatty acids (PUFA), and some essential amino acids. Linoleic acid and oleic acid are the main fatty acid components of sorghum lipids. The grains are usually eaten with the testa, which retain most of the nutrients. Sorghum is a good source of minerals; however, the wide spread variety of its mineral composition is generated by environmental settings [42].

Sorghum contains both essential and nonessential amino acids, including alanine (7.34–9.62 g/100 g), aspartic acid (4.83–7.06 g/100 g), glutamic acid (17.5–28.12 g/100 g), leucine (12.02–14.48 g/100 g), phenylalanine (4.03–5.62 g/100 g), proline (6.66–12.34 g/100 g), and valine (4.22–6, 86 g/100 g). It has reduced values for tryptophan and lysine. However, it has bioactive peptides and beneficial protein fractions—e.g., α-caffeine, kafirin, protease, amylase and xylanase inhibitors. Sorghum also contains cationic peroxidase, which has anticancer, antiviral, and antioxidant properties, and can lead to decreased cholesterol and reduced risk of hypertension.

Available studies have also shown that sorghum contains a number of minerals and vitamins, which are part of the vital nutrients that humans need in order to perform the functions that sustain life. Sorghum contains relatively high levels of potassium (K) (900–6957.67 mg/kg) and phosphorus (P) (1498–3787.25 mg/kg), minerals known to facilitate muscle movement, maintain a healthy nervous system and build strong bones and teeth. B vitamins (0.1–19.9 mg/100 g) and vitamin E (1.38 mg/100 g) are also among the main essential vitamins reported [43].

Sorghum grains are known for their toughness compared to other cereals. The hardness of the sorghum grain is owed to the increased protein and prolamin content (3.6–5.1%). The lysine quantity varies from 1.06 to 3.64%. Studies of proteins in sorghum have indicated that the allocations of albumin-globulin, prolamin, and glutelin are approximately 15%, 26%, and 44% of total nitrogen, respectively [43].

Pontieri et al. (2013) analyzed current knowledge on the development of functional non-gluten sorghum products, such as non-gluten bread, noodles, baby food, and beer [30]. The energetic value of 100 g of sorghum is around 400 calories, similar to corn and wheat. However, sorghum has a higher content of resistant starch. It also contains micronutrients like B-complex vitamins and minerals (potassium and phosphorus) [65].

2.3. The Use in Brewing

The use of maize and sorghum as adjuvants in brewing has been known for a long time. Recently, interest in and research on the use of these cereals and their malted forms to obtain new varieties of beer have increased.

2.3.1. Maize

Maize can be used as an adjuvant in brewing in many ways: as flour, groats, starch, expanded, extruded, cereals, corn syrup, etc. It is the most widely used adjuvant in Europe. In Brazil, it has been shown that the use of 30% maize as an adjuvant can reduce production costs by 8%, though this value can vary based on local raw material prices and other production costs [66]. In this country, the use of substituents may not exceed 45% of the primitive extract [67].

Maize starch is widely applied (due to its high fermentability) as an adjuvant in the production of high-gravity beer [68]. Corn flakes or pre-gelatinized maize can be used to significantly reduce mashing time. Corn kernels produce a somewhat lower extract compared to other raw adjuvants (such as rice) due to the lower amount of dextrin in the wort after mashing. They also contain higher levels of lipids and proteins. It should be noted that the addition of corn derivatives has an important impact on the organoleptic properties of beer.

Poreda et al. (2014) obtained a beer with 10% and 20% maize addition, and analyzed the influence of these adjuvants on the production process and beer quality. Both adjuvant levels influenced the color of the wort: the intensity decreased (11.1 EBC for 10% maize addition and 10.5 EBC for 20% maize addition). Dimethyl sulfide (DMS) and extract content were considerably lower in the wort produced with maize. Free amino nitrogen (FAN) content also decreased because—with the addition of corn–the content of hydrolysable proteins increased. The characteristics of the finished product were not considerably influenced [69]. Furthermore, as reported by Diakabana et al. (2013), corn beer has a low alcohol content, normal pH (about 4.5) and darker color due to the Maillard reaction [70].

In the future, it may be recommended that brewers/manufacturers use exogenous enzymes together with corn in order to enhance saccharification and amylolytic activity. The most convenient way to use corn in the beer industry is as an adjuvant because they are a source of carbohydrates. This is why maize granules are in common use. However, the process of obtaining malt from maize is quite expensive and difficult, though it is commonly used in brewing [7]. Rocha dos Santos Mathias et al. (2019) reported that the use of adjuvants and the exclusion of the proteolytic step from the mashing phase will lead to must with lower nitrogen compounds. Despite the fact that these are alternative steps to decrease production costs, they can influence yeast activity during brewing and affect the quality of the finished product [15].

Fumi et al. (2011) reported that when maize starch was used as an adjuvant, the phenolic content of the wort was lower compared to conventional wort [71]. Fumi et al. (2009) also reported that adjuvanted corn wort had a lower total nitrogen substance content than malt wort, and the free amine nitrogen was almost double that of malt [72]. However, Perez-Carrillo et al. (2012) reported that the content of free amine nitrogen in the wort increased due to the addition of protease, and the free nitrogen content in the corn wort (60%) was twice as high as that discovered in sorghum wort (30%) [73]. The starch granules in the maize endosperm were inserted in a protein matrix and enclosed in the cell walls; consequently, researchers are now looking for techniques to improve the conversion of maize starch into monoglucides [22].

He et al. (2018) analyzed the quality of adjuvants in the brewing of extruded maize starch and cooked maize starch. The aromatic substances of the beer obtained using extruded maize starch and cooked maize starch were examined via headspace solid-phase microextraction gas chromatography mass spectrometry. Eight volatile compounds in extruded maize starch and cooked maize starch beer were measured using gas chromatography. They concluded that both extruded maize starch and cooked maize starch could be used to produce beer—but the concentration of representative flavor substances was higher in the former than in the latter [22].

The potential use of maize in the beer industry as a malted cereal is low. Consequently, it is mainly utilized as an adjuvant. Zweytik and Berghofer (2009) obtained corn malt on a pilot scale to produce lower fermentation beer from 100% corn. They concluded that the finished product (the beer) was clear, with a light yellow color, and presented a good foam stability and flavor similar to normal beer [74].

Hernández-Becerra et al. (2020) reported that maize grains required fewer days to germinate compared to barley, which needed another day under the same experimental conditions. Maize grains are attractive for brewing because they germinate faster and have a high level of reducing sugars [17].

Table 2 summarizes data from the literature on the production of beer assortments based on maize or maize derivatives. For some varieties of beer, other ingredients are used as raw materials (for example, sorghum, wheat, and barley). To obtain these beer assortments, old manufacturing recipes and traditional processing are often used. However, technical and scientific advances in the brewin gindustry have optimized the process and increased the shelf life of finished products.

Table 2. Beer assortments in which maize is used as a raw material.

Beer Name (Origin Country)	Raw Materials	Tehnological Process	Finished Product Characteristics	References
Sendechó (Mexic)	Blue maize, chili Guajillo, pulque	Malting, grinding, mashing, brewing, fermentation	Fermented fruit flavor, smells of cooked vegetables, tortillas, bread, dried fruit and dried chili, amber-copper red color	[75,76]
Chicha de jora (Argentina, Euador, Peru)	Maize	Malting, grinding, brewing, lactic fermentation, alcoholic fermentation	Clear liquid, yellow color, effervescent drink, and a low alcohol content (1–3%)	[77,78]
Umqombothi (Africa de Sud)	Maize flour, sorghum malt	Mashing, brewing, fermentation, filtration	Opaque, pink in color, rich in B vitamins, with a distinct aroma, acid and a creamy consistency, shelf life of 2–3 days	[79,80]
Sesotho (Lesotho)	Maize, sorghum and/or wheat flour	Grinding, mashing, lactic fermentation, cooling, alcoholic fermentation	Opaque liquid, thin consistency, distinct sour taste, 3–5% (v/v) alcohol content, rich in B vitamins	[81,82]
Chibuku (Zimbabwe, Tanzania, Zambia, Ghana, Nigeria)	Maize, sorghum, sorghum malt, barley malt	Malting, grinding, brewing, acidification, lactic fermentation, alcoholic fermentation	Opaque brown-pink liquid containing suspended and dissolved solids (3.6% w/v), alcohol content of 3–5%, pH of 3–4 and lactic acid levels of approx. 0.5 g/L	[83,84]
Tella (Etiopia)	Maize, barley, wheat, *Rhamnus prinoides* L.	Malting, grinding, brewing, alcoholic fermentation	pH 3.87–4.67 alcohol content (%v/v) 3.04–3.75 CO_2 content (%) 0.24–0.034	[85,86]
Sekete (Nigeria)	Sprouted maize	Mashing, brewing, acidification, lactic fermentation, alcoholic fermentation	Dark brown color alcohol content of 1–3%	[14,87]

In Mexico, the country in which corn originated, it occurs in a wide variety of pigmented grains, including white, yellow, red, purple, blue, and black. Over time, the inhabitants of Mexico have created several fermented drinks obtained from specific varieties of pigmented maize. These are widely known as corn beers [75].

Sendechó is a representative fermented drink obtained by the Mazahuas inhabitants of the Valley of Mexico. Their production technique is very comparable to the conventional technologies in use today for beer production. Sendechó is made using local ingredients, such as blue pigmented maize, which go through a process of malting with Guajillo chili—itself a customary ingredient in Mexican gastronomy. The wort obtained is fermented with pulque, a thick, white alcoholic beverage from the moorlands of Mexico, which is made by fermenting mead or the juice of various types of agave (*Agave americana*, *Agave feroce Agave atrovirens*, *Agave salmiana* and *Agave mapisaga*). The fermentation process for agave begins with spontaneous microorganisms such as yeasts, lactic acid bacteria, ethanol-producing bacteria and exopolysaccharide-producing bacteria [8].

The beer obtained with these varieties of pigmented maize is mainly characterized by aromas of fermented fruit, boiled vegetables, bread, tortillas, dried fruit and dried chilis. Romero-Medina et al. (2020) was the first to prove that ketones, anthocyanins, terpenes and volatile phenolic compounds were pertinent criteria for differentiation of maize beers. Anthocyanins can also be used as an indicator to determine whether a beer was obtained with pigmented maize malt. This could potentially be used as a quality characteristic in future studies. Studying the relations between sensory properties and chemical characteristics via multiple factor analysis (MFA) helped explain the influence of each malt type (red maize, blue maize and barley malt) on chemical characteristics and sensory attributes. Over 100 volatile compounds were quantified by headspace solid-phase microextraction coupled with gas chromatography mass spectrometry (HS-SPME/GC-MS). Terpenes and phenols were the groups of volatile compounds that better characterized the beers containing maize. The anthocyanin content of maize beers ensures colors in the amber, copper, and red families and can prevent the development of unwanted tastes and aromas [75].

Flores-Calderón et al. (2017) analyzed the chemical characteristics and antioxidant activity in three separate stages of the blue corn malt beer manufacturing process. The influence of adding caramel malt and various amounts of Guajillo hops and chili was analyzed, creating several types of blue maize malt beer. Upon completing statistical analysis, using ANOVA and multivariate methods, the best concentration of anthocyanins and antioxidant capacity was found in two varieties of beer (obtained with 85% corn malt and 15% caramel malt) [8].

Chicha de jora is a type of traditional corn beer (chicha) from South America. It is still widespread in Andean countries, where it is consumed for its nutritional properties [78].

Obtaining this product begins with the process of preparing maize malt from corn morocho. The corn is soaked for 3–8 days. For germination, the water is removed, and the grains are set for one to two weeks on the leaves of *Baccharis latifolia* (Chilca), *Sambucus nigra* (Sauco) or *Alnus glutinosa* (Aliso) at room temperature. After that, the germinated grains are dried for one to two weeks. Stones are traditionally used for the grinding process. The resultant flour is introduced into the water and boiled for 1–2 h. At this stage, some manufacturing recipes provide for the addition of other ingredients, e.g., chancaca corn, barley, sugar, cinnamon, cloves, wheat flour, quinoa, fava beans, herbs, or fruits. Fermentation takes place in several containers, depending on the capacity of the chichería, and is frequently an uncontrolled procedure that can take anywhere from 24 h to 15 days. To start the fermentation process, according to ancient practices, ceramic vessels called tomin are used. These vessels are made of a porous material that helps the adhesion and multiplication of microorganisms. During fermentation, particles of corn from the grinding process are continuously drained to obtain borage which can then be reused as the inoculum for subsequent productions. The addition of chicha borra in the second stage of fermentation gives the beverage a higher alcohol content by volume.

Biochemical and physiological research has established the main role of the genus *Lactobacillus* and genus *Leuconostoc* in the fermentation process. Recent molecular studies utilizing next-generation sequencing methods have shown that lactic acid bacteria and yeast are the main populations accountable for the organoleptic properties of this drink. The data from the literature have expanded knowledge on the microbiota of this fermented beverage in different stages of the technological process, allowing the detection of minority species of bacteria or difficult-to-grow microorganisms. Although corn chicha is primarily brewed in northwestern Argentina, Brazilian rice chicha and cassava chicha from Ecuador have also been researched. Peruvian chicha has not been studied as closely or as often.

Basi et al. (2020) analyzed twenty-seven chicha samples from fourteen different chicherias in seven regions of Peru. They showed that fermentation was produced by a select group of microorganism species, dominated by lactic acid bacteria groups, which promote health and determine the hygienic and beneficial qualities of the finished product [77].

Umqombothi is a traditional South African beer. It is opaque, pink in color, and rich in B vitamins, with a distinct, acidic aroma and a creamy consistency. To produce it, corn flour is combined with sorghum malt and water. The mixture is then boiled to a soft consistency and cooled for approximately 6 h. After adding sorghum malt and umqombothi from a previous batch, the mash is left to ferment for about 18 h, after which the mixture is filtered. Umqombothi is characterized by a short shelf life (2–3 days) and is consumed as a product in which fermentation is not complete. It is common, especially among rural black populations in South Africa. Umqombothi is produced mainly by women, either for social events or for sale [79,80].

Sesotho is a well-liked beer, obtained through spontaneous fermentation, which originated from Lesotho (South Africa). It is produced from ground corn, wheat, sorghum flour, or a combination of these. This beverage is opalescent, has a low consistency and a distinct acidic taste. Sesotho is produced mainly in rural households or for small-scale commercial purposes. It is also used as a beverage for funerals, weddings, and cultural ceremonies. Lactic acid bacteri (such as *Lactobacillus*, *Enterococcus*, *Leuconostoc*, *Pediococcus* and *Wiesella*) and yeasts of the genus *Saccharomyces* are in common use in all beer factories where Sesotho is produced [81]. Some factories produce this beer using 50% barley malt, 34% unmalted corn and 16% unmalted wheat. Krstanovic et al. (2020) analyzed the colloidal stability and shelf life of this type of beer and concluded that it should be stabilized by methods that guarantee the elimination of haze inducers [88].

Sesotho was traditionally made using only sorghum and wheat flour. According to information reported from women who produced sesotho, sorghum and wheat flour are traditionally mixed together in equal amounts, after which cold water is added to obtain a rigid consistency. Hot water is then introduced to obtain a thin porridge, which is cooled to about 30–35 °C. After the addition of a traditional liquid starter known as tomoso, the mixture is left to ferment, either overnight or for 24–48 h, depending on the ambient temperature. It is then boiled for 2–3 h and cooled to 30–35 °C. A solid starter culture called moroko (using scraps from previous fermentations) is added and the mixture is fermented for another 24–48 h. It is then filtered to remove coarse particles to obtain the finished beer [82].

Chibuku is the best-known opaque beer made in Zimbabwe. Over 420 million L are produced in twenty breweries annually. Chibuku is also known as hwahwa, doro, utshwala or mhamba in various parts of Zimbabwe. Other brands are made locally, such as Pungwe, Simba, Go-beer and Ingwebu. This beer was first manufactured in Zimbabwe in 1962 by Delta Beaches Breweries, after originating in Zambia in the 1950s. Chibuku is currently brewed using advanced technology, unlike many traditional beers from rural regions of Zimbabwe, Zambia, Tanzania, Nigeria, Ghana and other African states. This beer is obtained from corn, sorghum, barley malt, sorghum malt, water, lactic acid and superior fermentation yeast of the species *Saccharomyces cerevisiae* [83].

Chibuku is an opaque brown-pink beverage containing dissolved and suspended solid substances (3.6% *w/v*), with an alcoholic strength of 3–5%, a pH of 3–4 and a lactic acid content of about 0.5 g/L. In essence, the process of obtaining the finished product consists of lactic fermentation and alcoholic fermentation. This beer is sold at retail while still microbiologically active, so lactic acid bacteria can spoil the product. Useful microorganisms continue to multiply even after acceptable levels of alcohol and lactic acid have been obtained. The product becomes unacceptable when lactic acid levels reach 0.5% (*v/v*) and other metabolic substances have accumulated [84].

Opaque beer has been a sociocultural drink in African for over a century. Depending on the geographical location, these beers bear different names. Opaque beers are called burukutu or pito in Ghana and Nigeria, chibuku in Zimbabwe, utshwala in South Africa, ikagage in Rwanda, mtama in Tanzania and tchakpalo in Ivory Coast and Togo. The technology used to produce these opaque beers differs slightly from country to country, depending on the local raw materials available to them. Some use sorghum, while others may use millet, cornmeal, or even bananas. These beverages are a completely natural and nutritionally balanced product and rightly considered a liquid food and an alcoholic drink. Perhaps this is the reason why Africans—especially in rural regions—consume them in great quantities. The consumption of these beers often occurs while the fermentation state is still active [83].

Tella is a beer that is widely brewed and consumed in both rural and urban regions of Ethiopia. It is mainly obtained from cereals e.g., maize (*Zea mays* L.), wheat (*Triticum aestivum* Z.) and barley (*Hordeum vulgare* L.). In addition to cereals, *Rhamnus prinoides* L., which is known as gesho in Amharic, is used to add a special flavor to the finished product. It is also used as an antiseptic agent against bacteria. Alcohol is produced by *Saccharomyces cerevisiae*, but it is not able to metabolize starch. However, Andualem and Gessesse (2013) showed that yeasts other than *Saccharomyces cerevisiae* should be avoided in the tella production process, as their role in converting starch into fermentable sugar is reduced compared to malt amylase [85].

In the process of preparing tella, various substrates, including malt, enkuro (roasted dry corn ground into fine flour, mixed with water and boiled), kita (bread made from grain flour such as corn, wheat, or barley, separately or in the mixture), and derokote (fried maize, wheat or barley grains) are used as a source of carbon in fermentation. Malt amylases may not be sufficient to break down starch molecules into fermentable sugars in different substrates, as kita, enkuro and derokote are produced from cereals with a high starch content. Thus, various microorganisms from environmental sources may contribute to amylases and increase the efficiency of the conversion of starch from malt and adjuvants into maltose and other fermentable sugars [86].

Sekete is a drink made from germinated corn grains. It has a low alcohol content. This beer is commonly consumed by native people in rural areas in western Nigeria. Very little scientific research has been done on this beer. Studies show that after fermentation, the mineral content (phosphorus, magnesium, potassium, and sodium) is decreased, while the content of some vitamins (riboflavin and niacin) is increased. This beer is a rich source of important amino acids like tyrosine, alanine, lysine and leucine, but it is deficient in arginine and proline. The drink exhibits high acidity. It contains species of bacteria of the genus *Lactobacillus* and yeasts of the genus *Saccharomyces,* in whose technological processes, lactic fermentation predominates—to the detriment of alcoholic fermentation. The commonly experienced relaxing, medicinal, and stimulating effects of the drink may be the result of its microbial profile, acid content and microorganism activity. Obatolu et al. (2016) recommended fortifying and sweetening sekete to increase the overall acceptability of the drink [89].

2.3.2. Sorghum

Outside of Mexico and Niger, sorghum has not been widely used as an adjuvant, although its potential has been promoted. Sorghum semolina offers several advantages in beer brewing, including short boiling time, easy filtration, high extract content and highly nutritious wort [90].

In 2017, sorghum dominated the non-gluten beer market; sorghum beer accounted for 37.9% of the total volume of non-gluten beers produced. Non-gluten beer is not only marketed to people with gluten intolerance. Many are consumed by people interested in new drinks and new products. In this way, the market for non-gluten beer consumers can be significantly expanded. It has been found that women are a special target group, because the incidence of celiac disease is higher among women (60% of adult patients) and because women are often more concerned with a healthy lifestyle, including in their diet. The global non-gluten-beer market is projected to grow to \$18.7 billion by 2025, with an annual growth rate of 16.3%, according to a report published by Fior Markets [15].

Goode and Arendt (2003) used unmalted sorghum as an adjuvant in 50% barley malt and found that a good quality finished product was obtained, comparable to beer obtained from 100% barley malt [91]. Ogbonna (2011) recommended the use of sorghum as an adjuvant to barley malt at 50%, and showed that it was possible to obtain beer from 100% sorghum malt [92]. In other research, it was argued that the use of 40% sorghum as an adjuvant in brewing did not significantly influence the quality of the finished product. Other researchers have produced lager beer with 25% sorghum adjuvant and concluded that a good quality beer was obtainable when barley malt was substituted with up to 25% sorghum [90].

Beta et al. (1995) analyzed the malting characteristics of several varieties of sorghum with characteristics similar to commercial barley malt, suggesting that they might produce different qualities of malt [93]. Chandra, Proudlove, & Baxter (1999) reported that the texture of the endosperm affected malting, influenced water absorption, and impacted the enzymatic process [94]. Oyewole & Agboola (2011) showed that malting losses were very high for tropical grains [95].

Eneje et al. (2004) and Oyewole & Agboola (2011) studied the qualities of rice, sorghum, millet and maize malt and reported that sorghum and millet malts were more suitable for use in food formulation, while maize—which bears some similarity to barley in compositional characteristics—could be used as a substitute for conventional malt in beer brewing [54,95,96].

The use of sorghum malt in beer manufacture has led to some difficulties, largely due its low amylolytic activity (which is insufficient for complete saccharification), high gelatinization temperature, and low content of free amino nitrogen. Sorghum malt exhibits higher α-amylase activity, but lower β-amylase activity, compared to barley malt. Reduced enzymatic activity can lead to insufficient production of fermentable carbohydrates, high dextrin content, and increased viscosity [97,98]. The gelatinization temperature of sorghum malt is limited by kafirin [99], and thus the hydrolysis of starch into fermentable sugars is only partially completed. Therefore, in order to avoid technological challenges, the use of sorghum in brewing requires an adequate malting process. Otherwise, it is recommended to use exogenous enzymes to produce sorghum beers [7].

Espinosa-Ramírez et al. (2014) studied the effect of the addition of β-amylase and amyloglucosidase during sorghum mashing. They produced a beer with higher alcohol content [100]. Urias-Lugo and Salvidar used amyloglucosidase to achieve improved production efficiency of wort and its filtration rate, resulting in a higher percentage of ethanol [101]. However, the alcohol content of sorghum beer was 1.1% lower than barley malt beer. The color, pH and content of free amino nitrogen were not affected by the addition of amyloglucosidase. To reduce these deficiencies, *Aspergillus oryzae* can be added; it has been shown to improve the malting characteristics of sorghum. By using this adjuvant, α-amylase was positively affected. There were no differences for β-amylase [7].

In Africa, sorghum is traditionally the most important raw material in beer production, both malted and as an adjuvant [102,103]. The presence of sorghum beers on the market is proof of the potential of this cereal in brewing. However, in order to improve the sensory properties of the finished product, it is necessary to obtain a quality sorghum malt or to use it in combination with other cereals [7,104].

In Africa, different types of traditional fermented beverages have been described, commonly named opaque beers or sorghum beers. These beverages have both sociocultural and nutritional value. They are known as dolo in Burkina Faso, ikagage in Rwanda, amgba in Cameroon, pito or burukutu in Nigeria and Ghana, merissa in Sudan, doro or chibuku in Zimbabwe, bili bili in Chad, mtama in Tanzania, tchapalo in Ivory Coast, Togo and Benin, and kaffir in South Africa. These beverages play a central role in these cultures and are a significant part of the diet for a growing part of the population.

The presence of nonspecific microorganisms in the traditional starter cultures of some of these beverages can make it more difficult to monitor the fermentation process—this, in turn, can lead to finished products of variable quality. That is why it is recommended to use pure starter cultures to reduce organoleptic variations and decrease the risk of contamination with pathogenic microorganisms. Controlled fermentation also increases the chances of preserving traditional sorghum beer, giving it a longer shelf life. Pasteurization would also solve the issue of sorghum beer's relatively short shelf life, which is a problem for the brewing industry. As a potential alternative to the use of synthetic preservatives, pasteurization of beer could be combined with plant extracts. It is well known that the main determinant factor in the deterioration of beverages is the presence of Gram-negative bacteria, Gram-positive bacteria, and molds [84].

In African countries, sorghum beer is generally consumed at festivals, weddings, prayers, rituals, birth ceremonies, and funeral rituals. For example, in Burundi and Rwanda, the consumption of this beer marks the beginning of the handing over of the dowry during traditional marriages. The families in question share pleasures around a clay pot of sorghum beer. Sorghum beer signifies the connection between the couple and their families. Meetings and community work also often end with the consumption of this beer. Sorghum beer contributes significantly to the diets of many persons in developing countries and is largely consumed by the poor. Sorghum beer has a high content of B vitamins, such as riboflavin, nicotinic acid, and folic acid. It is also rich in amino acids, and in mineral substances such as calcium, sodium, potassium, magnesium, iron, and zinc. In general, sorghum beer has a higher nutritional value than European barley beers, due to its high content of lactic acid bacteria, yeast and other suspended materials [13]. Sorghum beer is considered both a food (energizer) and an alcoholic drink. This may be why Africans, especially in rural regions, consume fairly lare quantities of sorghum beer [83].

Nigeria produces over 900 million liters of beer annually, and most of it is made from sorghum. This cereal has crossed to parts of Eastern and Southern Africa, the United States, Mexico, Cuba, and Israel (where sorghum beer is also produced) [30,105].

In Togo, about 60% of national sorghum production is used to produce two types of sorghum beer: Tchoukoutou and Tchakpalo. The manufacturing and marketing of this beer remains primarily the purview of Togolese women, from which they derive significant earnings. Sorghum beer is a key socioeconomic factor in northern Togo, where it is served in large quantities on holidays. Sorghum beer is also used in traditional rituals and religious ceremonies [84].

Clear sorghum lager beers have reportedly been manufactured in many parts of the world. Sorghum lager beers have been produced in Mexico and Cameroon. In Sri Lanka, specialists analyzed sorghum varieties in order to select the most appropriate type for obtaining a conventional lager. In the USA, sorghum has been used since the 1980s as an adjuvant in the preparation of lagers. The most significant progress in the production of sorghum beer has probably been achieved in Nigeria. Due to the government's 1988 ban on importing barley malt, local brewers were compelled to use alternative native cereals, such as corn and sorghum, as substitutes for malted barley [90].

Ogbonna and Adejemilu (1992), in a review of sorghum beer production, discussed the technological difficulties associated with sorghum malting and sorghum beer production [106]. Comprehensive reviews on the preparation of sorghum lager beer, especially sorghum malt, were conducted by Agu and Palmer (1998), Owuama (1999) Schnitzenbaumer, Arendt (2014), and Embashu et al. (2019) [21,104,107,108]. Unmalted sorghum grain and commercial enzymes are more sustainable for use in brewing—mainly because malted sorghum has several inconveniences, including limited protein modification, insufficient diastatic power, high malt losses, high malting costs, and the necessity to improve the mash with exogenous enzymes [90].

The main problems in the manufacture of sorghum beer are the reduced diastatic power of the resulting malt (especially the deficiency in β-amylase activity) and the high gelatinization temperature of sorghum starch, in contrast to barley starch [109]. Espinosa-

Ramirez et al. (2013) successfully produced lager beers from several varieties of sorghum malts and non-gluten adjuvants. These were also supplemented with amyloglucosidase and β-amylase [98,100].

Pilot-scale beer (1000 L) was produced using raw malted sorghum (50% of the total wet weight of the grains) and barley malt (50% of the total wet weight of the grains) as raw materials. The raw materials were subjected to the mashing operation with rests at temperatures of 50 °C, 95 °C and 60 °C. Organoleptic analysis showed that there were no noteworthy differences in terms of aroma, taste and clarity between sorghum beer, the control beer, and commercial barley beer. However, it was found that sorghum beer differed significantly from the control and barley beers in the color, taste and stability of the foam [90].

Table 3 summarizes data from the literature on obtaining beer assortments that use sorghum as a basic raw material.

Table 3. Beer assortments in which sorghum is used as basic raw material.

Beer Name (Origin Country)	Raw Materials	Tehnological Process	Finished Product Characteristics	References
Burukutu/Otika (Nigeria, Niger, Ghana)	Sorghum	Malting (steeping, germination), milling, mashing, boiling, fermentation, maturation	Viscous, opaque, light brown liquid, alcohol content approx. 4% (v/v), sour taste, pH = 3.3–3.5	[42,87]
Pito (Ghana, Togo, Nigeria)	Sorghum	Malting, grinding, mashing, brewing, lactic fermentation, alcoholic fermentation	Sour taste, characteristic, alcohol content 3–5% (v/v)	[110–113]
Tchapalo (Coasta de Fildeș, Togo, Benin)	Sorghum	Lactic fermentation, alcoholic fermentation	Non-alcoholic beer, turbid, shelf life 3 days	[114–118]
Bantu (Africa de Sud)	Sorghum	Malting, grinding, mashing, lactic fermentation, alcoholic fermentation	Turbid liquid, alcohol content 3–4% (v/v), sour taste, brown-pink color, rich in B vitamins	[119–122]
Dolo (Burkina Faso, Benin, Rwanda)	Sorghum	Malting of red sorghum grains, crushing, mashing, cooking, lactic fermentation, filtration, boiling, alcoholic fermentation	Turbid liquid, alcohol content 1–5% (v/v), sweet-sour taste, fruit flavor	[123–126]
Bili bili (Ciad)	Sorghum	Malting, mashing, boiling, souring, and fermenting	Turbid liquid, brown-pink color, sour taste, fruity, alcohol content 1–8% (v/v), low in carbohydrates and high in protein	[104,127–130]
Omalovu (Namibia)	Sorghum, millet	Malting, drying, milling, souring, boiling, mashing, straining, alcoholic fermentation	Unpasteurized beer, opaque, red-brown or cream color, pH = 3.06–4.34, alcohol content 0.18–4.05% (v/v)	[104,131]

Burukutu is another fermented alcoholic drink produced from sorghum grains, having an alcohol content of about 4% (v/v). This beer is a viscous, opaque liquid, due to the suspended solids and yeast. It is light brown in color, with an acidic taste caused by the presence of lactic acid and a pH of 3.3 to 3.5. It is frequently sold in Mammi markets (which are attached to soldiers' barracks in Nigeria) and in northern Niger [87]. Burukutu contains vitamins including potassium, magnesium, iron, manganese, and calcium, and contains about 26.7 g of starch and 5.9 g of protein per liter. This regional drink is known as tchoukoutou in Benin or Togo, burukutu or otika in Nigeria, bilibili in Chad, pito in Ghana, dolo in Burkina Faso, mtama in Tanzania, and kigage in Rwanda. Technological processes are highly inconsistent and dependent on the region in which the beverage is produced. In general, the brewing process involves malting, soaking, germination, grinding, mashing, boiling, fermentation and maturation [42].

Pito is part of many native African beers collectively called sorghum beer or opaque beer. It is a light alcoholic drink commonly found in many regions of Africa. It is produced from sorghum and is known under different names, depending on the tribe or locality. Ghana produces some of the most well-liked types of ethnic pito, such as Frafra pito, Dagarti pito, Kasena pito, Kusasi pito and Grushie pito. In general, pito has a unique sour taste and holds 3–5% alcohol by volume. The technological process consists of two types of fermentation: lactic fermentation and alcoholic fermentation. Pito brewing is still artisanal, done primarily by women, and the quality of the product differs greatly producer to producer and even batch to batch.

The process of obtaining pito varies depending on the ethnic group or tribe. Generally, sorghum grains are soaked in water, left to germinate for 4–6 days and dried in the sun for two days or more. The dried malt is coarsely ground and mixed with water. Clarifying agents (such as crushed okro stalk bark) are added, and mixed with the suspension, which is then left to stabilize and clear. The obtained mash is boiled, followed by cooling and filtration. the obtained must is subjected to boiling, cooling, and clarification. The clear must is then decanted and inoculated with yeast to ferment. Finally, pito usually has a low brewing yield of 47.28 ± 11.74% and an extract recovery yield of 62.21 ± 15.44%—compared to a brewing yield of 76% and an extract recovery yield of 97% for lager beer brewed in industrial beer factories [108].

Tchapalo is a traditional nonalcoholic, opaque beer that contains suspended solids and yeast and is typically obtained from sorghum in Ivory Coast. This traditional drink is produced in a two-step process comprising lactic fermentation to generate the sweet wort followed by alcoholic fermentation. This beer is typically consumed by women, children and those who want to drink beer without alcohol [114–116].

Tchapalo and sweet wort contain several nutritional substances that help improve consumer diets. Additionally, popular opinion has attributed medicinal benefits to this drink, including laxative, antimalarial and antihemorrhoidal effects, whilst scientific research has indicated that it may potentially work against diabetes, cardiovascular diseases, and cancer. This type of drink is consumed at different African ceremonies (e.g., marriage, birth, baptism, etc.) and festivals, and is also a source of revenue for female brewers. Nevertheless, tchapalo and sweet wort is often produced under deplorable hygienic conditions, using rudimentary equipment and a large amount of work. Consequently, the finished products have a short shelf life (3 days), and the quality of the product differs widely [117,118].

Bantu is a traditional sorghum beer made in South Africa. It is often turbid and contains yeast with a persistent sour taste. It is brownish-pink. Higher pH levels produce a more pronounced pink color. Most traditional sorghum beers have low alcohol content (3% to 4%). Many traditionally prepared sorghum beers also contain maltotriose, the last fermented sugar in yeast during fermentation. Some amino acids and peptides are also present. Cassava root is also used throughout Africa as an alternative to cereals, in addition to sorghum malt in brewing [119–122].

Dolo is a well-known traditional beer from Burkina Faso. This beer is prepared from the *Sorghum bicolor* variety of sorghum; in some localities, millet, and maize are used as adjuvants. In Burkina Faso, yeasts and lactic acid bacteria are the main microorganisms used in making dolo beer. Dolo yeast—called rabilé in Burkina Faso, kpètè-kpètè or otchè in Benin, and umusemburo in Rwanda—is used as a spice and an important source of protein for some communities in Burkina Faso. Dolo yeast is generally collected from the bottom of the fermentation vessel, after a period of 12–13 h of fermentation. This beer is frequently filtered but remains turbid and has a blend of sweet/acidic tastes and fruit flavors. It contains 1–4% *v/v* alcohol. The traditional production of this beer involves malting the red sorghum grain, followed by the crushing, mixing, boiling, lactic fermentation, filtration, boiling and alcoholic fermentation of the wort. The color of the cereals and sorghum flours used plays a significant role in dolo's acceptance by consumers [123–126].

Bili bili is a traditional Chadian beer that has a relatively low concentration of alcohol and is obtained without hops. This beer has a slightly acidic taste and is consumed unfiltered (containing particles), especially in rural areas. It is produced by lactic fermentation, boiling, mashing, filtration, and alcoholic fermentation. The lactic fermentation process is achieved by inoculating a suspension of ground malt in water with *Lactobacillus leichmannii*. Sorghum beer is more viscous than commercial beer. The color of this beer is intense, depending on the pH of the finished product [53,104].

Omalovu is one of the most popular beers in Namibia. Producing this type of beer is a traditional process and is contingent on the skills of the producer. It is brewed only locally. It is obtained from white or red sorghum, the latter being preferred, possibly due to the sensory properties of color, bitterness and astringency that it can bring to beer. Omalovu is not commercially manufactured, but it is freshly produced and used to supplement household income when sold on open markets. During important events and holidays, this beer is prepared in batches of 20–30 L. Omalovu can be produced with malted sorghum flour and water, though malted millet flour can also be used. The quality of this beer is inconsistent, due to the variable malting process and uncontrolled technological conditions. Nevertheless, with advancements in the malting and brewing process, the possibility of producing omalovu on an industrial scale is greatly increased [104].

2.4. Other Uses

In general, cereals have multiple uses, and maize and sorghum are not exceptions to this rule.

Maize or corn (*Zea mays* L.) is among the most significant crops in the world from an energy point of view. Historically, the demand for maize has been determined by the starch industry and the bird feed industry. More recently, research has studied the use of other parts of this cereal. Maize outperforms other ingredients in bioethanol production, with respect to generated starch content levels (over 72% of the U.S.), crop yield and ethanol yield [14,36].

Globally, corn is mainly used for animal feed (64%), with significant amounts used as human food (16%), industrial starch and beverages (19%), and seeds (1%). Corn has reached a significant position as an industrial crop; 83% of its products are used in the starch and feed industry [8,38].

Today, it is also considered a vital cereal in several countries, especially in Asia, Central/South America, and Africa. Maize serves as the basic ingredient for thousands of products, including: alcohol, oil, protein, beverages, pharmaceuticals, food sweeteners, food grains, cornflakes, corn flour, starch, syrups, dextrose, cosmetics, films, textiles, paper, biofuels, and more. As one of our fundamental food crops, corn can be used in various forms, including: corn flour for confectionery use, semolina (in soup), ground corn (in animal feed), and fried corn. Corn is also cooked or boiled as a form of porridge [132–134].

Around 40–75% of animal feed is maize; thus, maize grains are essential for meat, eggs, and dairy products, providing the animals with energy and other essential for survival. Several beverages, including alcoholic beverages, have been obtained from maize, both locally and industrially [135,136]. Edible oils obtained from maize seeds contain a high level of natural antioxidants, and are used in the preparation of salads or for cooking household foods. They provide nourishing nutrients for the human body [137,138]. Roasted seeds are used as a coffee substitute. Maize starch is also well known for its use in the cosmetics industry, and in the pharmaceutical industry, where it is used to create edible packaging [139,140]. Maize seeds are essential in the manufacture of alcohol and stem fibers for papermaking. Industrial uses of maize have grown and diversified over many years, ranging from mixed feed manufacturing, dry milling, wet milling, distillation, and fermentation. Fermentation industries produce butyl alcohol, ethyl alcohol, propyl alcohol, lactic acid, acetic acid, acetaldehyde, acetone, citric acid, glycerol, and whiskey, among other relevant products which are currently flooding the industrial market [34].

Sorghum is also utilized for many purposes. This cereal is traditionally grown as a food source for people and animals. Because it is grown in most African and Asian countries, it has been used in many types of food (often indigenous) as well as alcoholic and nonalcoholic drinks. People of the savannah regions of Africa and Asia generally consume sorghum wine, or soft beverages such as kunu-zaki.

Many developing countries (especially Africa) use over 78% of their sorghum crops for food, with approximately 14% used for animal feed and 7% for diverse uses [43].

For western countries, where 40% of the global sorghum crop is produced, this cereal is mostly used as fodder or for ethanol production. More recently, research has focused on the benefits of sorghum in human food, including its vast nutritional potential and its impact on the quality of life of people with celiac disease. In addition, the growing demand for gluten-free foods and drinks has led to numerous studies of sorghum as a crop for human consumption. Various sorghum-based products have been created, including bakery products, pasta, and biscuits. Moreover, innovative sorghum hybrids with tannin-free white cereals (often referred to as food-grade sorghum) have been developed for gluten-free food markets [141–145].

3. Perspectives

The development of brewing from unconventional raw materials is always a major challenge for researchers and specialists in the field. There is currently growing concern about the real and potential failures of industrial crops, including important brewing cereals such as barley and wheat. Climate change is adding to those concerns. Research is needed to identify new ingredients (e.g., maize and sorghum) in the brewing industry [45].

Beer is the most important part of the global alcoholic beverage market and is projected to grow at a compound annual growth rate (CAGR) of 5.2% during the next period (2021–2026). An upward trend has also been noted for consumers' preferences for beers obtained from new ingredients and with innovative flavors [146]. As technology advances, more economic and technical criteria will be taken into account—including the consumer response to new beverages. Maize and sorghum are destined for wider consumption in the future, in the form of various foods and nonalcoholic and alcoholic beverages.

In the beer industry, research into gluten-free beer assortments and new nonalcoholic or low-alcohol beers will continue. Additionally, research will continue on increasing beers' shelf lives while complying with current food safety regulations.

More research is needed on germination and the physicochemical properties and enzymatic activities of sorghum and maize in order to obtain malt comparable to barley malt in quality [17].

Future research is needed to develop a non-gluten beer with the technological and sensory characteristics—as well as the foam stability, wort fermentability, and final taste—of beers brewed using conventional raw materials. It is recommended to combine different gluten-free raw materials with the use of exogenous enzymes to optimize production recipes and obtain new varieties of beer with desirable qualities [30].

It is necessary to improve the technologies involved in the production and preservation of beer from unconventional raw materials in order to maintain their nutritional and marketable qualities.

4. Conclusions

The studies undertaken demonstrated the real potential benefits of using maize and sorghum in the brewing process, whether as simple adjuvants or via the brewing of beers made from 100% sorghum or maize malt.

Maize is a versatile money crop and is adaptable to various climatic conditions; globally, it is known as the queen of cereals. Sorghum is genetically close to corn, and is also called the camel plant due to its resistance to extreme drought conditions. Sorghum is also a vital staple food in many semi-arid areas of the developing world.

There are some limitations in the use of these two cereals: maize has bare grains and lacks a husk (which would act as an adjuvant for filtration). It also has a low level of enzymatic activity. The structure of the sorghum grain is similar to maize; it has no shell, and the aleurone layer inhibits the flow of enzymes. Moreover, the development of amylolytic enzymes during the germination of maize and sorghum is lower than in barley.

In countries around the world, craft and functional beer brewing has revived old varieties and created new ones. Specialist brewers have worked to advance novelty beers that exhibit a complete and rich taste through efficient processing. These new beverages are created using various ingredients, and often involve modifications of the brewing process.

In conclusion, industrial and scientific research can promote innovation by creating new assortments of beer using maize and sorghum. This, in turn, could have a significant impact on product quality improvements.

Author Contributions: A.D., M.E.C., A.C. and G.G.C. contributed equally to the collection of data and preparation of the paper. All authors have read and agreed to the published version of the manuscript.

Funding: This research received no external funding.

Conflicts of Interest: The authors declare no conflict of interest.

References

1. Fox, G. The brewing industry and the opportunities for real-time quality analysis using infrared spectroscopy. *Appl. Sci.* **2020**, *10*, 616. [CrossRef]
2. Albanese, L.; Ciriminna, R.; Meneguzzo, F.; Pagliaro, M. Gluten reduction in beer by hydrodynamic cavitation assisted brewing of barley malts. *LWT* **2017**, *82*, 342–353. [CrossRef]
3. Humia, B.V.; Santos, K.S.; Barbosa, A.M.; Sawata, M.; Mendonça, M.D.C.; Padilha, F.F. Beer molecules and its sensory and biological properties: A review. *Molecules* **2019**, *24*, 1568. [CrossRef]
4. Rošul, M.Đ.; Mandić, A.I.; Mišan, A.Č.; Đerić, N.R.; Pejin, J.D. Review of trends in formulation of functional beer. *Food Feed. Res.* **2019**, *46*, 23–35. [CrossRef]
5. Chetrariu, A.; Dabija, A. Pre-treatments used for the recovery of brewer's spent grain-a minireview. *J. Agroaliment. Process. Technol.* **2020**, *26*, 304–312.
6. Habschied, K.; Živković, A.; Krstanović, V.; Mastanjević, K. Functional beer—A review on possibilities. *Beverages* **2020**, *6*, 51. [CrossRef]
7. Cela, N.; Condelli, N.; Caruso, M.C.; Perretti, G.; Di Cairano, M.; Tolve, R.; Galgano, F. Gluten-free brewing: Issues and perspectives. *Fermentation* **2020**, *6*, 53. [CrossRef]
8. Flores-Calderón, A.M.D.; Luna, H.; Escalona-Buendía, H.B.; Verde-Calvo, J.R. Chemical characterization and antioxidant capacity in blue corn (*Zea mays* L.) malt beers. *J. Inst. Brew.* **2017**, *123*, 506–518. [CrossRef]
9. Donadini, G.; Fumi, M.D.; Kordialik-Bogacka, E.; Maggi, L.; Lambri, M.; Sckokai, P. Consumer interest in specialty beers in three European markets. *Food Res. Int.* **2016**, *85*, 301–314. [CrossRef]
10. Betancur, M.I.; Motoki, K.; Spence, C.; Velasco, C. Factors influencing the choice of beer: A review. *Food Res. Int.* **2020**, 109367. [CrossRef]
11. Chetrariu, A.; Dabija, A. Brewer's spent grains: Possibilities of valorization, a review. *Appl. Sci.* **2020**, *10*, 5619. [CrossRef]
12. Zhang, T.; Zhang, H.; Yang, Z.; Wang, Y.; Li, H. Black rice addition prompted the beer quality by the extrusion as pretreatment. *Food Sci. Nutr.* **2019**, *7*, 3664–3674. [CrossRef] [PubMed]
13. Evera, E.; Abedin Abdallah, S.H.; Shuang, Z.; Sainan, W.; Yu, H. Shelf life and nutritional quality of sorghum beer: Potentials of phytogenic-based extracts. *J. Agric. Food. Technol.* **2019**, *9*, 1–14.
14. Chaves-López, C.; Rossi, C.; Maggio, F.; Paparella, A.; Serio, A. Changes occurring in spontaneous maize fermentation: An overview. *Fermentation* **2020**, *6*, 36. [CrossRef]
15. Rocha dos Santos Mathias, T.; Moreira Menezes, L.; Camporese Sérvulo, E.F. Effect of maize as adjunct and the mashing proteolytic step on the brewer wort composition. *Beverages* **2019**, *5*, 65. [CrossRef]
16. Bogdan, P.; Kordialik-Bogacka, E. Alternatives to malt in brewing. *Trends Food Sci. Technol.* **2017**, *65*, 1–9. [CrossRef]
17. Hernández-Becerra, E.; Contreras-Jiménez, B.; Vuelvas-Solorzano, A.; Millan-Malo, B.; Muñoz-Torres, C.; Oseguera-Toledo, M.E.; Rodriguez-Garcia, M.E. Physicochemical and morphological changes in corn grains and starch during the malting for Palomero and Puma varieties. *Cereal Chem.* **2020**, *97*, 404–415. [CrossRef]
18. Dabija, A. *Biotehnologies in the Food Industries*; Performantica Press: Iasi, Romania, 2019.
19. Puligundla, P.; Smogrovicova, D.; Mok, C.; Obulam, V.S.R. Recent developments in high gravity beer-brewing. *Innov. Food Sci. Emerg. Technol.* **2020**, *64*, 102399. [CrossRef]

20. Goode, D.L.; Arendt, E.K. Developments in the supply of adjunct materials for brewing. In *Brewing*; Woodhead Publishing: Cambridge, UK, 2006; pp. 30–67.
21. Schnitzenbaumer, B.; Arendt, E.K. Brewing with up to 40% unmalted oats (*Avena sativa*) and sorghum (*Sorghum bicolor*): A review. *J. Inst. Brew.* **2014**, *120*, 315–330. [CrossRef]
22. He, Y.; Cao, Y.; Chen, S.; Ma, C.; Zhang, D.; Li, H. Analysis of flavour compounds in beer with extruded corn starch as an adjunct. *J. Inst. Brew.* **2018**, *124*, 9–15. [CrossRef]
23. Mellor, D.D.; Hanna-Khalil, B.; Carson, R. A review of the potential health benefits of low alcohol and alcohol-free beer: Effects of ingredients and craft brewing processes on potentially bioactive metabolites. *Beverages* **2020**, *6*, 25. [CrossRef]
24. Donadini, G.; Porretta, S. Uncovering patterns of consumers' interest for beer: A case study with craft beers. *Food Res. Int.* **2017**, *91*, 183–198. [CrossRef] [PubMed]
25. Morgan, D.R.; Thomas Lane, E.; Styles, D. Crafty Marketing: An Evaluation of Distinctive Criteria for "Craft" Beer. *Food Rev. Int.* **2020**, 1–17. [CrossRef]
26. Salanță, L.C.; Coldea, T.E.; Ignat, M.V.; Pop, C.R.; Tofană, M.; Mudura, E.; Zhao, H. Non-alcoholic and craft beer production and challenges. *Processes* **2020**, *8*, 1382. [CrossRef]
27. Kerpes, R.; Fischer, S.; Becker, T. The production of gluten-free beer: Degradation of hordeins during malting and brewing and the application of modern process technology focusing on endogenous malt peptidases. *Trends Food Sci. Technol.* **2017**, *67*, 129–138. [CrossRef]
28. Ciocan, M.; Dabija, A.; Codină, G.G. Effect of some unconventional ingredients on the production of black beer. *Ukr. Food J.* **2020**, *9*, 322–331. [CrossRef]
29. Gumienna, M.; Górna, B. Gluten hypersensitivities and their impact on the production of gluten-free beer. *Eur. Food Res. Technol.* **2020**, *247*, 2147–2160. [CrossRef]
30. Adiamo, O.Q.; Fawale, O.S.; Olawoye, B. Recent trends in the formulation of gluten-free sorghum products. *J. Culin. Sci. Technol.* **2018**, *16*, 311–325. [CrossRef]
31. Dlamini, B.C.; Taylor, J.R.N.; Buys, E.M. Influence of ammonia and lysine supplementation on yeast growth and fermentation with respect to gluten-free type brewing using unmalted sorghum grain. *Int. J. Food Sci. Technol.* **2020**, *55*, 841–850. [CrossRef]
32. Yorke, J.; Cook, D.; Ford, R. Brewing with Unmalted Cereal Adjuncts: Sensory and Analytical Impacts on Beer Quality. *Beverages* **2021**, *7*, 4. [CrossRef]
33. Ambra, R.; Pastore, G.; Lucchetti, S. The Role of Bioactive Phenolic Compounds on the Impact of Beer on Health. *Molecules* **2021**, *26*, 486. [CrossRef] [PubMed]
34. Adiaha, M.S.; Agba, O.A.; Attoe, E.E.; Ojikpong, T.O.; Kekong, M.A.; Obio, A.; Undie, U.L. Effect of maize (*Zea mays* L.) on human development and the future of man-maize survival: A review. *World Sci. News* **2016**, *59*, 52–62.
35. Steiner, E.; Auer, A.; Becker, T.; Gastl, M. Comparison of beer quality attributes between beers brewed with 100% barley malt and 100% barley raw material. *J. Sci. Food Agric.* **2012**, *92*, 803–813. [CrossRef] [PubMed]
36. Szambelan, K.; Nowak, J.; Szwengiel, A.; Jeleń, H. Comparison of sorghum and maize raw distillates: Factors affecting ethanol efficiency and volatile by-product profile. *J. Cereal Sci.* **2020**, *91*, 102863. [CrossRef]
37. Bussmann, R.W.; Batsatsashvili, K.; Kikvidze, Z.; Ghorbani, A.; Paniagua-Zambrana, N.Y.; Khutsishvili, M.; Tchelidze, D. Ethnobotany of Mountain Regions: Far Eastern Europe. *Ethnobot. Mt. Reg. Far East. Eur. Ural North. Cauc. Turk. Iran* **2020**, 3–43. [CrossRef]
38. Malhotra, S.K. Diversification in Utilization of Maize and Production. *Dep. Agric. Co-Oper. Farmers Welf. Minist. Agric. Farmers Welf. Gov. India A Compend.* **2017**, *5*, 49–57.
39. Singh, N.; Singh, S.; Shevkani, K. Maize: Composition, bioactive constituents, and unleavened bread. In *Flour and Breads and Their Fortification in Health and Disease Prevention*; Academic Press: London, UK, 2019; pp. 111–121.
40. Shah, T.R.; Prasad, K.; Kumar, P. Maize—A potential source of human nutrition and health: A review. *Cogent Food Agric.* **2016**, *2*, 1166995.
41. Eshetie, T. Review of quality protein maize as food and feed: In alleviating protein deficiency in developing countries. *Am. J. Food Nutr.* **2017**, 99–105. [CrossRef]
42. Abah, C.R.; Ishiwu, C.N.; Obiegbuna, J.E.; Oladejo, A.A. Sorghum Grains: Nutritional Composition, Functional Properties and Its Food Applications. *Eur. J. Nutr. Food Saf.* **2020**, *5*, 101–111. [CrossRef]
43. Adebo, O.A. African sorghum-based fermented foods: Past, current and future prospects. *Nutrients* **2020**, *12*, 1111. [CrossRef]
44. FAOSTAT Food and Agriculture Organization Statistics. Available online: http://www.fao.org/faostat/en/#data/QC (accessed on 25 February 2021).
45. Nnamchi, C.I.; Okolo, B.N.; Moneke, A.N. Grain and malt quality properties of some improved Nigerian sorghum varieties. *J. Inst. Brew.* **2014**, *120*, 353–359. [CrossRef]
46. Xiong, Y.; Zhang, P.; Warner, R.D.; Fang, Z. Sorghum grain: From genotype, nutrition, and phenolic profile to its health benefits and food applications. *Compr. Rev. Food Sci. Food Saf.* **2019**, *18*, 2025–2046. [CrossRef] [PubMed]
47. Perreta, M.; Ramos, J.; Tivano, J.C.; Vegetti, A. Descriptive characters of growth form in Poaceae—An overview. *Flora* **2011**, *206*, 283–293. [CrossRef]
48. Gupta, M.; Abu-Ghannam, N.; Gallaghar, E. Barley for Brewing: Characteristic Changes during Malting, Brewing and Applications of its By-Products. *Compr. Rev. Food Sci. Food Saf.* **2010**, *9*, 318–328. [CrossRef] [PubMed]

49. Watson, S.A. Description, development, structure, and composition of the corn kernel. In *Corn: Chemistry and Technology*, 2nd ed.; White, P.J., Johnson, L.A., Eds.; American Association of Cereal Chemists, Inc.: St. Paul, MN, USA, 2003; pp. 69–106.
50. Dicko, M.H.; Gruppen, H.; Traoré, A.S.; Voragen, A.G.; Van Berkel, W.J. Sorghum grain as human food in Africa: Relevance of content of starch and amylase activities. *Afr. J. Biotechnol.* **2006**, *5*, 384–395.
51. Fox, G.P. Chemical composition in barley grains and malt quality. In *Genetics and Improvement of Barley Malt Quality*; Zhang, G., Li, C., Eds.; Springer: Berlin/Heidelberg, Germany, 2010; pp. 64–99.
52. Baik, B.K.; Ullrich, S.E. Barley for food: Characteristics, improvement, and renewed interest. *J. Cereal Sci.* **2008**, *48*, 233–242. [CrossRef]
53. Šterna, V.; Zute, S.; Jākobsone, I. Grain composition and functional ingredients of barley varieties created in Latvia. In *Proceedings of the Latvian Academy of Sciences. Section B. Natural, Exact, and Applied Sciences*; Sciendo, 2015; Volume 69, pp. 158–162.
54. Le, T.A.T. Maize Malt Supplementation of Barley for the New Beer Production. Ph.D. Thesis, School of Biotechnology, Institute of Agricultural Technology, Suranaree University of Technology, Nakhon Ratchasima, Thailand, 2017.
55. Fišteš, A.; Došenovic, T.; Rakic, D.; Pajin, B.; Šereš, Z.; Simovic, Š.; Loncarevic, I. Statistical analysis of the basic chemical composition of whole grain flour of different cereal grains. *Acta Univ. Sapientiae Aliment.* **2014**, *7*, 45–53.
56. Alijošius, S.; Švirmickas, G.J.; Kliševičiūtė, V.; Gružauskas, R.; Šašytė, V.; Racevičiūtė-Stupelienė, A.; Dailidavičienė, J. The chemical composition of different barley varieties grown in Lithuania. *Vet. Zootech.* **2016**, *73*, 9–13.
57. Andersson, A.A.; Elfverson, C.; Andersson, R.; Regnér, S.; Åman, P. Chemical and physical characteristics of different barley samples. *J. Sci. Food Agric.* **1999**, *79*, 979–986. [CrossRef]
58. Ndife, J.; Nwokedi, C.U.; Ugwuona, F.U. Optimization of malting and saccharification in the production of malt beverage from maize. *Niger. J. Agric. Food Environ.* **2019**, *15*, 134–141.
59. Ignjatovic-Micic, D.; Vancetovic, J.; Trbovic, D.; Dumanovic, Z.; Kostadinovic, M.; Bozinovic, S. Grain nutrient composition of maize (*Zea mays* L.) drought-tolerant populations. *J. Agric. Food Chem.* **2015**, *63*, 1251–1260. [CrossRef]
60. Available online: http://www.fao.org/3/T0395E/T0395E03.htm (accessed on 1 March 2021).
61. Udachan, I.S.; Sahu, A.K.; Hend, F.M. Extraction and characterization of sorghum (*Sorghum bicolor* L. Moench) starch. *Int. Food Res. J.* **2012**, *19*, 315–319.
62. Singh, E.; Jain, P.K.; Sharma, S. Effect of different household processing on nutritional and anti-nutritional factors in Vigna aconitifolia and *Sorghum bicolour* (L.) Moench seeds and their product development. *J. Med. Nutr. Nutraceut.* **2015**, *4*, 95.
63. Mohapatra, D.; Patel, A.S.; Kar, A.; Deshpande, S.S.; Tripathi, M.K. Effect of different processing conditions on proximate composition, anti-oxidants, anti-nutrients and amino acid profile of grain sorghum. *Food Chem.* **2019**, *271*, 129–135. [CrossRef]
64. Nghiem, N.P.; Montanti, J.; Johnston, D.B. Sorghum as a renewable feedstock for production of fuels and industrial chemicals. *Bioengineering* **2016**, *3*, 75–91. [CrossRef]
65. Vilela, J.D.A.S.; de Figueiredo Vilela, L.; Ramos, C.L.; Schwan, R.F. Physiological and genetic characterization of indigenous Saccharomyces cerevisiae for potential use in productions of fermented maize-based-beverages. *Braz. J. Microbiol.* **2020**, *51*, 1297–1307. [CrossRef]
66. Piacentini, K.C.; Rocha, L.O.; Savi, G.D.; Carnielli-Queiroz, L.; Corrêa, B. Beer industry in Brazil: Economic aspects, characteristics of the raw material and concerns. *Kvas. Prum.* **2018**, *64*, 284–286. [CrossRef]
67. Estevão, S.T.; e Silva, J.B.D.A.; Lourenço, F.R. Development and optimization of beer containing malted and non-malted substitutes using quality by design (QbD) approach. *J. Food Eng.* **2021**, *289*, 110182. [CrossRef]
68. Zhu, L.; Ma, T.; Mei, Y.; Li, Q. Enhancing the hydrolysis of corn starch using optimal amylases in a high-adjunct-ratio malt mashing process. *Food Sci. Biotechnol.* **2017**, *26*, 1227–1233. [CrossRef] [PubMed]
69. Poreda, A.; Czarnik, A.; Zdaniewicz, M.; Jakubowski, M.; Antkiewicz, P. Corn grist adjunct–application and influence on the brewing process and beer quality. *J. Inst. Brew.* **2014**, *120*, 77–81. [CrossRef]
70. Diakabana, P.; Mvoula-Tsiéri, M.; Dhellot, J.; Kobawila, S.C.; Louembé, D. Physico-chemical characterization of brew during the brewing corn malt in the production of maize beer in Congo. *Adv. J. Food Sci. Technol.* **2013**, *5*, 671–677. [CrossRef]
71. Fumi, M.D.; Galli, R.; Lambri, M.; Donadini, G.; De Faveri, D.M. Effect of full–scale brewing process on polyphenols in Italian allzmalt and maize adjunct lager beers. *J. Food Compos. Anal.* **2011**, *4–5*, 568–573. [CrossRef]
72. Fumi, M.D.; Galli, R.; Lambri, M.; Donadini, G.; Faveri, D.M.D. Impact of full-scale brewing processes on lager beer nitrogen compounds. *Eur. Food Res. Technol.* **2009**, *230*, 209–216. [CrossRef]
73. Perez-Carrillo, E.; Serna-Saldivar, S.O.; Chuck-Hernandez, C.; Cortes-Callejas, M.L. Addition of protease during starch liquefaction affects free amino nitrogen, fusel alcohols and ethanol production of fermented maize and whole and decorticated sorghum mashes. *Biochem. Eng. J.* **2012**, *67*, 1–9. [CrossRef]
74. Zweytik, G.; Berghofer, E. Production of gluten-free beer. In *Gluten-Free Food Science and Technology*; Gallagher, E., Ed.; Wiley-Blackwell: Oxford, UK, 2009.
75. Romero-Medina, A.; Estarrón-Espinosa, M.; Verde-Calvo, J.R.; Lelièvre-Desmas, M.; Escalona-Buendía, H.B. Renewing traditions: A sensory and chemical characterisation of mexican pigmented corn beers. *Foods* **2020**, *9*, 886. [CrossRef]
76. Bernal-Gil, N.Y.; Favila-Cisneros, H.J.; Zaragoza-Alonso, J.; Cuffia, F.; Rojas-Rivas, E. Using projective techniques and Food Neophobia Scale to explore the perception of traditional ethnic foods in Central Mexico: A preliminary study on the beverage Sende. *J. Sens. Stud.* **2020**, *35*, e12606. [CrossRef]

77. Bassi, D.; Orrù, L.; Cabanillas Vasquez, J.; Cocconcelli, P.S.; Fontana, C. Peruvian chicha: A focus on the microbial populations of this ancient Maize-based fermented beverage. *Microorganisms* **2020**, *8*, 93. [CrossRef]

78. Williams, P.R.; Nash, D.J.; Henkin, J.M.; Armitage, R.A. Archaeometric Approaches to Defining Sustainable Governance: Wari Brewing Traditions and the Building of Political Relationships in Ancient Peru. *Sustainability* **2019**, *11*, 2333. [CrossRef]

79. Adekoya, I.; Obadina, A.; Adaku, C.C.; De Boevre, M.; Okoth, S.; De Saeger, S.; Njobeh, P. Mycobiota and co-occurrence of mycotoxins in South African maize-based opaque beer. *Int. J. Food Microbiol.* **2018**, *270*, 22–30. [CrossRef] [PubMed]

80. Hlangwani, E.; Adebiyi, J.A.; Doorsamy, W.; Adebo, O.A. Processing, Characteristics and Composition of *Umqombothi* (a South African Traditional Beer). *Processes* **2020**, *8*, 1451. [CrossRef]

81. Cason, E.D.; Mahlomaholo, B.J.; Taole, M.M.; Abong, G.O.; Vermeulen, J.G.; De Smidt, O.; Viljoen, B. Bacterial and fungal dynamics during the fermentation process of sesotho, a traditional beer of Southern Africa. *Front. Microbiol.* **2020**, *11*, 1451. [CrossRef]

82. Gadaga, T.H.; Lehohla, M.; Ntuli, V. Traditional fermented foods of Lesotho. *J. Microbiol. Biotechnol. Food Sci.* **2020**, *9*, 2387–2391.

83. Mawonike, R.; Chigunyeni, B.; Chipumuro, M. Process improvement of opaque beer (chibuku) based on multivariate cumulative sum control chart. *J. Inst. Brew.* **2018**, *124*, 16–22. [CrossRef]

84. Konfo, C.T.R.; Chabi, N.W.; Dahouenon-Ahoussi, E.; Cakpo-Chichi, M.; Soumanou, M.M.; Sohounhloue, D.C.K. Improvement of African traditional sorghum beers quality and potential applications of plants extracts for their stabilization: A review. *J. Microbiol. Biotechnol. Food Sci.* **2020**, *9*, 190–196. [CrossRef]

85. Andualem, B.; Gessesse, A. Isolation and identification of amylase producing yeasts in 'tella' (Ethiopian local beer) and their amylase contribution for 'tella'production. *J. Microbiol. Biotechnol. Food Sci.* **2020**, *9*, 30–34.

86. Fentie, E.G.; Emire, S.A.; Demsash, H.D.; Dadi, D.W.; Shin, J.-H. Cereal- and Fruit-Based Ethiopian Traditional Fermented Alcoholic Beverages. *Foods* **2020**, *9*, 1781. [CrossRef] [PubMed]

87. Adesulu-Dahunsi, A.T.; Dahunsi, S.O.; Olayanju, A. Synergistic microbial interactions between lactic acid bacteria and yeasts during production of Nigerian indigenous fermented foods and beverages. *Food Control* **2020**, *110*, 106963. [CrossRef]

88. Krstanović, V.; Habschied, K.; Lukinac, J.; Jukić, M.; Mastanjević, K. The Influence of Partial Substitution of Malt with Unmalted Wheat in Grist on Quality Parameters of Lager Beer. *Beverages* **2020**, *6*, 7. [CrossRef]

89. Obatolu, V.A.; Adeniyi, P.O.; Ashaye, O.A. Nutritional, Sensory and Storage Quality of Sekete from *Zea mays*. *Int. J. Food Sci. Nutr.* **2016**, *6*, 73–80.

90. Agu, R.C.A. comparison of maize, sorghum and barley as brewing adjuncts. *J. Inst. Brew.* **2002**, *108*, 19–22. [CrossRef]

91. Goode, D.L.; Arendt, E.K. Pilot Scale Production of a Lager Beer from a Grist Containing 50% Unmalted Sorghum. *J. Inst. Brew.* **2003**, *109*, 208–217. [CrossRef]

92. Ogbonna, A.C. Current Developments in Malting and Brewing Trials with Sorghum in Nigeria: A Review. *J. Inst. Brew.* **2011**, *117*, 394–400. [CrossRef]

93. Beta, T.; Rooney, L.W.; Waniska, R.D. Malting characteristics of sorghum cultivars. *Cereal* **1995**, *72*, 533–538.

94. Chandra, G.S.; Proudlove, M.O.; Baxter, E.D. The structure of barley endosperm an important determinant of malt modification. *J. Sci. Food Agric.* **1999**, *79*, 37–46. [CrossRef]

95. Oyewole, O.I.; Agboola, F.K. Comparative studies on properties of amylases extracted from kilned and unkilned malted sorghum and corn. *Int. J. Biotechnol. Mol. Biol. Res.* **2011**, *2*, 146–149.

96. Eneje, L.O.; Ogu, E.O.; Aloh, C.U.; Agu, R.C.; Palmer, G.H. Effects of steeping and germination on malting performance of Nigerian white and yellow maize varieties. *Process. Biochem.* **2004**, *39*, 1013–1016. [CrossRef]

97. Taylor, J.R.N.; Dlamini, B.C.; Kruger, J. 125th anniversary review: The science of the tropical cereals sorghum, maize and rice in relation to lager beer brewing. *J. Inst. Brew.* **2013**, *119*, 1–14. [CrossRef]

98. Espinosa-Ramírez, J.; Pérez-Carrillo, E.; Serna-Saldívar, S.O. Production of brewing worts from different types of sorghum malts and adjuncts supplemented with β-amylase or amyloglucosidase. *J. Am. Soc. Brew. Chem.* **2013**, *71*, 49–56. [CrossRef]

99. Heredia-Olea, E.; Cortés-Ceballos, E.; Serna-Saldívar, S.O. Malting sorghum with Aspergillus oryzae enhances gluten-free wort yield and extract. *J. Am. Soc. Brew. Chem.* **2017**, *75*, 116–121. [CrossRef]

100. Espinosa-Ramírez, J.; Pérez-Carrillo, E.; Serna-Saldívar, S.O. Maltose and glucose utilization during fermentation of barley and sorghum lager beers as affected by β-amylase or amyloglucosidase addition. *J. Cereal Sci.* **2014**, *60*, 602–609. [CrossRef]

101. Urias-Lugo, D.A.; Saldivar, S.O.S. Effect of amyloglucosidase on properties of lager beers produced from sorghum malt and waxy grits. *J. Am. Soc. Brew. Chem.* **2005**, *63*, 63–68. [CrossRef]

102. Shen, S.; Huang, R.; Li, C.; Wu, W.; Chen, H.; Shi, J.; Chen, S.; Ye, X. Phenolic Compositions and Antioxidant Activities Differ Significantly among Sorghum Grains with Different Applications. *Molecules* **2018**, *23*, 1203. [CrossRef]

103. Attchelouwa, C.K.; Aka-Gbézo, S.; N'guessan, F.K.; Kouakou, C.A.; Djè, M.K. Biochemical and Microbiological Changes during the Ivorian Sorghum Beer Deterioration at Different Storage Temperatures. *Beverages* **2017**, *3*, 43. [CrossRef]

104. Embashu, W.; Iileka, O.; Nantanga, K.K.M. Namibian opaque beer: A review. *J. Inst. Brew.* **2019**, *125*, 4–9. [CrossRef]

105. Zaukuu, J.L.Z.; Oduro, I.; Ellis, W.O. Processing methods and microbial assessment of *pito* (an African indigenous beer), at selected production sites in Ghana. *J. Inst. Brew.* **2016**, *122*, 736–744. [CrossRef]

106. Ogbonna, A.C. Developments in the malting and brewing trials with sorghum. *World J. Microbiol. Biotechnol.* **1992**, *8*, 87–91. [CrossRef] [PubMed]

107. Agu, R.C.; Palmer, G.H. A reassessment of sorghum for lager-beer brewing. *Bioresour. Technol.* **1998**, *66*, 253–261. [CrossRef]

108. Owuama, C.I. Brewing beer with sorghum. *J. Inst. Brew.* **1999**, *105*, 23–34. [CrossRef]
109. Rubio-Flores, M.; García-Arellano, A.R.; Perez-Carrillo, E.; Serna-Saldivar, S.O. Use of Aspergillus oryzae during sorghum malting to enhance yield and quality of gluten-free lager beers. *Bioresour. Bioprocess.* **2020**, *7*, 1–11. [CrossRef]
110. Sawadogo-Lingani, H.; Lei, V.; Diawara, B.; Nielsen, D.; Møller, P.; Traoré, A.; Jakobsen, M. The biodiversity of predominant lactic acid bacteria in dolo and pito wort for the production of sorghum beer. *J. Appl. Microbiol.* **2007**, *103*, 765–777. [CrossRef]
111. Djameh, C.; Saalia, F.K.; Sinayobye, E.; Budu, A.; Essilfie, G.; Mensah-Brown, H.; Sefa-Dedeh, S. Optimization of the sorghum malting process for pito production in Ghana. *J. Inst. Brew.* **2015**, *121*, 106–112. [CrossRef]
112. Adadi, P.; Kanwugu, O.N. Potential application of tetrapleura tetraptera and hibiscus sabdariffa (malvaceae) in designing highly flavoured and bioactive pito with functional properties. *Beverages* **2020**, *6*, 22. [CrossRef]
113. Djameh, C.; Ellis, W.O.; Oduro, I.; Saalia, F.K.; Haslbeck, K.; Komlaga, G.A. West African sorghum beer fermented with Lactobacillus delbrueckii and *Saccharomyces cerevisiae*: Fermentation by-products. *J. Inst. Brew.* **2019**, *125*, 326–332. [CrossRef]
114. Vázquez-Araújo, L.; Chambers IV, E.; Cherdchu, P. Consumer Input for Developing Human Food Products Made with Sorghum Grain. *J. Food Sci.* **2012**, *77*, S384–S389. [CrossRef]
115. N'Guessan, F.K.; Coulibaly, H.W.; Alloue-Boraud, M.W.A.; Cot, M.; Djè, K.M. Production of freeze-dried yeast culture for the brewing of traditional sorghum beer, tchapalo. *Food Sci. Nutr.* **2016**, *4*, 34–41. [CrossRef]
116. Ezekiel, C.N.; Ayeni, K.I.; Misihairabgwi, J.M.; Somorin, Y.M.; Chibuzor-Onyema, I.E.; Oyedele, O.A.; Abia, W.A.; Sulyok, M.; Shephard, G.S.; Krska, R. Traditionally Processed Beverages in Africa: A Review of the Mycotoxin Occurrence Patterns and Exposure Assessment. *Compr. Rev. Food Sci. Food Saf.* **2018**, *17*, 334–351. [CrossRef]
117. Aka, S.; Dridi, B.; Bolotin, A.; Yapo, E.A.; Koussemon-Camara, M.; Bonfoh, B.; Renault, P. Characterization of lactic acid bacteria isolated from a traditional Ivoirian beer process to develop starter cultures for safe sorghum-based beverages. *Int. J. Food Microbiol.* **2020**, *322*, 108547. [CrossRef] [PubMed]
118. Tano, M.B.; Aka-Gbezo, S.; Attchelouwa, C.K.; Koussémon, M. Use of Lactic Acid Bacteria as Starter Cultures in the Production of Tchapalo, a Traditional Sorghum Beer from Côte d'Ivoire. *Am. J. Food Sci. Health* **2020**, *6*, 23–31.
119. Shimotsu, S.; Asano, S.; Iijima, K.; Suzuki, K.; Yamagishi, H.; Aizawa, M. Investigation of beer-spoilage ability of *Dekkera/Brettanomyces* yeasts and development of multiplex PCR method for beer-spoilage yeasts. *J. Inst. Brew.* **2015**, *121*, 177–180. [CrossRef]
120. Rogerson, C.M. African traditional beer: Changing organization and spaces of South Africa's sorghum beer industry. *Afr. Geogr. Rev.* **2019**, *38*, 253–267. [CrossRef]
121. Aruna, C.; Visaradal, K.B.R.S. Sorghum grain in food and brewing industry. In *Breeding Sorghum for Diverse End Uses*; Woodhead Publishing: Cambridge, UK, 2019; pp. 209–228.
122. Tamang, J.P.; Cotter, P.; Endo, A. Fermented foods in a global age: East meets West. *Compr. Rev. Food Sci. Food Saf.* **2020**, *19*, 184–217. [CrossRef] [PubMed]
123. Pale, S.; Taonda, S.J.; Bougouma, B.; Mason, S.C. Sorghum malt and traditional beer (dolo) quality assessment in Burkina Faso. *Ecol. Food Nutr.* **2010**, *49*, 129–141. [CrossRef]
124. Hadebe, S.T.; Modi, A.T.; Mabhaudhi, T. Drought Tolerance and Water Use of Cereal Crops: A Focus on Sorghum as a Food Security Crop in Sub-Saharan Africa. *J. Agron. Crop. Sci.* **2017**, *203*, 177–191. [CrossRef]
125. Mogmenga, I.; Dadiré, Y.; Somda, M.K.; Keita, I.; Ezeogu, L.I.; Ugwuanyi, J.; Traoré, A.S. Isolation and Identification of Indigenous Yeasts from "Rabilé", a Starter Culture Used for Production of Traditional Beer "dolo", a Condiment in Burkina Faso. *Adv. Microbiol.* **2019**, *9*, 646. [CrossRef]
126. Disharoon, A.; Boyles, R.; Jordan, K.; Kresovich, S. Exploring diverse sorghum (*Sorghum bicolor* (L.) Moench) accessions for malt amylase activity. *J. Inst. Brew.* **2021**, *127*, 5–12. [CrossRef]
127. Nso, E.; Ajebesome, P.; Mbofung, C.; Palmer, G. Properties of Three Sorghum Cultivars Used for the Production of Bili-Bili Beverage in Northern Cameroon. *J. Inst. Brew.* **2003**, *109*, 245–250. [CrossRef]
128. Maoura, N.; Mbaiguinam, M.; Nguyen, H.V.; Gaillardin, C.; Pourquie, J. Identification and typing of the yeast strains isolated from bili bili, a traditional sorghum beer of Chad. *Afr. J. Biotechnol.* **2005**, *4*, 646–656.
129. Desobgo, Z.S.C.; Nso, E.J.; Tenin, D.; Kayem, G.J. Modelling and Optimizing of Mashing Enzymes—Effect on Yield of Filtrate of Unmalted Sorghum by Use of Response Surface Methodology. *J. Inst. Brew.* **2010**, *116*, 62–69. [CrossRef]
130. Kubo, R.; Funakawa, S.; Araki, S.; Kitabatake, N. Production of indigenous alcoholic beverages in a rural village of Cameroon. *J. Inst. Brew.* **2014**, *120*, 133–141. [CrossRef]
131. Embashu, W.; Nantanga, K.K.M. Malts: Quality and phenolic content of pearl millet and sorghum varieties for brewing nonalcoholic beverages and opaque beers. *Cereal Chem.* **2019**, *96*, 765–774. [CrossRef]
132. Soji, P.A. The Potential Importance of Maize, (*Zea mays* L), in Nigeria, [A Case Study of 2800 Farmers Sampled at Different Locations]. *Adv. Biochem.* **2020**, *8*, 1. [CrossRef]
133. Palanisamy, C.P.; Cui, B.; Zhang, H.; Jayaraman, S.; Kodiveri Muthukaliannan, G. A Comprehensive Review on Corn Starch-Based Nanomaterials: Properties, Simulations, and Applications. *Polymers* **2020**, *12*, 2161. [CrossRef] [PubMed]
134. Gálvez Ranilla, L. The Application of Metabolomics for the Study of Cereal Corn (*Zea mays* L.). *Metabolites* **2020**, *10*, 300. [CrossRef] [PubMed]
135. Green, D.I.G.; Agu, R.C.; Bringhurst, T.A.; Brosnan, J.M.; Jack, F.R.; Walker, G.M. Maximizing alcohol yields from wheat and maize and their co-products for distilling or bioethanol production. *J. Inst. Brew.* **2015**, *121*, 332–337. [CrossRef]

136. Onuki, S.; Koziel, J.A.; Jenks, W.S.; Cai, L.; Grewell, D.; van Leeuwen, J.H. Taking ethanol quality beyond fuel grade: A review. *J. Inst. Brew.* **2016**, *122*, 588–598. [CrossRef]
137. Ai, Y.; Jane, J. Macronutrients in Corn and Human Nutrition. *Compr. Rev. Food Sci. Food Saf.* **2016**, *15*, 581–598. [CrossRef]
138. Ortíz-Islas, S.; García-Lara, S.; Preciado-Ortíz, R.E.; Serna-Saldívar, S.O. Fatty acid composition and proximate analysis of improved high-oil corn double haploid hybrids adapted to subtropical areas. *Cereal Chem.* **2019**, *96*, 182–192. [CrossRef]
139. Shi, M.; Gao, Q.; Liu, Y. Corn, potato, and wrinkled pea starches with heat–moisture treatment: Structure and digestibility. *Cereal Chem.* **2018**, *95*, 603–614. [CrossRef]
140. Zhang, Y.; Li, Y. Comparison of physicochemical and mechanical properties of edible films made from navy bean and corn starches. *J. Sci. Food Agric.* **2021**, *101*, 1538–1545. [CrossRef]
141. Ari Akin, P.; Miller, R.; Jaffe, T.; Koppel, K.; Ehmke, L. Sensory profile and quality of chemically leavened gluten-free sorghum bread containing different starches and hydrocolloids. *J. Sci. Food Agric.* **2019**, *99*, 4391–4396. [CrossRef] [PubMed]
142. Galassi, E.; Taddei, F.; Ciccoritti, R.; Nocente, F.; Gazza, L. Biochemical and technological characterization of two C4 gluten-free cereals: Sorghum bicolor and Eragrostis tef. *Cereal Chem.* **2020**, *97*, 65–73. [CrossRef]
143. Awobusuyi, T.D.; Pillay, K.; Siwela, M. Consumer Acceptance of Biscuits Supplemented with a Sorghum–Insect Meal. *Nutrients* **2020**, *12*, 895. [CrossRef] [PubMed]
144. Dabija, A. *Unconventional Raw Materials for the Bakery Industry*; Performantica Publishing House: Iași, Romania, 2020; ISBN 978-606-685-731-4.
145. Rashwan, A.K.; Yones, H.A.; Karim, N.; Taha, E.M.; Chen, W. Potential processing technologies for developing sorghum-based food products: An update and comprehensive review. *Trends Food Sci. Technol.* **2021**, *110*, 168–182. [CrossRef]
146. Available online: https://www.mordorintelligence.com/industry-reports/beer-market (accessed on 25 March 2021).

 *applied sciences*

Article

Sorghum Flour: A Valuable Ingredient for Bakery Industry?

Livia Apostol [1], Nastasia Belc [1],*, Liviu Gaceu [2,3,4], Oana Bianca Oprea [2],* and Mona Elena Popa [5]

[1] National Research & Development Institute for Food Bioresources—IBA Bucharest, 6 Dinu Vintila Street, 0211202 Bucharest, Romania; apostol_livia@yahoo.com
[2] Faculty of Food and Tourism, Transilvania University of Brasov, 148 Castelului Street, 500014 Brasov, Romania; gaceul@unitbv.ro
[3] CSCBAS &CE-MONT Centre/INCE—Romanian Academy, Casa Academiei Române, Calea 13 Septembrie no. 13, 050711 Bucharest, Romania
[4] Academy of Romanian Scientists, Ilfov Street, no. 3, 050711 Bucharest, Romania
[5] Faculty of Biotechnology, University of Agronomic Sciences and Veterinary Medicine of Bucharest, 011464 Bucharest, Romania; monapopa@agral.usamv.ro
* Correspondence: nastasia.belc@bioresurse.ro (N.B.); oprea.oana.bianca@unitbv.ro (O.B.O.); Tel.: +40-727-171-083 (O.B.O.)

Received: 27 October 2020; Accepted: 27 November 2020; Published: 30 November 2020

Abstract: The information from this study may provide opportunities for industrial application of sorghum seed flour as a useful bakery ingredient and a suitable alternative source of functional compounds to whole wheat flour. The chemical composition of sorghum was evaluated compared to that of wheat whole flour, showing high contents of mineral and fibers. Next were evaluated the dough rheological properties of flour mixtures using Mixolab equipment, "Chopin+" protocol. Finally, six bread samples were obtained from wheat flour with addition of sorghum seed flour in various percentages, in which three samples were fortified with *Lactobacillus plantarum* compared to the other three bread samples without the addition of any *lactic acid bacteria*. All six bread sample were compared to a control bread sample with wheat flour type 550. The results show the fat and raw fiber were higher in sorghum compared to whole wheat flour. Also, the content of magnesium, potassium, and iron were much higher than in whole wheat flour. A significant improvement of the sensorial characteristics was observed in bread samples in which lactic acid bacteria was used.

Keywords: sorghum seeds; whole wheat flour; *Lactobacillus plantarum*

---

1. Introduction

Sorghum bicolor (L.), originated in Africa, commonly called sorghum is a specie cultivated for its grain, which is used for human food, animal feed, and ethanol production. Sorghum is the world's fifth-most important cereal crop after rice, wheat, maize, and barley. *Sorghum bicolor* is typically an annual, but some cultivars are perennial. Sweet sorghums are sorghum cultivars that are primarily grown for forage, syrup production, and ethanol.

Sorghum plays a decisive role in food security in developing countries. It is used in many types of food recipes such as breads, porridges, pastes, and griddlecake [1]. Due to the content of phenolic compounds, diet fibers and antioxidant activity, *Sorghum bicolor* has various applications in African traditional medicine and many of its uses have been mentioned in literature [1,2].

In India, a decoction of sorghum seeds is used as a demulcent and diuretic for treating kidney and urinary tract complaints. Numerous studies have shown that the biocompounds responsible for the red color of sorghum also have antimicrobial, antifungal, and anti-anemic properties [3].

Lately, in occidental countries, the use of sorghum in human consumption has increased due to its antioxidant potential with a role in reducing the risk of developing chronic diseases (obesity, cardiovascular diseases, hypertension, diabetes, and cancer) [4,5].

Thus, S. Ben Slima et al. extracted a soluble polysaccharide from *Sorghum bicolor* (L.) seeds and evaluated in vitro hemolytic and antioxidant activities as well as its in vivo wound healing ability to treat burns induced by fractional CO_2 laser. The obtained results have shown that this polysaccharide was efficient on wound closure and that it might be useful as a wound healing agent in modern medicine. Also, sorghum is a cereal recommended as a safe food for celiac patients because it is gluten-free [6]. Sorghum is a major source of energy, serving as a staple food to many of the world's poorest and least privileged population [7]. Some of the challenges associated with the use of sorghum, in the production of bread are associated to reduced volume, hard texture, and poor sensory attributes. These shortcomings of unconventional flours have been enhanced with sourdough fermentation, providing nutritive food with attractive flavor and texture [8–10].

One of the main trends in the market is the use of sourdough especially for bread because it has been shown that it has a positive influence on bakery products both in terms of sensory, nutritional texture, and shelf life [11,12]. Overall, the fermentation with sourdough showed new ways to improve the quality and acceptability of bread with the addition of gluten-free matrices. During fermentation with sourdough, acid appears, which improves the swelling of polysaccharides that could partially replace gluten and improve the structure of gluten-free bread [13]. According to Coda et al. [14], lactic acid bacteria are very important because this ferments gluten-free flours. The result is the production of functional bread enriched with bioactive compounds. *L. plantarum* was selected to synthesize γ-aminobutyric acid (GABA) by fermenting wheat sourdough, sorghum, rye, spelled, oats, buckwheat, rice, amaranth, millet, chickpeas, and soybeans.

The objective of the present study was to compare the nutritional composition of *Sorghum bicolor* (alimentary hybrid) seeds flour with whole wheat flour, on the future prospect of using this cereal in the bakery industry. In order to highlight the valuable chemical composition of sorghum vis-à-vis wheat, whole meal sorghum flour and whole wheat flour were studied comparatively (potassium, magnesium, zinc, iron, etc. contents).

To evaluate the ability to integrate sorghum flour into classical bread making technology, the rheology of sorghum flour mixtures with white wheat flour was studied, compared to a control sample of 100% type 550 wheat flour. Finally, sensorial analysis of breads was performed.

2. Materials and Methods

2.1. Raw Materials

Sorghum bicolor (Alimentary hybrid) seeds flour was supplied by Agricultural Research and Development Station Secuieni, Neamț county (Romania).

Whole wheat flour used in the study was provided by *Hofigal Export—Import SA* (Bucharest, Romania). Sorghum flour was compared with whole wheat flour (provided by Hofigal Export) to demonstrate the special value of sorghum. Subsequently, rheological analyzes were performed comparatively between mixtures of sorghum and type 550 wheat flour (provided by Titan S.A.), respectively compared to the control sample of 100% white wheat flour type 550.

2.2. Preparation of Flour Mixtures

Raw materials: Wheat flour type 550 (ash = 0.55%) used to prepare bread samples was provided by Titan S.A. (Bucharest, Romania), *Sorghum bicolor* (Alimentary hybrid) seeds flour—by Agricultural Research and Development Station Secuieni (Romania), *L. plantarum* (*Aurum Plantarum* Millbo srl— Via Bellaria s.n.—Novara/Italy), dried yeast, iodized salt for food use.

In Table 1 are presented the six samples of mixtures from wheat flour and different proportions of sorghum seed flour.

Table 1. Types of flours used in the experiments.

P1	70% wheat flour (type 550) + 30% sorghum flour without *L. plantarum*
P2	70% wheat flour (type 550) + 30% sorghum flour with *L. plantarum*
P3	65% wheat flour (type 550) + 35% sorghum flour without *L. plantarum*
P4	65% wheat flour (type 550) + 35% sorghum flour with *L. plantarum*
P5	60% wheat flour (type 550) + 40% sorghum flour without *L. plantarum*
P6	60% wheat flour (type 550) + 40% sorghum flour with *L. plantarum*
P7	100% wheat flour (type 550)

Note: The used amount of *L. plantarum* was 0.3 g/100 g flour mixture.

2.3. Chemical Analysis

Moisture content was determined according to ICC Standard No. 110/1 [15]. The ash content was determined by incineration at $525 \pm 25\,°C$ (ICC no. 104/1) [16]. Total nitrogen (N) and crude protein content (N·6.25, conversion factor) was determined by the Macro Kjeldahl Method (SR EN ISO 20483:2007). Total fat content was determined by extracting with petroleum ether at 40–65 °C, to the Romanian standard SR 91/2007 [17]. The subject to chemical analysis was sorghum flour and whole wheat flours.

2.4. Crude Fiber Content Analysis

Using FIBRETHERM–Gerhardt apparatus the content of crude fiber (cellulose, hemicellulose, and lignin) was determined for the above listed samples. The method for determining the crude fiber begins with treating the sample with an acid detergent solution (20 g *N*-cetyl-*N*,*N*,*N*-trimethylammonium bromide dissolved in 1 L sulfuric acid 0.5 M). In this solution, cellulose hemicellulose and lignin from the analysed material are insoluble, unlike all other components. Using special fiber bags, the dilution and filtration steps are simplified. The most important aspects of this method of fiber analysis are adherence to strict boiling times and to weighing procedures.

After treatment with the acid detergent solution, the insoluble residue is dried, weighed, and then calcinated. The acid detergent fiber (ADF) content represents the insoluble part of the sample that is left after boiling in acid detergent solution from which the ash obtained upon calcination is subtracted

$$\%ADF = \frac{((\chi - \alpha) - (\delta - \xi)) \times 100}{\beta} \tag{1}$$

$$Blank\ value\ (\xi) = \gamma - \psi \tag{2}$$

where:
α = mass of fiber bBag (g)
β = sample mass (g)
χ = mass of crucible and dried fiber bag, after digestion (g)
δ = mass of crucible and and ash (g)
ζ = blank value of empty fiber bag (g)
γ = mass of crucible and ash of the empty fiber bag (g)
Ψ = mass of crucible (g)

2.5. Mineral Content Analysis

Mineral content was determined using an atomic absorption spectrophotometer (ContrAA 700; Analitykjena). Total ash was determined by incineration at 550 °C, in an oven. Analysis was performed using an external standard (Merck, multi element standard solution) and calibration curves for all minerals were obtained using six different concentrations. Dried samples were digested in concentrated HCl.

2.6. Amino Acid Content Analysis

For the analysis of amino acid content, samples were hydrolyzed at 100–120 °C in 6 N hydrochloric acid for 22–24 h under vacuum.

After evaporation to dryness of hydrochloric acid, the dry residue was diluted using 4 mM stock solution of Norleucine. For the separation of amino acids by gradient anion exchange with pulsed electrochemical detection (PED) was used an ICS300 (Dionex, sSunnyvale, CA, USA) equipment with the following eluents: deionized water, 0.250 M NaOH and 1 M CH3COONa. Amino acids were expressed as g amino acid/100 g protein.

The chemical score of samples was calculated according to FAO/WHO (1985) [18] as

$$\text{Chemical Score} = \frac{\text{mg/g of essential amino acid in test protein}}{\text{mg/g of essential amino acid reference protein}} \times 100 \tag{3}$$

2.7. In Vitro Protein Digestibility Method

The in vitro protein digestibility of sorghum and wheat flour was achieved according to the previous methods [19,20] with small changes using the enzyme Trypsin from the porcine pancreas—type IX-S.

An aliquot of 50 mL of aqueous protein suspension (6.25 mg protein/mL) in double distilled water is placed in a water bath with the temperature set to 37 °C and to an adjusted pH of 8.0 with 0.1 N HCl. The enzyme solution (containing 1.6 mg trypsin/mL) was maintained in an ice bath and 5 mL of the solution was then added to the protein suspension.

The pH drop was automatically recorded after a 10 min period of time, using a WTW InoLab 7110 (Weilheim, Germany). The protein digestibility percentage of each protein sample was calculated using the regression equation predicted by Hsu et al. (1977) [19] as

$$\% \text{ Protein digestibility (Y)} = 210.46 - 18.1X \tag{4}$$

where X is the final pH value of each sample after a 10 min digestion.

2.8. Fatty Acid Content by Gas Chromatography Method

The oil content was analyzed using AOAC method, 920.85 [21] with Soxhlet apparatus. The lipid extracts from the sorghum and wheat flours were mixed with boron trifluoride (BF3)–methanol reagent (20%) and fatty acids were converted into the methyl ester derivatives [22].

Chromatographic analysis was performed on a Trace GC Ultra/TSQ Quantum XLS system (Thermo Fisher Scientific, Waltham, MA, USA), a gas chromatograph coupled with a mass spectrometer (MS) TSQ Quantum XLS, autosampler, TriPlus AS. FAMEs/FAs separation was realized on a high polarity capillary column, TR-FAME (60 m × 0.25 mm × 0.25 µm film thickness) of 70% cyanopropyl and 30% polysilphenyl-siloxane. Analysis of calibration solutions and the acquired extract samples were performed in the positive electron impact ionization (EI+) mode, selected ion monitoring (SIM) mode, using 24 segments. The ion source temperature was 250 °C, the oven temperature was programmed at 100 °C for 0.2 min, increased to 240 °C with 2 °C/min and hold for 15 min. The mobile phase was He of a 99.9995% (5.0) purity at a constant flow rate of 1 mL/min. A volume of 0.5 µL extract was injected at 240 °C in split mode with a 1:50 split ratio and a 50 mL/min splitting rate. Instrument control, data acquisition, and processing were performed using the Xcalibur Program. The total run time of a GC-MS chromatogram was 85.20 min.

Components identification was done by comparison of their recognition and their retention times and mass spectra with corresponding data from reference compounds.

2.9. Rheological Properties Testing

Dough rheological behavior was determined using the predefined "Chopin+" protocol on Mixolab, according to ICC no. 173 [23], a protocol for complete characterization of flours (water absorption, protein quality, amylase activity, and starch quality), and a simplified graphic interpretation of the results was performed (Mixolab device—Chopin, Tripette et Renaud, Paris, France) [24].

The rheological behavior analysis of the Mixolab procedure parameters were the following: tank temperature 30 °C, mixing speed 80 min^{-1}, heating rate 2 °C/min, total analysis time 45 min. Mixolab curves recorded are basically characterized by torque in five defined points ("C1"–"C5", N·m), temperatures and processing times corresponding to them (Figure 1).

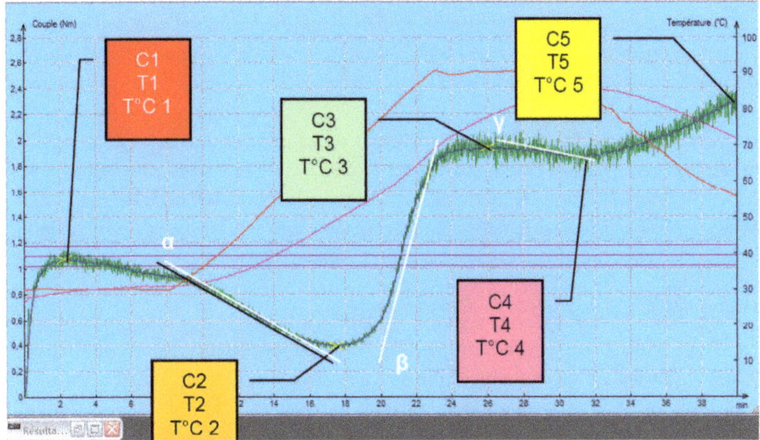

Figure 1. Mixolab parameters.

Relation of these features to physical state of tested dough during mixing and heating [24–27] are:

C1 represents maximum torque during mixing (used to determine water absorption);
C2 represents the weakening of the protein based on the mechanical work and the increasing temperature;
C3 expresses the rate of starch gelatinization;
C4 measures the stability of the hot-formed gel;
C5 represents starch retrogradation during the cooling period;
C1–C2 show the protein network strength under increasing heating;
C3–C4 denotes to starch gelatinization rate;
C4–C5 relate to the anti-staling effects (starch retrogradation at cooling phase), represents the shelf life of the end products.

The correlation between parameters (Table 2) is tested during dough mixing and heating processes by Mixolab.

Table 2. Mixolab parameters correlation and significance.

Parameter	Calculation Method	Significance
Water Absorption (%)	Quantity of water required to obtain C1 = 1.1 Nm +/− 0.05	Amount of water taken up by flour to achieve the desired consistency and create a quality end-product.
Time for C1 (min)	Time required to obtain C1	Dough formation time: the stronger the flour, the longer it takes.
Stability (min)	Time during which torque is > C1–11% (constant T° phase)	Dough resistance to kneading: The longer it takes the 'strengthen' the dough.
Amplitude (Nm)	Curve width at C1	Dough elasticity: The higher the value, the greater the dough elasticity.

Sourse: MIXOLAB APPLICATIONS HANDBOOK—May 2012 edition [28].

2.10. Bread Making

The recipe consists of the following raw materials: wheat + sorghum flour (1 kg), dried yeast (3.0 g/kg flour), sodium chloride (1.5 g/kg flour), and water according to the Mixolab water absorption parameter (52.7%...54.7%). Samples were coded according to Table 1. All ingredients were added into a mixer (Diosna, Osnabrück, Germany). The program starts with a 10 min kneading process, followed by a 5 min resting time before the rounding and fermentation procedures that take 60 min at a temperature of 28–30 °C.

The 600 g obtained dough parts were subjected to remodelling process, after which were placed in special bread baking form and fermented at a 30 °C temperature and 90% humidity for 60 min, followed by baking at 200 ± 5 °C in a baking oven (Mondial Forni, Verona, Italy). To make the measurements, the bread was left to cool at room temperature.

The procedure for preparing the dough with addition of *L. plantarum* was the following: one third of the flour mixtures was mixed with a part of water and *bacterial freeze–dried powder (Aurum Plantarum, Millbo)* (0.3 g/100 g flour). There were made two batches of two breads for each sample.

This dough was kept for 12 h at a temperature of 25 °C, stirring slowly from time to time. During kneading, the dough was added over the amount of flour remaining in the sample.

2.11. Physicochemical Characteristics of the Experimental Bread

The bread specific volume was determined, by the rape seed displacement method according to SR 91:2007, AACC 2000 [17,21]. The ratio of the obtained data was the average of three measurements of the fresh made bread loaf, expressing it in cm^3/g.

For porosity measurement, knowing the mass and density, the porosity was expressed % by measuring the total scale of holes, in a known crumb volume.

Elasticity content measurement consists in pressing a piece of bread crumb of a certain shape, a given time and measuring the return to the initial position/shape after removing the pressing force. Crumb elasticity is expressed in percent meaning the ratio between the height expressed in % by pressing and return, and the initial height of the cylinder crumb bread.

Moisture content measurement consists in drying approximately 5 g of bread crumb at 103 °C (±2 °C) to constant weight; reported data consists in mean of three measurements, each time performed on a freshly new bread loaf.

Acidity measurement, expressed in degrees (SR 91/2007) [17,21], was determined by titration of a fluid extract of bread with 0.1 N NaOH solution, in the presence of phenolphthalein as indicator.

2.12. Proximate Analysis

Proximate analysis of the sorghum and whole wheat flour samples were carried out on dry matter basis. Ash, crude protein, fat, and fiber contents were determined as described by ICC standard no. 173 [24].

2.13. Sensory Analysis

A group of 10 specially trained panelist, with ages between 25 and 61, evaluated the bread samples, giving grades from 1 (lowest intensity) to 5 (highest intensity), for the following sensory attributions: crust color (degree of perceived brown color characterizing the crust), crumb color (degree of color darkness in the crumb ranging from white to dark brown), crumb pore uniformity (size of pores on the surface; (small/big)), crumb softness (minimum force necessary to compress the sample), bitter taste (perceived by the back of the tongue and characterized by solutions of quinine, caffeine, and other alkaloids; usually caused by over-roasting), salty taste (fundamental taste sensation elicited by sodium chloride), sour taste (fundamental taste sensation evoked by acids, e.g., tartaric acid), specific aroma (aroma of fresh baked bread and odor associated with aromatic exchange from yeast fermentation), after-taste (flavor staying after tasting). Also, there has been made a consumer overall acceptability determination in a 9-point hedonic scale (from 9 = like extremely to 1 = dislike extremely) [10,29,30].

For the hedonic test, 35 untrained panelists with ages between 22 and 60 (70% females and 30% males) have tasted the samples that were coded with 3 random letters. Water was served to panelists for mouth cleaning between the sample's evaluation. The results are expressed as mean standard deviation; $n = 10$ (descriptive test) and $n = 35$ (hedonic). Differences were considered as significantly different at a value of $p < 0.05$.

2.14. Statistical Analysis

All analyses were executed in triplicate and the mean values with the standard deviations were related. For statistical analysis was used Microsoft Excel Program, with the level of significance set at 95%. Analysis of variance (ANOVA) and Tukey's test was used to estimated statistical differences between samples. Differences were considered significant for a value at $p < 0.05$.

3. Results and Discussion

3.1. Chemical Composition

The chemical characterization of sorghum and whole wheat flours used in this study are presented in Table 3. As it can be seen, the nutritional composition of *Sorgum bicolor* is like that of whole wheat flour. Moreover, sorghum has a higher fat and raw fiber content than whole wheat flour. It is easy to notice that, apart from the smaller protein and calcium levels, the potassium, magnesium, and iron contents of sorghum flour are higher than of whole wheat flour.

Table 3. Chemical composition of sorghum and whole wheat flours.

Parameter	Sorghum Seed Flour	Whole Wheat Flour	p-Value (t-Test)
Moisture content (g/100 g)	11.40 ± 0.10 [a]	11.51 ± 0.09 [a]	0.256
Ash (g/100 g)	1.80 ± 0.01 [a]	1.90 ± 0.02 [b]	0.0015
Protein (g/100 g)	11.80 ± 0.07 [a]	12.80 ± 0.07 [b]	<0.0001
Fat (g/100 g)	3.70 ± 0.02 [a]	2.35 ± 0.03 [b]	<0.0001
Raw fiber (g/100 g)	3.07 ± 0.08 [a]	2.07 ± 0.06 [b]	<0.0001
Starch (g/100 g)	64.74 ± 0.15 [a]	64.00 ± 0.44 [a]	0.052
Potassium (mg/100 g d.m.)	575 ± 1.73 [a]	305.70 ± 1.14 [b]	<0.0001
Magnesium (mg/100 g d.m.)	296 ± 2.37 [a]	78.70 ± 0.81 [b]	<0.0001
Calcium (mg/100 g d.m.)	3.50 ± 0.31 [a]	41.20 ± 0.61 [b]	<0.0001
Iron (mg/100 g d.m.)	13.90 ± 0.09 [a]	4.71 ± 0.10 [b]	<0.0001
Zinc (mg/100 g d.m.)	2.17 ± 0.03 [a]	2.14 ± 0.03 [a]	0.214
Manganese (mg/100 g d.m.)	1.95 ± 0.04 [a]	2.45 ± 0.07 [b]	0.0003

Note: [a,b] Values are the means of triplicate determinations. The results are presented as mean values ± standard deviation. Different letters in the same row indicate significant diferences ($p < 0.05$).

The results of the analyzes performed show that sorghum seed flour contains a high amount of minerals. 100 g of sorghum assures 21% of the daily reference intake for K, 81% of the daily reference intake for Fe and 91% of the daily reference intake for Mg, according to recommendation of the FDA [31]. The content of potassium from sorghum seed flour is almost double than that of whole wheat flour. The magnesium content level is almost 4 times higher for sorghum seed flour, and, the iron content is almost 3 times higher in sorghum seed flour than is wheat flour. For zinc and starch content level, statistical analyzes show no significant differences.

3.2. Amino Acid Composition

As the world's population rises quickly, and the stress of limited water and food resources is more and more oppressive, it is more important than ever to provide high quality protein, that meet the human nutritional needs. The amino acid composition of sorghum and whole wheat flour are presented in Table 4.

The results show that sorghum protein has a content of essential amino acids, approximately similar to that of wheat protein. Moreso, the leucine content of sorghum protein is 12.84% compared to content of wheat protein (6.69%). Leucine is known on having excellent effect in bone, skin, and tissue wound healing, and stimulates hormone synthesis.

Table 4. Amino acids composition of sorghum and whole wheat flours (g amino acid/100 g protein).

Type	Amino Acids	Sorghum Seed Flour	Whole Wheat Seed Flour	*p*-Value (*t*-Test)
Essential amino acids	Arginine	3.65 ± 0.18 [a]	4.30 ± 0.54 [a]	0.123
	Leucine	12.84 ± 0.66 [a]	6.69 ± 0.65 [b]	<0.0001
	Valine	4.76 ± 0.24 [a]	4.46 ± 0.43 [a]	0.347
	Lysine	2.1 ± 0.40 [a]	2.56 ± 0.73 [a]	0.391
	Phenylalanine	4.65 ± 0.18 [a]	4.63 ± 0.16 [a]	0.909
	Isoleucine	3.65 ± 0.14 [a]	3.14 ± 0.23 [b]	0.029
	Threonine	2.77 ± 0.20 [a]	2.73 ± 0.42 [a]	0.888
	Histidine	1.88 ± 0.54 [a]	2.40 ± 0.47 [a]	0.272
Dispensable amino acids	Glutamic ac.	19.93 ± 0.50 [a]	26.20 ± 0.46 [b]	<0.0001
	Glycine	3.1 ± 0.19 [a]	4.21 ± 0.27 [b]	0.004
	Serine	4.1 ± 0.43 [a]	4.63 ± 0.55 [a]	0.260
	Proline	8.41 ± 0.41 [a]	10.50 ± 0.43 [b]	0.003
	Aspartic ac.	5.54 ± 0.50 [a]	3.31 ± 0.31 [b]	0.002
	Alanine	8.41 ± 0.44 [a]	3.31 ± 0.37 [b]	<0.0001
	Tyrosine	2.99 ± 0.17 [a]	1.57 ± 0.30 [b]	0.0019
	Methionine	1.88 ± 0.14 [a]	1.65 ± 0.12 [a]	0.009
	Cysteine	1.33 ± 0.04 [a]	2.31 ± 0.07 [b]	<0.0001

Note: [a,b] The results are expressed as mean ± standard deviation. Different letters in the same row indicate significant diferences (*t*-Test, $p < 0.05$).

Regarding the nonessential amino acids content, sorghum protein has a lower content of glutamic acid, glycine, cysteine, and proline than wheat protein. However, sorghum protein has a higher content of alanine aspartic acid and tyrosine than wheat protein. Whitaker and Tannenbaum [32] have evaluated the capability of a protein source to meet human amino acid requirements, by developing a chemical score procedure. The procedure consists in calculating the percentage that each amino acid in the protein tested represents from the respective amount of amino acid in the standard protein, where the egg protein was established as a standard for the evaluation of food proteins.

Figure 2 represents chemical scores of amino acids from the sorghum seed flour vs. chemical scores of amino acids from whole wheat flour.

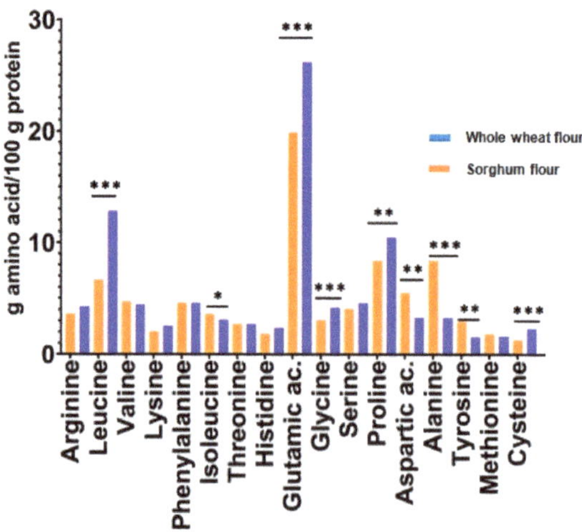

Figure 2. Chemical scores of amino acids of sorghum and whole wheat seeds. Note: *, **, *** Correlations significant at $p < 0.05$, $p < 0.01$, and $p < 0.001$, respectively.

An appreciable correlation was reported between the chemical scores determined in this way and the values obtained by biological tests for protein quality [33]. It can be seen in Figure 2 that both sorghum and wheat protein are a rich source of isoleucine, leucine, and valine. The most limited amino acid is lysine, followed by phenylalanine, threonine, and methionine.

3.3. In Vitro Protein Digestibility (IPD)

Trypsin from the porcine pancreas—type IX-S was used to simulate the gastrointestinal enzymatic process that took place in the normal human digestion of food proteins [20,34]. During protein digestion, hydrolysis of the peptide bond releases H+, which lowers the pH; rapid decreases mean higher digestion rates and can be used as an index of protein digestibility.

The sorghum flour pH value decreased from 7.90 to 7.72 about 60 s after the trypsin addition; compared to wheat flour values, that were each at a pH level of 7.90, respectively 7.64. After the first 60 s, the digestion of sorghum flour experienced a slow but continuous pH, decreasing up to 600 s for a final pH value of 7.53. The digestion of wheat flour experienced a slow, continuous pH decreases to 600 s for a final pH value of 7.43. The IPD from sorghum and wheat flour was calculated according to the formula defined in Section 2.6.

The values of Y (% Protein digestibility) obtained for the two samples were statistically different as

$$Y_{\text{whole wheat flour}} = 75.98 \tag{5}$$

$$Y_{\text{Sorghum}} = 74.17 \tag{6}$$

Note: standard deviation within in vitro meassurement = ±0.70; regression equation: Y = 210.46 − 16.10X, where X = pH at 10 min.

The IPD value obtained for whole wheat flour in this study is higher than the 74.17 reported for sorghum. The slightly lower digestibility of sorghum protein is due to the higher fiber content of sorghum. Fiber has been reported to have a negative effect on IPD because of nonspecific interactions between proteins and polysaccharide constituents in food [35].

The 1985 FAO/WHO [18] recommended that the amino acid score method corrected by protein digestibility to be the most appropriate approach for the routine assessment of the overall quality of proteins for humans and recommended that this method to be adopted as an official method internationally.

3.4. Fatty Acid Composition

In Table 5 is given fatty acid composition of seed oil of whole wheat and sorghum seed.

The total oil content in sorghum seeds was 3.70% (w/w), in favor of whole wheat which had a value of 2.35% (w/w). The results revealed that, unlike whole wheat, sorghum seeds had a higher oleic acid content (35.06% respectively 16.55%). It was observed that monounsaturated fatty acids (MUFA) content of sorghum was double that of whole wheat, and the polyunsaturated fatty acids (PUFA) content of sorghum was slightly lower. Besides, the (MUFA + PUFA) content of sorghum was higher than that of whole wheat.

In Figure 3, it is presented Spectra of GC–MS of fatty acid methyl esters of sorghum, and in Figure 4 it is presented the spectra of GC–MS for fatty acid methyl esters of whole wheat.

These results recommend that *Sorghum bicolor* seed oils can be used as a potential dietary source of MUFA and PUFA [36].

Table 5. Fatty acid composition of seed oil of whole wheat and sorghum seed.

	Fatty Acids	**Rt (Min.)**	**Sorghum**	**Whole Wheat**
C14:0	Tetradecanoic/Miristic	26.21	0.04 ± 0.00	0.10 ± 0.01
C16:0	Hexadecanoic/Palmitic	33.58	17.28 ± 0.92	21.23 ± 0.15
C16:1n7	*Cis*-9-hexadecenoic/Palmitoleic	35.00	0.67 ± 0.08	0.16 ± 0.02
C18:0	Octadecanoic/Stearic	40.47	1.83 ± 0.23	1.22 ± 0.15
C18:1n9	*Cis*-9-octadecenoic/Oleic	41.68	34.38 ± 0.74	16.55 ± 0.14
C18:1n11	*Cis*-11-octadecenoic/*Cis*-vaccenic	41.94	2.28 ± 0.08	1.34 ± 0.13
C18:2n6	*Cis*-9,12-octadecadienoic/Linoleic	43.83	42.29 ± 0.71	55.25 ± 0.67
C18:3n3	*Cis*-9,12,15-octadecatrienoic/α-linolenic	46.27	0.76 ± 0.08	3.03 ± 0.01
C20:0	Eicosanoic/Arahidic	46.9	0.14 ± 0.02	0.11 ± 0.01
C20:1n9	*Cis*-11-eicosenoic/Gondoic	48.06	0.17 ± 0.02	0.72 ± 0.05
C22:0	Docosanoic/Behenic	52.86	0.06 ± 0.01	0.12 ± 0.01
C24:0	Tetracosanoic/Lignoceric	58.38	0.10 ± 0.01	0.18 ± 0.01
	SFA, %		19.45	22.96
	MUFA, %		37.50	18.77
	PUFA, %		43.05	58.28
	MUFA + PUFA, %		80.55	77.04
	omega-3, %		0.76	3.03
	omega-6, %		42.29	55.25
	omega-9, %		34.55	17.27

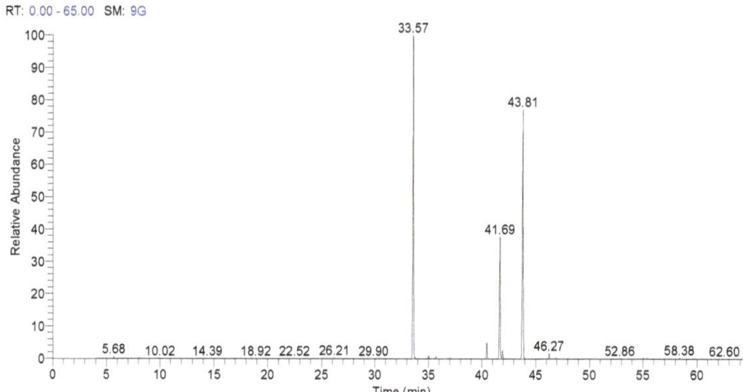

Figure 3. Spectra of GC–MS of fatty acid methyl esters of sorghum.

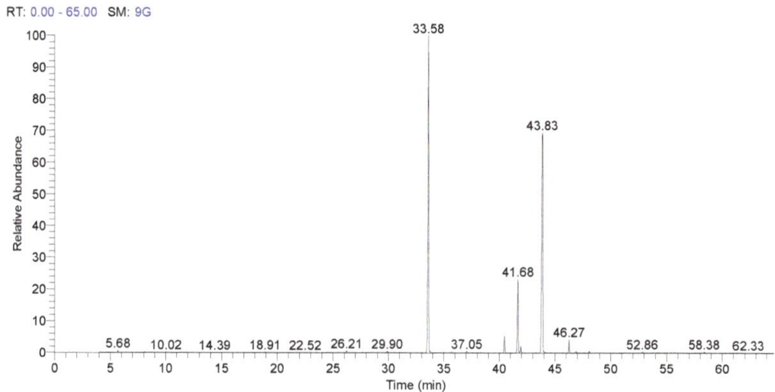

Figure 4. Spectra of GC–MS for fatty acid methyl esters of whole wheat.

3.5. Rheological Properties of Flour Mixtures

Table 6 presents the rheological properties of wheat and sorghum flour mixtures, versus the trial simple wheat flour.

Table 6. Rheological characteristics of flour mixtures and wheat flour.

	100% Wheat Flour	70% Wheat Flour + 30% Sorghum Flour	65% Wheat Flour + 35% Sorghum Flour	60% Wheat Flour + 40% Sorghum Flour
Water Absorption (%)	58.2 ± 0.06 [a]	54.7 ± 0.06 [b]	53.4 ± 0.1 [c]	52.7 ± 0.1 [c]
Stability (min)	9.15 ± 0.01 [a]	8.69 ± 0.01 [c]	8.82 ± 0.01 [b]	8.51 ± 0.01 [d]
Amplitude (Nm)	0.097 ± 0.001 [c]	0.107 ± 0.001 [b]	0.109 ± 0.001 [b]	0.124 ± 0.002 [a]
C1	1.098 ± 0.01 [a]	1.093 ± 0.01 [a]	1.097 ± 0.01 [a]	1.097 ± 0.01 [a]
C2	0.462 ± 0.01 [a]	0.469 ± 0.01 [a]	0.474 ± 0.01 [a]	0.483 ± 0.01 [a]
C3	1.924 ± 0.01 [c]	1.995 ± 0.01 [b]	2.040 ± 0.01 [a]	2.048 ± 0.02 [a]
C4	1.809 ± 0.01 [d]	1.939 ± 0.01 [c]	1.998 ± 0.02 [b]	2.090 ± 0.03 [a]
C5	3.049 ± 0.01 [c]	3.139 ± 0.01 [b]	3.138 ± 0.02 [b]	3.233 ± 0.02 [a]

Note: [a,b] The results are expressed as mean ± standard deviation. Different letters in the same row indicate significant diferences ($p < 0.05$).

Mixolab C1–C5 values of wheat dough were 1.098 N·m, 0.462 N·m, 1.924 N·m, 1.809 N·m and 3.049 N·m, respectively (Table 6). Similar Mixolab behavior was mentioned in other studies [27] for three wheat varieties with small differences.

From our results, it can be seen that as the amount of added sorghum seed flour increases, the water absorption capacity decreases from 58.2% (wheat) to 54.7% (30% sorghum), 53.4% (35% sorghum), and 52.7% (40% sorghum), respectively. For wheat flour for baking, normal CH values are between 55–62% [37].

It can be noticed that the addition of sorghum seed flour did not have a significant influence on the wheat flour dough's stability (Table 6).

As the percentage of sorghum flour increased, the amplitude the width of the curve during dough formation increase, which suggests a higher elasticity of the dough, due to higher content of fat. This increase in fat content is due to sorghum seeds, with a positive influence on the dough.

The increase of consistency (C2—the behavior of the gluten when heating the dough) reveal a higher resistance of dough as an effect of temperature. The increase in sorghum addition did not lead to changes in this parameter.

In phase 3, the starch gel formation (at temperature 50–55 °C) the lowest C3 was observed for wheat flour. The difference in C3 results between wheat flour and mixture of wheat + 40% sorghum was 0.124 N·m, so the influence on dough preparation recipe was low.

The C4 parameter corresponds to the stability of the hot-formed gel. The lowest C4 was found for wheat flour (Table 6). The difference for C4 results between wheat flour and wheat flour + 40% sorghum was only 0.281 N·m.

The retrogradation stage of starch (C5) for the tested wheat flour and wheat–sorghum flour mixtures demonstrated similar differences as for C4 parameter.

From those presented above, it can be determined that, with regard to their baking quality, the flour mixtures studied can be categorized as flours adequate for bakery products.

The mixolab curves of the wheat control sample and of all three mixtures of wheat and sorghum flour are shown in Figure 5.

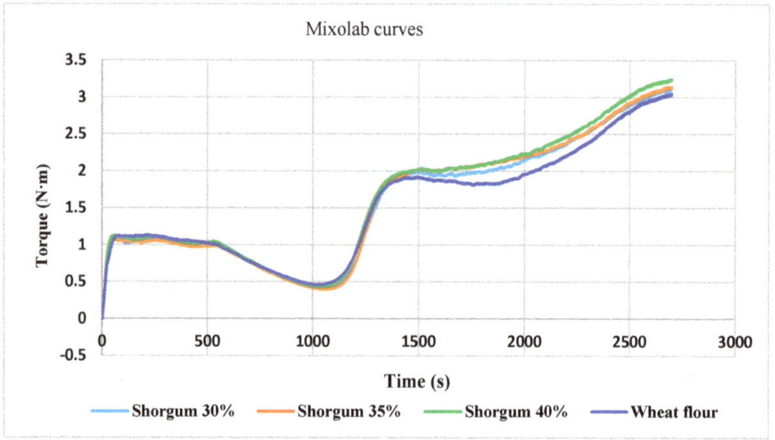

Figure 5. The influence of substitution level of sorghum flour on the Mixolab curves.

3.6. Bread Quality

The six bread samples obtained from the mixtures of wheat and sorghum flowers with and without *L. plantarum* and were analyzed compared to the control/trial sample of bread obtained only from wheat flour type 550 (ash = 0.55%), (Figure 6). The coding of the experimental bread samples was presented in Table 1. As it can be seen in Table 7, the porosity values of the bread samples obtained from the variants of mixtures are within limits of SR 878/1996 [38] (minimum 66%). The acidity of the bread samples does not exceed 1.6 degrees of acidity which is a value within the normal limits for wheat bread (max. 3.3%, SR 878-1996) [38]. The elasticity of the bread samples obtained from the flour mixtures was close, but significantly lower than that of the control sample [38].

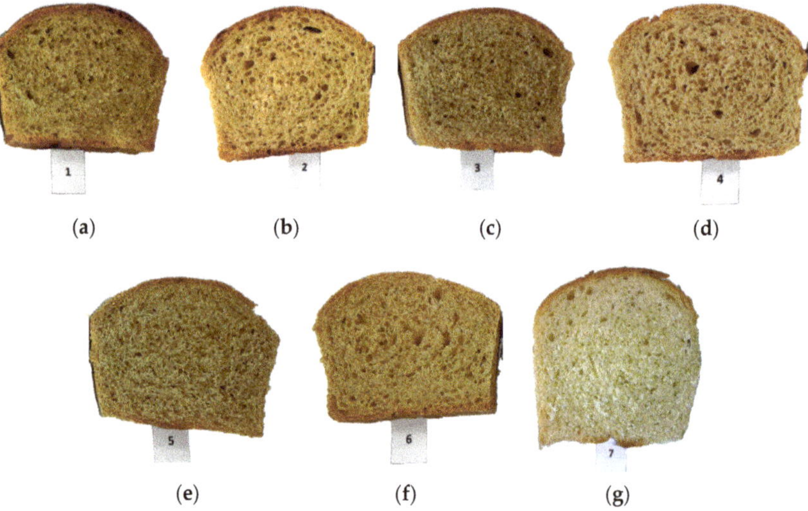

Figure 6. Bread samples ((a) P1 = 70% wheat + 30% sorghum without *L. plantarum*; (b) P2 = 70% wheat + 30% sorghum with *L. plantarum*; (c) 65% wheat + 35% sorghum without *L. plantarum*; (d) P4 = 65% wheat + 35% sorghum with *L. plantarum*; (e) P5 = 65% wheat + 35% sorghum without *L. plantarum*; (f) P6 = 65% wheat + 35% sorghum with *L. plantarum*; (g) P7 = 100% wheat flour (type 550)–control sample).

Table 7. Physico-chemical results for the experimental bread samples.

Sample	Mass (g)	Volume (cm³)	Porosity (%)	Elasticity (%)	Diameter (mm)	Humidity (%)	Acidity (Degree)
P1	620 ± 0.01 [e]	290 ± 2.87 [b]	75.7 ± 0.45 [b]	92 ± 0.94 [b]	15.75 ± 0.09 [a]	44.40 ± 0.10 [b]	0.6 ± 0.09 [e]
P2	620 ± 0.01 [e]	242 ± 2.63 [c]	71.3 ± 0.55 [c]	90 ± 0.82 [b]	15.50 ± 0.10 [a]	43.57 ± 0.10 [c,d]	1.2 ± 0.10 [b]
P3	624 ± 0.02 [d]	236 ± 2.68 [d]	71.8 ± 0.35 [c]	76 ± 0.45 [c]	15.75 ± 0.10 [a]	44.41 ± 0.10 [b]	0.4 ± 0.06 [f]
P4	625 ± 0.01 [c]	218 ± 2.68 [f]	66.7 ± 0.50 [e]	75 ± 0.82 [c,d]	15.40 ± 0.10 [b]	43.66 ± 0.16 [c]	1.2 ± 0.10 [b]
P5	626 ± 0.02 [b]	225 ± 2.69 [e]	68.6 ± 0.55 [d]	74 ± 0.47 [c,d]	15.65 ± 0.08 [a]	45.09 ± 0.12 [a]	0.8 ± 0.09 [d]
P6	624 ± 0.02 [d]	218 ± 2.67 [f]	64 ± 0.63 [f]	72 ± 0.82 [e]	15.55 ± 0.08 [a]	43.84 ± 0.11 [c]	1.6 ± 0.10 [a]
P7	628 ± 0.02 [a]	299 ± 2.89 [a]	77.8 ± 0.60 [a]	97 ± 0.94 [a]	15.85 ± 0.09 [a]	43.42 ± 0.12 [c]	1.0 ± 0.10 [c]

Note: [a,b,c,d,e] Each value represents a mean of three replicates. The results are expressed as mean ± standard deviation. Values followed by different letters in the same column are significantly different ($p < 0.05$).

It can be seen in Table 7 that the porosity decreases with increasing degree of sorghum replacement (porosity values for P1 (75.7 ± 0.45), P3 (71.8 ± 0.35), P5 (68.6 ± 0.55)). On the other hand, there is a negative influence of the addition of L. plantarum on the porosity, decreasing from P1 (75.7%) to P2 (71.3%), respectively P3 (71.8%) to P4 (66.7%), or from 68.6% (P5) to 64% (P6). In all cases of addition of sorghum, respectively *L. plantarum*, the porosity value was lower than that of the control sample (P7—77.8%). The same effects of adding sorghum can be seen on the elasticity since the value obtained in the case of the P7 control sample was considerably higher than all samples, with a value of 97%. The acidity was significantly influenced by the presence of *L. plantarum*, the values determined in samples P2 (1.2), P4 (1.2), and P6 (1.6) being significantly higher than the acidity of samples P1 (0.6), P3 (0.4), P5 (0.8), and P7 (1), respectively. The acidity of bread samples with the addition of *L. plantarum* was much higher than that of samples without *L. plantarum*, however within normal limits according to the Romanian standards in force (SR 878:1996). Taking into account these elements, we can conclude that the P4 sample ensures a balance between the benefits brought by a high intake of sorghum and *L. plantarum*, respectively the reduction of the organoleptic characteristics of the samples.

3.7. Sensory Evaluation

The sensory evaluation (Figure 7) of the bread samples was performed 24 h after baking. The bread samples received out of 9 points for overall acceptability the following scores: P1 = 7.17; P2 = 7.58; P3 = 7.03; P4 = 7.34; P5 = 6.95; P6 = 7.29 and control sample (P7) = 7.61. As it can be seen, all the tests got a score of 7 and 8, meaning between "I like it very much" and "I like it moderately". However, all the samples with the same percentage of sorghum flour, but which used *L. plantarum* in the preparation obtained a higher score than the samples without *L. plantarum* (P1 vs. P2; P3 vs. P4 and P5 vs. P6). In the case of sensory analysis, we can highlight the P4 sample with a score of 7.34, as a variant that combines the positive effects of the addition of sorghum (35%) and *L. plantarum*.

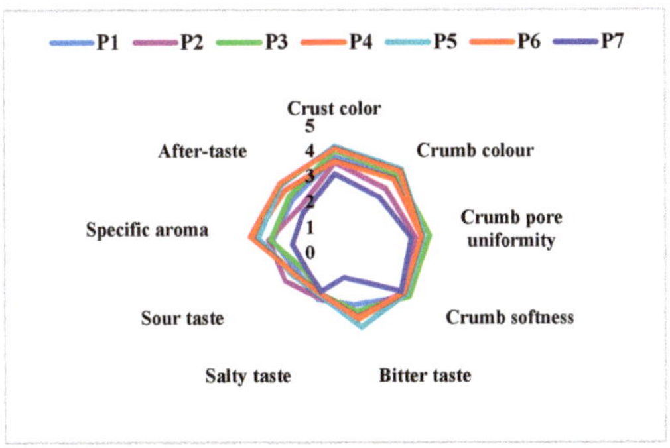

Figure 7. Sensorial properties of bread samples. P1 = 70% wheat + 30% sorghum without *L. plantarum*; P2 = 70% wheat + 30% sorghum with *L. plantarum*; P3 = 65% wheat + 35% sorghum without *L. plantarum*; P4 = 65% wheat + 35% sorghum with *L. plantarum*; P5 = 65% wheat + 35% sorghum without *L. plantarum*; P6 = 65% wheat + 35% sorghum with *L. plantarum*; P7 = 100% wheat flour (type 550).

4. Conclusions

The research presented in the paper demonstrates that sorghum flour is a cereal with a nutritional composition similar to that of whole wheat flour. Moreover, sorghum has a higher content of fat and crude fiber than whole wheat flour, and the content of monounsaturated fatty acids + polyunsaturated fatty acids (MUFA + PUFA,%) is higher than that of whole wheat.

Sorghum seeds have an important mineral content, 100 g of sorghum providing 21% of the daily reference intake for K, 81% of the daily reference intake for Fe and 91% of the daily reference intake for Mg, according to recommendation of the FDA. However, the in vitro protein digestibility (IPD) is lower than that of whole wheat flour.

The results of the rheological analyzes performed with the help of the Mixolab device showed that the addition of sorghum had a significant effect on water absorption, as well as small negative significant qualitative changes in protein composition and dough stability. There are no significant changes in starch degradation. From rheological point of view, it can be determined that, with regard to their baking quality, the flour mixtures studied can be categorized as flours adequate for bakery products.

Taking into consideration the physicochemical results and the sensorial ones, it can be concluded that the addition of 35% sorghum seeds and *L. plantarum* (0.3 g/100 g flour) is a variant that combines the benefits of high amounts of minerals (K, Mg, Fe) and is preferred by consumers.

The final conclusion of this study is that sorghum can be used to partially substitute wheat flour in the production of bakery products with acceptable physical and chemical characteristics.

Author Contributions: Conceptualization, L.A. and N.B.; Methodology, L.A., N.B. and O.B.O.; Software, L.G. and O.B.O.; Validation, L.A., N.B. and M.E.P.; Formal analysis, L.A., O.B.O. and L.G.; Writing—original draft preparation, L.A. and O.B.O.; Writing—review and editing, N.B. and L.G.; Supervision, M.E.P.; Project administration, L.A. and M.E.P.; Funding acquisition, L.G. and O.B.O. All authors have read and agreed to the published version of the manuscript.

Funding: This work was supported by a grant of the Romanian Ministry of Research and Innovation CCDI—UEFISCDI, "Complex system of integral capitalization of agricultural species with energy and food potential", project number PN-III-P1-1.2-PCCDI-2017-0566, contract no. 9PCCDI/2018, within PNCDI III.

Acknowledgments: Authors wish to acknowledge the support of the Transilvania University of Brasov, Romania.

Conflicts of Interest: The authors declare no conflict of interest.

References

1. Coulibaly, W.H.; Bouatenin, K.M.J.-P.; Boli, Z.B.I.A.; Alfred, K.K.; Bi, Y.C.T.; N'Sa, K.M.C.; Cot, M.; Djameh, C.; Djè, K.M. Influence of yeasts on bioactive compounds content of traditional sorghum beer (tchapalo) produced in Côte d'Ivoire. *Curr. Res. Food Sci.* **2020**, *3*, 195–200. [CrossRef] [PubMed]

2. Devi, S.P.; Saravanakumar, M.; Mohandas, S. Identification of 3-deoxyanthocyanins from red sorghum (Sorghum bicolor) bran and its biological properties. *Afr. J. Pure Appl. Chem.* **2011**, *5*, 181–193. [CrossRef]

3. Lim, T.K. *Edible Medicinal and Non-Medicinal Plants*; Springer: Dordrecht, The Netherlands, 2013; Volume 5, pp. 379–380. [CrossRef]

4. Slima, S.B.; Trabelsi, I.; Ktari, N.; Bardaa, S.; Elkaroui, K.; Bouaziz, M.; Abdeslam, A.; Salah, R.B. Novel *Sorghum bicolor (L.)* seed olysaccharide structure, hemolytic and antioxidant activities, and laser burn wound healing effect. *Int. J. Biol. Macromol.* **2019**, *132*, 87–96. [CrossRef]

5. Poquette, N.M.; Gu, X.; Lee, S.-O. Grain sorghum muffin reduces glucose and insulin responses in men. *Food Funct.* **2014**, *5*, 894–899. [CrossRef] [PubMed]

6. Kasarda, D.D. Grains in relation to celiac disease. *Cereal Foods World* **2001**, *46*, 209–210.

7. Chinedu, S.M.; Yusuf, S.O.; Maxwell, I.E. Fermentation of sorghum using yeast n(Saccharomyces cerevisiae) as a starter culture for burukutu production. *Cont. J. Biol. Sci.* **2010**, *3*, 63–74.

8. Montemurro, M.; Coda, R.; Rizzello, C.G. Recent Advances in the Use of Sourdough Biotechnology in Pasta Making. *Foods* **2019**, *8*, 129. [CrossRef]

9. Nionelli, L.; Montemurro, M.; Pontonio, E.; Verni, M.; Gobbetti, M.; Rizzello, C.G. Pro-technological and functional characterization of lactic acid bacteria to be used as starters for hemp (*Cannabis sativa* L.) sourdough fermentation and wheat bread fortification. *Int. J. Food Microbiol.* **2018**, *279*, 14–25. [CrossRef]

10. Ogunsakin, A.; Sanni, A.; Banwo, K. Effect of legume addition on the physiochemical and sensorial attributes of sorghum-based sourdough bread. *LWT* **2020**, *118*, 108769. [CrossRef]

11. De Vuyst, L.; Vrancken, G.; Ravyts, F.; Rimaux, T.; Weckx, S. Biodiversity, ecological determinants, and metabolic exploitation of sourdough microbiota. *Food Microbiol.* **2009**, *26*, 666–675. [CrossRef]

12. Gobbetti, M.; Rizzello, C.G.; Di Cagno, R.; De Angelis, M. How the sourdough may affect the functional features of leavened baked goods. *Food Microbiol.* **2014**, *37*, 30–40. [CrossRef]

13. Moroni, A.V.; Bello, F.D.; Arendt, E.K. Sourdough in gluten-free bread-making: An ancient technology to solve a novel issue? *Food Microbiol.* **2009**, *26*, 676–684. [CrossRef] [PubMed]

14. Coda, R.; Rizzello, C.G.; Gobbetti, M. Use of sourdough fermentation and pseudo-cereals and leguminous flours for the making of a functional bread enriched of γ-aminobutyric acid (GABA). *Int. J. Food Microbiol.* **2010**, *137*, 236–245. [CrossRef]

15. *ICC Standard Method No.110/1, Determination of Moisture Content of Cereals and Cereal Products (Practical Method)*; International Association for Cereal Science and Technology: Vienna, Austria, 1976.

16. *ICC Standard Method No.104/1, Determination of Ash in Cereals and Cereal Products*; International Association for Cereal Science and Technology: Vienna, Austria, 1990.

17. *SR 91:2007. Pâine şi Produse Proaspete de Patiserie. Metode de Analiză*; ASRO Publisher House: Bucharest, Romania, 2007.

18. Food and Agriculture Organization of the United Nations; World Health Organization; United Nations University. Energy and Protein Requirements. 1985. Available online: http://www.fao:3/aa040e/aa040e00.htm#TOC (accessed on 22 August 2020).

19. Hsu, H.W.; Vavak, D.L.; Satterlee, D.L.; Mill, G.A. A multienzyme technique for estimating protein digestibility. *J. Food Sci.* **1977**, *42*, 1269–1273. [CrossRef]

20. Malomo, S.A.; Aluko, R.E. Conversion of a low protein hemp seed meal into a functional protein concentrate through enzymatic digestion of fibre coupled with membrane ultrafiltration. *Innov. Food Sci. Emerg. Technol.* **2015**, *31*, 151–159. [CrossRef]

21. AACC International. *Approved Methods of the American Association of Cereal Chemists*, 10th ed.; AACC International: St. Paul, MN, USA, 2000.

22. AOAC. *Methods of the Association of Official Analytical Chemists*, 15th ed.; AOAC: Arlington, VA, USA, 1990; Volume II, p. 780.

23. Morrison, W.R.; Smith, L.M. Preparation of fatty acid methyl esters and dimethylacetals from lipids with boron fluoride—Methanol. *J. Lipid Res.* **1964**, *5*, 600–608. [PubMed]

24. *ICC Standard No.173, Determination of Rheological Behaviour as a Fuction of Mixing and Temperature Increase*; International Association for Cereal Sciences and Technology: Vienna, Austria, 2010.

25. Chopin Technologies. Available online: https://chopin.fr/ (accessed on 7 September 2020).

26. Dubat, A. A new AACC international approved method to measure rheological properties of a dough sample. *Cereal Food World* **2010**, *55*, 150–153. [CrossRef]

27. Papoušková, L.; Capouchova, I.; Kostelanska, M.; Skeríkova, A.; Prokinova, E.; Hajslova, J. Changes in baking quality of winter wheat with different intensity of fusarium spp. contamination detected by means of new rheological system Mixolab. *Czech J. Food Sci.* **2011**, *29*, 420–429. [CrossRef]

28. Mixolab Applications Handbook. Available online: http://concereal.net/wp-content/uploads/2017/03/2012-CHOPIN-Mixolab-Applications-Handbook-EN-SPAIN-3.pdf (accessed on 8 September 2020).

29. Lawless, H.T.; Heymann, H. *Sensory Evaluation of Food. Principles and Practices*, 2nd ed.; Springer: New York, NY, USA, 2010. [CrossRef]

30. Ogunsakin, O.A.; Banwo, K.; Ogunremi, O.R.; Sanni, A.I. Microbiological and physicochemical properties of sourdough bread from sorghum flour. *Int. Food Res. J.* **2015**, *22*, 2610–2618.

31. Food and Drug Administration (FDA). Available online: https://www.fda.gov/drugs/science-and-research-drugs/regulatory-science-action (accessed on 13 September 2020).

32. Whitaker, J.R.; Tannenbaum, S.R. *Food Proteins*; AVI: Westport, CT, USA, 1977; ISBN 0870052309. [CrossRef]

33. Hegsted, D.M. 'Assessment of protein quality', in Improvement of protein nutriture. *Natl. Acad. Sci.* **1974**, 64–88.

34. Pihlanto, A.; Mäkinen, S. *Antihypertensive Properties of Plant Protein Derived Peptides*; Hernandez-Ledesma, B., Hsieh, C., Eds.; INTECH: Rijeka, Croatia, 2013.

35. Marambe, H.K.; Shand, P.J.; Wanasundara, J.P.D. In vitro digestibility of flaxseed (*Linum sitatissimum* L.) protein: Effect of seed mucilage, oil and thermal processing. *Int. J. Food Sci. Technol.* **2013**, *48*, 628–635. [CrossRef]

36. Mehmood, S.; Orhan, I.; Ahsan, Z.; Aslan, S.; Gulfraz, M. Fatty acid composition of seed oil of different Sorghum bicolor varieties. *Food Chem.* **2008**, *109*, 855–859. [CrossRef] [PubMed]
37. Bordei, D. *Tehnologia Modernă a Panificaţiei*; Agir: Bucuresti, Romania, 2004; pp. 80–81.
38. *SR 878:1996. Pâine de Făină de Grâu*; ASRO Publisher House: Bucharest, Romania, 1996.

Publisher's Note: MDPI stays neutral with regard to jurisdictional claims in published maps and institutional affiliations.

Article

A Comparative Assessment of the Baking Quality of Hybrid and Population Wheat Cultivars

Marta Jańczak-Pieniążek [1,*], Jan Buczek [1], Joanna Kaszuba [2], Ewa Szpunar-Krok [1], Dorota Bobrecka-Jamro [1] and Grażyna Jaworska [2]

[1] Department of Crop Production, University of Rzeszow, Zelwerowicza 4 St., 35-601 Rzeszow, Poland; jbuczek@ur.edu.pl (J.B.); szpunar-krok@wp.pl (E.S.-K.); bobjamro@ur.edu.pl (D.B.-J.)

[2] Department of Food Technology and Human Nutrition, University of Rzeszow, Zelwerowicza 4 St., 35-601 Rzeszow, Poland; jkaszuba@ur.edu.pl (J.K.); rrgjawor@cyf-kr.edu.pl (G.J.)

* Correspondence: mjanczak@ur.edu.pl

Received: 14 September 2020; Accepted: 10 October 2020; Published: 13 October 2020

Abstract: The study assessed the quality parameters of grain and flour, the rheological properties of dough and the quality of bread prepared from flour of hybrid cultivars of wheat in comparison with population cultivars of wheat. As the interest in wheat hybrids cultivars from the agricultural and milling industry is growing, their technological value of grain and flour was evaluated at two levels of nitrogen fertilisation (N1—110 kg/ha, N2—150 kg/ha). Increasing the fertilisation (N2) produced a significant influence on the crude protein and gluten content in the flour, as well as the moisture of the crumb and the yield of the dough without impacting other rheological traits and parameters of bread baking process. The performed principal component analysis (PCA) allowed for identification of the best cultivars among the studied wheat cultivars (Hybery and Hyvento). The hybrid cultivar Hyvento was characterised by favourable qualitative traits of the grain (vitreousness, crude protein content) and rheological parameters of the dough (bread volume), however, it had lower baking quality parameters. Among the hybrid cultivars, the best applicability for baking purposes was Hybery due to the favourable values of the baking process parameters and bread quality (bread yield, bread volume, Dallmann porosity index of crumb). Hybrid cultivars of wheat can therefore be used for the production of bread and be an alternative in agricultural production for population cultivars, which will contribute to filling the knowledge gap for the hybrid wheat cultivars.

Keywords: hybrid wheat; rheological properties; bread-making quality; N fertilisation

1. Introduction

Wheat (*Triticum aestivum* L.) production for food purposes in Europe and around the world dictates the consumption model and determines the safety of food and human nutrition [1,2].

With its high content of complex carbohydrates, wheat grain is a significant source of the energy provided in food, particularly due to the high contribution of cereal products (flour, groats, flakes, pasta, bread) in the daily diet of people. What is more, it constitutes a basic and valuable source of protein, minerals (P, K, Ca, Mg and micronutrients), B group vitamins, dietary fibre, and antioxidants, being at the same time one of the major allergens [3,4].

In combination with water, wheat flour creates dough with unique viscoelastic properties (distinguishing it among the flours of other cereals), enabling processing the dough for bread, pasta and other food products, which stems from the presence of gluten proteins in its composition [5,6].

Baking quality of wheat flour depends largely on the amount of seed storage proteins, in particular glutenins, the content of which is dictated by, among other things, the climate and soil conditions and fertilisation, in particular with nitrogen, which is also a cultivar-related trait [7,8]. Thus, the increased glutenin content contributes to increased dough resistance to stretching and prolonged dough development time, whereas a higher HMW glutenin content increases dough resistance and bread volume [9].

What is more, every new cultivar possesses its own set of genes controlling the synthesis of gluten proteins and affecting the formation of a different quality of gluten, determining the specified value of quality traits of the wheat grain and flour. Nowadays, as a result of accelerated biological advances, attempts are being made to obtain wheat cultivars with more favourable grain quality properties enriched in terms of nutrition with the greatest possible amount of bioactive substances, while being tolerant of stress and variable climatic and soil conditions [10].

One strategy in wheat breeding is the improvement of yield stability and size under stress conditions created by biotic and abiotic conditions of the environment by means of hybrid vigour [11]. Interest of the public and private sector in hybrid cultivars of wheat has been increasing, not only in the countries with growing concerns about food safety (China, India, Pakistan), but also in Europe and the USA [2].

The surface area occupied by hybrid wheat is less than 1% of the global wheat cultivation area [1,12]. In Europe, hybrid wheat is cultivated on an area of over 560.000 ha, and 80% of hybrid cultivars are grown in France, while the remaining 20% are grown in Germany, Hungary, Italy, the Czech Republic, Slovakia, Romania and Portugal [2]. In hybrid wheat, the vigour obtained via cross-fertilisation differs genetically from the distinct parent lines, resulting in an improved phenotype and traits desired for plant production, such as: higher rate of growth and plant differentiation, higher and stable grain yield, better resistance to environmental stress conditions and greater competitiveness with regards to diseases and weeds [11,12].

Thus far, the literature data concerning the assessment of the technological value of grain and flour from hybrid winter wheat cultivars is scarce. The objective of the conducted study was assessment of the quality of grain and flour, baking quality and applicability for the production of bread of flour obtained from the grain of hybrid wheat cultivars as compared with population wheat cultivars. The wheat grain originated from an experiment utilising two levels of nitrogen fertilisation.

2. Materials and Methods

2.1. Experimental Material

The study material consisted of winter wheat grain (*Triticum aestivum* L.) from seven hybrid cultivars: Hybery, Hyena, Hyfi, Hyking, Hymalaya, Hypocamp, and Hyvento. The wheat grain of two population cultivars, Belissa and Hondia were used as control samples. The wheat grain originated from an experiment utilising two levels of nitrogen fertilisation (N_1—110 kg/ha, N_2—150 kg/ha), carried out in the 2017/2018 period at the Experimental Station for Cultivar Assessment in Przecław (50°11′ N, 21°29′ E, altitude 185 m asl) near Mielec (Poland). The doses of nitrogen fertilisation were applied according to the COBORU method [13] intended for strict field experiments with winter wheat cultivars, at two levels of agricultural technology: medium intensity (A1) with N_1 fertilisation—110 kg/ha and high intensity (A2) with N_2 fertilisation—150 kg/ha. From the grain of the investigated cultivars flour, was obtained, which was subject to testing in the field of baking quality assessment.

2.2. Methods

2.2.1. Grain Quality Assessment

Grain samples were analysed for bulk density, referred to as mass per hectolitre, by means of EN ISO 7971-3:2019 [14]. Thousand grain weight (TGW) was determined with a grain counter (Sadkiewicz Instruments, Poland). Vitreousness was determined according to ICC Standard Method No. 129. [15]. Vitreousness was determined by cutting 50 grains with a Farinotom (Sadkiewicz Instruments, Poland) and calculating the number of grains that have completely or partially vitreous flour on the cross-section surface, measured as a percentage of vitreous kernels (0–100%).

2.2.2. Grain Milling

Grain with a moisture level of 13.0 ± 0.1% was milled in a Quadrumat Junior mill (Brabender, Germany) with a cone screen, mesh size φ = 212 μm in accordance with AACC Method No. 26-50.01 [16]. Flour yield [%] was calculated as the amount of flour obtained via milling 100 g of the grain.

2.2.3. Grain and Flour Quality Assessment

Total ash was determined in accordance with ICC Standard Method No. 104/1 [17]. Nitrogen content was measured and calculated into crude protein content using the N × 6.25 conversion ratio based on AACC Method No. 46-11.02 [16]. Wet gluten and dry gluten content and gluten index (GI) were determined in the flour following AACC Method No. 38-12.02 [16] using a Gluten Index System device (Perten, Sweden). Finally, the flour falling number was determined with a Falling Number 1800 apparatus (Perten, Sweden) following AACC Method No. 56-81.03 [16].

2.2.4. Farinograph Parameters Testing

Water absorption of the flour at a maximum consistency of 500 FU (Farinograph Units) as well as the development time of the dough and dough stability, degree of softening and quality number were determined using a Farinograph-E (Brabender, Germany) following ICC Standard No. 115/1 [17].

2.2.5. Bread Baking

The baking procedure was carried out with a modified ICC Standard Method No. 131 [17]. The dough was prepared from 300 g flour, 9 g yeast, 4.5 g salt and water, allowing for production of dough with a consistency equal to 350 FU, determined with a Farinograph-E (Brabender, Germany). The ingredients were combined with the use of a laboratory mixer (Sadkiewicz Instruments, Poland) and afterwards the dough was kneaded by hands for 3 min. The dough was placed in a fermentation chamber under conditions of 80% humidity and 30 °C (Sveba Dahlen, Sweden) for 30 min, then it was kneaded by hands for 3 min and again placed in the fermentation chamber for another 30 min. Subsequently, 250 g pieces of dough were formed, which were placed in greased baking pans. Fermentation of the dough pieces was continued until their optimum rise in the fermentation chamber was obtained (as above). The dough pieces were baked in a Classic electric oven (Sveba Dahlen, Sweden) at a temperature of 230 °C for 30 min. The breads were weighed twice: right after removing from the oven and after cooling down.

2.2.6. Bread Quality Assessment

Weight and volume were measured approximately 24 h from baking. The basic indicators of the wheat bread baking process, such as dough yield, total baking loss, and bread yield were calculated [18]. Volume of 100 g bread was calculated after weighing and measuring the loaves with a Sa-Wa volumeter (Sadkiewicz Instruments, Poland) following AACC Method No. 10-05.01 [16]. The Dallmann porosity index of the crumb was determined according to the Dallmann scale [19], in which 30 points are given to the lowest quality and 100 to the highest quality crumb. The bread's crumb moisture was determined according to AACC Method No. 44-15.02 [16].

2.3. Statistical Analysis of Results

The results of three replicate analyses are presented as mean values ± standard deviations and the coefficient of variation (CV). The two-way ANOVA was used to analyse the data. The Duncan test was used to determine the statistically significant difference at the 95% level ($p = 0.05$). In addition, the Pearson's linear correlation coefficients were calculated between the analysed parameters describing grain, flour and bread at the significance level of $p = 0.05$. Additionally, the principal component analysis (PCA) was used to provide a ready means of visualizing the differences and similarities between the investigated wheat cultivars in different nitrogen fertilisation. Statistical analysis of the results was performed using TIBCO Statistica 13.3.0 (TIBCO Software Inc., Palo Alto, CA, USA).

3. Results and Discussion

3.1. Grain Quality and Flour Yield

Table 1 lists the results of the assessment of selected quality parameters of winter wheat grain. According to Morgan et al. [20] TGW is linked to the yield and quality of the obtained flour, and determines its colour and ash content. No significant impact of increasing the nitrogen dose from N_1 to N_2 on the TGW change could be found in the present study, for hybrid as well as population cultivars. The lowest value of the parameter was determined forHyking cultivar (N_1—38.6 g, N_2—41.6 g). The population Hondia cultivar and the Hypocamp hybrid cultivar fertilised with N_2 nitrogen dose exhibited TGW values higher by 17.8 and 14.4% in comparison to the Hyking hybrid cultivar. Klikocka et al. [21] were also unable to determine the impact of different nitrogen fertilisation levels on TGW formation, whereas Abedi et al. [22] demonstrated a significant increase of the values of this parameter values under the impact of elevated nitrogen doses.

Table 1. Selected qualitative indicators of wheat grain and flour from hybrid and population cultivars.

Cultivar	Nitrogen	TGW [g]	Bulk Density [kg/hl]	Vitreousness [%]	Crude Protein Content [%]	Flour Yield [%]
Hybery	N_1	43.6 [b,c] ± 2.5	74.5 [b–d] ± 0.7	44 [a,b] ± 9	11.8 [a] ± 0.4	75.3 [b] ± 0.8
	N_2	46.5 [c–e] ± 2.1	76.0 [d–f] ± 1.6	48 [a,b] ± 0	12.7 [b,c] ± 0.4	75.7 [b,c] ± 0.3
Hyena	N_1	44.6 [b,c] ± 1.3	72.2 [a,b] ± 0.4	54 [a–c] ± 6	11.6 [a] ± 0.1	77.9 [e–g] ± 0.9
	N_2	45.1 [c,d] ± 0.3	75.3 [c–f] ± 0.9	77 [d] ± 6	13.0 [c,d] ± 0.1	78.7 [g,h] ± 0.6
Hyfi	N_1	44.1 [b,c] ± 3.1	74.3 [b–d] ± 1.6	40 [a] ± 1	12.7 [b,c] ± 0.4	78.2 [f–h] ± 0.6
	N_2	46.9 [c–e] ± 0.4	73.1 [a–c] ± 0.4	53 [a,b] ± 2	14.1 [f] ± 0.1	78.5 [g,h] ± 0.4
Hyking	N_1	38.6 [a] ± 2.3	71.1 [a] ± 1.6	50 [a,b] ± 7	12.6 [b,c] ± 0.1	75.6 [b,c] ± 0.4
	N_2	41.6 [a,b] ± 1.9	74.6 [b–d] ± 0.4	75 [d] ± 8	13.7 [e,f] ± 0.3	77.5 [e–g] ± 0.1
Hymalaya	N_1	45.6 [c,d] ± 1.2	74.8 [b–e] ± 0.1	71 [d] ± 5	12.3 [b] ± 0.2	80.1 [I] ± 0.6
	N_2	46.1 [c,d] ± 1.5	77.1 [e,f] ± 0.7	75 [d] ± 7	13.5 [d,e] ± 0.1	79.5 [h,i] ± 0.0
Hypocamp	N_1	47.2 [c–e] ± 0.6	77.6 [f] ± 0.4	45 [a,b] ± 1	11.7 [a] ± 0.2	76.8 [c–e] ± 0.1
	N_2	48.6 [d–f] ± 0.6	77.6 [f] ± 1.5	51 [a,b] ± 10	12.8 [b,c] ± 0.1	81.9 [j] ± 0.5
Hyvento	N_1	46.0 [c,d] ± 2.0	75.4 [c–f] ± 2.3	67 [c,d] ± 4	12.4 [b] ± 0.3	77.7 [e–g] ± 0.8
	N_2	47.1 [c–e] ± 1.3	73.8 [b–d] ± 1.6	74 [d] ± 1	14.2 [f] ± 0.3	78.8 [g,h] ± 0.4
Belissa [1]	N_1	43.9 [b,c] ± 0.0	73.6 [a–d] ± 0.2	45 [a,b] ± 5	11.4 [a] ± 0.3	70.8 [a] ± 0.7
	N_2	46.9 [c–e] ± 0.3	74.3 [b–d] ± 0.4	56 [b,c] ± 10	12.7 [b,c] ± 0.2	75.1 [d–f] ± 0.9
Hondia [1]	N_1	49.9 [e,f] ± 0.1	73.3 [a–c] ± 0.3	50 [a,b] ± 7	12.5 [b] ± 0.2	74.7 [c–e] ± 0.3
	N_2	50.6 [f] ± 0.0	74.0 [b–d] ± 1.2	57 [b,c] ± 4	13.3 [d,e] ± 0.1	75.2 [b–d] ± 0.5
CV (%) **	N_1	7.23	2.68	21.56	4.20	3.45
	N_2	5.45	2.26	20.00	4.39	3.76

[1] Population cultivars. The results are presented as mean values ± standard deviation. Different letters in the same column indicate significant differences (p = 0.05), according to ANOVA followed by Duncan test. ** CV coefficient of variation; N_1—110 kg/ha. N_2—150 kg/ha.

The population cultivars significantly differed from the hybrids at both nitrogen treatments only in the case of the Hypocampcultivar. The bulk density is an important quality indicator of grain, determining its accuracy, the degree of grain development, their structure, and the thickness of the coat, and it also determines the milling value of the grain. However, bulk density is not always an ideal measure of grain quality, because this parameter is determined by environmental factors. A higher bulk density indicates a better technological value of the wheat, because smaller grains can have a much lower ratio of endosperm to coat and germ [23]. The highest grain bulk density value was found for the Hypocamp hybrid cultivar (77.6 kg/hl), without the impact of N_1 and N_2 doses on the value of this trait. An increasing tendency of bulk density was observed as a result of applying N_2 fertilisation. However, the Hypocamp hybrid cultivar was characterised by a higher value of bulk density in N_1 fertilisation than the Hyfi, Hyking, Hyvento, Hondia, and Belissa cultivars with N_2 fertilisation. Among hybrid cultivars, only Hyena and Hyking were characterised by a significant increase of grain bulk density after the application of the N_2 dose [22]. Harasim and Wesołowski [24] indicated that winter wheat grain, as a result of increasing the nitrogen fertilisation dose from 100 to 150 kg/ha, was characterised by a higher value of the analysed parameter. According to Dziki et al. [25] vitreousness grains are characterised by a higher degree of endosperm content, and higher grain hardness and protein content in comparison with non-vitreous grain. The grain vitreousness of the tested cultivars ranged between 40.0 and 77.0%. Significantly, the highest grain vitreousness for both doses: N_1 and N_2 were found in the Hymalaya and Hyventohybrid cultivars, which at N_1 fertilisation, were characterised by a higher value of this parameter than the Hybery, Hyfi, Hypocamp, Belissa, and Hondia cultivars fertilised by N_2. Compared to N_1, N_2 resulted in a significant increase of grain vitreousness, by 23.0 and 25.0% only in the Hyena and Hykinghybrid cultivars. An appropriate content of total protein, which is not only determined by nitrogen fertilisation, but also by genetic predispositions, is the basic factor determining the applicability of wheat cultivar grain for bread baking [10,26,27]. Crude protein content in the grain ranged between 11.8 and 14.2% (hybrid cultivars) and between 11.4 and 13.3% (population cultivars). Similarly to a study by Skudra and Ruza [3], the present study confirmed increased protein content in the grain with increased N dosage for all cultivars. However, the cultivars reacted differently to this effect [9]. The Hyfi and Hyking cultivars fertilised with N_1 had a crude protein content at the same level as the Hybery, Hypocamp and Belissa cultivars fertilised with N_2. Also Haque et al. [28] indicated the effect of the interaction of wheat cultivars and different levels of N fertilisation on protein content of the grain. Increasing the dose from N_1 to N_2 resulted in a significantly higher crude protein content in the grain of the Hyena and Hykinghybrid cultivars, at the level determined in the grain of the Hondiapopulation cultivar. The highest crude protein content (N_2) was determined in the Hyfi and Hyvento hybrid cultivars [10,26]. The grain of the hybrid cultivars was generally characterised by a higher flour yield than the population cultivars. The highest flour yield was obtained from the grain of the Hymalaya (N_1) and Hypocamp (N_2) hybrid cultivars. Increasing the fertilisation dosage to N_2 had a significant impact on increasing the flour yield in Hypocamp and Hyking hybrid cultivars and the Belissa population cultivar, by 9.5, 2.5, and 8.2%, respectively. A similar wheat flour yield without the impact of fertilisation variants was obtained by Warechowska et al. [29], whereas Metho et al. [30] indicated a significant interaction between cultivars and nitrogen fertilisation on the value of the flour yield. In this study it was shown that the Hymalaya hybrid cultivar fertilised with N_1 was characterised by higher flour yield than other cultivars fertilised with N_2 with the exception of Hypocamp.

The Belissa population cultivar (N_1, N_2) is concerned, the Hypocamp hybrid cultivar developed a grain with a lower TGW and vitreousness but with a higher bulk density, and the crude protein content in the grain of both cultivars was similar. Especially, for the Hypocamp hybrid cultivar grain with the N_2 dose, a higher flour yield was obtained than from the grains of the Belissa and Hondia population cultivars.

Among the analysed features, the lowest variability was found in bulk density and simultaneously the highest grain vitreousness.

3.2. Flour Quality Parameters

Table 2 presents selected parameters of the quality of the flour obtained from the grain of the tested wheat cultivars. The ash content in the grain depends mainly on the genotype, wheat class, and cultivar as well as the growing location, year, and grinding method [31,32]. The ash content in the flour of the hybrid cultivars ranged from 0.43 to 0.69%. Compared to the population cultivars and other hybrid cultivars, the flour of the Hypocamp and Hybery hybrids was characterised by the lowest ash content. The highest ash content was found in the flour of the Hyfi (0.59–0.73%) and Hyking (0.69–0.58%) hybrid cultivars. According to Bucsella et al. [33] and Hemery et al. [34], an increase in the ash content in the flour combined with an increase in nutrients (fibre, vitamins) is desirable, but the technology quality of the flour is then lower due to weakening of the protein matrix during dough formation. Increased N fertilisation resulted in a reduced total ash content in the flour of Hyking hybrid cultivar (by 15.9%)), and an increase for Hyfi cultivar (by 19.2%) and Hondia population cultivar (by 20.7%). A study by Bayoumi and El-Demardash [35] also indicated a beneficial effect on this feature of nitrogen fertilisation, although in a later study by Warechowska et al. [29], such a relationship was not found. Statistical analysis revealed a negative linear correlation between total ash content in flour and bulk density (r = −0.62, $p < 0.05$). The falling number indicates of activity of amylolytic enzymes contained in the flour, their capacity for hydrolysing the starch present in the flour to sugars, which are substrates for the dough fermentation process [10,36]. A study by of Linina and Ruza [37] shows that the falling number value depends significantly on the weather conditions, grain storage time and the applied nitrogen dose. In the present study, the increased N dose resulted in a reduced flour falling number (by approximately 20 s) in the Hyfi, Hybery, and Hymalaya hybrid cultivars and Hondiapopulation cultivar. Kindred et al. [36] demonstrated that the impact of nitrogen fertilisation on the falling number value relies on the interactions between the genotype and N doses, which has been confirmed in the present study. The application of the N2 dose increased the value of the falling number in the Belissa population cultivar (by 3.5%) and in the hybrid cultivars in the range from 0.3 (Hyking Hypocamp) to 1.6% (Hyvento). Most flours used in the baking industry requires adjustment of the falling number to the perfect range (250–320 s), and a reduced value of this parameter is obtained by adding α-amylase preparation [38]. In the presented study, this condition was fulfilled by flours of hybrid wheat cultivars, with the exception of the Hyena and Hyventocultivars and the Belissa and Hondia population cultivars. Among the hybrid cultivars, the lowest falling number value from 190 (N_1) to 210 s (N_2) was found in the Hyfihybrid cultivar and the highest from 387 (N_2) to 389 s (N_1) Hyena cultivar. Jaskulska et al. [8] state that wheat cultivars react in a natural manner to the N fertilisation level, and it is displayed by grain quality changes, including increased wet gluten content. Increasing nitrogen fertilisation with N_2 resulted in an increase in the wet gluten content in the hybrid cultivars in the range from 10.2 (Hyvento) to 15.6% (Hypocamp). In the population cultivars, the increase ranged from 7.9 (Hondia) to 11.4% (Belissa). In studies by Jaskulska et al. [8], increasing nitrogen fertilisation from 100 to 200 kg/ha resulted in a significant increase in gluten in the flour by 17.6%, while Wojtkowiak et al. [39] showed no effect of the cultivar genotype and nitrogen dose on the content of wet gluten for the tested wheat cultivars. The Hyfi and Hyvento hybrid cultivar were characterised by a significantly higher (N_1, N_2) value of wet gluten compared to the population cultivars and other hybrid cultivars. The lowest (N_1) wet gluten content among all cultivars was found in the flour of Hyking and Hybery hybrid cultivars. A similar trend was observed for the differences between the tested cultivars fertilised with N_1 and N_2 doses in terms of dry gluten, which ranged from 7.9 to 11.0 [40]. The range of wet and dry gluten content in the flour of the studied hybrid wheat cultivars was similar to the values of these parameters reported by Taner et al. [41] and Jaskulska et al. [8]. However, higher contents of wet gluten in the range from 28.3 to 37.0% were found by Šip et al. [42]. The gluten index is the parameter used to determine the quality of washed gluten. According to Šekularac et al. [26], the cultivar genotype is the main factor for the statistically significant variability of this parameter. In the conducted study, no significant influence of cultivar (with the exception for Hybery and Hyventohybrid cultivars) and N fertilisation (with the exception for the Hyvento hybrid cultivar and the Hondiapopulation cultivar) on the values of this parameter could be determined. According to Ćurić et al. [5] GI values between 75 and 90 ensure the optimum bread baking quality. The GI value in the flour of the majority of hybrid cultivars ranged from 94 to 99, indicating too strong flour to be applicable for baking. The flour of the Hyvento hybrid cultivar was assessed markedly more favourably, as its GI remained in the optimum range from 83 (N_1) to 92 (N_2). The study results show a negative linear correlation between GI and bread specific volume (r = −0.64, $p < 0.05$), agreeing with the results obtained by Šekularac et al. [26]. Compared to the population cultivars, the Hyfi hybrid cultivar (N_1, N_2) was characterized by a higher total ash content. The Hyfi and Hyvento hybrid cultivars (N_1, N_2) compared to the population cultivars were characterised by a significantly higher value of wet gluten content. The flour of the Hyvento hybrid cultivar was definitely more advantageous as compared to the population cultivars as the GI was in the optimal range from 83 (N_1) to 92 (N_2).

Table 2. Quality parameters of wheat flour from hybrid and population cultivars.

Cultivar	Nitrogen	Content in Flour [%]			Gluten Index [%]	Falling Number [s]
		Total Ash	**Wet Gluten**	**Dry Gluten**		
Hybery	N_1	0.45 [a–d] ± 0.00	24.1 [a,b] ± 1.4	8.4 [a,b] ± 0.4	99 [e] ± 0	314 [d] ± 4
	N_2	0.40 [a] ± 0.07	27.4 [d–g] ± 0.1	9.3 [d–g] ± 0.4	98 [c–e] ± 1	296 [c] ± 6
Hyena	N_1	0.52 [c–e] ± 0.04	25.7 [b–e] ± 2.5	8.7 [b–d] ± 0.6	96 [c–e] ± 2	389 [h] ± 5
	N_2	0.44 [a–c] ± 0.02	28.6 [f–h] ± 0.6	9.7 [f–h] ± 0.1	96 [b–e] ± 0	387 [g,h] ± 3
Hyfi	N_1	0.59 [e] ± 0.02	27.1 [d–g] ± 0.1	9.2 [c–g] ± 0.2	96 [c–e] ± 2	210 [b] ± 0
	N_2	0.73 [f] ± 0.03	30.3 [h,i] ± 0.5	10.2 [h,i] ± 0.2	94 [b,c] ± 1	190 [a] ± 1
Hyking	N_1	0.69 [f] ±0.02	22.2 [a] ± 0.1	7.9 [a] ± 0.1	99 [e] ± 0	309 [d] ± 2
	N_2	0.58 [e] ± 0.04	25.2 [b–d] ± 0.4	9.0 [b–f] ± 0.3	99 [e] ± 0	310 [d] ± 1
Hymalaya	N_1	0.46 [a–d] ± 0.02	24.6 [b,c] ± 1.2	8.5 [a–c] ± 0.3	98 [d,e] ± 1	346 [f] ± 3
	N_2	0.54 [d,e] ± 0.09	28.2 [f–h] ± 0.8	9.7 [f–h] ± 0.2	96 [c–e] ± 1	324 [e] ± 2
Hypocamp	N_1	0.43 [a,b] ± 0.04	24.4 [a–c] ± 0.4	8.3 [a,b] ± 0.0	97 [c–e] ± 2	291 [c] ± 3
	N_2	0.41 [a,b] ± 0.02	28.2 [f–h] ± 0.5	9.6 [f–h] ± 0.0	96 [c–e] ± 0	292 [c] ± 1
Hyvento	N_1	0.46 [a–d] ± 0.05	29.4 [g,h] ± 1.2	9.8 [g,h] ± 0.2	83 [a] ± 3	378 [g] ± 2
	N_2	0.46 [a–d] ± 0.05	32.4 [i] ± 0.4	11.0 [j] ± 0.4	92 [b] ± 3	384 [g,h] ± 4
Belissa [1]	N_1	0.53 [c–e] ± 0.04	25.7 [c–f] ± 1.6	8.9 [b–e] ± 0.4	95 [b–d] ± 2	385 [g,h] ± 1
	N_2	0.50 [b–e] ± 0.00	29.0 [I] ± 1.2	9.5 [i,j] ± 0.5	97 [c–e] ± 1	399 [i] ± 3
Hondia [1]	N_1	0.46 [a–d] ± 0.02	26.8 [e–g] ± 0.8	9.4 [e–g] ± 0.0	96 [c–e] ± 1	403 [i] ± 3
	N_2	0.58 [e] ± 0.03	29.1 [h,i] ± 0.8	9.2 [h,i] ± 0.1	93 [b] ± 1	384 [g,h] ± 4
CV (%) **	N_1	16.41	8.89	7.33	5.17	17.92
	N_2	20.53	8.48	6.31	2.39	19.98

[1] Population cultivars. The results are presented as mean values ± standard deviation. Different letters in the same column indicate significant differences ($p = 0.05$), according to ANOVA followed by Duncan test. ** CV coefficient of variation; N_1—110 kg/ha. N_2—150 kg/ha.

The values of the coefficient of variation (CV) for the N_1 and N_2 doses were similar for the wet gluten as well as the dry gluten and falling number. The use of a higher dose of N_2 increased the stability of the gluten index whereas the ash content was characterised by higher variability.

3.3. Flour Water Absorption and Rheological Properties of Dough

Determination of flour water absorption, as well as the results of the farinographic analysis of the flour play a key role in the assessment of wheat flour baking quality [36,43]. The results of determinations of these parameters are presented in Table 3.

Water absorption of the flour of the hybrid cultivars was considerably lower than for the Belissa population cultivar. However, the Hymalaya, Hyena, Hyfi, and Hyking hybrid cultivars were found to have similar or higher values of this parameter than the Hondia population cultivar. The influence of the cultivar factor on significant differentiation of the flour water absorption and other traits of the wheat dough was also confirmed by Warechowska et al. [29] and Silva et al. [44]. Increased water absorption of flour with the increased N dose was determined at the range between 0.9 and 2.0% in the population cultivars and between 1.5 and 2.1% in the hybrid cultivars. Only for the Hyfi hybrid cultivar, the increased water absorption of flour was 0.6% for N_2 compared to N_1. Park et al. [7] also recorded increased wheat water absorption of flour under the influence of increased nitrogen fertilisation, but such a relationship was not shown by Warechowska et al. [29]. A negative linear correlation was observed between water absorption of flour and specific volume of its bread ($r = -0.64$, $p < 0.05$) and positive between water absorption of flour and dough yield ($r = 0.62$, $p < 0.05$).

Table 3. Results of rheological analysis of wheat flour obtained from hybrid and population cultivars.

Cultivar	Nitrogen	Water Absorption of Flour [%]	Properties of Dough			
			Development Time [min]	Stability [min]	Degree of Softening [FU]	Quality Number
Hybery	N_1	54.0 [a] ± 0.3	2.1 [a–c] ± 0.1	4.5 [e,f] ± 1.3	74 [a,b] ± 4	70 [e–g] ± 4
	N_2	56.0 [c] ± 0.1	2.7 [a–c] ± 0.5	5.6 [f,g] ± 0.1	78 [a] ± 1	73 [i–j] ± 4
Hyena	N_1	57.5 [d,e] ± 0.1	2.5 [a–c] ± 0.3	4.1 [b–f] ± 0.1	97 [a,b] ± 4	56 [e–g] ± 1
	N_2	59.0 [g] ± 0.0	2.6 [a–c] ± 0.6	3.9 [a–e] ± 0.8	102 [b,c] ± 0	52 [c–e] ± 4
Hyfi	N_1	57.9 [e] ± 0.1	3.0 [a–c] ± 0.4	2.9 [a–d] ± 0.3	125 [e] ± 2	47 [a–d] ± 2
	N_2	58.5 [f] ± 0.2	3.3 [a–c] ± 0.6	2.7 [a,b] ± 0.2	129 [d,e] ± 11	46 [a–d] ± 4
Hyking	N_1	57.5 [d,e] ± 0.1	2.4 [a–c] ± 0.5	5.3 [e–g] ± 0.4	78 [a] ± 7	72 [i–j] ± 2
	N_2	58.5 [f] ± 0.0	2.9 [a–c] ± 0.5	6.6 [g] ± 0.7	75 [a] ± 4	76 [j] ± 1
Hymalaya	N_1	56.0 [c] ± 0.1	2.5 [a–c] ± 0.4	5.1 [e,f] ± 0.8	86 [a,b] ± 6	62 [b–f] ± 3
	N_2	57.8 [e] ± 0.1	3.9 [c] ± 0.7	5.2 [e–g] ± 0.1	93 [a,b] ± 2	68 [h–I] ± 3
Hypocamp	N_1	55.1 [b] ± 0.1	2.1 [a–c] ± 0.1	2.9 [a–d] ± 0.0	120 [c–e] ± 1	44 [a–c] ± 1
	N_2	57.2 [d] ± 0.4	1.9 [a] ± 0.3	2.5 [a] ± 0.2	128 [e] ± 3	43 [a,b] ± 1
Hyvento	N_1	54.8 [b] ± 0.2	2.9 [a–c] ± 0.3	5.1 [e,f] ± 0.4	87 [a,b] ± 1	63 [g,h] ± 3
	N_2	56.4 [c] ± 0.1	3.6 [b,c] ± 0.1	4.3 [d–f] ± 0.1	89 [a,b] ± 1	61 [f–h] ± 0
Belissa [1]	N_1	59.0 [g] ± 0.1	2.1 [a,b] ± 0.5	2.8 [a–c] ± 1.1	135 [e] ± 3	39 [a] ± 7
	N_2	61.0 [h] ± 0.4	2.9 [a–c] ± 0.5	2.8 [a–d] ± 0.4	134 [e] ± 2	45 [a–c] ± 8
Hondia [1]	N_1	56.2 [c] ± 0.1	1.8 [a] ± 0.4	4.3 [d–f] ± 0.1	95 [a,b] ± 4	54 [d–f] ± 1
	N_2	57.1 [d] ± 0.1	2.1 [a,b] ± 0.4	4.1 [b–f] ± 0.4	105 [b–d] ± 4	50 [b–e] ± 6
CV (%) **	N_1	2.82	23.76	27.98	22.80	20.12
	N_2	2.55	32.47	34.09	21.75	22.34

[1] Population cultivars. The results are presented as mean values ± standard deviation. Different letters in the same column indicate significant differences (p = 0.05), according to ANOVA followed by Duncan test. ** CV coefficient of variation; N_1—110 kg/ha. N_2—150 kg/ha.

The dough development time remained in the range from 1.8 to 3.9 min. Significantly longer dough development time characterised only the Hymalaya and Hyvento hybrid cultivars, whereas the lowest value of this parameter was determined in the flour of the Hondia population cultivar. Low dough stability from 2.5 to 2.9 min characterised the Hyfi and Hypocamp hybrid cultivars, as well as the Belissa population cultivar. The character of the dough was considerably more favourably assessed in the tested Hybery, Hyking and Hymalaya hybrid cultivars. Dough stability in these cultivars was at least one minute or twofold longer than for the Hondia and Belissa population cultivars, which corresponds to strong flour of the three mentioned hybrid wheat cultivars. It was noted that the dough stability parameter of the Hyking cultivar (N_1) was significantly higher than the value of this parameter in the tests of the Hyfi, Hypocamp and Belissa dough, made from flour obtained from the experiment with a higher level of nitrogen fertilisation (N_2). [40,45]. As stated by Silva et al. [44], extension of the dough development time and stability is caused by increased total protein content, and dough with such parameters is obtained from flour with high gluten and HMW glutenin content. In the present study, a positive linear correlation was observed between dough development time and crude protein content (r = 0.63, p < 0.05), wet gluten (r = 0.65, p < 0.05) and dry gluten (r = 0.60, p < 0.05). In terms of dough softening, the lowest score was assigned to the Belissa population cultivar and the Hyfi and Hypocamphybrid cultivars, whereas the value of this parameter for these cultivars was 130 FU, on average. Dough mixing qualities are considered satisfactory when the degree of softening is below 70 FU [27]. The Hybery and Hyking hybrid cultivars were characterised by a degree of softening favourable in terms of dough quality, which was lower by 50 FU. Dough from these cultivars was characterised by better stability and a lower degree of softening, with a higher quality number. The significant increase of dough stability and degree of softening under the impact of the increased N dose known from the study of Park et al. [7] and Ma et al. [10] could not be confirmed. The Hyfi, Hypocamp, and Belissa cultivars fertilised with N_1 had a higher degree of dough softening than Hybery, Hyking, Hymalaya and Hyvento fertilised with N_2. The flour of the Hyking cultivar, both in the N_1 and N_2 variants, was distinguished by the highest quality number. The grain of the Hyena, Hyfi, Hypocamp, Hyvento, Belissa, and Hondia cultivars obtained under conditions of higher nitrogen fertilisation (N_2) did not match the quality number of the Hyking cultivar (N_1), which indicates that the cultivar's factor allows the studied cultivars to be distinguished in terms of their potential suitability for food processing. Statistical analysis did not confirm significant differentiation of the remaining rheological traits of the dough, for hybrid as well as population cultivars with an increased N dose, which, in contrast, coincides with the results obtained by Silva et al. [44]. Strong negative linear correlations were observed between the quality number and degree of softening, as well as between the dough stability and degree of softening [40]. Both values of the correlation coefficient were r = −0.94, $p < 0.05$. Moreover, a strong positive linear correlation between quality number and dough stability was demonstrated (r = 0.96, $p < 0.05$). The use of the N_2 dose resulted in an increase in the dough development time in both hybrid (except the Hypocampcultivar) and population cultivars. The N_2 dose influenced the increase of dough stability only in the Hybery, Hyking and Hymalayahybrid cultivars.

The dough quality parameters were characterised by high volatility, except for the coefficient of variation (CV) for water absorption (N_1 = 2.82%, N_2 = 2.55%). The increase in nitrogen dose from 110 to 150 kg/ha resulted in greater differentials in the dough development time, dough softening, and quality number.

3.4. The Flour Baking Quality Assessment

The flour baking quality assessment was completed via the laboratory baking method, comparing the results of the course of individual dough production phases and the assessment of the obtained product [5,6]. The dough from Belissa population cultivar was characterised by a significantly higher dough yield, and the Hybery hybrid cultivar had the lowest compared to the remaining cultivars (Table 4). The difference in terms of this parameter in the assessment of the baking process of the dough from Belissa population cultivar and Hybery hybrid cultivar flour was 13.0%. In the conducted study, as in the study by Jaskulska et al. [8] on winter wheat, a positive impact of the N_2 dose on dough yield occurred. The total baking loss was similar in the majority of cultivars, and for the hybrid cultivars, the lowest value was observed for the Hyfi cultivar (N_1) and highest for the Hyvento cultivar (N_2). The dough yield obtained from the hybrid flour ranged between 142.3 and 148.5 cm^3. The lower the baking loss, the higher the bread yield that was observed for hybrid the Hybery (N_1, N_2), Hyfi (N_2) and Hymalaya cultivars (N_2), without an impact of the N doses on the value of this parameter. The bread yield and Dallmann porosity index of the crumb were positively correlated (r = 0.48 and r = 0.49, $p < 0.05$) with the gluten index. Bread volume is one of the basic traits taken into account for bread quality assessment and it was slightly variable for the hybrid wheat cultivars. The Hybery and Hyvento hybrid cultivars had significantly higher value of this parameter than the remaining cultivars. The Hyvento hybrid cultivar was characterised by the lowest bread yield and the concomitant highest gluten quality (GI < 95), which was confirmed by Ćurić et al. [5] who indicated that the GI value is strongly correlated to bread quality. The influence of the cultivar's factor was also noted in the comparison of the bread yield parameter from the flour of the Hypocamp hybrid cultivar obtained from grains cultivated with a higher nitrogen dose (N_2), it was much lower than that obtained with the flour of the Hybery hybrid cultivar cultivated with a lower level of nitrogen fertilisation (N_1). A significant impact of N dose on bread volume was determined only for the Hymalaya and Hyvento hybrid cultivars. However, a lower bread volume, by 12.5 and 13.5% respectively, characterised the breads made from the flour of these cultivars fertilised with the N_2 dose. Jaskulska et al. [8] also were unable to demonstrate a direct effect of N fertilisation on bread yield and volume, whereas the same effect was confirmed by Grahmann et al. [6]. The Dallmann porosity index of crumb was good or very good. Also, in the case of the Hybery and Hyvento hybrid cultivars, the crumb of breads with the highest volume received the best score. The amount of crumb moisture content depends on a variety of factors, including the recipe and baking parameters, which is confirmed in the presented results, as the mentioned cultivars were distinguished by higher water absorption of the flour, and this parameter determines the amount of water in the recipe for the bread dough. The crumb moisture content from wheat flour obtained from the grain (N_1) of the hybrid cultivars was undifferentiated, and a higher crumb moisture content was found for Belissa and Hondia population cultivars. However, in the case of the N_2 dose, a higher crumb moisture content was observed in the breads made from the flour from the Hybery, Hyena, Hymalaya, and Hypocamp hybrid cultivars. Statistical analysis revealed a positive linear correlation between crumb moisture content and dough yield (r = 0.65, $p < 0.05$). The lack of a significant impact of the N_2 dose as compared to the N_1 on the assessment of bread volume, the Dallmann porosity index of crumb, and the crumb moisture content from hybrid cultivars suggest the possibility of obtaining flour of very good baking value with a lower N dose used [4]. A study by Vazquez et al. [43] showed that it is possible to modulate the wheat grain and flour for bread according to the nitrogen regime, although it is necessary to understand the genotype-environment relationship. Increasing the content of protein in the grain under the impact of nitrogen fertilisation does not directly affect the increased volume of the bread made from flour obtained from such grain [4]. It is important to minimise the environmental issues associated with late N usage, thus it is significant to know what dosage and fertiliser form, as well as date of N fertilisation would be necessary to obtain the desired baking traits of wheat cultivars [43]. Among the assessed characteristics of the baking process, the application of the N_2 dose increased only in the case of the dough yield.

The assessed quality indicators for bread were characterised by high stability and the values of the coefficient of variation (CV) were the lowest for bread yield and crumb moisture content. The use of the N_1 dose increased the stability of baking loss while the N_2 dose increased the stability of bread volume and the Dallmann porosity index of crumb. Photographs of the bread crumb of selected cultivars are shown in Figure 1.

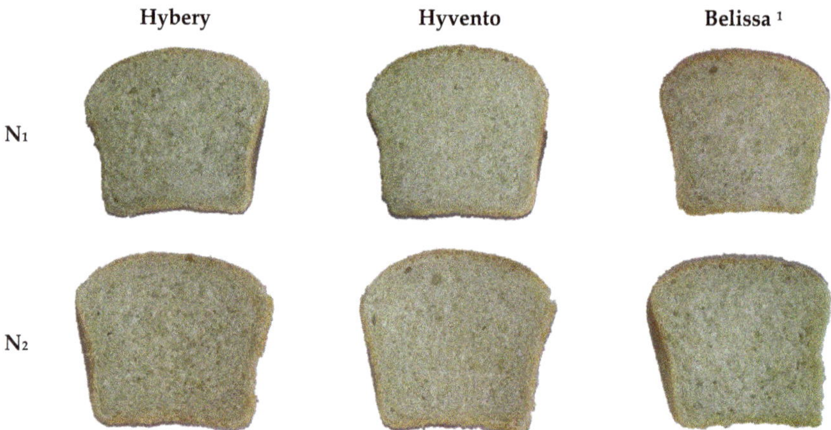

Figure 1. Crumb porosity of bread obtained from the flours of selected wheat cultivars with various levels of fertilisatio: N_1.—110 kg/ha, N_2—150 kg/ha. [1] Population cultivars.

Table 4. Baking process parameters of wheat flour and quality indicators of wheat bread obtained from hybrid and population cultivars.

Cultivar	Nitrogen	Dough Yield [%]	Baking Loss [%]	Bread Yield [%]	Bread Volume [cm³/g]	Dallmann Porosity Index of Crumb [Scores]	Crumb Moisture Content [%]
Hybery	N_1	143.8 ^{a} ± 0.3	15.1 ^{a,b} ± 0.9	148.1 ^{c,d} ± 1.6	3.2 $^{c-e}$ ± 0.2	100 ^{c} ± 0	43.9 $^{b-d}$ ± 0.4
	N_2	145.5 ^{b} ± 0.1	15.1 ^{a,b} ± 0.1	148.1 ^{c,d} ± 0.1	3.3 ^{e} ± 0.1	100 ^{c} ± 0	45.0 $^{e-g}$ ± 0.4
Hyena	N_1	148.6 ^{e,f} ± 0.0	15.9 $^{a-c}$ ± 0.5	146.5 $^{b-d}$ ± 0.8	2.9 ^{a,b} ± 0.0	95 ^{b,c} ± 7	43.1 ^{a,b} ± 0.6
	N_2	149.3 ^{f,g} ±0.5	16.2 $^{a-d}$ ± 0.9	146.2 $^{a-d}$ ± 1.6	2.9 ^{a,b} ± 0.0	90 $^{a-c}$ ± 0	44.3 $^{c-f}$ ± 0.3
Hyfi	N_1	146.2 ^{f,g} ±0.4	17.1 $^{a-d}$ ± 0.5	144.5 $^{a-d}$ ± 0.9	2.9 ^{a,b} ± 0.0	90 $^{a-c}$ ± 0	43.6 $^{a-c}$ ± 0.3
	N_2	149.6 $^{b-d}$ ± 1.2	14.8 ^{a} ± 0.5	148.5 ^{d} ± 1.2	3.0 $^{a-d}$ ± 0.1	90 $^{a-c}$ ± 0	44.2 $^{c-e}$ ± 0.6
Hyking	N_1	148.9 ^{f,g} ± 0.4	16.0 $^{a-d}$ ± 0.4	146.4 $^{a-d}$ ± 0.6	3.0 $^{a-d}$ ± 0.0	85 ^{a,b} ± 7	43.9 $^{b-d}$ ± 0.7
	N_2	150.0 ^{g} ± 0.4	16.4 $^{a-d}$ ± 0.7	145.8 $^{a-d}$ ± 1.3	3.0 $^{a-c}$ ± 0.2	90 $^{a-c}$ ± 0	44.5 $^{d-g}$ ± 0.2
Hymalaya	N_1	146.0 ^{b,c} ± 0.4	17.0 $^{a-d}$ ± 0.9	144.6 $^{a-d}$ ± 1.5	3.2 ^{d,e} ± 0.0	85 ^{a,b} ± 7	43.2 ^{a,b} ± 0.2
	N_2	147.0 ^{c,d} ± 0.2	15.3 ^{a,b} ± 0.2	147.7 ^{c,d} ± 0.4	2.8 ^{a} ± 0.0	90 $^{a-c}$ ± 0	44.1 ^{c,d} ± 0.5
Hypocamp	N_1	147.7 ^{d,e} ± 0.2	17.2 $^{b-d}$ ± 0.6	144.3 $^{a-c}$ ± 1.1	3.1 $^{b-d}$ ± 0.1	95 ^{b,c} ± 7	43.2 ^{a,b} ± 0.1
	N_2	154.3 ^{h} ± 0.5	17.8 ^{c,d} ± 0.1	143.5 ^{a,b} ± 0.1	3.1 $^{b-d}$ ± 0.1	95 ^{b,c} ± 7	44.7 $^{d-g}$ ± 0.2
Hyvento	N_1	145.6 ^{b} ± 0.1	18.4 ^{d} ± 0.2	142.3 ^{a} ± 0.4	3.7 ^{f} ± 0.1	96 ^{a} ± 0	43.1 ^{a,b} ± 0.3
	N_2	146.2 ^{b,c} ± 0.1	16.8 $^{a-d}$ ± 0.2	142.9 $^{a-d}$ ± 0.6	3.2 $^{c-e}$ ± 0.1	98 ^{a,b} ± 7	43.0 ^{a} ± 0.1
Belissa [1]	N_1	156.7 ^{i} ± 0.9	16.8 $^{a-d}$ ± 0.2	145.1 $^{a-d}$ ± 0.3	2.9 ^{a,b} ± 0.1	95 ^{b,c} ± 7	45.2 ^{g} ± 0.2
	N_2	158.3 ^{j} ± 0.7	16.7 $^{a-d}$ ± 0.1	145.3 $^{a-d}$ ± 0.2	3.0 $^{a-d}$ ± 0.0	93 ^{b,c} ± 4	45.1 ^{f,g} ± 0.1
Hondia [1]	N_1	154.5 ^{h} ± 0.6	16.7 $^{a-d}$ ± 0.2	145.4 $^{a-d}$ ± 0.4	3.0 $^{a-d}$ ± 0.0	95 ^{b,c} ± 7	44.3 $^{c-f}$ ±0.1
	N_2	155.8 ^{i} ± 0.5	16.6 $^{a-d}$ ± 0.1	145.5 $^{a-d}$ ± 0.1	3.1 $^{b-d}$ ± 0.1	95 ^{b,c} ± 7	44.9 $^{e-g}$ ± 0.1
CV (%) **	N_1	2.78	5.95	1.19	8.29	7.01	1.65
	N_2	3.05	7.98	1.69	5.12	4.85	1.47

[1] Population cultivars. The results are presented as mean values ± standard deviation. Different letters in the same column indicate significant differences ($p = 0.05$), according to ANOVA followed by Duncan test. ** CV coefficient of variation; N_1—110 kg/ha. N_2—150 kg/ha.

3.5. Principal Component Analysis

The relationship between the quality parameters measured in nine winter wheat cultivars fertilised with variable nitrogen dosage has been assessed by e principal component analysis (PCA) in order to estimate the source of variability. Among the twenty-one tested parameters, for the purpose of analysis eight extracted principal components were selected having eigenvalues higher than the average. The eigenvalues for the first component was 3.73, and the percentage of the explained variance was 46.56%. The second component explains 26.12% of the variance, and its own value was 2.09 (Table 5). The first two components transfer 72.69% of the variability of the original data. In order to determine the number of main components, the criterion of eigenvalue greater than the unity of the so-called Kaiser criterion was used [46]. It is therefore possible to approximate the original data set only in two dimensions. A scree plot confirms the significance of the first two components, which allows the reduction of the 8-dimensional space to two components (Figure 2).

Table 5. Individual values of correlation matrix.

Value Number	Eigenvalues (Correlations), Related Statistics			
	Eigenvalue	**% of Total Variance**	**Cumulative Eigenvalue**	**Cumulative%**
1	3.725399	46.56749	3.725399	46.5675
2	2.089800	26.12250	5.815199	72.6900
3	0.792959	9.91199	6.608158	82.6020
4	0.618540	7.73174	7.226698	90.3337
5	0.439531	5.49414	7.666229	95.8279
6	0.250196	3.12745	7.916425	98.9553
7	0.062311	0.77889	7.978736	99.7342
8	0.021264	0.26580	8.000000	100.0000

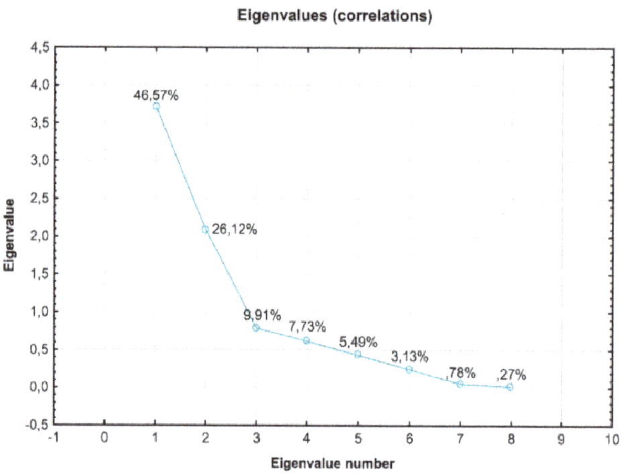

Figure 2. Scree plot for individual values of the matrix of correlation.

Figure 3a shows the projection of variables on the factor plane. The first principal component transfers the information contained in positively correlated variables, degree of softening, dough yield and negatively correlated values, dough stability, quality number and grain vitreousness (Table 6). The variable quality number is positively correlated with dough stability ($r = 0.96$, $p < 0.05$), while the variable dough yield is positively correlated with the variable dough degree of softening ($r = 0.58$, $p < 0.05$), which indicates the position of the vectors in close proximity. The second principal component comprises mostly of positively correlated variables, Dallmann porosity index of crumb, gluten index and bread yield. The variables of bread yield and Dallmann porosity index of crumb are positively correlated with the gluten index variable (r = 0.48 and r = 0.49, $p < 0.05$). The vectors of the degree of softening, dough quality number and grain vitreousness variables are longer and located closer to the circle, which indicates that these variables contribute the most information.

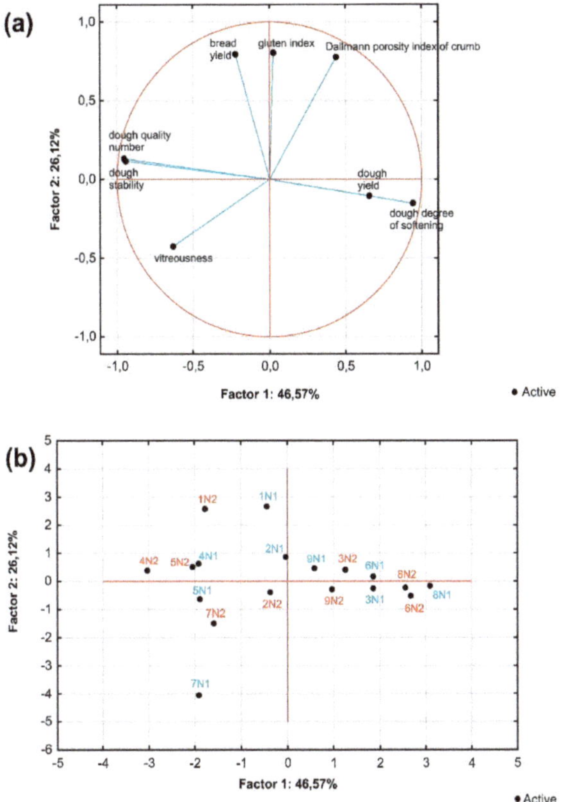

Figure 3. Principal components analysis of (**a**) distribution of the analyzed parameters (**b**) distribution of wheat cultivars and nitrogen fertilisation. 1—Hybery. 2—Hyena. 3—Hyfi. 4—Hyking. 5—Hymalaya. 6—Hypocamp. 7—Hyvento. 8—Belissa. 9—Hondia. N1—110 kg/ha. N2—150 kg/ha.

Table 6. Value of principal components coefficients.

Variable	Factor Coordinates of Variables, Based on Correlation	
	Factor 1	Factor 2
dough degree of softening	0.935932	−0.147075
dough stability	0.651097	−0.107195
Dallmann porosity index of crumb	0.433313	0.771696
gluten index	0.024504	0.797857
bread yield	−0.222052	0.788339
dough stability	−0.940939	0.117195
quality number	−0.951258	0.129446
vitreousness	−0.630539	−0.415473

Figure 3b presents the project of the analysed winter wheat cultivars fertilised with different nitrogen doses (N_1 and N_2) on the factor plane. Of the studied hybrid cultivars, the Hybery cultivar

(N_1 and N_2) had the best applicability for baking process, whereas the Hyvento cultivar (N_1) was characterised by favourable grain quality properties, dough rheological properties and bread volume, but its remaining baking quality parameters were less optimal. The proximity of the remaining cultivars confirms their similarity in terms of the analysed quality parameters.

4. Conclusions

Grain of the studied wheat cultivars was characterised by a variable technological property. Grain of the Hypocamp hybrid cultivar had a more favourable assessment in terms of grain bulk density (N_1 and N_2) and flour yield (N_2), the Hymalaya cultivar in terms of vitreousness (N_1 and N_2) and flour yield (N_1), whereas the Hyvento cultivar in terms of vitreousness (N_1 and N_2) and crude protein content (N_2). Increase of fertilisation to N_2 resulted in a significant increase of crude protein content in the case of all of the studied cultivars. Significant variability of falling number and water absorption of the flour was observed between cultivars fertilised with different nitrogen doses. The nitrogen dose increased to N_2 resulted in increased wet and dry gluten content, and the highest value of these parameters, similar to the population cultivars, was found in the Hyfi and Hyvento hybrid cultivars. The flour obtained from the Belissa population cultivar grain was distinguished by higher water absorption than in the remaining cultivars, whereas the Hyking hybrid cultivar flour had the highest quality number. The least favourable assessment of the rheological properties concerned the dough of the Hyfi and Hypocamp hybrid cultivars and the Belissa population cultivar.

The hybrid wheat flour was characterised by good quality, therefore it can be used for wheat bread production. No influence of the increased N_2 dose on the baking process and bread quality could be found with the exception of crumb moisture content and dough yield. Breads obtained from Hybery and Hyvento hybrid cultivars were characterised by greater volume than Belissa and Hondia population cultivars.

The greatest applicability for baking purposes was determined to be Hybery hybrid cultivar due to high values of the following parameters: -bread yield, specific bread volume, and low baking loss, which may contribute to increased interest in this cultivar by bread producers. Without a doubt, this may result in increased purchasing prices of hybrid cultivar grain, which ultimately justifies the purchase of more expensive grain of hybrid wheat cultivars. The research on hybrid wheat cultivars grown in different climatic and agronomic conditions have produced promising results, and considering the shrinking cultivation areas and climate change, such as prolonged droughts, it seems necessary to maintain the leading role of cereals in global plant production.

Author Contributions: Conceptualization, M.J.-P.; J.B. and J.K.; methodology, M.J.-P.; J.B. and J.K.; investigation, M.J.-P.; J.K.; writing—original draft preparation, M.J.-P.; J.K.; and J.B.; writing—review and editing, M.J.-P.; J.B.; J.K.; E.S.-K.; D.B.-J. and G.J. All authors have read and agreed to the published version of the manuscript.

Funding: The project was supported by the Minister of Science and Higher Education of Poland (Project No.026/RID/2018/19 "Regional Initiative of Excellence", 2019–2022).

Conflicts of Interest: The authors declare no conflict of interest.

References

1. Liu, G.; Zhao, Y.; Mirdita, V.; Reif, J.C. Efficient strategies to assess yield stability in winter wheat. *Theor. Appl. Genet.* **2017**, *130*, 1587–1599. [CrossRef] [PubMed]
2. Gupta, P.K.; Balyan, H.S.; Gahlaut, V.; Saripalli, G.; Pal, B.; Basnet, B.R.; Josh, A.K. Hybrid wheat: Past, present and future. *Theor. Appl. Genet.* **2019**, *132*, 2463–2483. [CrossRef] [PubMed]
3. Skudra, I.; Ruza, A. Winter wheat grain baking quality depending on environmental conditions and fertilizer. *Agron. Res.* **2016**, *14* (Suppl. S2), 1460–1466.
4. Rossmann, A.; Scherf, K.A.; Rühl, G.; Greef, J.M.; Mühling, K.H. Effects of a late N fertiliser dose on storage protein composition and bread volume of two wheat varieties differing in quality. *J. Cereal Sci.* **2020**, 102944. [CrossRef]

5. Ćurić, D.; Karlović, D.; Tušak, D.; Petrović, B.; Đugum, J. Gluten as a standard of wheat flour quality. *Food Technol. Biotechnol.* **2001**, *4*, 353–361.

6. Grahmann, K.; Govaerts, B.; Fonteyne, S.; Guzmán, C.; Galaviz Soto, A.P.; Buerkert, A.; Verhulst, N. Nitrogen fertilizer placement and timing affects bread wheat (*Triticum aestivum*) quality and yield in an irrigated bed planting system. *Nutr. Cycl. Agroecosyst.* **2016**, *106*, 185–199. [CrossRef]

7. Park, H.; Clay, D.E.; Hall, R.G.; Rohila, J.S.; Kharel, T.P.; Clay, S.A.; Lee, S. Winter wheat quality responses to water, environment, and nitrogen fertilization. *Commun. Soil Sci. Plant* **2014**, *45*, 1894–1905. [CrossRef]

8. Jaskulska, I.; Jaskulski, D.; Gałęzewski, L.; Knapowski, T.; Kozera, W.; Wacławowicz, R. Mineral composition and baking value of the winter wheat grain under varied environmental and agronomic conditions. *J. Chem.* **2018**, 1–7. [CrossRef]

9. Johansson, E.; Prieto-Linde, M.L.; Svensson, G. Influence of nitrogen application rate and timing on grain protein composition and gluten strength in Swedish wheat cultivars. *J. Plant Nutr. Soil Sci.* **2004**, *167*, 345–350. [CrossRef]

10. Ma, D.; Guo, T.; Wang, Z.; Wang, C.; Zhu, Y.; Wang, Y. Influence of nitrogen fertilizer application rate on winter wheat (*Triticum aestivum* L.) flour quality and Chinese noodle quality. *J. Sci. Food Agric.* **2009**, *89*, 1213–1220. [CrossRef]

11. Whitford, R.; Fleury, D.; Reif, J.C.; Garcia, M.; Okada, T.; Korzun, V.; Langridge, P. Hybrid breeding in wheat: Technologies to improve hybrid wheat seed production. *J. Exp. Bot.* **2013**, *64*, 5411–5428. [CrossRef] [PubMed]

12. Singh, S.P.; Srivastava, R.; Kumar, J. Male sterility systems in wheat and opportunities for hybrid wheat development. *Acta Physiol. Plant* **2015**, *37*, 1713. [CrossRef]

13. COBORU. *Cereals. Methodology Study of Varieties*; Polish Research Centre for Cultivar Testing: Słupia Wielka, Poland, 2002.

14. ISO 7971-3:2019. *Cereals: Determination of Bulk Density, Called Mass Per Hectoliter—Part 3: Routine Method*; ISO: Geneva, Switzerland, 2019.

15. ICC. *International Association for Cereal Science and Technology (ICC) Standard Method No. 129. Method for the Determination of the Vitreousness of Durum Wheat*; ICC: Vienna, Austria, 1980.

16. AACC. *American Association of Cereal Chemists (AACC) International Approved Methods of Analysis*, 11th ed.; AACC: St. Paul, MN, USA, 2009.

17. ICC. *International Association for Cereal Science and Technology (ICC) Standard Methods*; ICC: Vienna, Austria, 2005.

18. Alvarez-Jubete, L.; Auty, M.; Arendt, E.K.; Gallagher, E. Baking properties and microstructure of pseudocereal flours in gluten-free bread formulations. *Eur. Food Res. Technol.* **2010**, *230*, 437–445. [CrossRef]

19. Dallmann, H. *Porentabelle. Detmold*; Verlag Moriz Schafer: Moriz Schafer, Germany, 1981.

20. Morgan, B.C.; Dexter, J.E.; Preston, K.R. Relationship of kernel size to flour water absorption for Canada western red spring wheat. *Cereal Chem.* **2000**, *77*, 286–292. [CrossRef]

21. Klikocka, H.; Cybulska, M.; Barczak, B.; Narolski, B.; Szostak, B.; Kobiałka, A.; Nowak, A.; Wójcik, E. The effect of sulphur and nitrogen fertilization on grain yield and technological quality of spring wheat. *Plant Soil Environ.* **2016**, *62*, 230–236. [CrossRef]

22. Abedi, T.; Alemzadeh, A.; Kazemeini, S.A. Wheat yield and grain protein response to nitrogen amount and timing. *Aust. J. Crop Sci.* **2011**, *5*, 330–336.

23. Gaines, C.S.; Finney, P.L.; Andrews, L.C. Influence of kernel size and shriveling on soft wheat milling and baking quality. *Cereal Chem.* **1997**, *74*, 700–704. [CrossRef]

24. Harasim, E.; Wesołowski, M. The effect of retardant moddus 250 EC and nitrogen fertilization on yielding and grain quality of winter wheat. *Fragm. Agron.* **2013**, *30*, 70–77.

25. Dziki, D.; Cacak-Pietrzak, G.; Miś, A.; Jończyk, K.; Gawlik-Dziki, U. Influence of wheat kernel physical properties on the pulverizing process. *J. Food Sci. Technol.* **2014**, *54*, 2648–2655. [CrossRef]

26. Šekularac, A.; Torbica, A.; Živančev, D.; Tomić, J.; Knežević, D. The influence of wheat genotype and environmental factors on gluten index and the possibility of its use as bread quality predictor. *Genetika* **2018**, *50*, 85–93. [CrossRef]

27. García-Molina, D.M.; Barro, F. Characterization of changes in gluten proteins in low-gliadin transgenic wheat lines in response to application of different nitrogen regimes. *Front. Plant Sci.* **2017**, *8*, 1–13. [CrossRef] [PubMed]

28. Haque, A.; Hossain, M.; Haque, M.; Hasan, M.; Malek, M.; Rafii, M.; Shamsuzzaman, S. Response of yield, nitrogen use efficiency and grain protein content of wheat (*Triticum aestivum* L.) varieties to different nitrogen levels. *Bangladesh J. Bot.* **2017**, *46*, 389–396.

29. Warechowska, M.; Stępień, A.; Wojtkowiak, K.; Nawrocka, A. The impact of nitrogen fertilization strategies on selected qualitative parameters of spring wheat grain and flour. *Pol. J. Nat. Sci.* **2019**, *34*, 199–212.

30. Metho, L.A.; Taylor, J.R.N.; Hammes, P.S.; Randall, P.G. Effects of cultivar and soil fertility on grain protein yield, grain protein content, flour yield and breadmaking quality of wheat. *J. Sci. Food Agric.* **1999**, *79*, 1823–1831. [CrossRef]

31. Piironen, V.; Salmenkallio-Marttila, M. Micronutrients and Phytochemicals in Wheat Grain. In *Wheat: Chemistry and Technology*; American Association of Cereal Chemists: St. Paul, MN, USA, 2009; pp. 179–222. [CrossRef]

32. Cardoso, R.V.C.; Fernandes, A.; Heleno, S.A.; Rodrigues, P.; Gonzalez-Paramas, A.M.; Barros, L.; Ferreira, I. Physicochemical characterization and microbiology of wheat and rye flours. *Food Chem.* **2019**, *280*, 123–129. [CrossRef]

33. Bucsella, B.; Molnár, D.; Harasztos, A.H.; Tömösközi, S. Comparison of the rheological and end-product properties of an industrial aleurone-rich wheat flour, whole grain wheat and rye flour. *J. Cereal Sci.* **2016**, *69*, 40–48. [CrossRef]

34. Hemery, Y.; Holopainen, U.; Lampi, A.-M.; Lehtinen, P.; Nurmi, T.; Piironen, V.; Edelmann, M.; Rouau, X. Potential of dry fractionation of wheat bran for the development of food ingredients, part II: Electrostatic separation of particles. *J. Cereal Sci.* **2011**, *53*, 9–18. [CrossRef]

35. Bayoumi, T.Y.; El-Demardash, I.S. Influence of nitrogen application on grain yield and end use quality in segregating generations of bread wheat (*Triticum aestivum* L). *Afr. J. Biochem. Res.* **2008**, *2*, 132–140.

36. Kindred, D.R.; Gooding, M.J.; Ellis, R.H. Nitrogen fertilizer and seed rate effects on Hagberg falling number of hybrid wheats and their parents are associated with α-amylase activity, grain cavity size and dormancy. *J. Sci. Food Agric.* **2005**, *85*, 727–742. [CrossRef]

37. Liniṇa, A.; Ruza, A. Impact of agroecological conditions on the Hagberg falling number of winter wheat grain. *Res. Rural. Dev.* **2015**, *1*, 19–26.

38. Bueno, M.M.; Thys, R.C.S.; Rodrigues, R.C. Microbial enzymes as substitutes of chemical additives in baking wheat flour-part I: Individual effects of nine enzymes on flour dough rheology. *Food Bioprocess. Technol.* **2016**, *9*, 2012–2023. [CrossRef]

39. Wojtkowiak, K.; Stępień, A.; Orzech, K. Effect of nitrogen fertilisation on the yield components, macronutrient content and technological quality parameters of four winter wheat (*Triticum aestivum* ssp. *vulgare*) varieties. *Fragm. Agron.* **2018**, *35*, 146–155. [CrossRef]

40. Baljeet, S.Y.; Yogesh, S.; Ritika, B.Y. Physicochemical and rheological properties of Indian wheat varieties of *Triticum aestivum*. *Qual. Assur. Saf. Crop* **2017**, *9*, 369–381. [CrossRef]

41. Taner, A.; Arısoy, R.Z.; Kaya, Y.; Gültekin, I.; Partigöç, F. The effects of various tillage systems on grain yield, quality parameters and energy indices in winter wheat production under the rainfed conditions. *Fresenius Environ. Bull.* **2015**, *24*, 1463–1473. [CrossRef]

42. Šip, V.; Vavera, R.; Chrpova, J.; Kusa, P. Winter wheat yield and quality related to tillage practice, input level and environmental condition. *Soil Tillage Res.* **2013**, *132*, 77–85. [CrossRef]

43. Vazquez, D.; Berger, A.; Prieto-Linde, M.L.; Johansson, E. Can nitrogen fertilization be used to modulate yield, protein content and bread-making quality in Uruguayan wheat? *J. Cereal Sci.* **2019**, *85*, 153–161. [CrossRef]

44. Silva, R.R.; Zucareli, C.; de Batista Fonseca, I.C.; Gazola, D.; Riede, C.R. Influence of nitrogen, environment, and cultivars on the industrial quality of wheat. *Semin. Ciênc. Agrár.* **2019**, *40*, 2571–2580. [CrossRef]

45. Voicu, G.; Constantin, G.; Stefan, E.M.; Ipate, G. Variation of farinographic parameters of doughs obtained from wheat and rye flour mixtures during kneading. *Sci. Bull.–Univ. Politeh. Buchar. D* **2012**, *74*, 307–320.

46. Kaiser, H.F. The application of electronic computers to factor analysis. *Educ. Psychol. Meas.* **1960**, *20*, 141–151. [CrossRef]

applied
sciences

Article

Type and Amount of Legume Protein Concentrate Influencing the Technological, Nutritional, and Sensorial Properties of Wheat Bread

Nastasia Belc, Denisa Eglantina Duta *, Alina Culetu and Gabriela Daniela Stamatie

National Institute of Research & Development for Food Bioresources, IBA Bucharest, 6 Dinu Vintila Street, 021102 Bucharest, Romania; nastasia.belc@bioresurse.ro (N.B.); alinaculetu@gmail.com (A.C.); gabriela.stamatie@bioresurse.ro (G.D.S.)
* Correspondence: denisa.duta@bioresurse.ro

Abstract: Plant protein concentrates are used to enhance the nutritional quality of bread and to respond to the demand of consumers with respect to increased protein intake. In the present study, bread samples were produced using pea protein concentrate (PP) and soy protein concentrate (SP) substituting wheat flour by 5%, 10%, and 15%. The protein levels were between 1.2- and 1.7-fold (PP) and 1.1- and 1.3-fold (SP) higher than the control bread. The incorporation of 10% and 15% PP allowed for the achievement of a "high protein" claim. Water absorption was correlated with the protein contents of the breads ($r = 0.9441$). The decrease in bread volume was higher for the PP than SP incorporations, and it was highly negatively correlated with the protein content ($r = -0.9356$). Soy breads had a softer crumb than pea breads. The total change in crumb colour was higher in the PP than SP breads. The soy breads had an overall acceptability between 6.3 and 6.8, which did not differ ($p > 0.05$) from the control. PP breads were statistically less liked ($p < 0.05$). The results underlined that the choice of the type and amount of protein concentrates influenced the bread properties differently.

Keywords: soy protein concentrate; pea protein concentrate; bread; texture; sensory

Citation: Belc, N.; Duta, D.E.; Culetu, A.; Stamatie, G.D. Type and Amount of Legume Protein Concentrate Influencing the Technological, Nutritional, and Sensorial Properties of Wheat Bread. *Appl. Sci.* **2021**, *11*, 436. https://doi.org/10.3390/app11010436

Received: 14 December 2020
Accepted: 30 December 2020
Published: 4 January 2021

1. Introduction

Nowadays, people are more aware of what they consume and they want nutritionally improved products like bread with a high protein content, which can provide nitrogen and amino acids to the body. On the other hand, there is a general global trend towards reducing the use of animal proteins and increasing the consumption of plant proteins for protecting the environment and for reducing gas emissions. Bread is considered an important food in several countries, as part of the daily diet. The forecasted bread consumption volume in Europe is estimated to reach 18,797 million kg by 2021 [1].

The use of wheat flour in baking manufacturing has some limitations from a nutritional point of view. Replacement of wheat flour by different ingredients in bakery products brings not only health effects, but also changes the dough rheology and technological features [2].

Different types of protein ingredients were used for wheat flour supplementation like legumes flours, concentrates, or isolates from pea, chickpea, faba bean, and soy protein [3]. The manufacture of bread with high levels of incorporated legume proteins is very difficult from a technological aspect, because of the lack of gluten network needed to meet the viscoelastic, fermentative, and structure-forming requirements [4]. This is why it is more convenient to use for supplementation ingredients with a high content in proteins (80–90%) for processing, which can be added in a small amount, rather than using legume flours with a lower protein content, which should be added in sufficient quantity so as to obtain a final product with the same protein content.

Pea proteins are very popular in food matrices because of their lower content of anti-nutrients compared with soy protein, which contain phytates, tannins, or trypsin inhibitors [5]. The health benefits of peas are represented by the level of protein, starch, fibre, vitamins, minerals, and phytochemical components. It has been demonstrated that pea ingredients can influence the glycaemic response, insulin resistance, and cardiovascular and gastrointestinal health [6]. Research has been done for producing bread with a 10% protein content through the replacement of wheat flour with 30% raw, germinated, and toasted pea flour [7]. In addition, in bread formulations, pea protein isolate has been used to substitute 15% of the wheat flour, and the effect on the dough properties and bread quality have been analysed [8].

Ribotta et al. [9] showed that different types of soy products used in bread manufacturing prevented the formation of the gluten, and reduced dough strength and gas retention ability of the dough. The addition of soy protein to 30% impacted the dough rheology and the quality of the end products—the specific bread volume decreased and the hardness increased [10]. On the other hand, Ivanovski et al. [11] proved that soy flour at a 20% substitution level to wheat flour used as a protein supplement in bread manufacturing did not negatively affect the flavour and texture properties.

Studies on the supplementation of bread with 5% pea protein isolate showed little adverse effects on the bread volume and physical attributes of the crumb, and had a good overall sensory acceptance. Higher amounts of soy protein concentrate and pea isolate in spelt flour (20% and 10%, respectively) determined the decrease in the loaf volume [12].

The aim of this study was to obtain and analyse the quality of high-protein breads with the addition of plant-based proteins, which can be a "source" of protein or can be claimed to be "rich" in proteins, based on EFSA (European Food Safety Authority) recommendations.

2. Materials and Methods

2.1. Materials

The following protein concentrate powders were employed in the experiment: pea protein concentrate (PP; Paradisul Verde, Romania) and soy protein concentrate (SP; Solaris, Romania). For bread formulations, white wheat flour, yeast, and salt were commercially available.

2.2. Analysis of the Raw Materials and Breads

The raw materials and bread samples were analysed for the following characteristics: moisture using the drying method (AOAC 925.10), protein content using the Kjeldahl method with a conversion factor of nitrogen to protein of 6.25 (AOAC 979.09), fat content using Soxhlet extraction with petroleum ether (AOAC 963.15), and ash using the gravimetric method by burning at 550 °C in a furnace (AOAC 923.03) [13]. The total carbohydrate contents were assessed by subtracting the values of the moisture, protein, fat, and ash content from 100. The caloric values were calculated using the following conversion factors: nine for fat, four for carbohydrates, and four for protein [14].

2.3. Bread Dough Mixolab Analysis

The rheological properties of wheat flour and wheat flour–protein concentrate powder blends were studied using a Mixolab analyser (Chopin Technologies, Villeneuve-la-Garenne, France) [15]. For both types of protein concentrates, the following percentages were used: 5%, 10%, and 15% of the wheat flour. The samples were loaded into the Mixolab bowl and mixed with distilled water to produce 75 g dough, and the target consistency (torque) was 1.1 ± 0.05 Nm. The dough was subjected to dual-mixing (80 rpm) during a heating and cooling programme following the "Chopin+" protocol, namely: 8 min mixing at 30 °C, 4 °C/min heating to 90 °C, 7 min holding at 90 °C, 4 °C/min cooling to 50 °C, and 5 min holding at 50 °C. The Mixolab software (version 4.0.8) was used to record the curves and calculate the dough mixing parameters. The samples were analysed at adapted hydration (i.e., the initial torque C1 was kept constant at 1.1 Nm). The parameters from the Mixolab curves refer to the following: dough development time, the time needed to attain a torque of 1.1 Nm (min); stability, the dough mixing resistance (min); C2, the torque associated with protein weakening (Nm); C3, the degree of the gelatinisation of the starch

(Nm); C4, the stability of the starch gel formed (Nm); and C5, the retrogradation of the starch (Nm) [16]. The analysis was performed in triplicate.

2.4. Bread Preparation

The bread samples' formulation was based on wheat flour, which was replaced by different concentrations of pea protein concentrate, namely 5%, 10%, and 15% (5PP, 10PP, and 15PP, respectively) and of soy protein concentrate, namely, 5%, 10%, and 15% (5SP, 10SP, and 15SP, respectively). A control sample (100% wheat flour) was also considered. The other ingredients (as % of wheat flour/wheat flour–protein concentrates blends) were as follows: fresh yeast (3.0%), salt (1.5%), and water. The amount of water used to form the dough was the amount calculated by the Mixolab measurements (Table 1). The ingredients were mixed in a Diosna mixer (Germany) for 3 min at a low speed and 3–5 min at a high speed. The dough was allowed to rest for 15 min and then divided into pieces of 575–580 g (in order to have a final end product of around 500 g), rounded and left to rest 10 min, and then moulded in the aluminium bread pans and proofed in a proofer at 35 °C and 49% humidity for 40 min (MCE Meccanica, Italy). The samples were baked in an oven (Mondial Forni, Italy) for 30 min at 230–235 °C. In the first 4 s, steam was used inside oven. The breads were cooled down for 2 h at room temperature and were stored in polypropylene bags for analysis.

Table 1. Bread recipes.

Sample	Wheat Flour (g)	PP (g)	SP (g)	Yeast (g)	Salt (g)	Water (mL)
Control	1500	-	-	45	22	895.5
5PP	1425	75	-	45	22	927.0
10PP	1350	150	-	45	22	984.0
15PP	1275	225		45	22	1073.25
5SP	1425	-	75	45	22	904.5
10SP	1350	-	150	45	22	919.5
15SP	1275	-	225	45	22	924.0

PP—pea protein concentrate; SP—soy protein concentrate.

2.5. Physical Parameters of Fresh Bread Samples

2.5.1. Bread Volume, Porosity, and Elasticity Analysis

For the bread volume, the rapeseed displacement method was employed using the Fornet bread volumeter (Chopin, France) [17]. The bread porosity was determined with the help of a cylindrical sharpened brass perforator (internal diameter of 45 mm) from a 60 mm slice obtained from the middle of the loaf and then weighed [18]. For bread elasticity, the crumb cut for the porosity analysis was pressed to half of its height for 1 min using a screw-driven pressing device and, then, after removing the pressure, the height of the compressed piece of crumb was measured. The ratio between the height after compression and recovery, as well as the initial height, represented the elasticity (expressed as a percentage) [19]. All of the above parameters were the average of three determinations.

2.5.2. Bread Crumb Colour Analysis

The colour parameters of the bread crumb samples were analysed using a CM-5 spectrophotometer (Konica Minolta Sensing, Inc., Osaka, Japan) equipped with a D65 illuminant and a 10° observer angle. The colour was determined on 10 different points from a slice of 20 mm thickness, taken from the middle of the loaf. The colour parameters were as follows: lightness ($L^* = 0$ is black and $L^* = 100$ is white), a^* (-a = green and +a = red), and b^* (-b = blue and +b = yellow). Using the colour coordinates, the following parameters were calculated: chroma ($C^* = \sqrt{a^{*2} + b^{*2}}$) and hue angle ($h = \arctan(b^*/a^*)$). The total colour difference (ΔE) between the control bread and the different protein formulations was determined as follows: $\Delta E = \sqrt{(L_C^* - L_P^*)^2 + (a_C^* - a_P^*)^2 + (b_C^* - b_P^*)^2}$, where subscripts

C and P refer to the control bread and to the different bread formulations with protein concentrates, respectively.

2.5.3. Bread Crumb Texture Analysis

The texture properties of the bread samples were calculated based on a compression profile with the help of the Instron Texture Analyzer (model 5944, Illinois Tool Works Inc., USA) equipped with a cylinder probe with a 12 mm diameter and a load cell of 50 kg· m· s^{-2}. The slice of bread (20 mm) was compressed to up to 40% of its height at a speed rate of 1.667 mm/s, and the following parameters were measured: hardness, cohesiveness, elasticity, and chewiness [20]. The average values for three determinations were included.

2.6. Headspace-Electronic Nose

The global volatile composition of the bread samples was analysed using an electronic nose system (FOX 4000, Alpha M.O.S., France), according to a previous study [21]. Briefly, 2 g of bread crumb was put into vials that were incubated (35 °C, 600 s) in order to generate a volatile headspace; then, the bread headspace was injected into the measuring chamber of the electronic nose. Three replications were performed for each type of bread and a statistical analysis was applied to the recorded signals.

2.7. Sensory Analysis

Twelve trained panellists (nine females and three males, with an average age of 24–62 years old) analysed the bread samples through the quantitative descriptive test, focused on attribute scoring from a low intensity (1) to high intensity (5). The sensory attributes included the following: appearance of the crust and crumb colour, lighter–darker; bitter flavour, no bitter flavour–very intense bitter flavour; astringent flavour, no astringent flavour–very intense astringent flavour; crumb firmness at the first bite, soft–rough; crumb gumminess at first bite, less gummy–very gummy; and mass adherence during chewing, no adherence–high adherence.

For the bread overall acceptability, a nine-point hedonic scale (from 9 = like extremely to 1 = dislike extremely) was used [22]. Another method employed to evaluate the bread acceptability was the bread note, which was performed according to the patent RO 130586 A2 [23]. Accordingly, a maximum score was set for each of the following parameters: volume (24), outer appearance (7), crust colour (7), crumb colour (10), crumb porosity (20), crumb texture (20), and flavour (12), taking into consideration a standard volume of 400 cm^3/100 g bread and a standard porosity of 85%.

2.8. Statistical Analysis

A data analysis was carried out using ANOVA (one-way analysis of variance) with Tukey's test (Minitab®19, Minitab Ltd., Coventry, UK). Significant differences among the samples were considered at $p < 0.05$. Values were expressed as mean ± standard deviation.

3. Results and Discussion

3.1. Effect of Protein Concentrates on Bread Compositional Analysis

The wheat flour used in this study had the following characteristics: 13.28% moisture, 10.9% protein, 1.33% fat, 0.48% ash, and 64.74% starch. The protein, fat, ash, and moisture in PP and SP were 77.96%, 0.29%, 3.77%, and 7.43%, and 39.21%, 10.01%, 5.66%, and 8.84%, respectively.

The bread compositional analysis was performed in order to evaluate the changes in the macronutrient composition determined by the partial substitution of wheat flour with protein concentrate such as pea and soy (Table 2). The protein and carbohydrates content constitute the main differences in the compositional profile, because of the substitution of wheat flour (which has a high starch content of 74.65% dry matter (d.m.)) by protein concentrates with protein contents of 84.22% d.m. for PP and 43.01% d.m. for SP. In addition to these parameters, for SP bread, the fat content brought significant differences ($p < 0.05$) in the bread samples, because of the higher fat content in SP (10.98% d.m.) compared with PP (0.31% d.m.).

Table 2. Compositional analysis of the control and protein concentrate breads (expressed in % of the fresh bread).

Bread Samples	Moisture (%)	Protein (%)	Fat (%)	Ash (%)	Carbohydrates * (%)	Energy (kcal/100g)	Energy from Protein ** (%)
Control	45.19 ± 0.08 [d]	7.61 ± 0.10 [f]	0.77 ± 0.06 [d]	1.00 ± 0.10 [c]	45.44 ± 0.13 [a]	219.6 ± 0.58 [b]	13.89 ± 0.16 [f]
5PP	46.06 ± 0.05 [c]	9.34 ± 0.21 [d]	0.80 ± 0.03 [d]	0.93 ± 0.06 [c]	42.86 ± 0.19 [c]	216.04 ± 0.13 [d]	17.30 ± 0.39 [d]
10PP	46.99 ± 0.09 [b]	11.49 ± 0.15 [b]	0.80 ± 0.02 [d]	1.05 ± 0.03 [bc]	39.68 ± 0.13 [e]	211.82 ± 0.43 [e]	21.69 ± 0.27 [b]
15PP	48.76 ± 0.07 [a]	13.02 ± 0.08 [a]	0.81 ± 0.01 [d]	1.16 ± 0.02 [b]	36.25 ± 0.16 [f]	204.34 ± 0.37 [f]	25.48 ± 0.19 [a]
5SP	45.90 ± 0.10 [c]	8.33 ± 0.13 [e]	1.10 ± 0.05 [c]	1.14 ± 0.01 [b]	43.53 ± 0.23 [b]	217.33 ± 0.33 [c]	15.33 ± 0.26 [e]
10SP	45.26 ± 0.05 [d]	9.28 ± 0.09 [d]	1.43 ± 0.06 [b]	1.16 ± 0.02 [b]	42.87 ± 0.11 [c]	221.47 ± 0.27 [a]	16.75 ± 0.18 [d]
15SP	45.25 ± 0.05 [d]	10.20 ± 0.010 [c]	1.60 ± 0.10 [a]	1.33 ± 0.02 [a]	41.62 ± 0.25 [d]	221.68 ± 0.31 [a]	18.40 ± 0.16 [c]

* Calculated by difference. ** Calculated based on the energy content, protein content, and 4 kcal/g protein. The results are mean ± standard deviation (n = 3). Values followed by different superscript letters in the same column are significantly different ($p < 0.05$).

In general, the calculated caloric value of the supplemented breads had a lower value ($p < 0.05$) than the control bread, with the exception of the 10SP and 15SP breads, where a small increase of 0.85% and 0.95%, respectively, was noted. Comparing the percentage of calories provided by the protein, the formulations for 10PP and 15PP reached values higher than 20, and the nutritional claim of being "high protein" can be applied according to regulation (EC) no. 1924/2006 [14].

3.2. Dough Mixing Properties

The influence of the protein concentrate additions on the dough mixing properties is summarized in Table 3.

Table 3. Mixolab rheological parameters of wheat flour and wheat flour–protein concentrate blends.

Sample	Water Absorption (%)	Development Time (min)	Stability (min)	C2 (Nm)	C3 (Nm)	C4 (Nm)	C5 (Nm)
Control	59.70 ± 0 [d]	3.05 ± 0.04 [c]	8.50 ± 0.06 [b]	0.45 ± 0 [bc]	1.88 ± 0.01 [b]	1.75 ± 0.01 [a]	2.91 ± 0.01 [a]
5PP	61.80 ± 0 [c]	6.27 ± 0.23 [a]	10.02 ± 0.14 [a]	0.49 ± 0 [a]	2.00 ± 0.01 [a]	1.82 ± 0.04 [a]	2.76 ± 0.05 [b]
10PP	65.60 ± 0 [b]	1.58 ± 0.21 [d]	9.88 ± 0.18 [a]	0.47 ± 0.01 [ab]	1.74 ± 0.04 [c]	1.19 ± 0.05 [d]	2.24 ± 0.01 [c]
15PP	71.55 ± 0.49 [a]	1.51 ± 0.01 [d]	7.75 ± 0.25 [b]	0.44 ± 0.02 [c]	1.50 ± 0.01 [e]	1.02 ± 0.02 [e]	1.54 ± 0.03 [f]
5SP	60.30 ± 0 [d]	7.0 ± 0.47 [a]	10.54 ± 0.27 [a]	0.48 ± 0 [ab]	1.91 ± 0 [b]	1.82 ± 0.04 [a]	2.67 ± 0.01 [b]
10SP	61.30 ± 0 [c]	6.18 ± 0.14 [a]	10.16 ± 0.04 [a]	0.43 ± 0.01 [c]	1.75 ± 0.01 [c]	1.53 ± 0 [b]	2.08 ± 0.01 [d]
15SP	61.60 ± 0 [c]	4.71 ± 0.27 [b]	9.80 ± 0.25 [a]	0.44 ± 0 [c]	1.66 ± 0.01 [d]	1.32 ± 0.01 [c]	1.80 ± 0.01 [e]

The results are mean ± standard deviation (n = 3). Values followed by different superscript letters in the same column are significantly different ($p < 0.05$).

Water absorption increased as the level of protein concentrates increased. PP significantly increased ($p < 0.05$) dough water absorption by almost 3.5% (5PP), 9.9% (10PP), and 19.8% (15PP), while in the case of SP, the maximum increase was by 3.2% (15SP).

Proteins are one of the main components involved in water adsorption, and the type of protein is responsible for the increase in water absorption [24]. Taherian et al. [25]

showed that the supplementation of wheat flour with 10% pea protein isolate (with a protein content of 96.1% dry matter basis) determined an increase in water absorption, a fact that was explained through the capacity of proteins to absorb a high quantity of water. In the present study, a high correlation between water absorption and the protein contents of the breads was found ($r = 0.9441$). The dough development time of the SP breads were significantly higher than the control bread ($p < 0.05$). In addition, the stability time was significantly increased ($p < 0.05$) by SP addition compared with the control, which denotes a stronger dough structure. However, there was no concentration-dependent tendency in the stability time, as there were no significant differences ($p > 0.05$) between the three levels of SP addition. The same trend was observed for the 5PP and 10PP samples.

On the other hand, the 15% PP addition had the effect of decreasing the dough stability compared with lower levels of addition (5% and 10%), but it did not significantly differ ($p > 0.05$) from the control. A similar behaviour with an increase in the dough stability through the addition of soy protein was reported by Zhou et al. [10], while in contrast, the incorporation of whey protein brought a decrease in the dough stability time.

The addition of 5% PP and 5% SP showed higher values for the C2 parameter, which denotes a dough with strong mixing resistance, while higher addition levels lead to a reduced strength in the protein network. The highest weakening of the gluten network was found for the samples 15PP, 10SP, and 15SP. The results are similar to Hoehnel et al. [8], who pointed out that a weakened gluten network was obtained when 15% wheat flour was substituted with pea and potato protein. It was also proven that the gluten network was loosened as a consequence of the interference effect of the soy proteins, and the lesser availability of water in the build-up of the gluten network [26].

The stability of the starch gel (C4) for 5PP and 5SP was similar to that of the control, while at higher concentrations it was significantly decreased, being more prominent in the case of pea protein blends.

The addition of both types of protein concentrates in wheat dough contributed to a decrease in C3 and C5 parameters, which correspond to a lower degree of starch gelatinization and starch retrogradation, respectively, towards the control, which is accounted by the higher protein content provided by the addition of protein concentrates, subsequently decreasing the starch content. Hadnadev et al. [27] attributed a lower starch retrogradation to a higher protein and fat content and lipid–amylose complex forming capacity. The parameters C3 and C5 were highly positively correlated ($r = 0.9418$).

3.3. Physical Characteristics of Bread Samples

Figure 1 shows the bread obtained with different levels of protein concentrates in comparison with the control bread. The physical characteristics of the bread samples are presented in Table 4.

A significant decrease ($p < 0.05$) in bread volume, porosity and elasticity was noted with the addition of protein concentrates in the bread receipt; the decrease being more obvious in the case of PP than SP additions. The use of PP significantly decreased ($p < 0.05$) the bread volume in a concentration-dependent trend; the PP bread volume was between 1.5- and 2.1-fold lower towards the wheat bread control. In the case of the SP breads, the volume was between 1.2- and 1.3-fold lower than the control. However, no significant difference ($p > 0.05$) regarding bread volume between the 10SP and 15SP breads was observed. The physical and chemical interactions between gluten and the components from the protein concentrates contributed to the volume lowering. The bread volume was highly negatively correlated with the protein content ($r = -0.9356$). The results are in line with other studies reported previously on the impact of pea protein isolate [3] or soy protein [10] on bread volume.

Figure 1. Comparation between bread obtained with different additions of pea protein concentrate (5PP, 10PP, and 15PP) and soy protein concentrate (5SP, 10SP, and 15SP), respectively, with the control bread.

Table 4. Physical parameters of the bread samples.

Bread Sample	Volume (cm³/100 g)	Crumb Porosity (%)	Elasticity	L*	a*	b*	C*	h°	ΔE
Control	324 ± 2 ª	80.7 ± 0.4 ª	97 ± 1 ª	76.75 ± 0.03 ª	0.96 ± 0.01 ᵍ	19.82 ± 0.03 ᵍ	19.85 ± 0.03 ᵍ	87.22 ± 0.03 ª	-
5PP	222 ± 2 ᵈ	68.0 ± 0.3 ᵈ	95 ± 0 ᵇ	73.91 ± 0.05 ᶜ	1.64 ± 0.01 ᶠ	21.16 ± 0.03 ᵉ	21.22 ± 0.03 ᵉ	85.58 ± 0.02 ᵇ	3.21 ± 0.05 ᵉ
10PP	174 ± 1 ᵉ	60.6 ± 0.1 ᵉ	89 ± 1 ᵈ	72.14 ± 0.05 ᶠ	2.47 ± 0.01 ᶜ	21.44 ± 0.04 ᵈ	21.58 ± 0.04 ᵈ	83.43 ± 0.03 ᵉ	5.12 ± 0.04 ᶜ
15PP	154 ± 2 ᶠ	54.8 ± 0.1 ᶠ	85 ± 1 ᵉ	71.23 ± 0.04 ᵍ	2.87 ± 0.03 ᵇ	22.06 ± 0.04 ᵇ	22.24 ± 0.04 ᵇ	82.60 ± 0.07 ᶠ	6.26 ± 0.07 ª
5SP	265 ± 4 ᵇ	73.4 ± 0.1 ᵇ	95 ± 0 ᵇ	74.69 ± 0.09 ᵇ	1.76 ± 0.01 ᵉ	20.37 ± 0.06 ᶠ	20.44 ± 0.06 ᶠ	85.05 ± 0.03 ᶜ	2.28 ± 0.07 ᶠ
10SP	247 ± 1 ᶜ	71.7 ± 0.7 ᶜ	93 ± 1 ᶜ	73.68 ± 0.04 ᵈ	2.42 ± 0.01 ᵈ	21.95 ± 0.05 ᶜ	22.08 ± 0.05 ᶜ	83.72 ± 0.03 ᵈ	4.01 ± 0.03 ᵈ
15SP	249 ± 1 ᶜ	71.7 ± 0.1 ᶜ	93 ± 0 ᶜ	72.25 ± 0.03 ᵉ	3.24 ± 0.01 ª	22.51 ± 0.02 ª	22.74 ± 0.02 ª	81.82 ± 0.03 ᵍ	5.71 ± 0.03 ᵇ

The results are mean ± standard deviation (n = 3 determinations for volume, porosity, and elasticity; n = 10 determinations for colour parameters). Values followed by different superscript letters in the same column are significantly different ($p < 0.05$).

The incorporation of protein concentrates decreased the crumb porosity by up to 47.3% (for 15PP bread) and 12.5% (for 15SP bread). There was a negative correlation between the crumb porosity and protein content ($r = -0.9559$). Another similar finding was observed in relation to the elasticity and protein content ($r = -0.8782$).

Measurements of the bread colour parameters showed that all supplemented breads were darker than the control bread. L* value (lightness parameter) significantly decreased ($p < 0.05$) with the inclusion of concentrates. The highest lightness was obtained for 5SP bread, whereas the lowest lightness was in the 15PP bread. For each of the three levels of substitution used, the PP breads were darker than the SP breads ($p < 0.05$). Protein supplemented breads showed significantly ($p < 0.05$) higher redness (a* parameter) and yellowness (b* parameter) values than the bread control. Both parameters significantly increased ($p < 0.05$) relative to the level of additions. Bread 15SP had the highest a* and b* values, showing a more reddish and yellowish crumb colour. Among the protein supplemented breads, sample 5SP had the smallest C* value, a lower chroma value indicating a less strong colour. The highest value for the hue angle (h°) for the control bread indicated yellowish tones. The hue angle was significantly reduced in the bread crumb of the supplemented breads ($p < 0.05$).

In order to establish the colour differences between the samples, ΔE was calculated. The total change in crumb colour was higher in the PP than SP breads. Only sample 5SP resulted in colour differences towards the control that were not noticeable to the human eye (ΔE < 3). It is stated that values for ΔE higher than 3 are visible to the human eye [28]. The texture properties of the bread crumbs are exhibited in Figure 2.

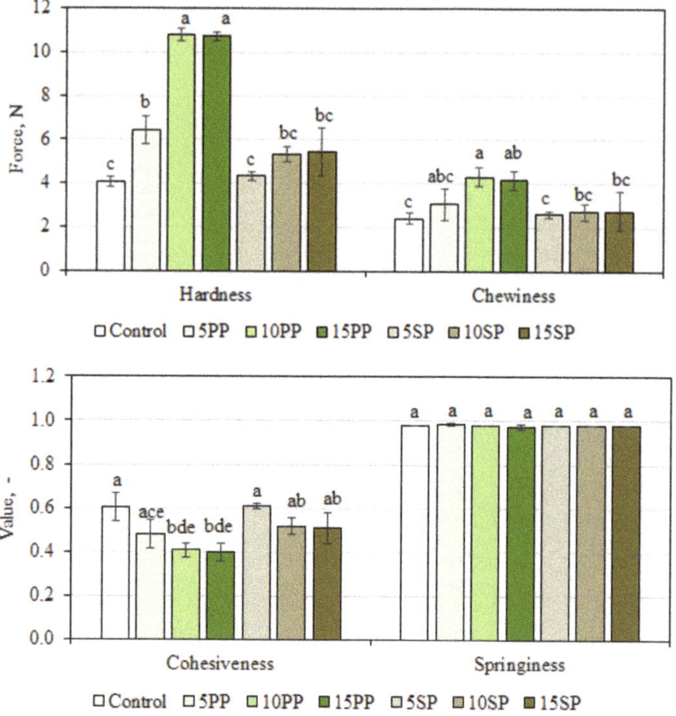

Figure 2. Texture properties of the bread samples. The results are mean ± standard deviation (n = 3). The different letters show significance difference ($p < 0.05$).

Crumb hardness for the control and 5SP breads were significantly lower ($p < 0.05$) compared with the other formulations, which indicates an adequate quality of the bread made with a lower concentration of SP. Pea protein showed a stronger effect on bread hardness than soy protein; so, soy breads had a softer crumb. The results were in agreement with a previous study [29], where the addition of soy protein into gluten-free bread formulations resulted in the lowest bread volume compared with other protein sources such as carob, lupin, or pea. Bread prepared with PP at a high concentration (10% and 15%) showed the highest hardness, which was almost 2.6-fold higher than the control wheat bread. Similar results of an increase in bread hardness were shown by Garcia-Segovia et al. [30] when wheat flour was substituted with 5% and 10% pea protein, respectively. The crumb hardness was highly negatively correlated with bread volume ($r = -0.9282$). In addition, a correlation between the hardness and protein content was noted ($r = 0.9223$).

Regarding bread chewiness, the same trend was observed as for hardness. Bread chewiness increased slowly with SP addition, but the difference was not significant

($p > 0.05$) compared with control bread. The higher increase in chewiness for PP breads points out that a longer time is needed for chewing the samples before swallowing. However, no significant differences ($p > 0.05$) in chewiness between the 10PP and 15PP samples were noted.

The difference in crumb cohesiveness between the control, SP breads, and 5PP was not significant ($p > 0.05$). At a higher level of PP used in the bread formulations, the crumb was less cohesive (lower values of cohesiveness registered for the 10PP and 15PP breads). Regarding the springiness parameter, no significant difference ($p > 0.05$) between the control and protein-based breads were observed.

3.4. Discrimination of Bread Sample by Electronic Nose System

Principal component analysis (PCA) was employed to detect the differences in volatile composition between the bread samples based on the calculation of a discrimination index. Thus, the higher the discrimination index, the better the discrimination. Figure 3 depicts the PCA plot that represents a map of the discrimination of the bread samples. A discrimination index of 85 was obtained, which is explained by the very different odour of the samples. The contribution rates for the two principal components (PC1 and PC2) were 99.18% and 0.67%, respectively.

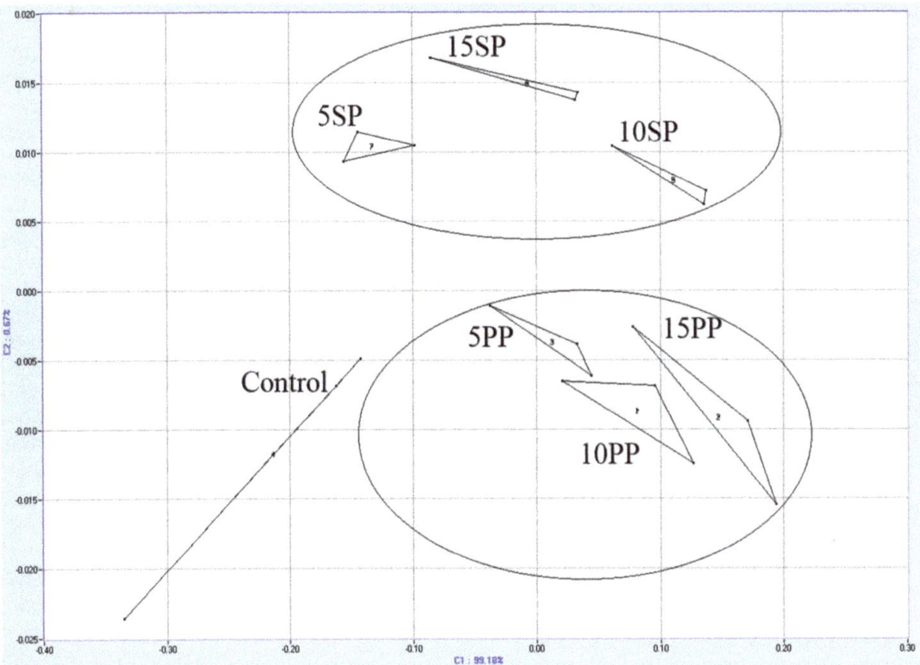

Figure 3. Principal component analysis (PCA) plot for the bread samples: control, pea (5PP, 10PP, and 15 PP), and soy (5SP, 10 SP, and 15SP) breads.

Considering only the same level of addition and comparing this with the control, a good discrimination was obtained: 93 (between the control, 5PP, and 5SP breads), 91 (between the control, 10PP, and 10SP breads), and 91 (between the control, 15PP, and 15SP breads), respectively. In all of the cases, the control bread was significantly separated from the protein bread samples, positioned on the left side of the PCA map.

3.5. Sensory Analysis

The sensory attributes of the bread samples are included in Table 5. Higher levels of protein concentrate additions led to a stronger intensity in the colour of the crumb and crust. There was no statistically significant difference ($p > 0.05$) in the bitter and astringent flavours between the bread samples. Crumb firmness increased with protein concentrate addition and was positively correlated with the values of hardness provided by the texture analyser ($r = 0.9493$). As the addition of protein concentrates increased, mass adherence during chewing increased in comparison with the bread. However, this increase was not significantly different ($p > 0.05$) between the control and soy protein breads and 5PP. Only higher levels of PP (10% and 15%) addition brought about a significant increase in mass adherence. Hence, a good correlation between the protein content in the bread and mass adherence was found ($r = 0.8544$).

Table 5. Sensory profile of bread samples using a descriptive method.

Samples	Control	5PP	10PP	15PP	5SP	10SP	15SP
Colour of crust	2.21 ± 0.84 [d]	2.50 ± 0.71 [cd]	2.83 ± 0.62 [cd]	2.21 ± 0.54 [cd]	3.00 ± 0.43 [bc]	3.21 ± 0.78 [b]	4.08 ± 0.51 [a]
Colour of crumb	1.75 ± 0.97 [b]	2.29 ± 0.78 [b]	2.54 ± 0.86 [b]	2.50 ± 1.00 [b]	2.13 ± 0.96 [b]	2.50 ± 1.07 [b]	3.04 ± 1.05 [a]
Bitter flavour	1.00 ± 0 [a]	1.08 ± 0.29 [a]	1.08 ± 0.29 [a]	1.08 ± 0.29 [a]	1.08 ± 0.29 [a]	1.00 ± 0 [a]	1.17 ± 0.58 [a]
Astringent flavour	1.00 ± 0 [a]	1.08 ± 0.29 [a]	1.00 ± 0 [a]	1.33 ± 0.89 [a]	1.08 ± 0.29 [a]	1.00 ± 0 [a]	1.17 ± 0.58 [a]
Crumb firmness at first bite	1.33 ± 0.49 [d]	2.17 ± 0.39 [c]	2.92 ± 0.76 [ab]	3.21 ± 0.75 [a]	1.63 ± 0.57 [cd]	2.17 ± 0.39 [c]	2.50 ± 0.52 [bc]
Crumb gumminess at first bite	1.96 ± 1.01 [d]	2.25 ± 1.06 [cd]	3.33 ± 1.15 [ab]	3.50 ± 1.22 [ab]	1.96 ± 0.75 [d]	2.42 ± 0.70 [bcd]	2.67 ± 0.89 [bcd]
Mass adherence during chewing	1.83 ± 1.03 [b]	2.50 ± 1.00 [ab]	3.21 ± 1.44 [a]	3.42 ± 1.49 [a]	1.83 ± 0.83 [b]	1.75 ± 0.66 [b]	2.21 ± 0.89 [ab]

The results are mean $\pm$ standard deviation (n = 12). Values followed by different superscript letters in the same row are significantly different ($p < 0.05$).

According to the hedonic test (Figure 4), the soy breads had a good overall acceptability, with a score varying from 6.3 to 6.8. Moreover, the SP breads did not significantly differ ($p > 0.05$) from the wheat control bread. On the other hand, breads with a pea protein concentrate addition were less liked compared with the soy breads. The overall acceptability noted for the 10PP and 15PP breads decreased significantly ($p < 0.05$) compared with the control bread and SP samples.

The results from the bread note are presented in Table 6. These results were in line with those obtained through other methods. Thus, positive correlations were observed between the physical methods and bread note methods for volume ($r = 0.9969$), porosity ($r = 0.9578$), and elasticity ($r = 0.8356$). In addition, a highly positive correlation was found between the overall acceptability performed by the panelists and the total score calculated through the bread note procedure ($r = 0.9770$). The L* parameter was positive correlated with the crumb colour in the bread note ($r = 0.8392$). There was also a negative correlation between L* and crumb colour in the sensory analysis ($r = -0.8254$).

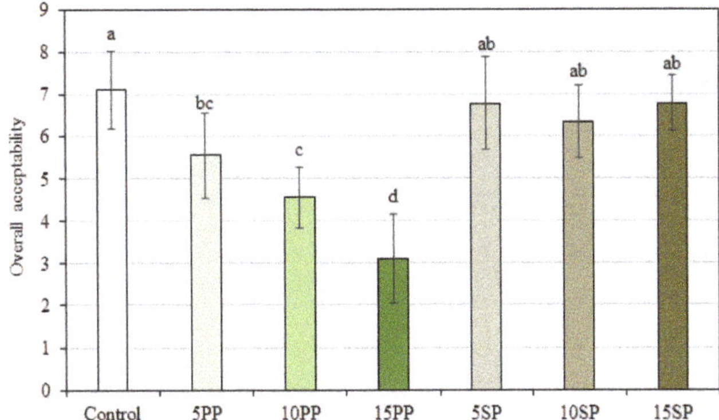

Figure 4. Mean of the overall acceptability of the breads. The different letters show a significance difference ($p < 0.05$).

Table 6. Bread note.

Samples	Volume	Outer Appearance	Crust Colour	Crumb Colour	Porosity	Crumb Elasticity	Flavour	Total
Control	19	6	6	10	18	17	12	88
5PP	13	4	6	10	14	13	11	71
10PP	10	3	7	9	13	6	10	58
15PP	9	3	7	8	10	5	10	52
5SP	16	5	7	10	17	17	12	84
10SP	15	4	7	9	17	18	12	82
15SP	15	4	6	9	17	18	11	80

4. Conclusions

In the present study, bread samples were produced using pea protein concentrate and soy protein concentrate, with the aim of enhancing the nutritional properties of wheat bread. SP breads showed a stronger dough structure, higher volume, and softer crumb than the bread obtained with PP. The total change in crumb colour was higher in the PP than SP breads. The bread samples were different in the odour profile according to the PCA map provided by the electronic nose system. Sensory analysis showed that SP allowed for obtaining a more acceptable bread than PP incorporation into bread formulation. The amount and type of protein have a substantial role in particular applications referring to protein fortified bakery products.

Author Contributions: Conceptualization, N.B.; methodology, N.B. and D.E.D.; formal analysis, A.C. and D.E.D.; investigation, D.E.D. and G.D.S.; resources, D.E.D.; data curation, A.C.; writing—original draft preparation, A.C. and D.E.D.; writing—review and editing, A.C. and D.E.D.; visualization, A.C.; supervision, N.B. and D.E.D.; project administration, D.E.D.; funding acquisition, N.B. and D.E.D. All authors have read and agreed to the published version of the manuscript.

Funding: This work was achieved through the Core programme (PN 19 02) supported by the Ministry of Education and Research (project number 19 02 01 01).

Institutional Review Board Statement: The study was conducted according to the guidelines of the Declaration of Helsinki, and approved by the Ethics Committee of the National Institute of Research & Development for Food Bioresources IBA Bucharest (08/28.02.2011)

Informed Consent Statement: Informed consent was obtained from all subjects involved in the study.

Data Availability Statement: Data is contained within the article.

Conflicts of Interest: The authors declare no conflict of interest.

References

1. Statistica. Bread and Bakery Consumption Volume in Europe from 2010 to 2021, by Category. Available online: https://www.statista.com/statistics/806319/europe-bread-and-bakery-production-volume-by-category/ (accessed on 29 November 2020).
2. Schmiele, M.; Ferrari Felisberto, M.H.; Silva Clerici, M.T.P.; Chang, Y.K. Mixolab™ for rheological evaluation of wheat flour partially replaced by soy protein hydrolysate and fructooligosaccharides for bread production. *LWT Food Sci. Technol.* **2017**, *76*, 259–269. [CrossRef]
3. Des Marchais, M.F.; Mercier, S.; Villeneuvea, S.; Mondor, M. Bread-making potential of pea protein isolate produced by a novel ultrafiltration/diafiltration process. *Procedia Food Sci.* **2011**, *1*, 1425–1430. [CrossRef]
4. Collar, C. Use of Mixolab on Formulated Flours. In *Mixolab: A New Approach to Rheology*; Dubat, A., Rosell, C.M., Gallagher, E., Eds.; AACC International, Inc.: St. Paul, MN, USA, 2013; pp. 33–44.
5. Shah, N.N.; Umesh, K.V.; Singhal, R.S. Hydrophobically modified pea proteins: Synthesis, characterization and evaluation as emulsifiers in eggless cake. *J. Food Eng.* **2019**, *255*, 15–23. [CrossRef]
6. Dahl, W.J.; Foster, L.M.; Tyler, R.T. Review of the health benefits of peas (*Pisum sativum L.*). *Br. J. Nutr.* **2012**, *108*, 3–10. [CrossRef] [PubMed]
7. Millar, K.A.; Barry-Ryan, C.; Burke, R.; McCarthy, S.; Gabllagher, E. Dough properties and baking characteristics of white bread, as affected by addition of raw, germinated and toasted pea flour. *Innov. Food Sci. Emerg. Technol.* **2019**, *56*, 102189. [CrossRef]
8. Hoehnel, A.; Axel, C.; Bez, J.; Arendt, E.K.; Zannini, E. Comparative analysis of plant-based high-protein ingredients and their impact on quality of high-protein bread. *J. Cereal Sci.* **2019**, *89*, 102816. [CrossRef]
9. Ribotta, P.D.; Arnulphi, S.A.; León, A.E.; Añón, M.C. Effect of soybean addition on the rheological properties and breadmaking qualityof wheat flour. *J. Sci. Food Agric.* **2005**, *85*, 1889–1896. [CrossRef]
10. Zhou, J.; Liu, J.; Tang, X. Effects of whey and soy protein addition on bread rheological property of wheat flour. *J. Texture Stud.* **2018**, *49*, 38–46. [CrossRef] [PubMed]
11. Ivanovski, B.; Seetharaman, K.; Duizer, L.M. Development of soy-based bread with acceptable sensory properties. *J Food Sci.* **2012**, *77*, S71–S76. [CrossRef] [PubMed]
12. Šimurina, O.; Filipčev, B.; Belić, Z.; Jambrec, D.; Krulj, J.; Brkljača, J.; Pestorić, M. The influence of plant protein on the properties of dough and the quality of wholemeal spelt bread. *Croat. J. Food Sci. Technol.* **2016**, *8*, 107–111. [CrossRef]
13. AOAC. *Official Methods of Analysis of AOAC International*; Association of Official Analytical Chemists: Gaithersburg, MD, USA, 2005.
14. Regulation (EC) No. 1924/2006 of the European Parliament and of the Council of 20 December 2006 on Nutrition and Health Claims Made on Foods. Available online: https://eur-lex.europa.eu/legal-content/EN/TXT/PDF/?uri=CELEX:32006R1924&from=EN (accessed on 27 November 2020).
15. AACC. *Method 54–60.01 Determination of Rheological Behavior as a Function of Mixing and Temperature Increase in wheat FLour and whole Wheat Meal by Mixolab*, 11th ed.; AACC International: St. Paul, MN, USA, 2010.
16. Dubat, A. The Mixolab. In *Mixolab: A New Approach to Rheology*; Dubat, A., Rosell, C.M., Gallagher, E., Eds.; AACC International, Inc.: St. Paul, MN, USA, 2013; pp. 3–13.
17. AACC. *Method 10–05.01 Guidelines for Measurement of Volume by Rapeseed Displacement*, 11th ed.; AACC International: St. Paul, MN, USA, 2010.
18. *SR 91 Bread and Fresh Pastry Products. Methods of Analysis*; Romanian Standard Association (ASRO) Publisher House: Bucharest, Romania, 2007. (In Romanian)
19. Lazaridou, A.; Duta, D.; Papageorgiou, M.; Belc, N.; Biliaderis, C.G. Effects of hydrocolloids on dough rheology and bread quality parameters in gluten-free formulations. *J. Food Eng.* **2007**, *79*, 1033–1047. [CrossRef]
20. Bourne, M.C. Texture profile analysis. *Food Technol.* **1978**, *32*, 62–66.
21. Culetu, A.; Duta, D.E.; Andlauer, W. Influence of black tea fractions addition on dough characteristics, textural properties and shelf life of wheat bread. *Eur. Food Res. Technol.* **2018**, *244*, 1133–1145. [CrossRef]
22. Lawless, H.T.; Heymann, H. *Sensory Evaluation of Food: Principles and Practices*, 2nd ed.; Springer: New York, NY, USA, 2010.
23. Belc, N.; Ghencea, S. Procedure for Evaluating the Bread Quality by Scoring (in Romanian). Patent RO 130586 A2. 2015. Available online: http://pub.osim.ro/publication-server/pdf-document?PN=RO130586%20RO%20130586&iDocId=7515&iepatch=.pdf (accessed on 29 November 2020).
24. Collar, C.; Angioloni, A. High-legume wheat-based matrices: Impact of high pressure on starch hydrolysis and firming kinetics of composite breads. *Food Bioproc. Tech.* **2017**, *10*, 1103–1112. [CrossRef]
25. Taherian, A.R.; Mondor, M.; Labranche, J.; Drolet, H.; Ippersiel, D.; Lamarche, F. Comparative study of functional properties of commercial and membrane processed yellow pea protein isolates. *Food Res. Int.* **2011**, *44*, 2505–2514. [CrossRef]
26. Roccia, P.; Ribotta, P.D.; Pérez, G.T.; Leon, A.E. Influence of soy protein on rheological properties and water retention capacity of wheat gluten. *LWT Food Sci Technol.* **2009**, *42*, 358–362. [CrossRef]
27. Hadnađev, T.D.; Torbica, A.; Hadnađev, M. Rheological properties of wheat flour substitutes/alternative crops assessed by Mixolab. 11th International Congress on Engineering and Food. *Procedia Food Sci.* **2011**, *1*, 328–334. [CrossRef]
28. Francis, F.J.; Clydesdale, F.M. *Food Colorimetry: Theory and Applications*; AVI Publishing Company Inc.: Westport, CT, USA, 1975.

29. Horstmann, S.W.; Foschia, M.; Arendt, E.K. Correlation analysis of protein quality characteristics with gluten-free bread properties. *Food Funct.* **2017**, *8*, 2465–2474. [CrossRef] [PubMed]
30. Garcia-Segovia, P.; Igual, M.; Martinez-Monzo, J. Physicochemical properties and consumer acceptance of bread enriched with alternative proteins. *Foods* **2020**, *9*, 933. [CrossRef] [PubMed]

Review

Perspectives on the Use of Germinated Legumes in the Bread Making Process, A Review

Denisa Atudorei * and Georgiana Gabriela Codină *

Faculty of Food Engineering, Stefan cel Mare University of Suceava, 720229 Suceava, Romania
* Correspondence: denisa.atudorei@fia.usv.ro (D.A.); codina@fia.usv.ro (G.G.C.); Tel.: +40-758-600-035 (D.A.); +40-745-460-727 (G.G.C.)

Received: 17 August 2020; Accepted: 6 September 2020; Published: 8 September 2020

Abstract: Nowadays, it may be noticed that there is an increased interest in using germinated seeds in the daily diet. This high interest is due to the fact that in a germinated form, the seeds are highly improved from a nutritional point of view with multiple benefits for the human body. The purpose of this review was to update the studies made on the possibilities of using different types of germinated legume seeds (such as lentil, chickpea, soybean, lupin, bean) in order to obtain bakery products of good quality. This review highlights the aspects related to the germination process of the seeds, the benefits of the germination process on the seeds from a nutritional point of view, and the effects of the addition of flour from germinated seeds on the rheological properties of the wheat flour dough, but also on the physico–chemical and sensory characteristics of the bakery products obtained. All these changes on the bread making process and bread quality depend on the level and type of legume seed subjected to the germination process which are incorporated in wheat flour.

Keywords: germination process; legumes; technological process; bread quality

1. Introduction

Seed germination is defined as the process of developing a new plant. From a mechanical point of view, the germination process can be described as an interaction between the protective coating of the seeds and the force of pushing on it by the developing seed embryo [1,2]. Germination includes three stages. The first one consists of the absorption of water by the dried seeds and the onset of the mRNA biosynthesis of the seed. The second stage is characterized by the coleoptile elongation of the seed. This is the most important phase because here all the metabolic and physiological processes of the seed are reactivated. The success of this stage is influenced by the external germination conditions, but also by the seed phytohormones. The third stage consists of the continuous absorption of water and the emergence of the radicle due to the development of the seed embryo axis [3].

Lately, interest in the consumption of germinated cereals has grown a lot due to several reasons. Mainly, it is related to the improving of the human health, correlated to the fact that the germination process is a relatively simple one because it does not require special working techniques. The germination process presents a number of advantages related to the fact that it improves the nutritional composition of the seeds and its content in bioactive substances. Different studies have shown that the germination process, if performed correctly, may increase the nutrients' availability (amino acids, minerals, vitamins, etc.) [4–7] and at the same time decrease the antinutrient compounds from the seeds [8–11]. The international literature has highlighted that the germination process increases the amount of phenolic compounds which are chemical compounds that have antioxidant action. At the same time, the germination process activates the hydrolytic enzymes from seeds, which favors the digestion process of some compounds such as starch and proteins [12]. Some researchers have highlighted that germination leads to the activation of endo-enzymes in seeds, such as proteases and amylases, which help break down macromolecular substances such as proteins and carbohydrates and facilitate the digestibility of nutritional compounds from seeds. At the same time, the germination process contributes to the decrease in the antinutritional factors content and to the activation of some enzymes which exist in the seeds. Thus, germination is a process that allows improving the nutritional composition of seeds [4]. Some studies emphasize that the germination process increases the content of flavonoids [13–17]. Studies in the field also indicates that the germination process leads to the decrease in some antinutritional factors, which favors a higher bioavailability of the nutritional compounds from the seeds subjected to the germination process [18–20].

Due to the health benefits of germinated seeds, different studies have reported on the incorporation of germination seeds into various food manufacturing recipes, such as: biscuits [21,22], beverages [23], baby purees [24], muffins [25,26], and yogurt [27,28]. Given the fact that bread is one of the main foods from the daily diet of the population, attempts have been made to improve its quality with flour obtained from germinated legumes. The addition of the germinated legumes in bread recipe led to changes in the dough rheological properties, but also on the physico–chemical and sensory properties of the bread samples obtained. The purpose of this review was to highlight the current state of studies in the field related to the possibility of using germinated seeds in the bread making recipe, underlining the changes brought by germinated seed addition on the technological process and the bread quality characteristics.

2. Description of the Germination Process

Germination is the first and foremost important process for the plants' cultivation and the production of vegetables and cereals. Germination is a process influenced by many genetic, endogenous, but also environmental factors, and consists of a series of cellular processes which produce various changes in the structure of the cell, some being visible to the naked eye. The germination process is being studied even today to fully elucidate its mechanism [29]. Some researchers have reported that the germination process, used for plant reproduction, is influenced not only by abiotic factors, but also by biotic stressors, such as pathogens and infestants [30]. Other studies have shown that the influence of environmental factors, such as temperature, oxygen content, soil moisture and salinity has an important impact on the germination process [31].

The germination process used for plant reproduction is a complex process which consists of several stages. The first stage corresponds to the absorption of water by seeds. Water absorption depends on seed composition and the permeability of the outer shell. The absorption of water exerts a pressure on the outer layer of seed protection leading to its deterioration in order to develop the root. Along with these physical changes in the structure of the seed, a series of metabolic changes also take place. Thus, the substances stored in the endosperm are solubilized to give rise to new tissues. Thus, starch is broken into simpler forms, namely simple sugars (glucose and maltose), and the proteins, to amino acids and amides. At the same time, the enzymes are activated, presenting an important role in the germination process. In addition, hormones also play important roles in the germination process, which help nutrients transport and the formation of new compounds that are needed for the seed development. Studies have shown that in the case of some seeds, after 12 h from the beginning of the germination process, the process of cell division begins in order to start the development of the component parts of the future plant. The germination process is influenced by a number of factors, such as: seed integrity, water used for germination, parameters of air from germination space, lighting, soil characteristics. Each type of seed has a different germination period and requires particular germination conditions [32].

Numerous studies have shown that different artificial treatments applied to seeds have the effect of improving the germination process. For example, Rifna et al. [33] have shown that ultrasonic treatment and UV radiation have a positive effect on germination. The ultrasound treatment applied to the seeds consisted of frequencies between 20 and 100 kHz and an intensity of 10–1000 W/cm^2. The positive influence of ultrasound treatment is attributed to the improvement of water absorption by seeds. Ding et al. [34] have shown that an exposure to 25 kHz ultrasound treatment for 5 min of red rice and brown rice seeds leads to an improved germination rate and improved functional properties of the flour obtained from these seeds. Porto et al. [35] also concluded that the ultrasound treatment of dry seeds improves the germination process.

The artificially induced germination process was used to obtain germinated seeds with superior properties in order to be incorporated into various food products being described in different studies. Studies have shown that when the artificial germination process was applied to the seeds, some preliminary treatments must be used in order to assure the optimal conditions to the germination process. For example, Kaczmarska et al. [36] obtained the germinated seeds after the following process described below. First, the seeds must be selected to eliminate any impurities or degenerated seeds. Then, the seeds need to be sanitized with a 3% hydrogen peroxide solution. Then, the seeds need to be rinsed with pure water to obtain a neutral pH. After that, the seeds are soaked in pure water for 8 h. After this, the seeds are placed in special containers for germination and covered with a special paper for germination. The germination time depends on the type of seed, and the germinating parameters should be carefully monitored (temperature of 22 °C, relative humidity of air 50–60%, access to daylight for 12 h and in the dark for 12 h). After seed germination, a treatment for drying the seeds is applied in order to stop the evolution of the germination process.

3. Germination Changes on the Sensory and Nutritional Characteristics of the Seeds

3.1. Sensory Characteristics

Numerous studies have shown that germination has beneficial effects on the nutritional profile of the seeds subjected to this process. It is a low-cost and an effective technology to improve the nutritional quality of vegetables, by its antioxidant capacity, digestibility of proteins, by increasing its vitamin C and E content and by reducing its anti-nutritional factors [37]. Kaczmarska et al. [36] showed that the germination process had the effect of changing the flavor characteristics of seeds as lupine and soybean ones. Thus, they showed that after the germination process, the amount of volatile organic compounds such as 2-methylbutanal and dimethyltrisulfite increased, which had the effect of intensifying the seeds' flavor. At the same time, the sweet taste of the seeds was also intensified.

Xu et al. [38] have shown that in the case of the chickpea, the changes in the flavor profile became undesirable after a period of 4 days when the seeds were subjected to the germination process. It seems that a longer period of germination intensifies the specific flavor of beans in the case of lentils and chickpeas, and in the case of the chickpea the specific flavor of beans decreased and unpleasant aromas appeared. Additionally, the study showed that the aroma profile of the lentil and pea, after germination, was similar, and in the case of the chickpea flour the specific flavor of beans decreased, and unpleasant aromas appeared. The study of changes in the flavor profile during the germination process is desirable to successfully achieve the improvement of the nutritional profile of the seeds, without adversely affecting their flavor.

In general, with the increasing period of the germination time, the sensory profile of the cereals/legumes subjected to the germination process became more and more pronounced. Xu et al. [39] highlighted that the specific smell of beans became more pronounced in the case of chickpeas, lentils and peas, with the increase in the period in which the legumes were subjected to the germination process. They showed that three compounds involved in the formation of the sensory profile decreased in concentration with increasing germination time. This is due to the fact that germination degrades the endosperm, favoring the elimination of flavor compounds. A solution to reduce the specific aroma of beans would be the addition of antioxidants in the germination process.

Dueñas et al. [40] reported that the seeds' germination increased the amount of phenolic compounds and dietary fiber of the seeds, affecting its taste. These results were in agreement with those reported by Xu et al. [41] which concluded that phenolic compounds influenced the bread sensory quality in a significant way, taste being one of the main characteristics affected by them. They also concluded that bread samples with a high quantity of phenolic compounds presented a more astringent taste than the control sample. However, according to Gębski et al. [42] it seems that the dietary fiber did not affect in such a significant way the taste of the bread samples and therefore it may be concluded that the phenolic compounds has a higher impact on bread taste than the dietary fibers.

3.2. Nutritional Characteristics

The main purpose of using the germinated seeds or of the flour obtained from them in the recipes of the food manufacture products is, as we mentioned above, the improvement of the nutritional characteristics of foodstuffs. The germination process leads, along with the physical changes of the seeds, to improvements in their nutritional profile.

For example, studies have shown that the germination process influences the fat content of the seeds. Thus, Xu et al. [39] concluded that the decrease in the fat content might be due to the increase in lipolytic activity during the germination process. Thus, lipid compounds are hydrolyzed to ensure seed development. However, reducing the amount of fat depends on the species undergoing the germination process. Some studies show that the ash content increased during the germination process, in the case of amaranth and rice. This was due to the decrease in the amount of soluble solid substances (such as starch and different types of sugars). Germination does not significantly affect the carbohydrate content, seeming that the amylose content decreases and the total sugar content increases. Due to the fact that by the germination process, the nutritional characteristics of seeds are improved, the germinated seeds can be successfully used in bakery products [18].

Bueno et al. [43] concluded that the germination process, in the case of soybeans, increases the content of free amino acids, sugars, phenolic compounds and its antioxidant potential. After 32 h of germination, the maximum antioxidant capacity was highlighted. The antioxidant capacity began to increase after 8 h of germination. Thus, it demonstrated the possibility of using germinated soy beans for the development of innovative foods with special properties, such as soy milk.

In the case of buckwheat seeds, germination has been shown to increase the total content of flavonoids, amino acids and reducing sugars. In contrast, the total protein and sugar content increased. At the same time, the content of vitamin C and B_1 decreased. In the case of vitamins B_2 and B_6 there were no significant changes. Moreover, the value of the free radical-scavenging activities increased due to the increase in its antioxidant activity [44].

Lentil, soybean, lupine and beans flour are increasingly used to improve the nutritional properties of food. For this reason, current studies often highlight the effect of their germination, in order to see the possibilities of improving the sensory profile of food through germination. Lentils are a legume whose consumption has been shown to have many health benefits, including reducing the risk of cancer due to its content in anticarcinogenic substances such as lectins, glycosidic saponins, bioactive peptides which include protease inhibitors, fermentable fibers and oligosaccharides. The high polyphenol content in lentils has anti-tumor action. [45]. Lentil are of high interest to be incorporated into various foods because it contains no gluten. It has a high protein content, namely 30.65 g/100 g. This content increased during germination to 33.60 g/100 g. The total starch content in the case of lentils germinated decreased during this process. At the same time, the germination process had the effect of increasing the viscosity of the lentil flour to a value of 1061 cP even after the first 2 days of germination. After germination, the water absorption capacity was highly improved.

In the case of lentils, the following changes were noticed after six days of germination: an increase in the amount of ash and a decrease in the amount of starch, lipids and amylose. The decrease in the amount of starch is due to the release of enzymes from the outer shell of seed (α-amylase, glucosidase, dextranase) and from endosperm (β-amylase). The presence of hydrolytic enzymes causes the starch conversion to oligosaccharides or monosaccharides, which, leads to a decrease in the amount of starch. The starch amount was reduced from 41.02 g/100 g to 34.96 g/100 g, after six days of germination, according to a study by Xu et al. [46]. The decreases in the amount of lipids content are due to their use by the seeds during the development process. Thus, it was noted that during germination, triglycerol is hydrolyzed by enzymes to release fatty acids. They will be oxidized at the level of the cytosol and mitochondria to release the energy needed in the germination process [47]. Fouad and Rehab [48] also have highlighted that the germination of lentil seeds at a temperature of 25 °C for 3–6 days, in dark conditions, leads to a decrease in lipid content.

In relation to the lentil flour properties, some changes due to the germination process were also reported. The viscosity of lentil flour increased after three days of germination. After the germination time continued, the viscosity registered a downward trend [49]. This may be due to the fact that the enzymes had hydrolyzed a significant amount of amylose and amylopectin, which resulted in the beginning of a damaged network starch. This also could lead to an increase in the starch digestibility [50]. Studies have shown that germination improves the water absorption capacity of lentil flour, namely the flour ability to absorb water, after 2–4 days of germination. This is probably explained by the macronutrient structural changes, but also by the fact that the starch in the germinated seeds changes its structure. At the same time, during the germination process the amount of melatonin in lentils increases. This increase occurs with the increase in root size [37]. Therefore, it seems that the germination process significantly improves lentil's nutritional characteristics, such as its antioxidant capacity, leading to the obtaining of an ingredient with a functional role which may be used to obtain innovative and functional food products. Taking into account that lentil is a gluten-free legume, it could be also successfully used to obtain foodstuffs for people suffering from celiac disease [46].

Another legume that is of interest in the food industry because it has special properties for improving human health is chickpeas. Recently, there has been a special interest in chickpeas because they are an ideal source of proteins, fibers, carbohydrates as well as minerals and thus help maintain a balanced diet. For vegetarians, chickpeas are an ideal source of proteins. Moreover, the content of allergens from it is low. At the same time, chickpeas could be considered an alternative to soy. Studies have shown an in vitro digestibility for chickpeas between 48% and 89.01%, higher than for soybeans and beans [51].

In the case of chickpeas, germination has been shown to increase the total flavonoid content, polyphenolic compounds that have antioxidant action. The total flavonoid content increases after germination, from 0.22 to 0.42 g/kg, at a germination temperature of 30 °C. At a germination temperature of more than 10 °C, the total flavonoids content was 0.38 g/kg. Thus, it can be concluded that the germination temperature influences the increase in the flavonoid content. This increase depends on the seed response to different stressors, biotic and abiotic [52]. However, the change in the total flavonoid content during germination is different for each type of seed. This change is based on the seed characteristics, the class of which it belongs to and their hardness [53].

Chickpea germination, as outlined by studies to date, does not significantly influence its content in proteins, lipids, fiber, ash and carbohydrates. However, this process increases the amount of ascorbic acid. This content increases from 1.9 mg/100 g to 9.4 mg/100 g (after a germination period of 24 h) and to 15.6 mg/100 g (after a germination period of 48 h). At the same time, an increase in some essential amino acids was noticed after germination. For example, the contents of threosine, lysine, leucine, valine and isoleucine increased after 24 h of germination, and in the case of some of them their contents decreased with increasing germination time to 48 h. The highest increase was in the case of lysine [54]. However, studies have shown that only the germination of chickpeas in optimal conditions leads to the nutritional improvement of chickpeas. Thus, the monitoring of the temperature and germination time is absolutely necessary in order to achieve the desired nutritional improvement. A study carried out by Domínguez-Arispuro et al. [55] concluded that the optimal parameters for the germination process in the case of chickpeas were: 27.5–35 °C and 125–240 h. The parameters of the germination process in which the nutritional compounds considered (polyphenol content, the total content of flavonoids, antioxidant activity) recorded the highest values were of 33.7 °C and 171 h. Under these conditions, it was shown that the content of dietary fiber significantly increased in the case of chickpeas. Moreover, after the germination process, a significant increase in ferulic and ellagic acids was noticed, which have antioxidant effects. At the same time, the germination conducted led to a significant increase in protein content (germination took place at 33.7 °C for 171 h) [55]. The increase in protein content has been attributed to the loss of dry matter, especially in carbohydrates, during the process of seed respiration during germination. In addition, a decrease in lipid content was observed, due to its consumption as an energy source in the germination process [56].

Another legume that occupies a special place among vegetarians is soy. At the same time, soy consumption has various health benefits. Watanabe and Uehara [57] have shown that isoflavones from soy contents are nonsteroidal phytoestrogenic and antioxidative diphenolic compounds that have a positive role in preventing diseases such as osteoporosis, cardiovascular disease, postmenopausal syndrome [57,58]. Soy protein is also an important source of amino acids [59]. A study by Miglani and Sharma [60] indicated that the content of soy phenolic constituents is significantly influenced by germination time and temperature. The germination process changes the structure of the polysaccharide wall of the seed cell wall. Megat et al. [61] studied the influence of the germination process on soybean fibers and showed that the total content of dietary fiber (soluble and insoluble) increased significantly compared to ungerminated seeds. Thus, the germination process can be used as a method to improve the dietary fiber content of soybeans, this having favorable health implications. In the case of soybeans, it has also been shown that the germination process leads to a significant increase in the content of phenols and flavonoids. Its maximum amount was determined at seven days of germination [62]. Regarding green soybeans, after germination, there was a decrease of 14% in protein, 37% in lipids, 22% in carbohydrates and 16% in ash, according to a study by Chen and Chang [63]. Furthermore, in some studies, following the germination of soybean seeds, there was a decrease in lipid content and the activity of lipoxygenase-1 and lipoxygenase-3. The variation of the last two depended on the germination time. The longer the germination time, the lower their value was due to the degradation of lipoxygenase which plays an important role in the oxidation of unsaturated fatty acids, such as linoleic, linolenic, arachidonic. Following germination, the content of trypsin inhibitors also decreased. They played an important role in decreasing the digestibility of proteins by inactivating trypsin. The highest decrease in trypsin inhibitors was recorded after two days of germination [64]. At the same time, it has been shown that the germination process has the effect of reducing the glycemic index. This is because arabinose, the dominant polysaccharide in soy, is used in the germination process as an energy source or for macromolecule biosynthesis. Thus, it can be concluded that germination decreases the activity of lipoxygenase and trypsin inhibitors, which leads to a decrease in the formation of unpleasant flavor compounds and an increase in the proteins' digestibility [65].

Another seed that was subjected to the germination process to be incorporated into the manufacturing recipes of various foods was lupine. Lupine belongs to the genus *Lupinus*, and three species of this genus are used in human consumption. Consumers should be aware that its consumption can cause allergic reactions [66]. Lately, the interest in using lupine flour in different foods has increased a lot due to its nutritional value. Lupine is high in protein and fiber and low in fat [67]. Studies show that lupine seeds have anti-inflammatory action [68]. Olkowski [69] reported that lupine seeds contain a higher amount of protein than all other grain legumes. Lupine seeds have a high lysine content [70], which contributes to extending the shelf life of bakery products in which lupine flour has been used [71]. The germination of lupine seeds leads to increased melatonin, total phenol and flavonoid contents and antioxidant activity [72].

The positive effect of the germination process on the nutritional value of seeds has been highlighted by various studies also in the case of beans. Beans are of interest because they have a high nutritional value and their consumption has positive effects on human health. Common beans are rich in protein and various macronutrients, especially zinc and iron [73]. Iron has an important role in the human body because it is used by hemoglobin to transport oxygen to the body [74]. Zinc has an important role in cell division, immune system, wound healing process, and carbohydrate metabolism [75]. Beans contain a significant amount of fiber [76–78]. Epidemiological studies have shown that eating beans reduces the risk of cardiovascular disease, diabetes, cancer, helps control blood sugar levels and have antioxidant action [79]. The germination process of beans has the effect of improving the bioavailability and potential health benefits of starch and the protein fractions contained in them. Germination causes a decrease in the starch content of beans. This reduction can be attributed to the enzymatic hydrolysis of this polymer by the enzymes activity during germination [80]. The amylose content of black beans was determined to be 26 g/100 g, and during germination this content decreases to 20 g/100 g [81]. This decrease is due to the amylolytic capacity of the beans' own enzymes [82].

4. The Use of Germinated Seeds in the Bread Making Recipe

In order to improve the bread quality from the nutritional point of view, an attempt was made to incorporate flours from different germinated seeds in its recipe. At the same time, the effect brought by this addition on the sensory and physico–chemical profile of the bread was analyzed, as were the changes made on the rheological characteristics of the dough samples. However, there are few studies on how the addition of germinated seeds influences the dough characteristics and bread quality.

The addition of flour from germinated seeds is recommended to be used in bread making recipes in order to improve its nutritional value [83]. This addition also has an influence on the characteristics of the dough, but also on the quality characteristics of the bread [84]. In general, the studies showed a positive influence of the addition of germinated seeds flour on the rheological characteristics of the dough samples. For example, in general, the addition increases the water absorption values [85]. However, there are also some types of germinated seeds flour whose incorporation into the bread making recipe has led to a decrease in the water absorption value. Therefore, for the use of a type of germinated seed flour in the bread making recipe, it is important to monitor its influence on this characteristic because the water absorption value influences in a significant way the dough behavior during the technological process of bread making.

4.1. The Influence of the Addition of Germinated Seeds on the Wheat Flour Dough

Regarding germinated soybean flour, Shin et al. [86] highlighted the fact that its addition in the dough recipe increases the water absorption value. In addition, when the flour obtained from germinated soybean was incorporated in wheat flour dough, its volume increased during the fermentation process compared to the addition of ungerminated soybean flour, steamed soybean flour and roasted soybean flour. Rasales-Juárez et al. [87] concluded that the farinographic values such as dough development time and dough stability slightly increased when germinated soybean flour was incorporated in the dough recipe. This was attributed to the increase in the enzymatic activity from the dough system or to the hydrophilic components due to the germination process. Thus, the authors pointed out the possibility of using flour from germinated soybean in wheat flour which contains a low amount of protein which may improve the characteristics of the wheat flour dough. Additionally, it seems that the germinated seed flour addition influences the viscosity of the wheat flour dough. This may be due to the fact that soy contains a significant amount of carbohydrates (including pentosans) that could influence the flour's ability to absorb water. Some studies have shown that a high level of pentosans led to a sticky dough [88] and to a delay in the development of gluten [89]. However, more studies on the possible interfering effects between soy carbohydrates and wheat flour on dough rheology would be necessary [87].

In the case of the addition of germinated bean flour, Morard et al. [90] have shown that the water absorption increased with the increasing amount of its addition in wheat flour. Aprodu et al. [91] highlighted that the addition of germinated bean flour led to an increase in the water solubility index of the flour mix (wheat and germinated beans). It seems that the addition of germinated bean flour improved the dough rheological properties due to the fact that its viscoelasticity was improved. An improvement in the dough volume was also observed during the fermentation in which germinated bean flour was incorporated.

Guardanelli et al. [92] highlighted the fact that the addition of germinated seed flour led to an increase in the dough moisture. This increase was also reported in a previous study by Bojnanská and Smitalová [93]. The authors noticed that the dough consistency increased for the samples in which germinated seed flour was incorporated compared to the control sample in which no germinated seed flour was added. In addition to the increase in the germinated seed flour addition, the dough adhesiveness and cohesiveness increased. The viscosity of the dough with the addition of germinated seed flour was also increased probably due to the depolymerization process during germination. Therefore, the addition of germinated seed flour influenced dough rheology and the physicochemical properties of the dough, also improving dough elasticity [94]. Therefore, germinated seed flour can be used to obtain dough with good rheological properties, especially if it is also considered to improve it from a nutritional point of view [95].

In the literature, there are studies that show that the addition of germinated wheat flour in the bread making recipe has the effect of increasing the mixing time of the dough obtained and also strengthens it. Furthermore, it was noticed that an increase in the lipolytic activity from the dough system may lead to a decrease in the shelf life of the mix samples. In general, an addition of less than 10% germinated seed flour has a positive effect on the dough rheological properties. A higher level of germinated seed flour addition leads to a decrease in the volume of the dough samples which also leads to a decrease in the bread samples' loaf volume [96]. Moreover, some studies have shown that subjecting wheat grains to germination reduces the quantity of peptides which play an important role in triggering celiac disease. This fact highlights another benefit of the germination process on the human body [97].

In general, different studies have reported that an amount of germinated seeds of up to 10% addition in wheat flour has no negative effects on the dough's rheological properties. According to these studies, it seems that a 5% addition of germinating seeds has led to optimum wheat flour dough rheological properties [96,98,99].

4.2. The Influence of the Addition of Germinated Seeds on the Bread Quality

Millar et al. [99] highlighted that the addition of germinated pea flour decreases the loaf volume of the bread samples. Fernandez and Berry [100] also reported that the addition of germinated chickpea flour influences in a negative way the loaf volume of the bread samples. However, Levent et al. [101] reported that the addition of 10% germinated chickpea flour has improved the characteristics of the bread samples obtained probably due to the increase in the hydrolytic enzymatic activity from the dough system and its soluble components. An addition of 20% germinated chickpea flour had the effect of decreasing the loaf volume of bread samples. In addition, the color of the bread samples was not affected by a 10% addition, but was significantly influenced by a 20% addition [102]. Guardado-Félix et al. [99] highlighted the fact that the addition of germinated chickpea flour decreases the loaf volume of bread to a level of a 7–13% addition. At the same time, the bread obtained with germinated chickpea flour has a darker crust, darker crumb and a higher density. The textural parameters are not influenced in a significant way. All chickpea flours increased with more than 2% the water absorption values and consequently the bread yield decreases by between 7 and 13% as does the loaf volume of the bread samples.

Shin et al. [86] highlighted the fact that the one with the addition of germinated soybean flour had a higher loaf volume than the bread obtained without germinated seed flour addition. This is due to the fact that, as a result of the germination process, the solubility of the proteins increased, which led to a better emulsification and foaming capacity [102]. Therefore, bread samples made with germinated soybean flour presented better quality characteristics.

The influence of the addition of germinated soybean flour in wheat flour on the bread crumb and bread crust was highlighted by Rasales-Juárez et al. [87], who reported that the bread crust was less yellow when germinated soybean flour was added in the bread recipe. It seems that compared with the addition of ungerminated soybean flour, the addition of germinated flour in a bread recipe had the effect of improving the color of the samples, making the crumb whiter and without changing in a negative way the color of the crust samples. Rosales-Juárez et al. [87] have also reported that the addition of germinated soybean flour improved the loaf volume of bread samples.

In general, studies have shown a negative influence of a high addition (higher than 10%) on the quality of bread [103] whilst an addition of up to 3% of germinated soy flour has a positive effect. The positive effect was due to the enzymes from the germinated flour (lipoxygenase, amylase, lipase, alpha galactosidase), but also to the content of lecithin and ascorbic acid which considerably increases during the germination process. It was also reported that germinated soybean seeds can be successfully used as substitutes for chemical additives often used in the bakery industry, due to its high content in ascorbic acid, lecithin and enzymes, often used as additives in bread making [104].

Al Omari et al. [105] studied the influence of wheat flour substitution with germinated lupine flour on the characteristics of the bread obtained and reported that a maximum substitution of 20% germinated lupine flour does not significantly influence the sensory profile of the bread samples. From all the sensory characteristics, the color parameter was the most influenced due to the presence of lupine seeds in yellow pigments, lutein and zeaxanthin, but also to the high lysine content from germinated lupine, lysine being the most reactive amino acid of the Maillard reaction. These results were similar to those reported by Obeidat et al. [106]. In addition, it was reported that as the amount of germinated lupine flour increased, the loaf volume of bread samples decreased. However, it seems that a level of 5% germinated lupine flour addition in the bread recipe does not significantly influence the loaf volume of bread samples. Moreover, in order to preserve the quality characteristics of the bread samples, in terms of freshness characteristics, a maximum substitution of wheat flour with germinated lupine flour of 10% was allowed. These results were similar with those reported by Abdul Hussain et al. [107]. According to the reported studies, it seems that germinated lupine flour can be used in bread making recipes in order to improve its quality from the nutritional and technological point of view. In order to obtain bread samples of a good quality, the recommended levels of germinated lupine flour addition in wheat flour was up to 10% [108].

Although the beans germination has led to a significant improvement in its nutritional profile [109], few studies have been made on the addition of them in bread making and the effect on bread quality. Morard et al. [90] highlighted that the addition up to a level of 5% germinated bean flour does not affect in a negative way the loaf volume of the bread. The loaf volume of the bread decreased, however, when high levels of germinated bean flour were incorporated in wheat flour. Moreover, the color of the bread crust decreased in intensity as the addition level of germinated bean flour increased.

The addition of germinated wheat flour up to a level of 50% addition in the wheat flour of a strong quality for bread making improves the loaf volume and porosity of the bread samples, as highlighted by Marti et al. [109]. This improvement was probably due to the increase in α-amylase content from the dough system due to the germination process of wheat flour. The addition also led to bread samples with a darker crust and a reddish tint. This was due to the Maillard reaction which is more intense in bread samples with germinated wheat flour. The higher level of α-amylase from the dough system has also been conducted for bread samples with a higher loaf volume.

It is well known that through the germination process, the amount of dietary fibers and polyphenolic compounds from the vegetables increases [108]. From the bread quality point of view, it seems that these compounds from the germinated vegetables have a high influence on its characteristics. It has been reported that the dietary fibers do not significantly affect the bread sensory characteristics such as taste and flavor but it had a significant effect on loaf volume [110,111]. Almeida et al. [110] reported that even at 20% fiber addition in wheat flour, the sensory characteristics of bread such as taste and flavor were not significantly affected. Moreover, they reported that the addition of fibers in wheat flour led to an improvement in bread quality by an increase in freshness due to the increase in the water absorption capacity and moisture retention. From the color point of view, it seems that depending on its type, fibers addition may influence this characteristic. For example, the wheat bran led to a decrease in the bread samples' brightness. In general, studies have shown that there was no significant negative influence of the addition of fiber (in the form of carob fiber, inulin and pea fiber type, etc.) on the overall quality and acceptability of the bread obtained [112]. Fendri et al. [113] concluded that the addition of fibers extracted from the pea and bean improved the textural profile of the bread samples and changes the dough's textural properties. They reported a decrease in the bread samples hardness and an increase in the cohesiveness and adhesiveness of the wheat flour dough. However, regarding the phenol effect point of view on bread making, it seems that its addition affects the bread quality in a more significant way. It was highlighted that the phenolic compounds' addition decreased the loaf volume and increased the bread hardness value [41,114]. It has also been reported that the polyphenols' addition affects starch gelatinization process leading to an increase in the gelatinization temperature [115] influencing starch retrogradation and increasing bread freshness [116]. According to these studies, it has been concluded that the addition of different ingredients in bread making with a high amount of polyphenol content may be a good alternative for substituting synthetic chemical additives, these being considered natural dyes, preservatives and antioxidant materials.

5. Conclusions and Perspectives

Germinated legumes can be successfully used in various food manufacturing recipes, especially in order to improve the nutritional value of these products. This way, the germination process can be successfully used to improve wheat flour in order to obtain healthier bakery products. In addition, the flour obtained from germinated seeds can be used as a substitute for different chemical additives that are often used in the bakery industry due to its high content of enzymes, ascorbic acid, lecithin, etc. However, even though there are many publications in the international literature on the influence of the addition of different types of legume flour in the bread making recipe, there are few studies on the addition of legume flour in a germinated form on bread making. The present study sought new alternatives for bread making improvement, underlining how different types of legumes in a germinated form can improve the dough's rheological characteristics and the final quality of bread.

Author Contributions: D.A. and G.G.C. contributed equally to the study design, collection of data, development of the sampling, analyses, interpretation of results, and preparation of the paper. All authors have read and agreed to the published version of the manuscript.

Funding: This research received no external funding.

Acknowledgments: This work was supported by a grant of the Romanian Ministry of Research and Innovation, CNCS–UEFISCDI, project number PN-III-P1-1.1-TE-2019-0892, within PNCDI III.

Conflicts of Interest: The authors declare no conflict of interest.

References

1. Steinbrecher, T.; Leubner-Metzger, G. The biomechanics of seed germination. *J. Exp. Bot.* **2017**, *68*, 765–783. [CrossRef] [PubMed]
2. Nonogaki, M.; Nonogaki, H. Germination. *Encycl. Appl. Plant Sci.* **2017**, *1*, 509–512.
3. Li, Q.F.; Zhou, Y.; Xiong, M.; Ren, X.Y.; Han, L.; Wang, J.D.; Zhang, C.Q.; Fan, X.L.; Liu, Q.Q. Gibberellin recovers seed germination in rice with impaired brassinosteroid signalling. *Plant Sci.* **2020**, *293*, 110435. [CrossRef] [PubMed]
4. Ma, M.; Zhang, H.; Xie, Y.; Yang, M.; Tang, J.; Wang, P.; Yang, R.; Zhenxin, G. Response of nutritional and functional composition, anti-nutritional factors and antioxidant activity in germinated soybean under UV-B radiation. *LWT Food Sci. Technol.* **2020**, *118*, 108709. [CrossRef]
5. Ohanenye, I.C.; Tsopmo, A.; Ejike, C.E.C.C.; Udenigwe, C.C. Germination as a bioprocess for enhancing the quality and nutritional prospects of legume proteins. *Trends Food Sci. Technol.* **2020**, *101*, 213–222. [CrossRef]
6. EL-Suhaibani, M.; Ahmed, M.A.; Osman, M.A. Study of germination, soaking and cooking effects on the nutritional quality of goat pea (*Securigera securidaca* L.). *J. King Saud Univ. Sci.* **2020**, *32*, 2029–2033. [CrossRef]
7. Xu, L.; Chen, L.; Yang, N.; Chen, Y.; Wu, F.; Jin, Z.; Xu, X. Impact of germination on nutritional and physicochemical properties of adlay seed (Coixlachryma-jobi L.). *Food Chem.* **2017**, *229*, 312–318. [CrossRef]
8. Singh, A.; Sharma, S. Bioactive components and functional properties of biologically activated cereal grains: A bibliographic review. *Crit. Rev. Food Sci.* **2017**, *57*, 3051–3071. [CrossRef]

9. Sokrab, A.M.; Mohamed-Ahmed, I.A.; Babiker, E.E. Effect of germination on antinutritional factors, total, and extractable minerals of high and low phytate corn (*Zea mays* L.) genotypes. *J. Saudi Soc. Agric. Sci.* **2012**, *11*, 123–128. [CrossRef]

10. Sangronis, E.; Machado, C.J. Influence of germination on the nutritional quality of *Phaseolus vulgaris* and *Cajanus cajan*. *LWT Food Sci. Technol.* **2007**, *40*, 116–120. [CrossRef]

11. Ma, Z.; Boye, J.Y.; Hu, X. Nutritional quality and techno-functional changes in raw, germinated and fermented yellow field pea (*Pisum sativum* L.) upon pasteurization. *LWT Food Sci. Technol.* **2018**, *92*, 147–154. [CrossRef]

12. Han, A.; Arijaje, E.O.; Jinn, J.R.; Mauromoustakos, A.; Wang, Y.J. Effects of germination duration on milling, physicochemical, and textural properties of medium- and long-grain rice. *Cereal Chem.* **2016**, *93*, 39–46. [CrossRef]

13. Chu, C.; Du, Y.; Yu, X.; Shi, J.; Yuan, X.; Liu, X.; Liu, Y.; Zhang, H.; Zhang, Z.; Yan, N. Dynamics of antioxidant activities, metabolites, phenolic acids, flavonoids, and phenolic biosynthetic genes in germinating Chinese wild rice (*Zizania latifolia*). *Food Chem.* **2020**, *318*, 126483. [CrossRef] [PubMed]

14. Aisyah, S.; Vincken, J.P.; Andini, S.; Mardiah, Z.; Gruppen, H. Compositional changes in (iso)flavonoids and estrogenic activity of three edible *Lupinus* species by germination and *Rhizopus*-elicitation. *Phytochemistry* **2016**, *122*, 65–75. [CrossRef]

15. Ujiroghene, O.J.; Liu, L.; Zhang, S.; Lu, J.; Zhang, C.; Lv, J.; Pang, X.; Zhang, M. Antioxidant capacity of germinated quinoa-based yoghurt and concomitant effect of sprouting on its functional properties. *LWT Food Sci. Technol.* **2019**, *116*, 108592. [CrossRef]

16. Agdel-Aty, A.M.; Salama, W.H.; Fahmy, A.S.; Mohamed, S.A. Impact of germination on antioxidant capacity of garden cress: New calculation for determination of total antioxidant activity. *Sci. Hortic.* **2019**, *246*, 155–160. [CrossRef]

17. Rasera, G.B.; Hilkner, M.H.; de Soares-Castro, R.J. Free and insoluble-bound phenolics: How does the variation of these compounds affect the antioxidant properties of mustard grains during germination? *Food Res. Int.* **2020**, *133*, 109115. [CrossRef]

18. Cornejo, F.; Novillo, G.; Villacrés, E.; Rosello, C.M. Evaluation of the physicochemical and nutritional changes in two amaranth species (*Amaranthus quitensis* and *Amaranthus caudatus*) after germination. *Food Res. Int.* **2019**, *121*, 933–939. [CrossRef]

19. Gong, K.; Chen, L.; Li, X.; Sun, L.; Liu, K. Effects of germination combined with extrusion on the nutritional composition, functional properties and polyphenol profile and related in vitro hypoglycemic effect of whole grain corn. *J. Cereal Sci.* **2018**, *83*, 1–8. [CrossRef]

20. Padmashree, A.; Negi, N.; Handu, S.; Khan, M.A.; Semwal, A.D.; Sharma, G.K. Effect of Germination on Nutritional, Antinutritional and Rheological Characteristics of Quinoa (*Chenopodium quinoa*). *Def. Life Sci. J.* **2019**, *4*, 55–60. [CrossRef]

21. Polat, H.; Capar, T.D.; Inanir, C.; Ekici, L.; Yalcin, H. Formulation of functional crackers enriched with germinated lentil extract: A Response Surface Methodology Box-Behnken Design. *LWT Food Sci. Technol.* **2020**, *123*, 109056. [CrossRef]

22. Patel, M.M.; Venkateswara Rao, G. Effect of untreated, roasted and germinated black gram (phaseolus mungo) flours on the physico-chemical and biscuit (cookie) making characteristics of soft wheat flour. *J. Cereal Sci.* **1995**, *22*, 285–291. [CrossRef]

23. Chavan, M.; Gat, Y.; Harmalkar, M.; Wagnmare, R. Development of non-dairy fermented probiotic drink based on germinated and ungerminated cereals and legume. *LWT Food Sci. Technol.* **2018**, *91*, 339–344. [CrossRef]

24. Jiménez, D.; Lobo, M.; Irigaray, B.; Grompone, M.A.; Sammán, N. Oxidative stability of baby dehydrated purees formulated with different oils and germinated grain flours of quinoa and amaranth. *LWT—Food Sci. Technol.* **2020**, *127*, 109229. [CrossRef]

25. Kaczmarska, K.T.; Chandra-Hioe, M.V.; Frank, D.; Arcot, J. Enhancing wheat muffin aroma through addition of germinated and fermented Australian sweet lupin (*Lupinus angustifolius* L.) and soybean (*Glycine max* L.) flour. *LWT Food Sci. Technol.* **2018**, *96*, 205–214. [CrossRef]

26. Kaur, A.; Kaur, R.; Bhise, S. Baking and sensory quality of germinated and ungerminated flaxseed muffins prepared from wheat flour and wheat atta. *J. Saudi Soc. Agric. Sci.* **2020**, *19*, 109–120. [CrossRef]

27. Cáceres, P.J.; Peñas, E.; Martínez-Villaluenga, C.; García-Mora, P.; Frías, J. Development of a multifunctional yogurt-like product from germinated brown rice. *Food Sci. Technol.* **2019**, *99*, 306–312. [CrossRef]

28. Park, K.M.; Oh, S.H. Production of yogurt with enhanced levels of gamma-aminobutyric acid and valuable nutrients using lactic acid bacteria and germinated soybean extract. *Bioresour. Technol.* **2007**, *98*, 1675–1679. [CrossRef]

29. Zhang, H.; Zhou, K.Z.; Wang, W.Q.; Liu, S.J.; Song, S.Q. Proteome analysis reveals an energy-dependent central process for Populus × canadensis seed germination. *J. Plant Physiol.* **2017**, *213*, 134–147. [CrossRef]

30. Eckelmann, D.; Kusari, S.; Spiteller, M. Spatial profiling of maytansine during the germination process of *Maytenus senegalensis* seeds. *Fitoterapia* **2017**, *119*, 51–56. [CrossRef]

31. Dai, W.; Wang, T.; Wang, C. Effects of interspecific interactions on seed germination between dominant species in the Yangtze River Estuary. *Estuar Coast Shelf S.* **2020**, *232*, 106483. [CrossRef]

32. Bubel, N.; Nick, J. *The New Seed-Starters Handbook*; Rodale Wellness: Allentown, PA, USA, 2018.

33. Rifna, E.J.; Ratish-Ramanan, K.; Mahendran, R. Emerging technology applications for improving seed germination. *Trends Food Sci. Tech.* **2019**, *86*, 95–108. [CrossRef]

34. Ding, J.; Hou, G.G.; Dong, M.; Xiong, S.; Zhao, S.; Feng, H. Physicochemical properties of germinated dehulled rice flour and energy requirement in germination as affected by ultrasound treatment. *Ultrason. Sonochem.* **2018**, *41*, 484–491. [CrossRef] [PubMed]

35. Porto, C.L.; Ziuzina, D.; Los, A.; Boehm, D.; Palumbo, F.; Favia, P.; Tiwari, B.; Bourke, P.; Cullen, P.J. Plasma activated water and airborne ultrasound treatments for enhanced germination and growth of soybean. *Innov. Food Sci. Emerg.* **2018**, *49*, 13–19. [CrossRef]

36. Kaczmarska, K.T.; Chandra-Hioe, M.V.; Frank, D.; Arcot, J. Aroma characteristics of lupin and soybean after germination and effect of fermentation on lupin aroma. *LWT Food Sci. Technol.* **2018**, *87*, 225–233. [CrossRef]

37. Aguilera, Y.; Herrera, T.; Liébana, R.; Rebollo-Hernanz, M.; Sanchez-Puelles, C.; Martín-Cabrejas, M.A. Impact of melatonin enrichment during germination of legumes on bioactive compounds and antioxidant activity. *J. Agric. Food Chem.* **2015**, *63*, 7967–7974. [CrossRef]

38. Xu, M.; Jin, Z.; Lan, Y.; Rao, J.; Chen, B. HS-SPME-GC-MS/olfactometry combined with chemometrics to assess the impact of germination on flavor attributes of chickpea, lentil, and yellow pea flours. *Food Chem.* **2019**, *280*, 83–95. [CrossRef]

39. Xu, M.; Jin, Z.; Gu, Z.; Rao, J.; Bingcan, C. Changes in odor characteristics of pulse protein isolates from germinated chickpea, lentil, and yellow pea: Role of lipoxygenase and free radicals. *Food Chem.* **2020**, *314*, 126184. [CrossRef]

40. Dueñas, M.; Sarmento, T.; Aguilera, Y.; Benitez, V.; Mollá, E.; Esteban, R.M.; Martín-Cabrejas, M.A. Impact of cooking and germination on phenolic composition and dietary fiber fractions in dark beans (*Phaseolus vulgaris* L.) and lentils (*Lens culinaris* L.). *LWT Food Sci. Technol.* **2016**, *66*, 72–78. [CrossRef]

41. Xu, J.; Wang, W.; Li, Y. Dough properties, bread quality, and associated interactions with added phenolic compounds: A review. *J. Funct. Foods* **2019**, *52*, 629–639. [CrossRef]

42. Gębski, J.; Jezewska-Zychowicz, M.; Szlachciuk, J.; Kosicka-Gębska, M. Impact of nutritional claims on consumer preferences for bread with varied fiber and salt content. *Food Qual. Prefer.* **2019**, *76*, 91–99. [CrossRef]

43. Bueno, D.B.; Silva-Júnior, S.I.; Seriani Chiarotoo, A.B.; Cardoso, T.M.; Neto, J.A.; Lopes dos Reis, G.C.; Abreu Glória, M.B.; Tavano, O.L. The germination of soybeans increases the water-soluble components and could generate innovations in soy-based foods. *LWT Food Sci. Technol.* **2020**, *117*, 108599. [CrossRef]

44. Yiming, Z.; Hong, W.; Linlin, C.; Xiaoli, Z.; Wen, T.; Xinli, S. Evolution of nutrient ingredients in tartary buckwheat seeds during germination. *Food Chem.* **2015**, *186*, 244–248. [CrossRef] [PubMed]

45. Faris, M.I.E.; Mohammad, M.G.; Soliman, S. Lentils (Lens culinaris L.): A candidate chemopreventive and antitumor functional food. *Funct. Foods Cancer Prev. Ther* **2020**, 99–120. [CrossRef]

46. Xu, M.; Jin, Z.; Simsek, S.; Hall, C.; Rao, J.; Chen, B. Effect of germination on the chemical composition, thermal, pasting, and moisture sorption properties of flours from chickpea, lentil, and yellow pea. *Food Chem.* **2019**, *295*, 579–587. [CrossRef] [PubMed]

47. Cornejo, F.; Caceres, P.J.; Martínez-Villaluenga, C.; Rosell, C.M. Effects of germination on the nutritive value and bioactive compounds of brown rice breads. *Food Chem.* **2015**, *173*, 298–304. [CrossRef]

48. Fouad, A.A.; Rehab, F.M.A. Effect of germination time on proximate analysis, bioactive compounds and antioxidant activity of lentil (*Lens culinaris Medik*) sprouts. *Acta Sci. Pol. Technol. Aliment.* **2015**, *14*, 233–246. [CrossRef]

49. Ghumman, A.; Kaur, A.; Singh, N. Impact of germination on flour, protein and starch characteristics of lentil (*Lens culinari*) and horsegram (*Macrotyloma uniflorum* L.) lines. *LWT Food Sci. Technol.* **2016**, *65*, 137–144. [CrossRef]

50. Acevedo, B.A.; Thompson, C.M.B.; González-Foutel, N.S.; Chaves, M.G.; Avanza, M.V. Effect of different treatments on the microstructure and functional and pasting properties of pigeon pea (*Cajanus cajan* L.), dolichos bean (*Dolichos lablab* L.) and jack bean (*Canavalia ensiformis*) flours from the north-east Argentina. *Int. J. Food Sci. Tech.* **2017**, *52*, 222–230. [CrossRef]

51. Rachwa-Rosiak, D.; Nebesny, E.; Budryn, G. Chickpeas—Composition, Nutritional Value, Health Benefits, Application to Bread and Snacks: A Review. *Crit. Rev. Food Sci.* **2015**, *8*, 1137–1145. [CrossRef] [PubMed]

52. Mamilla, R.K.; Mishra, V.K. Effect of germination on antioxidant and ACE inhibitory activities of legumes. *LWT Food Sci. Technol.* **2017**, *75*, 51–58. [CrossRef]

53. Kigel, J.; Rosenthal, L.; Fait, A. Seed physiology and germination of grain legumes, Grain legumes. *Handb. Plant Breed.* **2015**, *10*, 327–363.

54. Fernandez, M.L.; Berry, J.M. Nutritional evaluation of chickpea and germinated chickpea flours. *Plant Food Hum. Nutr.* **1988**, *38*, 127–134. [CrossRef] [PubMed]

55. Domínguez-Arispuro, D.M.; Cuevas-Rodríguez, E.O.; Milán-Carrillo, J.; León-López, L.; Gutiérrez-Dorado, R.; Reyes-Moreno, C. Optimal germination condition impacts on the antioxidant activity and phenolic acids profile in pigmented desi chickpea (*Cicer arietinum* L.) seeds. *J. Food. Sci. Technol.* **2018**, *55*, 638–647. [CrossRef]

56. El-Adawy, T.A. Nutritional composition and antinutritional factors of chickpeas (*Cicer arietinum* L.) undergoing different cooking methods and germination. *Plant Food Hum. Nutr.* **2002**, *57*, 83–97. [CrossRef]

57. Watanabe, S.; Uehara, M. Health Effects and Safety of Soy and Isoflavones. *Role Funct. Food Secur. Glabal Health* **2019**, 379–394. [CrossRef]

58. Cao, Z.H.; Green-Johnson, J.M.; Buckley, N.D.; Lin, Q.Y. Bioactivity of soy-based fermented foods: A review. *Biotechnol Adv.* **2019**, *37*, 223–238. [CrossRef]

59. Ju, M.; Huang, G.; Shen, X.; Zhang, Y.; Jiang, L.; Sui, X. A novel pickering emulsion produced using soy protein-anthocyanin complex nanoparticles. *Food Hydrocolloid.* **2020**, *99*, 105329. [CrossRef]

60. Miglani, H.; Sharma, S. Impact on germination time and temperature on phenolics, bioactive compounds and antioxidant activity of different coloured soybean. *Proc. Natl. Acad. Sci. India Sect. B Boil. Sci.* **2016**, *88*, 175–184. [CrossRef]

61. Megat, R.M.R.; Azrina, A.; Norhaizan, M.E. Effect of germination on total dietary fibre and total sugar in selected legumes. *Int. Food. Res. J.* **2016**, *23*, 257–261.

62. Lee, A.L.; Yu, Y.P.; Hsieh, J.F.; Kuo, M.I.; Ma, Y.S.; Lu, C.P. Effect of germination on composition profiling and antioxidant activity of the polysaccharide-protein conjugate in black soybean [*Glycine max* (L.) Merr.]. *Int. J. Biol. Macromol.* **2018**, *113*, 601–606. [CrossRef] [PubMed]

63. Chen, Y.; Chang, S.K.C. Macronutrients, Phytochemicals, and Antioxidant Activity of Soybean Sprout Germinated with or without Light Exposure. *J. Food Sci.* **2015**, *80*, S1391–S1398. [CrossRef] [PubMed]

64. Joshi, P.; Varma, K. Effect of germination and dehulling on the nutritive value of soybean. *Nutr. Food Sci.* **2016**, *46*, 595–603. [CrossRef]

65. Maetens, E.; Hettiarachchy, N.; Dewettinck, K.; Horax, R.; Moens, K.; Moseley, D.O. Physicochemical and nutritional properties of a healthy snack chip developed from germinated soybeans. *LWT Food Sci. Technol.* **2017**, *84*, 505–510. [CrossRef]

66. Koeberl, M.; Sharp, M.F.; Tian, R.; Buddhadasa, S.; Clarke, D.; Roberts, J. Lupine allergen detecting capability and cross-reactivity of related legumes by ELISA. *Food Chem.* **2018**, *256*, 105–112. [CrossRef]

67. Cabello-Hurtado, F.; Keller, J.; Ley, J.; Sanchez-Lucas, R.; Jorrin-Novo, J.; Aïnouche, A. Proteomics for exploiting diversity of lupin seed storage proteins and their use as nutraceuticals for health and welfare. *J. Proteom.* **2016**, *143*, 57–68. [CrossRef]

68. Cruz-Chamorro, I.; Álvarez-Sánchez, N.; Millán-Linares, M.C.; Yust, M.M.; Pedroche, J.; Millán, F.; Lardone, P.J.; Carrera-Sánchez, C.; Guerrero, J.M.; Carrillo-Vico, A. Lupine protein hydrolysates decrease the inflammatory response and improve the oxidative status in human peripheral lymphocytes. *Food Res Int.* **2019**, *126*, 108585. [CrossRef]

69. Olkowski, B. Feeding high lupine based diets for broiler chickens: Effect of soybean meal substitution with yellow lupine meal at various time points of growth cycle. *Livest. Sci.* **2018**, *218*, 114–118. [CrossRef]

70. El-Adawy, T.A.; Rahma, E.H.; El-Bedawey, A.A.; Gafar, A.F. Nutritional potential and functional properties of sweet and bitter lupin seed protein isolates. *Food Chem.* **2001**, *74*, 455–462. [CrossRef]

71. López, E.P.; Goldner, M.C. Influence of storage time for the acceptability of bread formulated with lupine protein isolate and added brea gum. *LWT Food Sci. Technol.* **2015**, *64*, 1171–1178. [CrossRef]

72. Saleh, H.M.; Hassan, A.A.; Mansour, E.H.; Fahmy, H.A.; El-Fath, A.; El-Bedawey, A.E. Melatonin, phenolics content and antioxidant activity of germinated selected legumes and their fractions. *J. Saudi Soc. Agric. Sci.* **2019**, *18*, 294–301. [CrossRef]

73. Philipo, M.; Ndakidemi, P.A.; Mbega, E.R. Environmental and genotypes influence on seed iron and zinc levels of landraces and improved varieties of common bean (*Phaseolus vulgaris* L.) in Tanzania. *Ecol. Genet Genom.* **2020**, *15*, 100056. [CrossRef]

74. UCSF Medical Center. Hemoglobin and Functions of Iron. *Ucsf Heal.* 2019. Available online: https://www.ucsfhealth.org/education/hemoglobin-and-functions-of-iron (accessed on 2 May 2020).

75. Lokuruka, M.N.I. Role of zinc in human health with reference to African elderly: A review. *AJFAND* **2012**, *12*, 6646–6664.

76. Reverri, E.J.; Randolph, J.M.; Kappagoda, C.T.; Park, E.; Edirisinghe, I.; Burton-Freeman, B.M. Assessing beans as a source of intrinsic fiber on satiety in men and women with metabolic syndrome. *Appetite* **2017**, *118*, 75–81. [CrossRef]

77. Xia, Q.; Gu, N.; Liu, J.; Niu, Y.; Yu, L.L. Novel composite gels of gelatin and soluble dietary fiber from black bean coats with interpenetrating polymer networks. *Food Hydrocolloid.* **2018**, *83*, 72–78. [CrossRef]

78. Kutoš, T.; Golob, T.; Kač, M.; Plestenjak, A. Dietary fibre content of dry and processed beans. *Food Chem.* **2003**, *80*, 231–235. [CrossRef]

79. Hayat, I.; Ahmad, A.; Masud, T.; Ahmed, A.; Bashir, S. Nutritional and Health Perspectives of Beans (*Phaseolus vulgaris* L.): An Overview. *Crit. Rev. Food Sci.* **2014**, *54*, 580–592.

80. Yin, Y.; Yang, R.; Gu, Z. Organ-specific proteomic analysis of NaCl-stressed germinating soybeans. *J. Agric. Food Chem.* **2014**, *62*, 7233–7244. [CrossRef]

81. Rosa-Millán, J.; Heredia-Olea, E.; Perez-Carrillo, E.; Guajardo-Flores, S.; Serna-Saldívar, S.R.O. Effect of decortication, germination and extrusion on physicochemical and in vitro protein and starch digestion characteristics of black beans (*Phaseolus vulgaris* L.). *LWT Food Sci. Technol.* **2019**, *102*, 330–337. [CrossRef]

82. Menon, L.; Majumdar, S.D.; Ravi, U. Development and analysis of composite flour bread. *J. Food Sci. Tech.* **2015**, *52*, 4156–4165. [CrossRef]

83. Ertaș, N.; Bilgiçli, N. Effect of different debittering processes on mineral and phytic acid content of lupin (*Lupinus albus* L.) seeds. *J. Food Sci. Tech.* **2014**, *51*, 3348–3354. [CrossRef]

84. Boukid, F.; Zannini, E.; Carini, E.; Vittadini, E. Pulses for bread fortification: A necessity or a choice? *Trends Food Sci. Tech.* **2019**, *88*, 416–428. [CrossRef]

85. Patrascu, L.; Aprodu, I.; Garnai, M.; Vasilean, I. Effect of germination and fermentation on the functionality of wheat-pulses flour mixtures. *J. Biotechnol.* **2018**, *280*, S52–S53. [CrossRef]

86. Shin, D.J.; Kim, W.; Kim, Y. Physicochemical and sensory properties of soy bread made with germinated, steamed, and roasted soy flour. *Food Chem.* **2013**, *141*, 517–523. [CrossRef]

87. Rosales-Juárez, M.; Beatriz-González-Mendoza, B.; López-Guel, E.C.; Lozano-Bautista, F.; Chanona-Pérez, J.; Gutiérrez-López, G.; Farrera-Rebollo, R.; Calderón-Domínguez, G. Changes on Dough Rheological Characteristics and Bread Quality as a Result of the Addition of Germinated and Non-Germinated Soybean Flour. *Food Bioprocess. Tech.* **2008**, *1*, 152–160. [CrossRef]

88. Ribotta, P.D.; Arnulphi, S.A.; León, A.E.; Añón, M.C. Effect of soybean addition on the rheological properties and breadmaking quality of wheat flour. *J. Sci. Food Agric.* **2005**, *85*, 1889–1896. [CrossRef]

89. Labat, E.; Rouau, X.; Morel, M.H. Effect of Flour Water-Extractable Pentosans on Molecular Associations in Gluten During Mixing. *LWT Food Sci. Technol.* **2002**, *35*, 185–189. [CrossRef]

90. Morard, M.M.; Leung, H.K.; Hsu, D.L. și Finney, P.L. Effect of Germination on Physicochemical and Bread-Baking Properties of Yellow Pea, Lentil, and Faba Bean Flours and Starches. *Cereal Chem.* **1980**, *57*, 390–396.

91. Aprodu, I.; Vasilean, I.; Muntenită, C. și Patrasu, L. Impact of broad beans addition on rheological and thermal properties of wheat flour based sourdoughs. *Food Chem.* **2019**, *293*, 520–528. [CrossRef]

92. Guardianelli, L.M.; Salinas, M.V.; Puppo, M.C. Hydration and rheological properties of amaranth-wheat flour dough: Influence of germination of amaranth seeds. *Food Hydrocolloid.* **2019**, *97*, 105242. [CrossRef]

93. Bojňanská, T.; Šmitalová, J. Impact of amaranth (amaranth sp.) on technological quality of bakery products during frozen storage. *JMBFS* **2014**, *3*, 187–189.

94. Houben, A.; Götz, H.; Mitzscherling, M.; Becker, T. Modification of the rheological behavior of amaranth (*Amaranthus hypochondriacus*) dough. *J. Cereal Sci.* **2010**, *51*, 350–356. [CrossRef]

95. Poudel, R.; Finnie, S.; Rose, D.J. Effects of wheat kernel germination time and drying temperature on compositional and end-use properties of the resulting whole wheat flour. *J. Cereal Sci.* **2019**, *86*, 33–40. [CrossRef]

96. Boukid, F.; Prandi, B.; Vittadini, E.; Francia, E.; Sforza, S. Tracking celiac disease-triggering peptides and whole wheat flour quality as function of germination kinetics. *Food Res. Int.* **2018**, *112*, 345–352. [CrossRef]

97. Millar, K.A.; Barry-Ryan, C.; Burke, R.; McCarthy, S.; Gallagher, E. Dough properties and baking characteristics of white bread, as affected by addition of raw, germinated and toasted pea flour. *Innov. Food Sci. Emerg.* **2019**, *56*, 102189. [CrossRef]

98. Ouazib, M.; Garzon, R.; Zaidi, F.; Rosell, C.M. Germinated, toasted and cooked chickpea as ingredients for breadmaking. *J. Food Sci. Tech.* **2016**, *53*, 2664–2672. [CrossRef] [PubMed]

99. Guardado-Félix, D.; Lazo-Vélez, M.A.; Pérez-Carrillo, E.; Panata-Saquicili, D.E.; Serna-Saldívar, S.O. Effect of partial replacement of wheat flour with sprouted chickpea flours with or without selenium on physicochemical, sensory, antioxidant and protein quality of yeast-leavened breads. *LWT Food Sci. Technol.* **2020**, *129*, 109517. [CrossRef]

100. Farnandez, M.L.; Berry, J.W. Rheological properties of flour and sensory characteristics of bread made from germinated chickpea. *Int. J. Food Sci. Tech.* **1989**, *24*, 103–110. [CrossRef]

101. Levent, H.; Bilgiçli, N.; Ertaş, N. The assessment of leavened and unleavened flat breads properties enriched with wheat germ. *Qual. Assur. Saf. Crop* **2014**, *7*, 321–326. [CrossRef]

102. Mostafa, M.M.; Rahma, E.H.; Rady, A.H. Chemical and nutritional changes in soybean during germination. *Food Chem.* **1987**, *23*, 257–275. [CrossRef]

103. Diowksz, A.; Kordialik-Bogacka, E.; Ambroziak, W. Se-enriched sprouted seeds as functional additives in sourdough fermentation. *LWT Food Sci. Technol.* **2014**, *56*, 524–528. [CrossRef]

104. López-Guel, E.C.; Lozano-Bautista, F.; Mora-Escobedo, R.; Farrera-Rebollo, R.R.; Chanona-Pérez, J.; Gutiérrez-López, G.F.; Calderón-Domínguez, G. Effect of Soybean 7S Protein Fractions, Obtained from Germinated and Nongerminated Seeds, on Dough Rheological Properties and Bread Quality. *Food Bioprocess Tech.* **2012**, *5*, 226–234. [CrossRef]

105. al Omari, D.Z.; Abdul-Hussain, S.S.; Ajo, R.Y. Germinated lupin (Lupinus albus) flour improves Arabic flat bread properties. *Qual. Assur. Saf. Crop* **2016**, *8*, 57–63. [CrossRef]

106. Obeidat, B.A.; Abdul-Hussain, S.S.; al Omari, D.Z. Effect of addition of germinated lupin flour on the physiochemical and organoleptic properties of cookies. *J. Food Process. Pres.* **2013**, *37*, 637–643. [CrossRef]

107. Abdul-Hussain, S.S.; Ajo, R.Y.; Obeidat, B.A. Acceptability and chemical composition of thick flat bread supplemented with chickpea flour and isolated soy protein. In Proceedings of the 5th International Congress, Flour-Bread '09, 7th Croatian Congress of Cereal Technologists, Opatija, Croatia, 21–23 October 2009; pp. 280–287.

108. Setia, R.; Dai, Z.; Nickerson, M.T.; Sopiwnykb, E.; Malcolmsonc, L.; Ai, Y. Impacts of short-term germination on the chemical compositions, technological characteristics and nutritional quality of yellow pea and faba bean flours. *Food Res. Int.* **2019**, *122*, 263–272. [CrossRef] [PubMed]

109. Marti, A.; Cardone, G.; Ambrogina-Pagani, M.; Casiraghi, M.C. Flour from sprouted wheat as a new ingredient in bread-making. *LWT Food Sci. Technol.* **2018**, *89*, 237–243. [CrossRef]

110. Almeida, E.L.; Chang, Y.K.; Steel, C.J. Dietary fiber sources in bread: Influence on technological quality. *LWT Food Sci. Technol.* **2013**, *50*, 545–553. [CrossRef]

111. Ragaee, S.; Guzar, I.; Dhull, N.; Seetharaman, K. Effects of fiber addition on antioxidant capacity and nutritional quality of wheat bread. *LWT Food Sci. Technol.* **2011**, *44*, 2147–2153. [CrossRef]

112. Wang, J.; Rosell, C.M.; Benedito de Barber, C. Effect of the addition of different fibres on wheat dough performance and bread quality. *Food Chem.* **2002**, *79*, 221–226. [CrossRef]

113. Fendri, L.B.; Chaari, F.; Maaloul, M.; Kallel, F.; Abdelkafi, L.; Chaabouni, S.E.; Ghribi-Aydi, D. Wheat bread enrichment by pea and broad bean pods fibers: Effect on dough rheology and bread quality. *LWT Food Sci. Technol.* **2016**, *73*, 584–591. [CrossRef]

114. Xu, K.; Guo, M.; Roman, L.; Pico, J.; Martinez, M.M. Okra seed and seedless pod: Comparative study of their phenolics and carbohydrate fractions and their impact on bread-making. *Food Chem.* **2020**, *317*, 126387. [CrossRef] [PubMed]

115. Sivam, A.S.; Sun-Waterhouse, D.; Perera, C.O.; Waterhouse, G.I.N. Exploring the interactions between blackcurrant polyphenols, pectin and wheat biopolymers in model breads; a FTIR and HPLC investigation. *Food Chem.* **2012**, *131*, 802–810. [CrossRef]

116. Ning, J.; Hou, G.G.; Sun, J.; Wan, X.; Dubat, A. Effect of green tea powder on the quality attributes and antioxidant activity of whole-wheat flour pan bread. *LWT Food Sci. Technol.* **2017**, *79*, 342–348. [CrossRef]

Article

Impact of Germination and Fermentation on Rheological and Thermo-Mechanical Properties of Wheat and Triticale Flours

Iuliana Banu [1,†], Livia Patraşcu [2,†], Ina Vasilean [1], Georgiana Horincar [1] and Iuliana Aprodu [1,*]

[1] Faculty of Food Science and Enginnering, Dunarea de Jos University of Galati, 111 Domneasca Str, Galati 800201, Romania; iuliana.banu@ugal.ro (I.B.); ina.vasilean@ugal.ro (I.V.); georgiana.parfene@ugal.ro (G.H.)

[2] Cross-Border Faculty, Dunarea de Jos University of Galati, 111 Domneasca Str, Galati 800201, Romania; livia.patrascu@ugal.ro

* Correspondence: iuliana.aprodu@ugal.ro

† Authors I.B. and L.P contributed equally.

Received: 14 October 2020; Accepted: 28 October 2020; Published: 29 October 2020

Abstract: Common cereal processing through germination and fermentation usually has an important impact on the technological performance of the flours, mainly because of the activation of endogenous enzymes acting on macromolecules. The aim of the present study is to estimate the effect of germination and fermentation, using a mixture of *Lactobacillus casei* and *Kluyveromyces marxianus* subsp. *marxianus*, on the rheological properties of different wheat and triticale varieties. Moreover, the thermo-mechanical behaviour of the white wheat flour-based dough, including germinated grain flour or sourdough was also tested. Grains germination and sourdough fermentation exerted a high influence on the rheological behaviour of the flour-based suspensions. Germination affected the structure and stability of the suspensions, resulting in samples with viscous behaviour prevailing over the elastic one. The temperature ramp tests revealed that germination together with fermentation lead to higher resistance to temperature changes. In agreement with the results of the rheological investigations on rheometer, the Mixolab test performed on flour obtained from germinated grains revealed lower dough stability and protein weakening at temperature increase. On the other hand, a significant improvement of the pasting properties of the dough was obtained when adding sourdoughs to the wheat flour.

Keywords: wheat; triticale; sourdough; rheological properties; Mixolab

---

1. Introduction

Bread products obtained from refined wheat flour have relatively poor nutritional value, being low in mineral, vitamin and fibre content [1]. Recently, many studies focused on different processing techniques, aiming to enhance functional properties of wheat flour and quality of the final products. Besides fortification by addition of pseudocereals or leguminous flours to the wheat flour, fermentation and germination were proved to be simple techniques for improving the nutritive and functional properties of grains [2–4]. In particular, Sibian et al. [5] showed that germination improved functional properties of both wheat and triticale flours.

During germination, the main storage biopolymers of cereal grains, namely carbohydrates, proteins and lipids are hydrolysed to lower molecular weight compounds, because of the activity of the hydrolytic enzymes, which are inactive in raw seeds [6]. These processes result in improved digestibility, bioavailability and bioaccessibility of various nutrients and biologically active molecules. In addition, the use of sourdough in bread-making technology leads to increasing levels of numerous bioactives [3,4,7–9]. In particular, the mineral bioavailability is improved through the activity of phytate degrading enzymes, which exist in yeast and lactic acid bacteria isolated from sourdough, whereas the digestibility of starch is reduced in the presence of lactic acid produced during sourdough fermentation, resulting in low glycemic index products [1,7].

Both cereal grains germination and sourdough fermentation were reported to significantly influence the technological performance of the flours and doughs. Many studies were focused on the effect of addition of germinated legumes or pseudocereals, such as pea [10], soy [11], broadbean [12], amaranth [6] etc., on the rheological properties of the wheat flour-based dough. Anyway, there are fewer studies regarding the effect of wheat germination and/or fermentation on dough rheology. Singh et al. [13] reported that wheat germination resulted in a softened dough structure, characterized by lower values of the elastic modulus and viscous modulus. A similar effect on the rheological properties caused by sourdough addition to the wheat flour was reported by Clarke et al. [14]. They showed that sourdough addition allowed obtaining more easily extensible doughs. On the other hand, Angioloni et al. [15] reported less elastic and firmer doughs, most probably as a result of physicochemical changes occurring in the protein network induced by the addition of sourdough. The addition of sourdough lowers the pH of the dough, affecting the structure and properties of different flour components, such as gluten and starch, which influences to a high extent the rheological behaviour of the dough [15].

The aim of the present study is to investigate the impact of the changes induced by wheat and triticale germination and sourdough addition on the rheological properties and thermomechanical behaviour of the dough. In this respect, sourdough samples were obtained through fermentation of the wholemeal flour suspensions using a mixture of lactic acid bacteria and yeast.

2. Materials and Methods

2.1. Materials

Three wheat cultivars bred in Romania, namely Gabriela (G) and Trivale (T), that are winter type common wheat (*Triticum aestivum* L.), and spelt (S) (*Triticum aestivum* subsp. *spelta*) and one triticale (X *Triticosecale* Wittmack) cultivar (TT) were used in the study.

The commercial white wheat flour (moisture content of 13.18%, protein content of 9.40%, and ash content of 0.48%) was purchased from the local market (Galati, Romania). All chemicals used in the experiment were of analytical grade.

2.2. Germination of Wheat and Triticale Grain

The grain samples used in the study were first washed with tap water and then sanitised using aqueous ethanol solution (70%). After a final rinsing with distilled water, the grains were evenly distributed on sterile cotton layers, moistened and allowed to germinate for four days at 23 °C ± 1. The germinated Gabriela (gG), Trivale (gT), spelt (gS) and triticale (gTT) samples further used in the experiment were obtained by drying at 55 °C ± 2 for 48 h in a convection oven (LabTech LDO-030E, Daihan LabTech Co., LTD, Kyonggi-Do, Korea).

2.3. Flour Obtaining

The raw and germinated grain samples were milled with a laboratory mill (WZ-2, Sadkiewicz Instruments, Bydgoszcz, Poland) into flours with particles size lower than 315 µm. The obtained flours were stored in glass containers at ~4 °C for further tests. The flour samples obtained by milling the raw grains were used as control for the germination experiment.

2.4. Proximate Composition

The proximate composition of the flour samples was determined as follows: the moisture content with the AACC 44-15.02 method [16]; the crude protein content through the Kjeldahl method (AACC 46-11.02;

Raypa Trade, R. Espinar, S.L., Barcelona, Spain) [16], using a nitrogen conversion factor of 5.83 and ash content using AACC method 08-01 [16].

2.5. Sourdoughs Preparation

Inoculum consisted of a mixture *of Lactobacillus casei* (Nutrish®), and yeast *Kluyveromyces marxianus* subsp. *marxianus* (LAF4) from Chr Hansen, Brasov, Romania. Producer recommendations were followed to obtain an inoculum size of 10^8 cfu/100 g sourdough. The fermentation was initiated by mixing the wholemeal flour obtained from raw and germinated grains with tap water and inoculum, such as to get a final dough yield (100 × mass of dough/mass of flour) of 300. The fermentation of the sourdoughs was carried out in large beakers covered with aluminium foil at 40 °C for 20 h, using a laboratory incubator (Pol-Eko Aparatura, Wodzisław Śląski, Poland).

For each type of flour used in the fermentation experiment, a control sample consisting of a flour suspension of the same concentration as in case of sourdoughs preparations, was prepared without inoculum addition.

2.6. Rheological Properties

The rheological properties of the control suspensions prepared using flours from raw (G, T, S and TT) or germinated grains (gG, gT, gS and gTT) and of the corresponding sourdough samples (G_SD, T_SD, S_SD and TT_SD prepared out of G, T, S and TT flours, and gG_SD, gT_SD, gS_SD and gTT_SD prepared out of gG, gT, gS and gTT flours) were measured using a controlled-stress rheometer (AR2000ex, TA Instruments, Ltd, New Castle, DE, USA) equipped with a Peltier temperature control system, and a plate–plate geometry (diameter of 40 mm). The closing gap was set to 2 mm and a solvent trap was used to avoid moisture loss during rheological tests.

For each sample the linear viscoelastic region (LVR) yield point was first identified through running stress sweep tests under oscillatory flow in small amplitude conditions, by increasing the stress values from 0.01 to 100 Pa, at an oscillation frequency of 1 Hz. The obtained data were used to determine the stress domain where the structure of the sample remained unaffected, known as the LVR. The intersection of the storage and loss moduli (G′ and G″), corresponding to the minimal stress value required for the proper beginning of flow, was also estimated [17]. Further dynamic frequency sweep tests were performed within LVR, at a constant oscillatory stress of 0.1 Pa. The working temperature was set at 20 °C for the stress and frequency sweep tests.

In order to study the structure changes induced by heating in the suspensions and sourdoughs, temperature ramp tests were carried out in quasi-static conditions (frequency of 1Hz, oscillatory stress of 0.1 Pa) by increasing the temperature from 20 to 90 °C by 1.5 °C/min. The samples were further maintained at 90 °C for 10 min. The temperature-induced gelatinization of the studied samples was estimated using the procedure proposed by Ding et al. [18]. The gelatinization temperature range was determined by calculating the first derivative of the third order equation used to fit the G′ vs. temperature values, in the area where inflection was observed. Only equations with r^2 approaching unity were considered proper for estimating the gelatinization temperature domain.

2.7. Thermo-Mechanical Properties

In order to estimate the influence of germination and fermentation processes on the thermo-mechanical properties of the wheat flour-based doughs, the Mixolab Chopin (Tripette & Renaud Chopin, Villeneuve La Garenne, France) was used.

The impact of germination was determined by performing measurements on mixtures consisting of 15% wholemeal flour obtained by milling the investigated raw and germinated grains and 85% white wheat flour. The flours obtained from raw (G, T, S and TT) and germinated (gG, gT, gS and gTT) grains were mixed with the white wheat flour directly into the Mixolab tank.

In the second part of the experiment the white wheat flour (85%) was supplemented with sourdoughs, prepared with raw and germinated wheat and triticale flours, in such ratio to provide the same amount of wholemeal flour (15%) as in the case of the first part of the experiment, when the influence of grains germination was tested.

The standard Chopin+ protocol was employed for the empirical rheological measurements, and the typical Mixolab parameters were registered for estimating the properties of the main flour constituents during mixing and thermal treatment. The amount of water added to the mixtures in the Mixolab tank was decided such as to reach the standard torque value of 1.1 Nm (C1) at initial mixing. In case of the samples including sourdough, the amount of added water was accordingly reduced. The following torque values were used to gather information on the impact of germination and fermentation on the thermo-mechanical performance of main components of the flours: C2 (Nm) which is a measure of protein weakening while applying double constraints to the dough, consisting of mechanical work and temperature raise, C3 (Nm) which gives indication on starch gelatinization, C4 (Nm) related to the stability of the starch gel at high temperature and C5 (Nm) which provides evidence on starch retrogradation in the cooling stage. Moreover, the cocking stability (C34) was estimated as the difference between C3 and C4.

2.8. Statistical Analysis

Two independent germination and fermentation experiments were conducted and the measurements were performed at least in triplicate. The average values are reported together with standard deviations. Analysis of variance (ANOVA) was used to identify significant differences among results. The one-way ANOVA and Tukey's test with a 95% confidence interval were applied using the Minitab 18 software. Regression analysis was performed to identify any potential relationships between the studied properties of the flour samples.

3. Results and Discussion

3.1. Influence of Germination and Fermentation on the Rheological Properties of Wheat and Triticale Flours

Wheat and triticale samples were subjected to germination for four days and were then dried to 93.09–93.91% d.w. The germination process caused a slight increase in the protein content (from 15.04% to 15.75% in case of G, from 13.01% to 13.81% in case of T, from 12.52% to 13.80% in case of S and from 12.42% to 12.80% in case of TT), and had no significant influence on the ash content, which varied between 1.30% (for G) and 1.73% (for TT).

Rheological behaviour of flour suspensions and sourdoughs over the oscillatory stress sweep test is presented in Figure 1. Analysing the evolution of the viscoelastic moduli as a function of increasing stress, one can see that, regardless of the germination status, the control samples of all studied grain varieties presented the highest G′ and G″ values. A strong solid-like behaviour was noticed for all control samples up to stress values in the 0.7–7.9 Pa range, depending on the flour type. In the LVR, delta values were lower than 24°. Germination affected the structure and stability of the suspension, resulting in formulations having poor rheological quality, with viscous behaviour prevailing over the elastic one. When compared to the raw grains, flour suspensions obtained from germinated samples presented significantly lower values of G′ and G″ moduli, with narrower but still present linear viscoelastic regions in case of G and TT (Figure 1a,d). For the other two germinated wheat varieties, flour suspensions entered directly into the transition phase toward flowing, with rapid decrease of both storage and loss moduli. The control samples yielded much higher stress values in comparison to flour suspensions obtained from the corresponding germinated grains (Table 1). The maximum yield stress value was obtained in the case of G variety, which also presented the highest protein content (15.04%). Significant correlation was established between protein content and yield stress (r^2 of 0.9009) in case of non-germinated wheat.

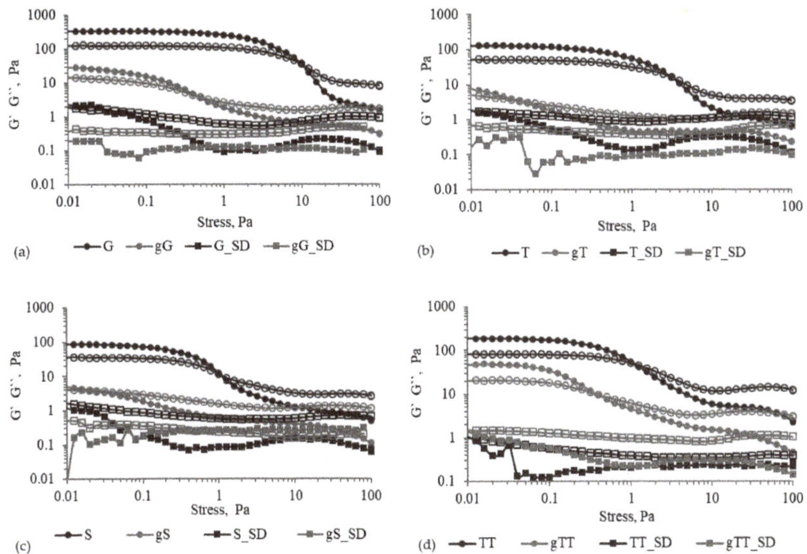

Figure 1. The evolution of the storage and loss moduli (G′ and G″) during the oscillatory stress sweep test for the suspensions and sourdoughs prepared with the wholemeal flours obtained by milling Gabriela (**a**), Trivale (**b**), Spelt (**c**) and Triticale (**d**) grains. Tests were performed on suspensions prepared using flours from raw (G, T, S and TT) or germinated grains (gG, gT, gS and gTT) and on corresponding sourdough samples (G_SD, T_SD, S_SD and TT_SD prepared out of G, T, S and TT flours and gG_SD, gT_SD, gS_SD and gTT_SD prepared out of gG, gT, gS and gTT flours). Full and open symbols are used for presenting the evolution of G′ and G″, respectively.

Table 1. Yield stress, corresponding to the crossover point of storage modulus and loss modulus, for wheat (Gabriela, Trivale, Spelt) and triticale-based suspensions and sourdoughs.

Sample		Yield Stress, Pa	
		Suspension	Sourdough
Gabriela	Raw	7.96 ± 4.39	0.03 ± 0.00
	Germinated	0.37 ± 0.19	-
Trivale	Raw	3.78 ± 0.00	-
	Germinated	0.04 ± 0.00	-
Spelt	Raw	0.71 ± 0.40	-
	Germinated	0.02 ± 0.00	-
Triticale	Raw	1.08 ± 0.16	-
	Germinated	0.23 ± 0.21	-

The sourdoughs obtained from raw and germinated wheat and triticale samples displayed viscous behaviour (G″ > G′) from the beginning of the test. Song and Zheng [19] presented the involvement of important molecules from flours, such as starch and gluten proteins, in defining the rheological behaviour of the dough. Among the main components of the flours, starch was shown to exhibit mainly non-linear viscoelastic response [20,21], while gluten was reported to be responsible for dough elasticity, being highly resistant to stress sweep [17,20]. Our results indicated that germination and fermentation interfered with gluten performance, most probably as a result of increased protease activity. The decrease of the molecular weight of proteins and peptides, as well as the potential discontinuities in the gluten network caused by the protease assisted cleavage of the peptide bonds affected dough resistance to the applied stress.

The rheological behaviour of the studied flour suspensions and sourdoughs under temperature constraint is shown in Figure 2. Higher G′ values were registered for the control samples obtained from raw grain flours at low temperatures, suggesting that the changes occurring in the grains during germination and in the flour matrix during fermentation, affected the rheological stability of the suspensions and sourdoughs.

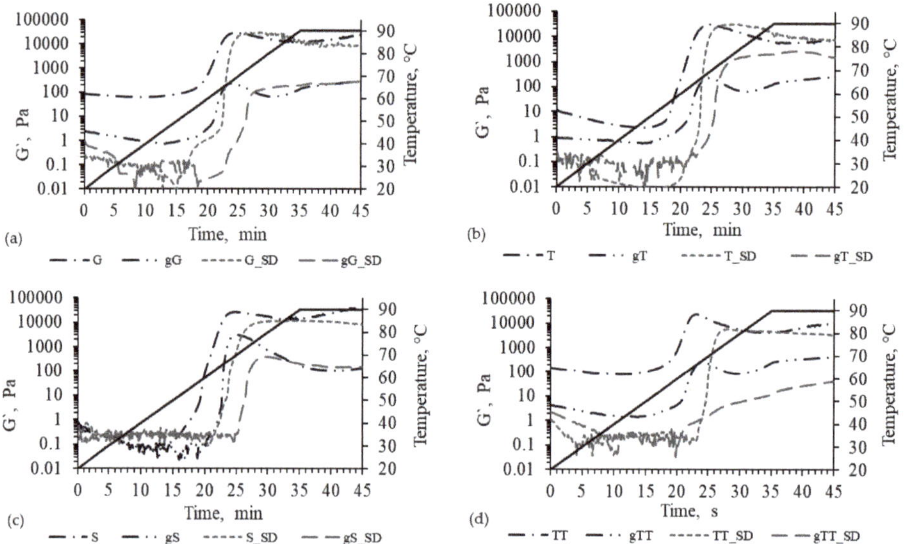

Figure 2. The evolution of the storage modulus of the suspensions and sourdoughs prepared with the wholemeal flours obtained by milling Gabriela (**a**), Trivale (**b**), Spelt (**c**) and Triticale (**d**) grains, during the temperature ramp test. Tests were performed on suspensions prepared using flours from raw (G, T, S and TT) or germinated grains (gG, gT, gS and gTT) and on corresponding sourdough samples (G_SD, T_SD, S_SD and TT_SD prepared out of G, T, S and TT flours, and gG_SD, gT_SD, gS_SD and gTT_SD prepared out of gG, gT, gS and gTT flours). The evolution of the temperature during the rheological test is represented with black full line.

Regardless of the investigated samples, in the first part of the heating stage, a slight decreasing trend of G′ was observed (Figure 2). This phenomenon could be associated to the heat-induced weakening of the proteins, and to a lower extent to the starch swelling process which tends to compensate the decrease of G′ values, caused by gradual proteins weakening. Gänzle et al. [22] stated that degradation of gluten protein structures in sourdoughs affects the viscoelastic properties of the final dough, which explains the lower G′ values in case of sourdoughs. Except for spelt, at low temperature all sourdough samples had significant lower G′ values compared to the corresponding suspensions. However, after completion of the gelling process, at 80–90 °C, no significant differences were observed between suspensions and sourdoughs. The curves showing the evolution of the elastic modulus (G′) were mainly grouped as a function of germination. As a result of the germination process, lower G′ values were obtained after achieving the gelling point (Figure 2). This behaviour might be attributed to starch, which competes with proteins for binding the water molecules available in the system. Over the upper limit of protein denaturation temperature range, the starch swelling prevails, further grabbing all available water. Jane et al. [23] stated that the amylose-to-amylopectin ratio highly influences the starch functional properties. Germination was found to increase the amount of amylopectin within starch granules [23,24]. Moreover, the high amount of dextrose and maltose formed through starch hydrolysis during germination as a result of increased amylolytic activity will interfere with gel formation. In this respect, Chinma et al. [25] observed that germination determined the increase of the least gelling concentration of *Cyperus esculentus* flours from 12 to 20%. Therefore, higher dosage of germinated flour is needed to achieve a dough consistency similar to the non-germinated wheat flour.

In order to study the temperature-induced gelatinization of the suspensions and sourdoughs, the experimental results in the inflexion area of G′ vs. temperature curves (Figure 2) were fitted using a third-order equation, and the gelatinization temperature domains were estimated by calculating the first derivative of these equations. Analysing the results presented in Figure 3 one can observe that germination together with fermentation determined the increase of the gelatinization temperature domains. Our results comply with Frias et al. [24], who showed that germination altered lentil starch granule surface, leading to higher resistance to temperature changes.

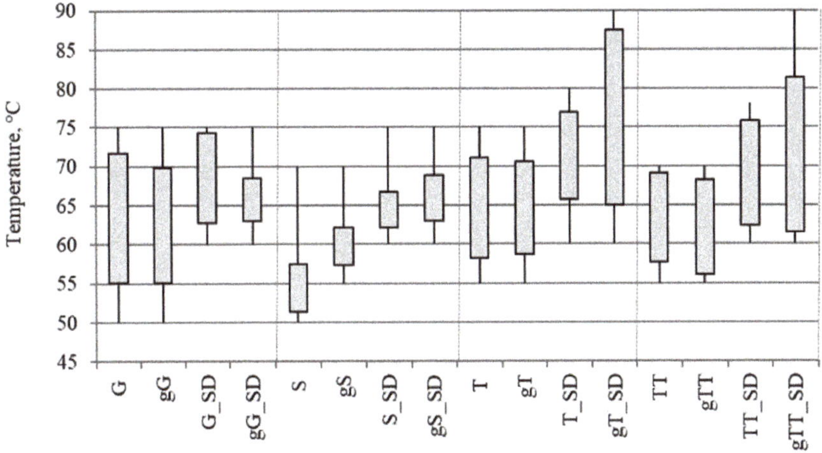

Figure 3. Temperature domains corresponding to the gelling process for the studied suspensions and sourdoughs during the temperature ramp test. The analysed temperature ranges corresponding to the inflection area of G′ vs. t curves are represented with lines, whereas the estimated gelling temperature domains are indicated as columns. Tests were performed on suspensions prepared using flours from raw (G, T, S and TT) or germinated grains (gG, gT, gS and gTT) and on corresponding sourdough samples (G_SD, T_SD, S_SD and TT_SD prepared out of G, T, S and TT flours, and gG_SD, gT_SD, gS_SD and gTT_SD prepared out of gG, gT, gS and gTT flours).

3.2. Influence of Germination and Fermentation on the Thermo-Mechanical Properties of Doughs

The impact of grains germination and sourdough fermentation on the bread making properties of the white wheat flour, was estimated by monitoring the thermo-mechanical behaviour of the doughs by means of Mixolab device. The main Mixolab parameters registered when supplementing the white wheat flour with 15% wholemeal flours from raw or grains or corresponding sourdoughs are presented in Table 2.

3.2.1. Influence of Grain Germination

The addition of 15% wholemeal flour obtained through grinding the germinated G, T, S and TT grain samples to the white wheat flour resulted in significant shortening of dough stability (DS) compared to the corresponding raw samples. The DS values ranged from 8.82 to 9.93 min in case of the doughs prepared with raw grain flour, and from 2.68 to 4.03 min for the samples with germinated grain flour (Table 2). Our results are in agreement with Rosales-Juarez et al. [11] who tested the influence of germinated and non-germinated soybean flour on the rheological behaviour of the doughs and, regardless of the addition level, reported shorter stability of the wheat flour-based dough samples supplemented with germinated soy flour. On the other hand, Sadowska et al. [10] reported dough stability prolongation for the wheat flour-based samples, including different percentages of germinated pea flour, compared to the samples with raw pea flour. When the dough was subjected to the dual constraint of mechanical shear and temperature, the consistency decreased to a minimum torque (C2) varying with the flour mixture and type of processing (Table 2). Higher C2 values were registered for dough samples with raw grain flour addition (0.42–0.47 Nm), compared to the samples prepared with corresponding germinated grain flour (0.12–0.17 Nm). The decrease of both C2 and DS values observed when testing flour from germinated grains, can be explained by the advanced disruption of the protein network caused by peptidase formed during germination. The protein aggregates stabilised through the disulfide or dityrosine linking and hydrogen bonds are essentially responsible for dough consistency and stability [26]. According to Bigiarini et al. [27], gliadins are degraded during germination into small peptides, the enzymes mainly responsible for this activity being cysteine proteinase and carboxypeptidases, which are highly active during wheat germination.

Table 2. Effect of the addition of flours from raw (G, T, S and TT) or germinated grains (gG, gT, gS and gTT) and of corresponding sourdough samples (G_SD, T_SD, S_SD and TT_SD prepared out of G, T, S and TT flours, and gG_SD, gT_SD, gS_SD and gTT_SD prepared out of gG, gT, gS and gTT flours). On the thermo-mechanical properties of the white wheat flour (WF) dough. Measurements were performed in triplicate by means of Mixolab using the Chopin+ protocol.

Samples	DS, min	C2, Nm	C3, Nm	C4, Nm	C5, Nm
WF + G	9.93 ± 0.12 [a]	0.46 ± 0.01 [a]	1.99 ± 0.01 [a]	1.76 ± 0.01 [a]	2.64 ± 0.02 [a]
WF + gG	3.23 ± 0.10 [c]	0.17 ± 0.01 [c]	0.51 ± 0.01 [c]	0.05 ± 0.00 [c]	0.02 ± 0.00 [c]
WF + G_SD	5.42 ± 0.10 [b]	0.31 ± 0.01 [b]	1.85 ± 0.01 [b]	1.74 ± 0.01 [a]	2.64 ± 0.01 [a]
WF + Gg_SD	5.32 ± 0.10 [b]	0.12 ± 0.01 [d]	0.40 ± 0.01 [d]	0.35 ± 0.01 [b]	1.09 ± 0.02 [b]
WF + T	8.83 ± 0.09 [a]	0.47 ± 0.01 [a]	2.20 ± 0.01 [a]	1.88 ± 0.01 [a]	2.83 ± 0.01 [a]
WF + gT	2.68 ± 0.10 [d]	0.16 ± 0.01 [c]	0.42 ± 0.01 [d]	0.03 ± 0.00 [d]	0.01 ± 0.00 [d]
WF + T_SD	5.23 ± 0.11 [b]	0.32 ± 0.01 [b]	1.82 ± 0.01 [b]	1.65 ± 0.01 [b]	2.56 ± 0.02 [b]
WF + Tg_SD	4.68 ± 0.10 [c]	0.12 ± 0.01 [d]	0.55 ± 0.01 [c]	0.18 ± 0.01 [c]	0.18 ± 0.01 [c]
WF + S	9.33 ± 0.10 [a]	0.43 ± 0.01 [a]	1.88 ± 0.01 [a]	1.52 ± 0.01 [b]	2.26 ± 0.02 [b]
WF + gS	4.03 ± 0.11 [d]	0.14 ± 0.01 [c]	0.94 ± 0.01 [c]	0.26 ± 0.01 [d]	0.32 ± 0.01 [d]
WF + S_SD	6.10 ± 0.09 [b]	0.37 ± 0.01 [b]	1.84 ± 0.01 [b]	1.80 ± 0.01 [a]	2.72 ± 0.02 [a]
WF + Sg_SD	4.83 ± 0.09 [c]	0.12 ± 0.01 [c]	0.96 ± 0.01 [c]	0.53 ± 0.01 [c]	0.75 ± 0.01 [c]
WF + TT	8.82 ± 0.11 [a]	0.42 ± 0.01 [a]	1.84 ± 0.01 [a]	1.33 ± 0.01 [b]	2.00 ± 0.02 [b]
WF + gTT	3.12 ± 0.09 [d]	0.12 ± 0.01 [c]	0.31 ± 0.01 [c]	0.01 ± 0.00 [d]	0.00 ± 0.00 [d]
WF + TT_SD	6.32 ± 0.10 [b]	0.33 ± 0.01 [b]	1.77 ± 0.01 [b]	1.50 ± 0.01 [a]	2.25 ± 0.02 [a]
WF + TTg_SD	5.32 ± 0.12 [c]	0.09 ± 0.01 [d]	0.29 ± 0.01 [c]	0.06 ± 0.00 [c]	0.10 ± 0.00 [c]

[a–d]: average values in columns signed by the same superscript letter are not statistically different ($p = 95\%$).

As shown in Table 2 and Figure 4, the use of gG, gT, gS and gTT flours resulted in changes of the pasting properties compared to the corresponding dough samples with raw G, T, S and TT flour (controls). It should be noted that addition of raw TT and S flour had a higher influence on pasting properties compared to G and T. The highest values of the maximum torque during heating (C3) were 1.99 and 2.20 Nm in case of the samples with raw G and T flour addition. Weaker starch gel stability (C4) was registered for dough samples with S and TT, most probably because of the high amylase activity in these samples, compared to G and T. Triticale and spelt tend to have higher α-amylase activity compared to wheat, depending on the cropping environment [28,29].

Different tendency of the Mixolab curves was registered when germinated grain flours were added to the white flour to obtain dough samples (Figure 4b). Germination induced the increase of amylase activity and consequently starch hydrolysis became very evident after the gelatinization. Therefore, significantly lower C4 values were registered for the dough samples including germinated grain flours. The cooking stability (C34) ranged from 0.3 Nm to 0.68 Nm, suggesting that the amylase activity continued even during the Chopin+ stage when the dough is heated at 90 °C. According to

Dubat and Boinot [30], the (C34) values increased from 0.11 Nm to 0.46 Nm when adding exogenous maltogenic amylases to the wheat flour. Similar reduction was observed for the torque values, which gives indication about starch retrogradation (C5). When studying the influence of brown rice and soy germination on the bread-making performance of the flours, Cornejo and Rosell [31] and Patraşcu et al. [32] registered lower torque values while heating the dough up to 90 °C and at further cooling down to 50 °C. They explained this change in the thermo-mechanical behaviour of the dough through the extensive hydrolysis of the starch by the amylases-activated during germination.

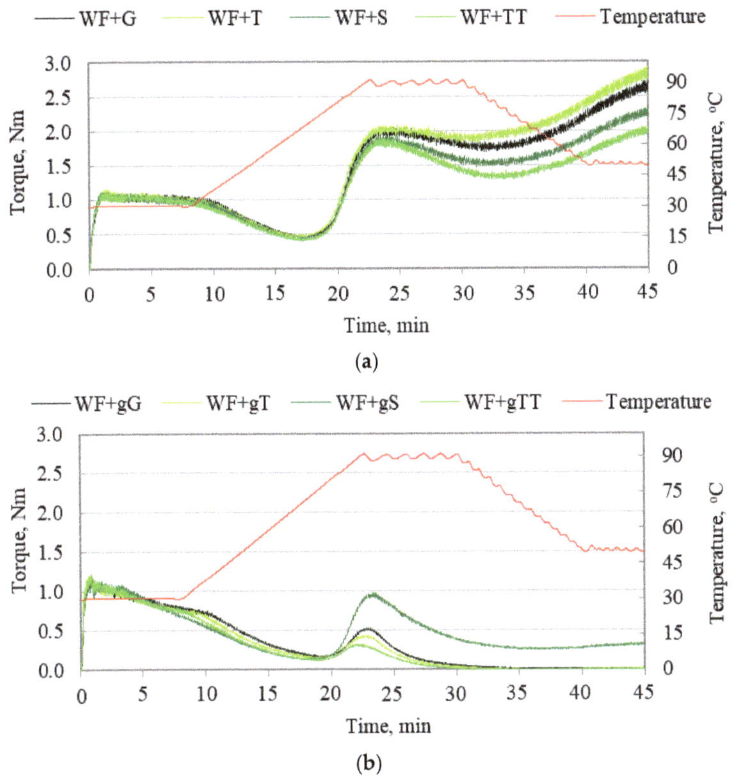

Figure 4. Mixolab curves of the white wheat flour (WF) supplemented with 15% wholemeal flours obtained by milling (**a**) raw (G, T, S and TT) and (**b**) germinated grains (gG, gT, gS and gTT).

3.2.2. Influence of Sourdough Addition

Regardless of the investigated grain sample, the addition of sourdoughs prepared with raw grain flours (G_SD, T_SD, S_SD and TT_SD) resulted in doughs with lower DS and C2 values (Table 2). The sourdough fermentation caused the DS shortening from 8.82–9.93 min to 5.23–6.32 min, and the C2 diminishing from 0.42–0.47 Nm to 0.31–0.37 Nm. On the other hand, the addition of sourdough prepared with germinated grain flour resulted in higher DS and lower C2 values compared to the corresponding samples with raw grain flour. This trend registered for the dough samples including sourdoughs might be the result of the low pH value. In agreement with Bottari et al. [33], the optimal pH for gliadin hydrolysis is 4.25. Moreover, Banu and Aprodu [26] and Clarke et al. [34] showed that dough behaviour during mixing at 30 °C and during temperature increase from 30 to 55–60 °C is influenced by pH, affecting both DS and C2 values. Similar results for DS and C2 were obtained when preparing dough with raw and germinated soybean flour added to the wheat flour [32]. The authors explained the reduction of C2 values by protein weakening during dough mixing and heating by 4 °C/min, but also by the action of proteolytic enzymes that are activated during germination.

The dough samples prepared with sourdoughs made of raw G, S and T flours displayed similar pasting properties, the torque values being significantly higher with respect to the TT-based sample (Figure 5). For instance, the values of C5 that defines starch retrogradation, ranged from 2.25 Nm, corresponding to the sample with TT-based sourdough, to 2.72 Nm in case of the sample with S, suggesting different optimal end-use for each sample. According to Dubat and Boinot [30] the flours suitable for bread making, brioche and biscuits have C5 of 2.05, 2.43 and 2.75 Nm, respectively.

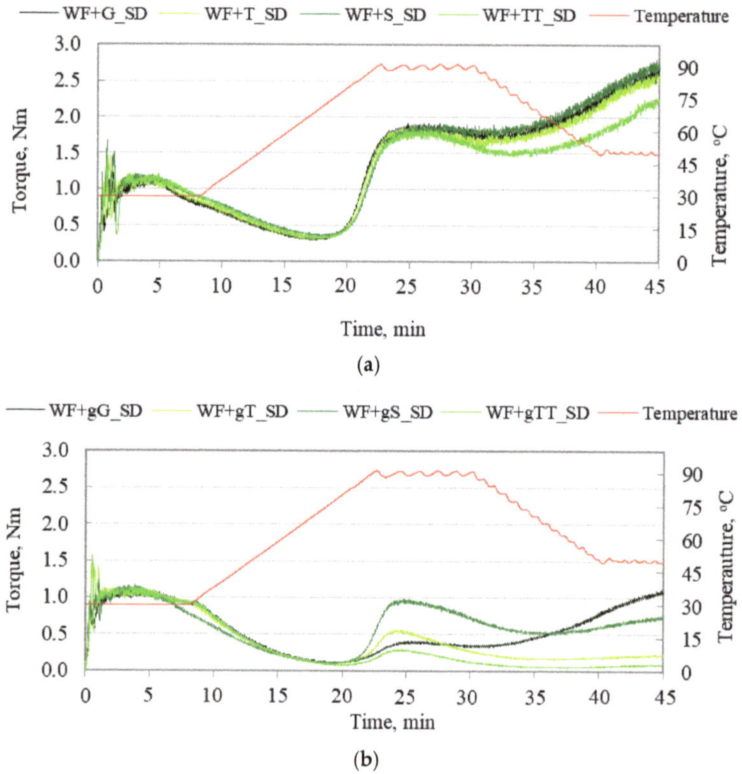

Figure 5. Mixolab curves of the white wheat flour (WF) supplemented with sourdoughs prepared with flours from (**a**) raw (G_SD, T_SD, S_SD and TT_SD) and (**b**) germinated grains (gG_SD, gT_SD, gS_SD and gTT_SD).

Grains germination improved starch gelatinization, starch stability and starch retrogradation properties of the dough samples prepared with sourdough (Table 2). The decrease of the pH during sourdough fermentation limited the activity of α-amylase, and consequently improved starch stability. The most significant improvement of the pasting properties was observed in case of the dough sample including sourdough prepared with germinated G flour.

4. Conclusions

Processing wheat and triticale through germination and further sourdough fermentation affected the rheological and thermo-mechanical properties of the flours. Regardless of the investigated grain, a narrower viscoelastic domain was obtained for samples subjected to germination and/or fermentation. The partial hydrolysis of the starch and proteins during germination as a result of increased hydrolytic activity of the enzymes interfered with gel formation at thermal treatment. The rheological tests indicated that combining grains germination and sourdough fermentation results in increased gelatinization temperature domains. The addition of 15% flour obtained from germinated wheat or triticale significantly affected protein weakening, starch gelatinization, the stability of the starch gel at high temperature and starch retrogradation, causing significant shortening of dough stability.

Author Contributions: Conceptualisation, I.A. and I.B.; methodology, L.P. and I.B.; software, I.A., L.P. and I.B.; validation, I.V. and G.H.; formal analysis, L.P., I.B. and I.A.; investigation, I.B., L.P., I.V. and G.H.; resources, I.A. and G.H.; writing—original draft preparation, I.B. and L.P.; writing—review and editing, I.A.; supervision, I.A.; project administration, I.A.; funding acquisition, I.A. and G.H. All authors have read and agreed to the published version of the manuscript.

Funding: This work was funded by a grant of the Romanian National Authority for Scientific Research and Innovation, CNCS–UEFISCDI, project number PN-II-RU-TE-2014-4-0618.

Acknowledgments: This work was supported by the project "EXPERT", financed by the Romanian Ministry of Research and Innovation, Contract no. 14PFE/17.10.2018. The Integrated Center for Research, Expertise, and Technological Transfer in Food Industry is acknowledged for providing technical support.

Conflicts of Interest: The authors declare no conflict of interest.

References

1. Dewettinck, K.; Van Bockstaele, F.; Kühne, B.; Van de Walle, D.; Courtens, T.M.; Gellynck, X. Nutritional value of bread: Influence of processing, food interaction and consumer perception. *J. Cereal Sci.* **2008**, *48*, 243–257. [CrossRef]

2. Katina, K.; Liukkonen, K.H.; Kaukovirta-Norja, A.; Adlercreutz, H.; Heinonen, S.M.; Lampi, A.M.; Pihlava, J.-M.; Poutanen, K. Fermentation-induced changes in the nutritional value of native or germinated rye. *J. Cereal Sci.* **2007**, *46*, 348–355. [CrossRef]

3. Banu, I.; Vasilean, I.; Aprodu, I. Effect of lactic fermentation on antioxidant capacity of rye sourdough and bread. *Food Sci. Technol. Res.* **2010**, *16*, 571–576. [CrossRef]

4. Patrascu, L.; Vasilean, I.; Turtoi, M.; Garnai, M.; Aprodu, I. Pulse germination as tool for modulating their functionality in wheat flour sourdoughs. *Qual. Assur. Saf. Crop.* **2019**, *11*, 269–282. [CrossRef]

5. Sibian, M.S.; Saxena, D.C.; Riar, C.S. Effect of germination on chemical, functional and nutritional characteristics of wheat, brown rice and triticale: A comparative study. *J. Sci. Food Agric.* **2017**, *97*, 4643–4651. [CrossRef] [PubMed]

6. Guardianelli, L.M.; Salinas, M.V.; Puppo, M.C. Hydration and rheological properties of amaranth-wheat flour dough: Influence of germination of amaranth seeds. *Food Hydrocoll.* **2019**, *97*, 105242. [CrossRef]

7. Katina, K.; Arendt, E.; Liukkonen, K.H.; Autio, K.; Flander, L.; Poutanen, K. Potential of sourdough for healthier cereal products. *Trends Food Sci. Technol.* **2005**, *16*, 104–112. [CrossRef]

8. Gobbetti, M.; De Angelis, M.; Di Cagno, R.; Calasso, M.; Archetti, G.; Rizzello, C.G. Novel insights on the functional/nutritional features of the sourdough fermentation. *Int. J. Food Microbiol.* **2019**, *302*, 103–113. [CrossRef]

9. Gobbetti, M.; De Angelis, M.; Di Cagno, R.; Polo, A.; Rizzello, C.G. The sourdough fermentation is the powerful process to exploit the potential of legumes, pseudo-cereals and milling by-products in baking industry. *Crit. Rev. Food Sci.* **2020**, *60*, 2158–2173. [CrossRef]

10. Sadowska, J.; Blaszczak, W.; Fornal, J.; Vidal-Valverde, C.; Frias, J. Changes of wheat dough and bread quality and structure as a result of germinated pea flour addition. *Eur. Food Res. Technol.* **2003**, *216*, 46–50. [CrossRef]

11. Rosales-Juarez, M.; Gonzalez-Mendoza, B.; Lopez-Guel, E.C.; Lozano-Bautista, F.; Chanona-Perez, J.; Gutierrez-Lopez, G.; Farrera-Rebollo, R.; Calderon-Domínguez, G. Changes on dough rheological characteristics and bread quality as a result of the addition of germinated and non-germinated soybean flour. *Food Bioprocess Technol.* **2008**, *1*, 152–160. [CrossRef]

12. Aprodu, I.; Vasilean, I.; Muntenită, C.; Patrascu, L. Impact of broad beans addition on rheological and thermal properties of wheat flour based sourdoughs. *Food Chem.* **2019**, *293*, 520–528. [CrossRef] [PubMed]

13. Singh, H.; Singh, N.; Kaur, L.; Saxena, S.K. Effect of sprouting conditions on functional and dynamic rheological properties of wheat. *J. Food Eng.* **2001**, *47*, 23–29. [CrossRef]

14. Clarke, C.I.; Schober, T.J.; Arendt, E.K. Effect of single strain and traditional mixed strain starter cultures on rheological properties of wheat dough and on bread quality. *Cereal Chem.* **2002**, *79*, 640–647. [CrossRef]

15. Angioloni, A.; Romani, S.; Pinnavaia, G.G.; Dalla Rosa, M. Characteristics of bread making doughs: Influence of sourdough fermentation on the fundamental rheological properties. *Eur. Food Res. Technol.* **2006**, *222*, 54–57. [CrossRef]

16. AACC International. *Approved Methods of Analysis*, 11th ed.; Methods 44-15.02, 46-11.02, 08-01; American Association of Cereal Chemists International: St. Paul, MN, USA, 2000.

17. Patraşcu, L.; Banu, I.; Vasilean, I.; Aprodu, I. Effect of gluten, egg and soy proteins on the rheological and thermo-mechanical properties of wholegrain rice flour. *Food Sci. Technol. Int.* **2016**, *23*, 142–155. [CrossRef] [PubMed]

18. Ding, Y.; Liu, R.; Rong, J.; Liu, Y.; Zhao, S.; Xiong, S. Rheological behavior of heat-induced actomyosin gels from yellowcheek carp and grass carp. *Eur. Food Res. Technol.* **2012**, *235*, 245–251. [CrossRef]

19. Song, Y.; Zheng, Q. Dynamic rheological properties of wheat flour dough and proteins. *Trends Food Sci. Technol.* **2007**, *18*, 132–138. [CrossRef]

20. Khatkar, B.S.; Fido, R.J.; Tatham, A.S.; Schofield, J.D. Functional properties of wheat gliadins. II. Effects on dynamic rheological properties of wheat gluten. *J. Cereal Sci.* **2002**, *35*, 307–313. [CrossRef]

21. Patraşcu, L.; Banu, I.; Vasilean, I.; Aprodu, I. Rheological and thermo-mechanical characterization of starch—Protein mixtures. *Agric. Agric. Sci. Proc.* **2016**, *10*, 280–288. [CrossRef]

22. Gänzle, M.G.; Loponen, J.; Gobbetti, M. Proteolysis in sourdough fermentations: Mechanisms and potential for improved bread quality. *Trends Food Sci. Technol.* **2008**, *19*, 513–521. [CrossRef]

23. Jane, J.; Chen, Y.Y.; Lee, L.F.; McPherson, A.E.; Wong, K.S.; Radosavljevic, M.; Kasemsuwan, T. Effects of amylopectin branch chain length and amylose content on the gelatinization and pasting properties of starch 1. *Cereal Chem.* **1999**, *76*, 629–637. [CrossRef]

24. Frias, J.; Fornal, J.; Ring, S.G.; Vidal-Valverde, C. Effect of germination on physico-chemical properties of lentil starch and its components. *LWT Food Sci. Technol.* **1998**, *31*, 228–236. [CrossRef]

25. Chinma, C.E.; Adewuyi, O.; Abu, J.O. Effect of germination on the chemical, functional and pasting properties of flour from brown and yellow varieties of tigernut (*Cyperus esculentus*). *Food Res. Int.* **2009**, *4*, 1004–1009. [CrossRef]

26. Banu, I.; Aprodu, I. Studies concerning the use of *Lactobacillus helveticus* and *Kluyveromyces marxianus* for rye sourdough fermentation. *Eur. Food Res. Technol.* **2012**, *234*, 769–777. [CrossRef]

27. Bigiarini, L.; Pieri, N.; Grilli, I.; Galleschi, L.; Capocchi, A.; Fontanini, D. Hydrolysis of gliadin during germination of wheat seeds. *J. Plant Physiol.* **1995**, *147*, 161–167. [CrossRef]

28. Hidalgo, A.; Brusco, M.; Plizzari, L.; Brandolini, A. Polyphenol oxidase, alpha-amylase and beta-amylase activities of *Triticum monococcum*, *Triticum turgidum* and *Triticum aestivum*: A two-year study. *J. Cereal Sci.* **2013**, *58*, 51–58. [CrossRef]

29. Zhu, F. Triticale: Nutritional composition and food uses. *Food Chem.* **2018**, *241*, 468–479. [CrossRef]

30. Dubat, A.; Boinot, N. Mixolab Applications Handbook. In *Rheological and Enzymes Analyses*; Chopin Technology: Villenueve, France, 2012; pp. 10–15, 114–115.

31. Cornejo, F.; Rosell, C.M. Influence of germination time of brown rice in relation to flour and gluten free bread quality. *J. Food Sci. Technol.* **2015**, *52*, 6591–6598. [CrossRef]

32. Patraşcu, L.; Banu, I.; Vasilean, I.; Aprodu, I. Effects of germination and fermentation on the functionality of whole soy flour. *Bull. USAMV Sci. Biotechnol.* **2016**, *73*, 126–134. [CrossRef]

33. Bottari, A.; Capocchi, A.; Fontanini, D.; Galleschi, L. Major proteinase hydrolyzing gliadin during wheat germination. *Phytochem* **1996**, *43*, 39–44. [CrossRef]

34. Clarke, C.I.; Schober, T.J.; Angst, E.; Arendt, E.K. Use of response surface methodology to investigate the effects of processing conditions on sourdough wheat bread quality. *Eur. Food Res. Technol.* **2003**, *217*, 23–33.

Publisher's Note: MDPI stays neutral with regard to jurisdictional claims in published maps and institutional affiliations.

Article

Nutritional Improvement of Bean Sprouts by Using Chitooligosaccharide as an Elicitor in Germination of Soybean (*Glycine max* L.)

Wenzhu Tang, Xinting Lei, Xiaoqi Liu and Fan Yang *

School of Biological Engineering, Dalian Polytechnic University, Dalian 116034, China;
tangwz@dlpu.edu.cn (W.T.); dalei970910@gmail.com (X.L.); liuxiaoqifighting@gmail.com (X.L.)
* Correspondence: yang_fan@dlpu.edu.cn; Tel./Fax: +86-411-86323646

Abstract: Soybean sprouts are among the healthiest foods consumed in most Asian countries. Their nutritional content, especially bioactive compounds, may change according to the conditions of germination. The purpose of this study was to test the effect of chitooligosaccharide with different molecular weight and dosage on nutritional quality and enzymatic and antioxidant activities of soybean sprouts. The chitooligosaccharide elicitor strongly stimulated the accumulation of vitamin C, total phenolics, and total flavonoid. The stimulation effect was correlated with the molecular weight and concentration of chitooligosaccharide. With treatment of 0.01% of 1 kDa chitooligosaccharide, the nine phenolic constituents and six isoflavone compounds were significantly increased. The antioxidant capacity (DPPH radical and hydroxyl radical scavenging activity) and antioxidase activities (catalase and peroxidase) of soybean sprouts were also enhanced after treatment with chitooligosaccharide. The degree of chitooligosaccharide-induced elicitor activity increased as the molecular weight of chitooligosaccharide decreased. These results suggest that soaking soybean seeds in a solution of chitooligosaccharide, especially in 0.01% of 1 kDa chitooligosaccharide, may effectively improve the nutritional value and physiological function of soybean sprouts.

Keywords: soybean sprouts; chitooligosaccharide; phytochemicals; antioxidant activity; catalase; peroxidase

Citation: Tang, W.; Lei, X.; Liu, X.; Yang, F. Nutritional Improvement of Bean Sprouts by Using Chitooligosaccharide as an Elicitor in Germination of Soybean (*Glycine max* L.). *Appl. Sci.* **2021**, *11*, 7695. https://doi.org/10.3390/app 11167695

Academic Editor: Georgiana Gabriela Codină

Received: 15 July 2021
Accepted: 19 August 2021
Published: 21 August 2021

Publisher's Note: MDPI stays neutral with regard to jurisdictional claims in published maps and institutional affiliations.

1. Introduction

As a major source of protein, soybean has been one of the most important crops in China since more than five thousand years and nowadays often appears in Western diets [1]. Soybean has high nutritional value and is recognized as a functional food that both promotes health and has therapeutic effects [2]. For example, soybean contains a high concentration of isoflavones, which are indicated to have benefits in reducing risk of multiple diseases, such as hormone-dependent cancers, and age-related cognitive decline [3]. Soybean is also believed to have antioxidant activity, due to its high amount of total phenolics, that may help neutralize harmful free radicals and reduce the risk of degenerative diseases [4].

Germination technology, which is an inexpensive and effective method, was applied to improve the taste and enhance the nutritional value of soybean [5]. During germination, the endogenous enzymes in soybean are activated and storage nutritional compounds are converted to bioactive components. As a result, the nutrient levels, bioavailability, and palatability are improved [6]. For example, vitamin C, which is famous for its physiological functions in animals and plants, barely exists in soybeans and significantly increases in soybean sprouts. Most importantly, germination is conducive to the accumulation of phenolics and flavonoids, which have antioxidant activity [7]. Furthermore, the sprouts of soybeans have been shown to be effective as an anti-hypertensive diet [8]. At the same time, germination can reduce or remove anti-nutritional factors in soybean, such as a-galactoside, trypsin inhibitor, phytic acid, lectin, and lipase inhibitor [9]. Germination technology expands the application of soybean in food processing and improves its utilization value.

Most of the functional ingredients in plants are secondary metabolites, which are biosynthesized as a defense response to different stresses. Hence, the accumulation of secondary metabolites may be stimulated by various elicitors [10]. Chitosan (poly [β-(1-4)-2-amino-2-deoxy-d-glucopyranose]) is an environmentally-friendly and biocompatible cationic polysaccharide, which is obtained by partially deacetylated chitin extracted from naturally occurring crustacean shells [11]. Studies have shown that chitosan can elicit defense responses with respect to callose formation, phytoalexin induction, and lignification [12]. Chitosan can also be employed in the treatment of soybean to enhance the yield and quality of soybean sprouts [13,14]. According to Lee et al., chitosan effectively increased the growth of soybean sprouts without adverse effects on the nutritional and postharvest characteristics [15]. Recently, Yang et al. reported that chitosan pre-soaking could improve the growth and quality of yellow soybean sprouts during germination, dependent on concentration [16]. However, application of chitosan is limited due to its poor solubility under physiological conditions. Recent studies showed that the hydrolytic product of chitosan, a water-soluble chitooligosaccharide, also has elicitor activity in seed germination [17]. Furthermore, chitooligosaccharide has a great advantage when utilized as antimicrobial agents, antioxidants, and enhancers of the nutritional quality of food, while molecular weight is identified as a main characteristic of chitooligosaccharide that is closely associated with the biological activity of the latter [18]. However, few reports to date have investigated the role of chitooligosaccharide as an elicitor in improving the quality of soybean sprouts, especially their phytochemicals, enzymatic, and antioxidant activities.

In the present study, we evaluated the elicitor effects of chitooligosaccharide on phytochemicals, enzymatic, and antioxidant activities of germinated soybean. To the best of our knowledge, this is the first report on the application of chitooligosaccharide as an elicitor in improving the quality of soybean sprouts. The results show that chitooligosaccharide could increase the concentration of bioactive compounds including vitamin C, phenolics, and isoflavones. In addition, the activities of catalase and peroxidase as well as free radical scavenging-linked antioxidant activities of soybean sprouts were improved significantly. Data from this study would be valuable for the production of high-quality and chemical-free soybean sprouts to satisfy consumers' demands.

2. Materials and Methods

2.1. Plant Material, Chitooligosaccharide, and Chemicals

Commercial soybeans (*Glycine max* L.) were purchased from the Walmart supermarket, Dalian city, China. Chitooligosaccharide with different molecular weights (1, 2, and 3 kDa) and a deacetylation degree of 70% were purchased from Golden-Shell Pharmaceutical Co. (Wenzhou, China). Gallic acid, protocatechuic acid, syringic acid, p-coumaric acid, vanillic acid, ferulic acid, ellagic acid, cinnamic acid, p-hydroxybenzoic acid, genistein, daidzein, glycitein, genistin, daidzin, and glycitin were purchased from Yuanye Bio-Technology Co. (Shanghai, China).

2.2. Soybean Germination

Soybean seed (40 g) was washed three times and soaked in 100 mL distilled water at room temperature for 3 h. After soaking, seeds were placed into the sprouter (Fumin, modelFMC-30B, Qingzhou, China) and kept in the dark at 25 °C for germination. The soybean sprouts were rinsed with tap water every 12 h. The sprouts were harvested after 3 days of treatment.

2.3. Chitooligosaccharide Treatment

Three different molecular chitooligosaccharides (1 kDa, 2 kDa, and 3 kDa) were dissolved in distilled water to give a final concentration of 0.001%, 0.01%, 0.1%, and 1.00%

(w/v), and they were used as soaking solution treatments. After 3 h of soaking, seeds were washed thoroughly with distilled water and placed into the sprouter for germination.

2.4. Determination of Vitamin C

Vitamin C content of soybean sprouts was determined using 2, 6-dichloroindophenol titrimetric method [19]. Briefly, 20 g of fresh sprouts were ground in a mortar with 40 mL of chilled 2% (w/v) metaphosphoric acid buffer, and then adjusted to 100 mL. After filtration, 10 mL filtrate was titrated with 0.1% (w/v) 2, 6-dichloroindophenol, and the color of the titration end-point was light red. Ascorbic acid was used as a quantitative standard. The content of vitamin C was expressed as mg per 100 g of fresh weight.

2.5. Phytochemical Extraction

After harvest, the sprouts were freeze-dried and ground with a sample mill (Jiuyang, modelJYL-C16V, Jinan, China) and passed through an 80-mesh sieve. An amount of 5 g of ground sprout powder was added to 100 mL of an 80% (v/v) methanol solution and extracted at 25 °C and 260 rpm for 3 h. The homogenate was centrifuged at 10,000 rpm for 10 min at room temperature. The supernatant was collected and used for determination of phytochemical content and antioxidant tests.

2.5.1. Determination of Total Phenolics Content

A volume of 100 μL of the sample extract was mixed with 400 μL of distilled water. An amount of 100 μL of a Folin–Ciocalteu reagent was added and left for 6 min. Next, an aliquot of 1 mL of a 7% (w/v) Na_2CO_3 solution was added, and then the mixture adjusted to 3 mL with distilled water. Samples were incubated for 90 min at room temperature, and measured at 760 nm against the blank using a spectrophotometer (Molecular Devices, model SpectraMax Plus 384, San José USA). A standard curve was prepared with known gallic acid concentrations. The content of total phenolics was expressed as mg/g of dry weight.

2.5.2. Determination of Total Flavonoid Content

An amount of 0.25 mL of the sample extract was added to 1.25 mL of distilled water, followed by the addition of 75 μL of a 5% (w/v) $NaNO_2$ solution. After 6 min, 150 μL of a 10% (w/v) $Al(NO_3)_3 \cdot 9H_2O$ solution was added and incubated for another 5 min before 0.5 mL of 1 M NaOH was added. Distilled water was added to adjust the final volume to 2.5 mL. The absorbance of the mixture was measured immediately at 510 nm and compared with standards prepared with known (+)-catechin concentrations. Results were expressed as mg/g of dry weight.

2.5.3. HPLC Analysis of Phenolic Components and Isoflavones

Phenolic components were determined by HPLC with a Poroshell 120 EC-C18 column (4.6 × 250 mm, 4 μm particle size) (Waldbronn, Germany) and ultraviolet detector. The mobile phase consisted of acetonitrile (A) and 0.3% (v/v) acetic acid (B). Flow rate was 0.6 mL/min, column temperature was 40 °C, and detection wavelength was 254 nm. The gradient elution process began with the mobile phase A at 20% in the first 5 min, then increased from 20% to 90% for the next 5 min, while in the following 5 min it reduced from 90% to 10%, and increased to 20% A in the last 10 min. The peaks of the samples were identified based on the retention times of the standards and the concentrations of phenolics were calculated from standard curves. The concentration was expressed as mg/g of dry weight.

Isoflavones were quantified using the same column and chromatographic conditions as described above, while the two solvents used in the mobile phase were solvent A, 0.1% (v/v) acetic acid, and solvent B, methanol. The gradient elution process began with the mobile phase A from 5% to 30% in the first 50 min, and then increased from 30% to 100% for the next 15 min, and the last 15 min was 100% A. The concentrations of isoflavones in the sample were calculated from standard curves of different isoflavone standards, and it was expressed as mg/g of dry weight.

2.5.4. Analysis of DPPH Radical Scavenging Activity

The 2, 2-diphenyl-1-picrylhydrazyl (DPPH) radical scavenging activity was examined using the DPPH method [20]. Briefly, 0.1 mL of the sample extract was mixed with 2.9 mL DPPH in ethanol (100 μM) and incubated in the dark at 33 °C for 30 min. The absorbance of the mixture was measured at 517 nm and denoted Ai. A control sample was obtained by mixing 0.1 mL ethanol with 2.9 mL DPPH solution (100 μM) and its absorbance value was denoted A0. A reference test was carried out by adding 0.1 mL of the sample extract to 2.9 mL ethanol, and Aj referred to the absorbance of the resultant solution. The DPPH radical scavenging activity was expressed as a scavenging rate and calculated as follows: DPPH scavenging rate (%) = [1 − (Ai − Aj)/A0] × 100%.

2.5.5. Analysis of Hydroxyl Radical Scavenging Activity

1.0 mL H_2O_2 (7.5 mM) was mixed with a solution containing 0.2 mL ammonium ferric sulfate (0.75 mM), 0.2 mL 1, 10-phenanthroline (0.5 mM), 1.0 mL sample extract, and 1.0 mL Tris buffer (500 mM, pH 7.4). The reaction mixture was adjusted to 10 mL with distilled water and incubated at 37 °C for 1 h. After cooling, the absorbance of the samples was measured at 508 nm and denoted As. The absorbance of the control without sample extract was denoted Ac. The absorbance of the reference without the sample extract and H_2O_2 addition was denoted A0. The hydroxyl radical scavenging activity was expressed as a scavenging rate and calculated as follows: hydroxyl scavenging rate (%) = (As − Ac)/(A0 − Ac) ×100%.

2.6. Enzyme Assays

Catalase was determined by following the consumption of H_2O_2 [21]. An amount of 5 g of fresh sprouts was suspended in 20 mL phosphate buffer (0.1 M, pH 7.8) ground into homogenate and adjusted to a final volume of 100 mL. The homogenate was centrifuged at 10,000 rpm and 4 °C for 15 min and the supernatant was collected and used for enzyme assay. An amount of 0.3 mL of H_2O_2 (0.1 M) was added to a solution containing 0.2 mL enzyme extract and 2.5 mL phosphate buffer (0.1 M, pH 7.8). Immediately after mixing, the concentration of H_2O_2 was measured spectroscopically at 240 nm at 10 s intervals for up to 3 min. One unit of activity was defined as the amount of enzyme that catalyzes the degradation of H_2O_2 leading to a decrease of 0.1 in the absorbance at 240 nm per minute under the above experimental conditions. The control experiment was carried out with boiled enzyme under the same conditions.

Peroxidase was evaluated with guaiacol in the presence of H_2O_2 [21]. The enzyme extraction from soybean sprouts was performed as described above for catalase assay. An amount of 0.1 mL of enzyme extract was added to a solution containing 2.9 mL of 0.4% H_2O_2 and 1 mL of 0.05 M guaiacol in 0.1 M phosphate buffer (pH 5.5). The increase in the absorbance of the mixture was monitored at 470 nm for 3 min. The control experiment was carried out with boiled enzyme. One unit of enzyme activity was defined as the amount of enzyme that catalyzes the oxidation of guaiacol leading to a 0.01 increase in the absorbance at 470 nm per minute under the above experimental conditions.

2.7. Statistical Analysis

All tests were conducted in triplicate and data were reported as the mean ± standard deviation. Significant differences among means were tested by one-way analysis of variance using SPSS software (SPSS, Chicago, IL, USA). Duncan's multiple range test was used to compare the differences of means among different groups, and differences with $p < 0.05$ were considered statistically significant.

3. Results and Discussion

3.1. Vitamin C Content

Vitamin C plays an important role in the process of collagen synthesis in humans. Because humans are unable to synthesize vitamin C, fresh fruit and vegetables are needed as a supplement of vitamin C [1]. As a result, soybean sprouts have attracted people's attention for their high content of vitamin C. Figure 1 presents the vitamin C content in soybean sprouts that have been treated with chitooligosaccharides of different molecular weights at different concentrations. It can be found that chitooligosaccharide could effectively promote the vitamin C accumulation in soybean sprouts, and all the different chitooligosaccharide treatments revealed higher vitamin C content than the control. The vitamin C content of soybean sprouts was increased by 45.84% with 0.001% of 1 kDa chitooligosaccharide. When the concentration of chitooligosaccharide increased to 0.01%, the vitamin C content reached the maximum, 33.33 mg/100 g fresh weight, about 2.11-fold in comparison with the control of 15.83 mg/100 g fresh weight. However, after treatment with 0.1% and 1% of 1 kDa chitooligosaccharide, the vitamin C content decreased to 29.17 mg/100 g and 24.67 mg/100 g, which were still more than the control. There was a similar trend for the treatment with 2 kDa and 3 kDa chitooligosaccharide. After treatment with 0.01% of 2 kDa and 3 kDa chitooligosaccharide, the vitamin C content of soybean sprouts increased to 30.83 mg/100 g fresh weight and 27.35 mg/100 g fresh weight, respectively.

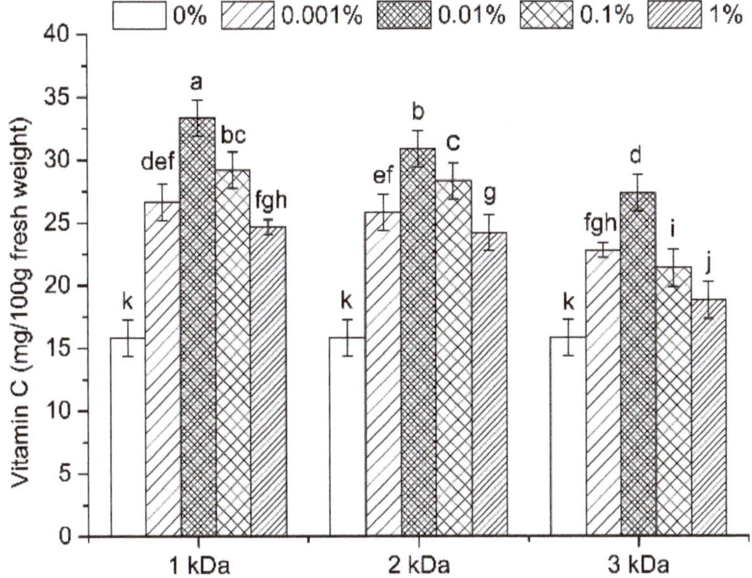

Figure 1. Vitamin C content of soybean sprouts under different chitooligosaccharide treatments. The treatments were conducted with chitooligosaccharides of three molecular weights (1, 2, and 3 kDa) at four different concentrations (0.001 to 1%). The control treatment (0%) included the same solutions without chitooligosaccharides. Vertical bars represent the standard error of three replicates. Different letters indicate statistically significant difference ($p < 0.05$).

As an important bioactive component, vitamin C does not exist in soybeans but increases after germination. This is owing to a response in the environmental factors [16]. Chitosan and chitooligosaccharide are well documented plant defense elicitors and are recognized as stress signals by plant cells [22]. It has been discovered that the size and dosage of chitooligosaccharide are crucial in determining their elicitor activity [6].

In the present study, the vitamin C content (15.83 mg/100 g fresh weight) of the control was similar to that (15.82 mg/100 g fresh weight) reported by NO and others. However, after treatment with 0.01% of 1 kDa chitooligosaccharide, the highest vitamin C content of soybean sprouts increased to 33.33 mg/100 g fresh weight, which was significantly higher than their treatment with 0.1% of 493 kDa chitosan (17.01 mg/100 g fresh weight) [13]. This result was in accordance with previous studies, which showed that the antiviral activity of chitosan increased as its molecular weight decreased. The highest activity was exhibited by the fractions with Mw = 2.2 and 1.2 kDa [23].

3.2. Total Phenolics Content

It has been proven that tremendous changes in functional substances are closely linked to germination of soybeans, such as high phenolics content [4]. Additionally, chemical elicitors can significantly increase the phenolics content in bean sprouts [24]. In this study, total phenolics content of soybean sprouts in all treatments with chitooligosaccharide was shown in Figure 2. Overall, the phenolics content of soybean sprouts was significantly increased by the chitooligosaccharide in three molecular weights (1 kDa, 2 kDa, and 3 kDa). The smallest molecular weight led to the highest increase in phenolics under the same treatment concentration. Total phenolics content increased gradually along with the increase in chitooligosaccharide until 0.01%, and after that it gradually reduced. Soybean sprouts treated with 0.01% of 1 kDa chitooligosaccharide exhibited the highest phenolics content of 25.60 mg/g dry weight, which was approximately 28% higher than that of the control. The total phenolics content of soybean sprouts treated with 0.01% of 2 kDa and 3 kDa chitooligosaccharide increased by 23% and 20%, respectively. These results were similar to the increase in phenolics as a result of ethephon treatment [25]. Compared with ethephon, chitooligosaccharide is safe and environmentally friendly, so it can be a useful alternative in increasing phenolic secondary metabolites in soybean sprouts.

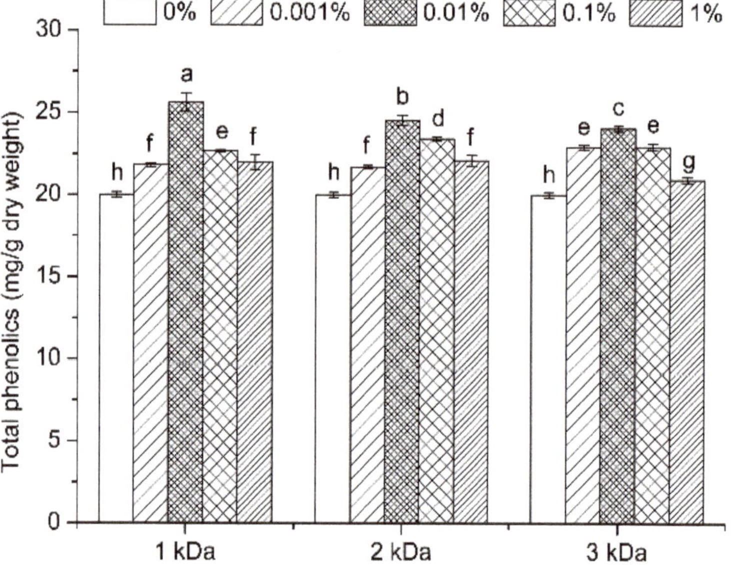

Figure 2. Total phenolics content of soybean sprouts under different chitooligosaccharide treatments. The treatments were conducted with chitooligosaccharides of three molecular weights (1, 2, and 3 kDa) at four different concentrations (0.001 to 1%). The control treatment (0%) included the same solutions without chitooligosaccharides. Vertical bars represent the standard error of three replicates. Different letters indicate statistically significant difference ($p < 0.05$).

3.3. Total Flavonoid Content

Germination could increase the flavonoid content, which might be owing to the activation of enzymes involved in the biosynthesis of flavonoids from large molecular weight polyphenols such as tannins [26]. Flavonoids contain a diverse group of phytochemicals with antioxidant and anticancer activity, and they have been detected in fruits, vegetables, tea, and wine [27,28]. In order to clarify the effects of chitooligosaccharide on the flavonoid's accumulation of soybean sprouts, soybeans were treated with chitooligosaccharide before germination. As shown in Figure 3, similar to the changing trend of vitamin C and total phenolics, the total flavonoid content of soybean sprouts was efficiently increased by all chitooligosaccharide elicitors with a molecular-weight-dependent mode. Among them, 1 kDa chitooligosaccharide was the most effective elicitor, and when using it in the concentration of 0.01%, the flavonoids content of soybean sprouts was 1.25 mg/g dry weight, about 22% more than that of the control (1.03 mg/g dry weight). Chitooligosaccharide with the molecular weight of 2 kDa and 3 kDa had a similar promoting action, and they improved the accumulation of flavonoids to 1.21 mg/g dry weight at a concentration of 0.01%. These results are consistent with earlier findings which showed that chitosan could induce the stimulation of secondary metabolism of plants [29].

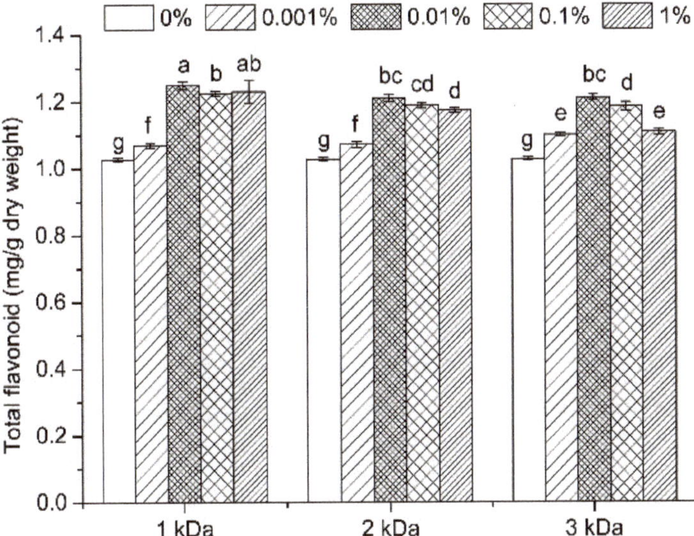

Figure 3. Total flavonoid content of soybean sprouts under different chitooligosaccharide treatments. The treatments were conducted with chitooligosaccharides of three molecular weights (1, 2, and 3 kDa) at four different concentrations (0.001 to 1%). The control treatment (0%) included the same solutions without chitooligosaccharides. Vertical bars represent the standard error of three replicates. Different letters indicate statistically significant difference ($p < 0.05$).

3.4. Quantification of Phenolic Components

Phenolic compounds are generally secondary metabolites with high antioxidative and antiaging properties, and they are beneficial to human health [30]. Khang and others have found several phenolic compounds in soybean sprouts including gallic acid, protocatechuic acid, syringic acid, p-coumaric acid, vanillic acid, ferulic acid, ellagic acid, cinnamic acid, p-hydroxybenzoic acid, and sinapic acid [4]. The impacts of chitooligosaccharide on nine of these phenolic compounds in soybean sprouts were studied (Table 1). Concentrations of phenolic components from different treatments differed significantly ($p < 0.05$). The contents of gallic acid ranged from 18.698 to 32.274 mg/g dry weight, protocatechuic acid from 0.539 to 1.444 mg/g dry weight, syringic acid from 0.073 to 0.116 mg/g dry weight, p-coumaric acid from 0.771 to 1.460 mg/g dry weight, vanillic acid from 1.021 to 1.844 mg/g dry weight, ferulic acid from 4.845 to 7.398 mg/g dry weight, ellagic acid from 0.459 to 0.775 mg/g dry weight, cinnamic acid from 0.053 to 0.254 mg/g dry weight, and p-hydroxybenzoic acid from 0.251 to 2.646 mg/g dry weight. All the phenolics were significantly increased in the presence of chitooligosaccharide, and the most effective treatment was 1 kDa chitooligosaccharide with the concentration of 0.01%. These results were in accordance with the total phenolics content measurement. Among them, p-hydroxybenzoic acid showed the highest proportion of increase, which was about 10.54-fold over the control. The contents of cinnamic acid and protocatechuic acid increased to 4.79-fold and 2.68-fold in comparison with that of the control, respectively. In addition, the contents of gallic acid, syringic acid, p-coumaric acid, vanillic acid, ferulic acid, and ellagic acid in sprouts treated with the same chitooligosaccharide increased by 73%, 59%, 89%, 81%, 53%, and 69%, respectively. This clearly indicated that chitooligosaccharide was extremely effective in increasing phenolic secondary metabolites in soybean sprouts, and the increase in these phenolic compounds was the result of resisting oxidative stress [25]. However, in contrast with research by Khang [4], the contents of these phenolic compounds were much higher in our study. These discrepancies might be due to differences in soybean cultivars and germination methods.

3.5. Quantification of Isoflavone Components

Numerous studies have confirmed that soybean isoflavones are effective cancer-preventive agents for lowering risks of various cancers [31]. Soybean isoflavones mainly exist in two forms: aglycon (genistein, daidzein and glycitein) and different types of glucosides (genistin, daidzin and glycitin). The aglycon form is the most biologically active isoflavone for mammal metabolism [6]. In the present study, six kinds of isoflavones were compared between different treatments (Table 2). Overall, chitooligosaccharide treatments induced statistically significant changes in their contents. Regardless of the molecular weight, all the promoting effects of chitooligosaccharide were dependent on its concentration. Treatment with an amount of 0.01% chitooligosaccharide increased the content of these six isoflavones in the soybean sprouts extract significantly. Among these elicitation treatments, 0.01% of 1 kDa chitooligosaccharide was the most effective one. The highest content of glycitein was 0.251 mg/g dry weight, which was about 20.92-fold compared with the control (0.012 mg/g dry weight) when the seeds were treated with 0.01% of 1 kDa chitooligosaccharide. After treatment with 0.01% of 2 kDa and 3 kDa chitooligosaccharide, the concentration of glycitein increased to about 18.33-fold and 19.67-fold in comparison with that of the control, respectively. There were also significant differences in the concentrations of genistein, daidzein, genistin, daidzin, and glycitin between the control and 0.01% of 1 kDa chitooligosaccharide treatments, and the contents of them increased by 16%, 6%, 62%, 90%, and 68%, respectively.

The observed impact of chitooligosaccharide on isoflavone concentrations was in accordance with previous studies, which reported that seed and foliar treatments of soybean plants with chitosan could bring about a significant increase in the concentration of daidzein, genistein, glycitein, as well as total isoflavone in soybean when compared with untreated control plants [32]. Furthermore, it is known that the biosynthesis of isoflavone is submetabolic, via the malonate and phenylpropanoid pathways [33]. Dixon and Paiva have reported that various biotic and abiotic stresses could stimulate the phenylpropanoid metabolism [34]. As a result, it is made clear that chitooligosaccharide has the eliciting activity in inducing the stress responses and secondary metabolism of soybean sprouts, and can be used as a harmless alternative to produce high quality soybean sprouts with a good market potential [35].

Table 1. Phenolic components concentration (mg/g dry weight) of soybean sprouts under different chitooligosaccharide treatments.

Treatment	COS Concentration (%)	Gallic Acid	Protocatechuic Acid	Syringic Acid	p-Coumaric Acid	Vanillic Acid	Ferulic Acid	Ellagic Acid	Cinnamic Acid	p-Hydroxybenzoic Acid
Control	0	18.698 ± 0.311 e	0.539 ± 0.023 e	0.073 ± 0.005 e	0.771 ± 0.015 g	1.021 ± 0.020 e	4.845 ± 0.047 g	0.459 ± 0.001 e	0.053 ± 0.004 e	0.251 ± 0.030 e
	0.001	30.462 ± 0.212 b	0.915 ± 0.017 d	0.106 ± 0.002 cd	1.261 ± 0.008 de	1.268 ± 0.014 d	7.296 ± 0.052 a	0.652 ± 0.007 cd	0.121 ± 0.003 d	0.913 ± 0.044 d
	0.01	32.274 ± 0.547 a	1.444 ± 0.032 a	0.116 ± 0.001 a	1.460 ± 0.019 a	1.844 ± 0.105 a	7.398 ± 0.078 a	0.775 ± 0.012 a	0.254 ± 0.003 a	2.646 ± 0.014 a
1 kDa COS	0.1	25.660 ± 0.305 d	0.809 ± 0.010 d	0.100 ± 0.002 d	1.194 ± 0.013 f	1.307 ± 0.020 d	6.477 ± 0.061 de	0.710 ± 0.016 b	0.126 ± 0.004 d	1.061 ± 0.026 cd
	1	26.855 ± 0.251 d	0.988 ± 0.034 c	0.105 ± 0.008 cd	1.191 ± 0.022 f	1.404 ± 0.024 d	7.050 ± 0.024 a	0.645 ± 0.0207 d	0.185 ± 0.006 c	1.170 ± 0.201 bc
	0.001	25.486 ± 0.481 d	0.860 ± 0.027 d	0.101 ± 0.002 d	1.231 ± 0.009 e	1.445 ± 0.016 d	6.291 ± 0.087 f	0.730 ± 0.056 ab	0.195 ± 0.002 bc	1.344 ± 0.037 bc
	0.01	31.020 ± 0.465 b	1.116 ± 0.034 b	0.110 ± 0.002 ab	1.372 ± 0.031 a	1.754 ± 0.037 a	7.155 ± 0.117 a	0.669 ± 0.009 bc	0.239 ± 0.000 a	1.660 ± 0.846 b
2 kDa COS	0.1	25.291 ± 0.540 d	1.144 ± 0.018 b	0.109 ± 0.001 bc	1.286 ± 0.015 cd	1.551 ± 0.031 c	6.559 ± 0.029 e	0.628 ± 0.008 d	0.172 ± 0.006 c	0.978 ± 0.518 cd
	1	29.939 ± 0.260 c	0.971 ± 0.012 c	0.103 ± 0.004 d	1.315 ± 0.008 bc	1.549 ± 0.021 c	6.682 ± 0.045 cd	0.637 ± 0.008 cd	0.142 ± 0.003 d	1.083 ± 0.021 cd
	0.001	29.026 ± 0.078 c	0.958 ± 0.010 a	0.109 ± 0.001 bc	1.331 ± 0.029 b	1.626 ± 0.045 b	6.839 ± 0.170 bc	0.671 ± 0.005 bc	0.197 ± 0.007 b	1.074 ± 0.083 cd
	0.01	32.066 ± 0.585 a	1.207 ± 0.006 b	0.112 ± 0.002 ab	1.381 ± 0.012 a	1.619 ± 0.051 b	6.973 ± 0.021 b	0.736 ± 0.003 a	0.230 ± 0.005 a	1.546 ± 0.319 b
3 kDa COS	0.1	28.318 ± 0.672 c	1.060 ± 0.028 c	0.109 ± 0.003 bc	1.259 ± 0.031 de	1.600 ± 0.048 b	6.307 ± 0.099 ef	0.658 ± 0.014 bc	0.151 ± 0.010 cd	1.056 ± 0.120 cd
	1	31.068 ± 0.323 b	1.012 ± 0.014c	0.105 ± 0.003 cd	1.266 ± 0.014 de	1.619 ± 0.024 b	6.440 ± 0.036 e	0.690 ± 0.010 b	0.148 ± 0.002 cd	0.914 ± 0.077 d

COS: chitooligosaccharide. Values represent mean ± standard deviation ($n = 3$). Values in a column with different letters are significantly different ($p < 0.05$).

Table 2. Isoflavone components concentration (mg/g dry weight) of soybean sprouts under different chitooligosaccharide treatments.

Treatment	COS Concentration (%)	Genistein	Daidzein	Glycitein	Genistin	Daidzin	Glycitin
Control	0	0.497 ± 0.007 e	0.461 ± 0.001 e	0.012 ± 0.004 cd	0.089 ± 0.001 h	0.079 ± 0.001 h	0.019 ± 0.001 h
	0.001	0.521 ± 0.007 c	0.463 ± 0.001 d	0.088 ± 0.026 b	0.122 ± 0.003 d	0.088 ± 0.003 g	0.025 ± 0.001 d
	0.01	0.575 ± 0.003 a	0.490 ± 0.001 a	0.251 ± 0.006 a	0.144 ± 0.001 a	0.150 ± 0.001 a	0.032 ± 0.002 a
1 kDa COS	0.1	0.551 ± 0.013 b	0.462 ± 0.000 de	0.102 ± 0.007 b	0.138 ± 0.004 b	0.132 ± 0.003 c	0.026 ± 0.001 d
	1	0.507 ± 0.012 cd	0.466 ± 0.001 c	0.041 ± 0.014 cd	0.102 ± 0.002 g	0.137 ± 0.001 bc	0.020 ± 0.001 gh
	0.001	0.511 ± 0.007 cd	0.461 ± 0.001 e	0.043 ± 0.007 c	0.118 ± 0.003 e	0.095 ± 0.004 fg	0.022 ± 0.001 ef
	0.01	0.555 ± 0.003 b	0.467 ± 0.001 c	0.220 ± 0.008 ab	0.136 ± 0.001 b	0.135 ± 0.003 c	0.029 ± 0.001 c
2 kDa COS	0.1	0.519 ± 0.008 c	0.466 ± 0.001 c	0.096 ± 0.041 b	0.119 ± 0.002 de	0.122 ± 0.002 e	0.021 ± 0.001 fg
	1	0.513 ± 0.007 cd	0.462 ± 0.002 de	0.019 ± 0.003 cd	0.099 ± 0.002 g	0.104 ± 0.003 f	0.019 ± 0.001 h
	0.001	0.515 ± 0.003 cd	0.466 ± 0.001 c	0.023 ± 0.002 cd	0.110 ± 0.000 f	0.105 ± 0.002 f	0.023 ± 0.001 e
	0.01	0.569 ± 0.010 a	0.484 ± 0.001 b	0.236 ± 0.021 a	0.139 ± 0.001 b	0.140 ± 0.001 b	0.031 ± 0.001 b
3 kDa COS	0.1	0.503 ± 0.004 de	0.466 ± 0.001 c	0.073 ± 0.029 b	0.131 ± 0.002 c	0.128 ± 0.002 de	0.025 ± 0.001 d
	1	0.518 ± 0.006 c	0.462 ± 0.001 de	0.009 ± 0.005 d	0.102 ± 0.001 g	0.102 ± 0.002 f	0.020 ± 0.001 gh

COS: chitooligosaccharide. Values represent mean ± standard deviation ($n = 3$). Values in a column with different letters are significantly different ($p < 0.05$).

3.6. Antioxidant Activity Evaluation

It is believed that free radicals are directly related to various diseases, and dietary antioxidants, which are able to scavenge free radicals, may be used to reduce the risk of these diseases. Therefore, it is important to investigate the radical scavenging effect of antioxidants in soybean sprouts [2]. The most commonly used method is based upon the stable free radical DPPH [36]. In the present study, antioxidant activity was evaluated by scavenging capability of DPPH (Figure 4). Regardless of molecular weight and concentration, antioxidant activity in chitooligosaccharide-treated soybean sprouts was higher than the control, which could be the result of high total phenolics accumulation observed in the preceding section [37]. Among them, the smallest molecular weight led to the highest DPPH radical scavenging activity under the same treatment concentration. DPPH radical scavenging activity increased gradually along with the increase in chitooligosaccharide until 0.01%, and after that it gradually reduced. The soybean sprouts treated with 0.01% of 1 kDa chitooligosaccharide showed the strongest DPPH scavenging activity of 28.14%, which is about 1.83-fold compared with the control of 15.40%. The scavenging activity of soybean sprouts treated with 0.01% of 2 kDa and 3 kDa chitooligosaccharide were 23.71% and 21.83%, respectively. This clearly indicated that chitooligosaccharide-treated soybean sprouts have better free radical scavenging ability to protect the plant from oxidative damage [38]. However, the antioxidant activity of soybean sprouts was continuously decreased when the concentration of chitooligosaccharide was higher than 0.01%. These changes in antioxidant activity may confirm that if the concentrations of chitooligosaccharides were too high and oxidative stress was too high, then its stimulatory effect was weakened.

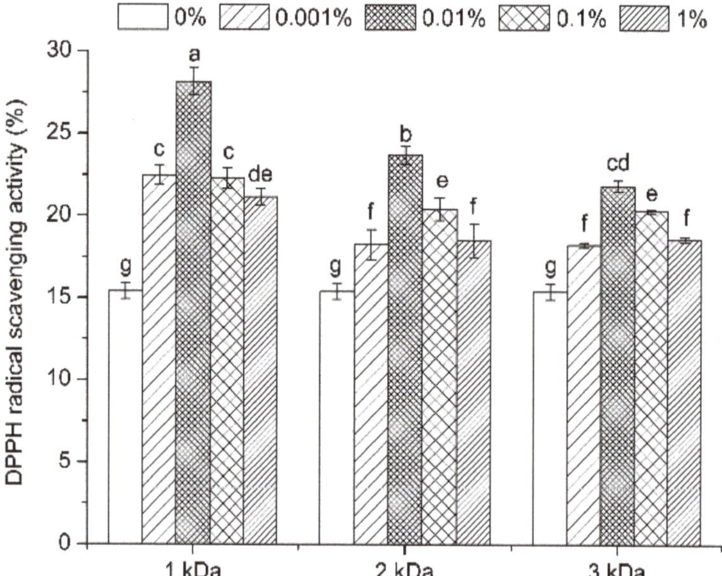

Figure 4. DPPH radical scavenging activity of soybean sprouts under different chitooligosaccharide treatments. The treatments were conducted with chitooligosaccharides of three molecular weights (1, 2, and 3 kDa) at four different concentrations (0.001 to 1%). The control treatment (0%) included the same solutions without chitooligosaccharides. Vertical bars represent the standard error of three replicates. Different letters indicate statistically significant difference ($p < 0.05$).

Because hydroxyl radicals are so reactive that they can cause potential damage to cells [39], the hydroxyl radicals scavenging activity in chitooligosaccharide-treated soybean sprouts was also investigated. As shown in Figure 5, the hydroxyl radical scavenging activity of soybean sprouts was similar to the changing trend of the DPPH radical scavenging activity. An amount of 0.01% of 1 kDa chitooligosaccharide was the most effective one, which increased the hydroxyl radical scavenging activity of soybean sprouts to 64.33%, which is about 1.43-fold compared with the control. Treatments with 0.01% of 2 kDa and 3 kDa chitooligosaccharide also improved the hydroxyl radical scavenging activity of soybean sprouts, representing 33% and 28% increases, respectively. The enhancement in antioxidant activity of soybean sprouts treated with chitooligosaccharide demonstrated an improvement in their biological function, and these soybean sprouts may meet the needs of people looking for new drugs from food materials [40].

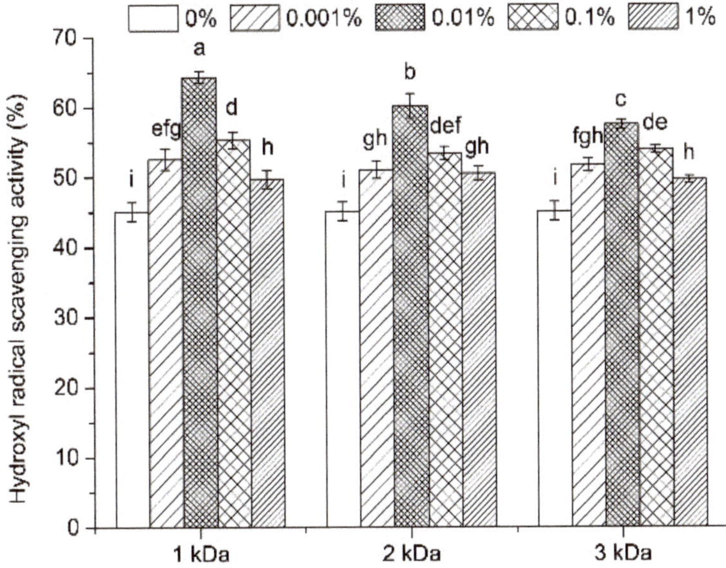

Figure 5. Hydroxyl radical scavenging activity of soybean sprouts under different chitooligosaccharide treatments. The treatments were conducted with chitooligosaccharides of three molecular weights (1, 2, and 3 kDa) at four different concentrations (0.001 to 1%). The control treatment (0%) included the same solutions without chitooligosaccharides. Vertical bars represent the standard error of three replicates. Different letters indicate statistically significant difference ($p < 0.05$).

3.7. Enzymatic Activities

It is known that chitosan has the potential for inducing defense-related enzymes [41], so the enzymatic activities, including catalase and peroxidase, in soybean sprouts after treatments of chitooligosaccharide were studied. Results showing catalase activity of soybean sprouts in different treatments were presented in Figure 6. In general, the treatments of chitooligosaccharide depicted higher catalase activity, among which, the highest increase was treatment with 0.01% of 1 kDa chitooligosaccharide, and resulted with 185.55 U/g fresh weight. Treatments with 2 kDa and 3 kDa chitooligosaccharide had similar stimulation effects on the activity of catalase, and at a concentration of 0.01%, they improved the activity of catalase to 156.67 U/g fresh weight and 159.64 U/g fresh weight, respectively. Regardless of the molecular weight, chitooligosaccharide with a concentration higher or lower than 0.01% has less elicitor activity. Catalase can detoxify H_2O_2 to water, thereby reducing the damage of reactive oxygen species to plants. The increase in the activity of catalase was most likely a response to the increased generation of reactive oxygen species and indicative of oxidative stress [42].

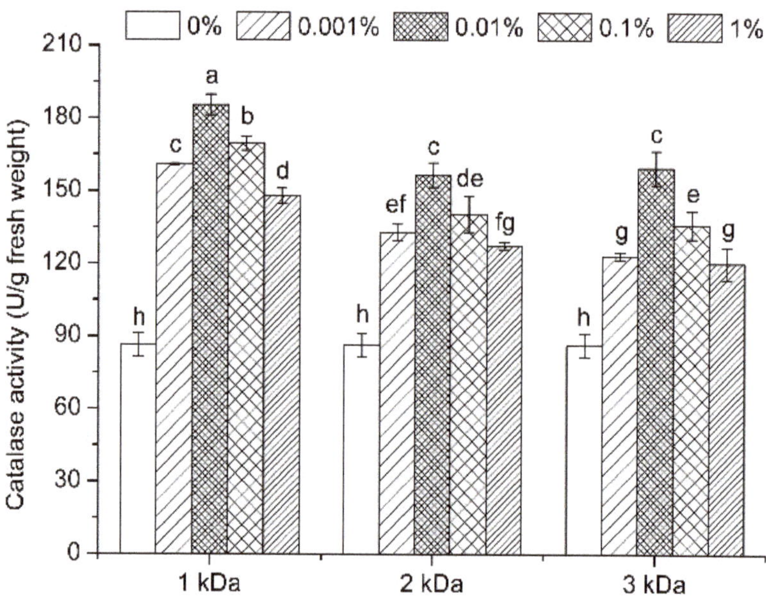

Figure 6. Catalase activity of soybean sprouts under different chitooligosaccharide treatments. The treatments were conducted with chitooligosaccharides of three molecular weights (1, 2, and 3 kDa) at four different concentrations (0.001 to 1%). The control treatment (0%) included the same solutions without chitooligosaccharides. Vertical bars represent the standard error of three replicates. Different letters indicate statistically significant difference ($p < 0.05$).

Peroxidase is involved in a variety of physiological functions in plants, and because of its importance in coloration, flavor development, nutritional properties, and texture of fruits and vegetables, it can change the consumer's acceptability of plant-based foods [19]. In the current study, soybean sprouts treated with chitooligosaccharide had higher levels of peroxidase than the control, which coincided with the trend of catalase (Figure 7). Regardless of the molecular weight, peroxidase activity first increased and then decreased with the concentration increase in chitooligosaccharide, and the best concentration of chitooligosaccharide was 0.01%. Among them, the highest peroxidase activity was shown in the soybean sprouts when they were soaked in 0.01% of 1 kDa chitooligosaccharide—this resulted in 2183.76 U/g fresh weight. Peroxidase activity in soybean sprouts treated with 0.01% of 2 kDa and 3 kDa chitooligosaccharide increased significantly as well.

Therefore, chitooligosaccharide treatment could enhance resistance to oxidative stress. Moreover, it also suggested that chitooligosaccharide could be recognized as an effective biotic elicitor, triggering a specific induction of catalase and peroxidase [43].

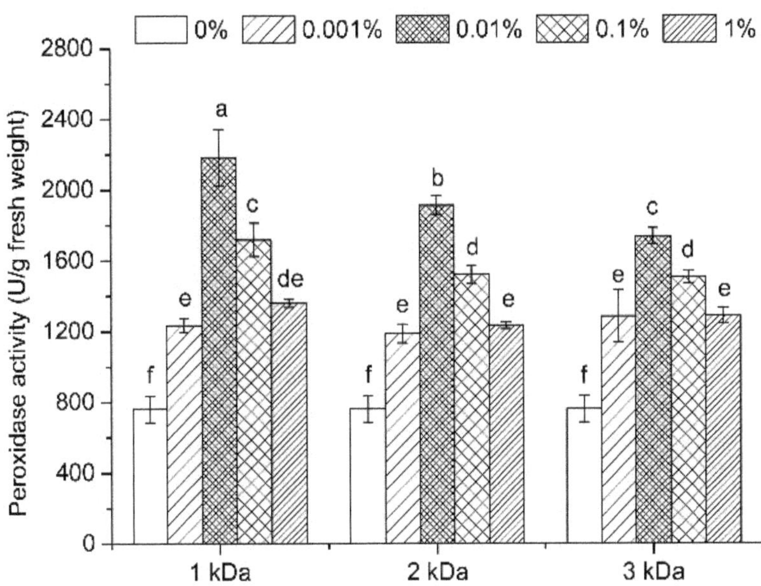

Figure 7. Peroxidase activity of soybean sprouts under different chitooligosaccharide treatments. The treatments were conducted with chitooligosaccharides of three molecular weights (1, 2, and 3 kDa) at four different concentrations (0.001 to 1%). The control treatment (0%) included the same solutions without chitooligosaccharides. Vertical bars represent the standard error of three replicates. Different letters indicate statistically significant difference ($p < 0.05$).

4. Conclusions

Chitooligosaccharide can significantly increase the content of phytochemicals and antioxidant properties of soybean sprouts. The highest accumulation of vitamin C, total phenolics, and total flavonoid occurred when the seeds were treated with 0.01% of 1 kDa chitooligosaccharide. Furthermore, the biosynthesis of nine phenolics and six isoflavones was also notably stimulated by chitooligosaccharide treatment. Moreover, the chitooligosaccharide strongly increased the antioxidant activities of soybean sprouts with the increase in the activities of catalase and peroxidase. The obtained data suggested that chitooligosaccharide could improve the physiological function of soybean sprouts by increasing their bioactive compounds content as well as antioxidant activity.

Author Contributions: Conceptualization, W.T. and F.Y.; methodology, W.T. and X.L. (Xinting Lei); validation, W.T., X.L. (Xinting Lei), and X.L. (Xiaoqi Liu); formal analysis, X.L. (Xiaoqi Liu); investigation, X.L. (Xinting Lei) and X.L. (Xiaoqi Liu); resources, W.T. and F.Y.; data curation, W.T. and F.Y.; writing—original draft preparation, W.T., X.L. (Xinting Lei), and X.L. (Xiaoqi Liu); writing—review and editing, W.T. and F.Y.; visualization, W.T., X.L. (Xinting Lei), and X.L. (Xiaoqi Liu); supervision, F.Y.; project administration, W.T. and F.Y.; funding acquisition, W.T. and F.Y. All authors have read and agreed to the published version of the manuscript.

Funding: This study was supported by Basic Research Projects of Liaoning Higher Education Institutions (2017J030) and Dalian High-level Talent Innovation Support Program (2018RQ24).

Institutional Review Board Statement: Not applicable.

Informed Consent Statement: Not applicable.

Data Availability Statement: Data is contained within the article.

Conflicts of Interest: No potential conflict of interest was reported by the authors.

References

1. Xu, M.J.; Dong, J.F.; Zhu, M.Y. Effects of germination conditions on ascorbic acid level and yield of soybean sprouts. *J. Sci. Food Agric.* **2005**, *85*, 943–947. [CrossRef]
2. Xu, B.J.; Chang, S.K. A comparative study on phenolic profiles and antioxidant activities of legumes as affected by extraction solvents. *J. Food Sci.* **2007**, *72*, 159–166. [CrossRef]
3. Pilšáková, L.; Riečanský, I.; Jagla, F. The physiological actions of isoflavone phytoestrogens. *Physiol. Res.* **2010**, *59*, 651–664. [CrossRef]
4. Khang, D.T.; Dung, T.N.; Elzaawely, A.A.; Xuan, T.D. Phenolic profiles and antioxidant activity of germinated legumes. *Foods* **2016**, *5*, 27–36. [CrossRef] [PubMed]
5. Vidal-Valverde, C.; Frias, J.; Sierra, I.; Blazquez, I.; Lambein, F.; Kuo, Y.H. New functional legume foods by germination: Effect on the nutritive value of beans, lentils and peas. *Eur. Food Res. Technol.* **2002**, *215*, 472–477. [CrossRef]
6. Shi, H.L.; Nam, P.K.; Ma, Y.F. Comprehensive Profiling of Isoflavones, Phytosterols, Tocopherols, Minerals, Crude Protein, Lipid, and Sugar during Soybean (*Glycine max*) Germination. *J. Agric. Food Chem.* **2010**, *58*, 4970–4976. [CrossRef] [PubMed]
7. Plaza, L.; Ancos, B.D.; Cano, M.P. Nutritional and health-related compounds in sprouts and seeds of soybean (*Glycine max*), wheat (*Triticum aestivum.* L) and alfalfa (*Medicago sativa*) treated by a new drying method. *Eur. Food Res. Technol.* **2003**, *216*, 138–144. [CrossRef]
8. Mastropasqua, L.; Dipierro, N.; Paciolla, C. Effects of Darkness and Light Spectra on Nutrients and Pigments in Radish, Soybean, Mung Bean and Pumpkin Sprouts. *Antioxidants* **2020**, *9*, 558. [CrossRef]
9. Bau, H.M.; Villaume, C.; Nicolas, J.P.; Méjean, L. Effect of germination on chemical composition, biochemical constituents and antinutritional factors of soya bean (*Glycine max*) seeds. *J. Sci. Food Agric.* **1997**, *73*, 1–9. [CrossRef]
10. Zhong, L.Y.; Niu, B.; Tang, L.; Chen, F.; Zhao, G.; Zhao, J.L. Effects of polysaccharide elicitors from endophytic *Fusarium oxysporum* Fat9 on the growth, flavonoid accumulation and antioxidant property of *Fagopyrum tataricum* sprout cultures. *Molecules* **2016**, *21*, 1590. [CrossRef] [PubMed]
11. Mao, S.R.; Shuai, X.T.; Unger, F.; Simon, M.; Bi, D.Z.; Kissel, T. The depolymerization of chitosan: Effects on physicochemical and biological properties. *Int. J. Pharm.* **2004**, *281*, 45–54. [CrossRef]
12. Shibuya, N.; Minami, E. Oligosaccharide signalling for defence responses in plant. *Physiol. Mol. Plant Pathol.* **2001**, *59*, 223–233. [CrossRef]
13. No, H.K.; Lee, K.S.; Kim, I.D.; Park, M.J.; Kim, S.D.; Meters, S.P. Chitosan treatment affects yield, ascorbic acid content, and hardness of soybean sprouts. *J. Food Sci.* **2003**, *68*, 680–685. [CrossRef]
14. No, H.K.; Kim, S.D.; Prinyawiwatkul, W.; Meyers, S.P. Growth of soybean sprouts affected by chitosans prepared under various deproteinization and demineralization times. *J. Sci. Food Agric.* **2006**, *86*, 1365–1370. [CrossRef]
15. Lee, Y.S.; Kim, Y.H.; Kim, S.B. Changes in the respiration, growth, and vitamin C content of soybean sprouts in response to chitosan of different molecular weights. *HortScience* **2005**, *40*, 1333–1335. [CrossRef]
16. Rui, Y.; Yu, J.; Xiu, L.; Huang, J. Effect of chitosan pre-soaking on the growth and quality of yellow soybean sprouts. *J. Sci. Food Agric.* **2019**, *99*, 1596–1603.
17. Lan, W.Q.; Wang, W.; Yu, Z.M.; Qin, Y.X.; Luan, J.; Li, X.Z. Enhanced germination of barley (*Hordeum vulgare* L.) using chitooligosaccharide as an elicitor in seed priming to improve malt quality. *Biotechnol. Lett.* **2016**, *38*, 1935–1940. [CrossRef]
18. Kim, S.K.; Rajapakse, N. Enzymatic production and biological activities of chitosan oligosaccharides (COS): A review. *Carbohydr. Polym.* **2005**, *62*, 357–368. [CrossRef]
19. Wang, J.H.; Ye, Y.T.; Li, Q.; Abbasi, A.M.; Guo, X.B. Assessment of phytochemicals, enzymatic and antioxidant activities in germinated mung bean (*Vigna radiata* L. Wilezek). *Int. J. Food Sci. Technol.* **2017**, *52*, 1276–1282. [CrossRef]
20. Yang, F.; Luan, B.; Sun, Z.; Yang, C.; Yu, Z.M.; Li, X.Z. Application of chitooligosaccharides as antioxidants in beer to improve the flavour stability by protecting against beer staling during storage. *Biotechnol. Lett.* **2016**, *39*, 305–310. [CrossRef]
21. Zhang, S.; Tang, W.Z.; Jiang, L.L.; Hou, Y.M.; Yang, F.; Chen, W.F.; Li, X.Z. Elicitor activity of algino-oligosaccharide and its potential application in protection of rice plant (*Oryza saliva* L.) against *Magnaporthe grisea*. *Biotechnol. Biotechnol. Equip.* **2015**, *29*, 646–652. [CrossRef]
22. Hamel, L.P.; Beaudoin, N. Chitooligosaccharide sensing and downstream signaling: Contrasted outcomes in pathogenic and beneficial plant-microbe interaction. *Planta* **2010**, *232*, 787–806. [CrossRef]
23. Kulikov, S.N.; Chirkov, S.N.; Il'ina, A.V.; Lopatin, S.A.; Varlamov, V.P. Effect of the molecular weight of chitosan on its antiviral activity in plants. *Appl. Biochem. Microbiol.* **2006**, *42*, 200–203. [CrossRef]
24. Mendoza-Sanchez, M.; Guevara-Gonzalez, R.G.; Castano-Tostado, E.; Mercado-Silva, E.M.; Acosta-Gallegos, J.A.; Rocha-Guzman, N.E. Effect of chemical stress on germination of cv Dalia bean (*Phaseolus vularis* L.) as an alternative to increase antioxidant and nutraceutical compounds in sprouts. *Food Chem.* **2016**, *212*, 128–137. [CrossRef] [PubMed]
25. Liu, H.K.; Cao, Y.; Huang, W.N.; Guo, Y.D.; Kang, Y.F. Effect of ethylene on total phenolics, antioxidant activity, and the activity of metabolic enzymes in mung bean sprouts. *Eur. Food Res. Technol.* **2013**, *237*, 755–764. [CrossRef]
26. Kumari, S.; Krishnan, V.; Sachdev, A. Impact of soaking and germination durations on antioxidants and anti-nutrients of black and yellow soybean (*Glycine max.* L.) varieties. *J. Plant Biochem. Biotechnol.* **2015**, *24*, 355–358. [CrossRef]
27. Guo, X.B.; Li, T.; Tang, K.X.; Liu, R.H. Effect of germination on phytochemical profiles and antioxidant activity of mung bean sprouts (*Vigna radiata*). *J. Agric. Food Chem.* **2012**, *60*, 11050–11055. [CrossRef] [PubMed]

28. Jiratanan, T.; Liu, R.H. Antioxidant activity of processed table beets (*Beta vulgaris var, conditiva*) and green beans (*Phaseolus vulgaris* L.). *J. Agric. Food Chem.* **2004**, *52*, 2659–2670. [CrossRef] [PubMed]

29. Ramirez-Estrada, K.; Vidal-Limon, H.; Hidalgo, D.; Moyano, E.; Golenioswki, M.; Cusidó, R.M.; Palazon, J. Elicitation, an effective strategy for the biotechnological production of bioactive high-added value compounds in plant cell factories. *Molecules* **2016**, *21*, 182. [CrossRef]

30. Kim, E.H.; Kim, S.H.; Chung, J.I.; Chi, H.Y.; Kim, J.A.; Chung, I.M. Analysis of phenolic compounds and isoflavones in soybean seeds (*Glycine max* (L.) Merill) and sprouts grown under different conditions. *Eur. Food Res. Technol.* **2006**, *222*, 201. [CrossRef]

31. Lee, A.H.; Su, D.D.; Pasalich, M.; Tang, L.; Binns, C.W.; Qiu, L.Q. Soy and isoflavone intake associated with reduced risk of ovarian cancer in southern Chinese women. *Nutr. Res.* **2014**, *34*, 302–307. [CrossRef]

32. Al-Tawaha, A.M.; Seguin, P.; Smith, D.L.; Beaulieu, C. Biotic elicitors as a means of increasing isoflavone concentration of soybean seeds. *Ann. Appl. Biol.* **2005**, *146*, 303–310. [CrossRef]

33. Chen, Y.M.; Chang, S.K.C. Macronutrients, phytochemicals, and antioxidant activity of soybean sprout germinated with or without light exposure. *J. Food Sci.* **2015**, *80*, 1391–1398. [CrossRef]

34. Dixon, R.A.; Paiva, N.L. Stress-induced phenylpropanoid metabolism. *Plant Cell* **1995**, *7*, 1085–1097. [CrossRef] [PubMed]

35. Algar, E.; Ramos-Solano, B.; García-Villaraco, A.; Sierra, M.D.S.; Gómez, M.S.M.; Gutiérrez-Mañero, F.J. Bacterialbioeffectors modify bioactive profile and increase isoflavone content in soybean sprouts (*Glycine max var Osumi*). *Plant Foods Hum. Nutr.* **2013**, *68*, 299–305. [CrossRef]

36. Molyneux, P. The use of the stable free radical diphenylpicrylhydrazyl (DPPH) for estimating antioxidant activity. *Songklanakarin J. Sci. Technol.* **2004**, *26*, 211–219.

37. Yun, J.; Li, X.H.; Fan, X.T.; Li, W.L.; Jiang, Y.Q. Growth and quality of soybean sprouts (*Glycine max* L. Merrill) as affected by gamma irradiation. *Radiat. Phys. Chem.* **2013**, *82*, 106–111. [CrossRef]

38. Zou, P.; Li, K.C.; Liu, S.; Xing, R.E.; Qin, Y.K.; Yu, H.H.; Zhou, M.M.; Li, P.C. Effect of chitooligosaccharides with different degrees of acetylation on wheat seedlings under salt stress. *Carbohydr. Polym.* **2015**, *126*, 62–69. [CrossRef]

39. Smirnoff, N.; Cumbes, Q.J. Hydroxyl radical scavenging activity of compatible solutes. *Phytochemistry* **1988**, *28*, 1057–1060. [CrossRef]

40. Chung, S.K.; Osawa, T.; Kawakishi, S. Hydroxyl radical-scavenging effects of spices and scavengers from brown mustard (*Brassica nigra*). *Biosci. Biotechnol. Biochem.* **1997**, *61*, 118–123. [CrossRef]

41. Liu, J.; Tian, S.P.; Meng, X.H.; Xu, Y. Effects of chitosan on control of postharvest diseases and physiological responses of tomato fruit. *Postharvest Biol. Technol.* **2007**, *44*, 300–306. [CrossRef]

42. Syeed, S.; Khan, N.A. Activities of carbonic anhydrase, catalase and ACC oxidase of mung bean (*Vigna radiata*) are differentially affected by salinity stress. *J. Food Agric. Environ.* **2004**, *2*, 241–249.

43. Day, R.B.; Okada, M.; Ito, Y.; Tsukada, K.; Zaghouani, H.; Shibuya, N.; Stacey, G. Binding site for chitin oligosaccharides in the soybean plasma membrane. *Plant Physiol.* **2001**, *126*, 1162–1173. [CrossRef] [PubMed]

Article

Influence of Buckwheat and Buckwheat Sprouts Flours on the Nutritional and Textural Parameters of Wheat Buns

Alina Sturza [1], Adriana Păucean [1], Maria Simona Chiş [1,*], Vlad Mureşan [1],
Dan Cristian Vodnar [2,3], Simona Maria Man [1], Adriana Cristina Urcan [4], Iulian Eugen Rusu [1],
Georgiana Fostoc [1] and Sevastiţa Muste [1]

[1] Department of Food Engineering, Faculty of Food Science and Technology, University of Agricultural
Sciences and Veterinary Medicine of Cluj-Napoca, 3–5 Mănăştur Street, 400372 Cluj-Napoca, Romania;
alina.sturza@usamvcluj.ro (A.S.); adriana.paucean@usamvcluj.ro (A.P.); vlad.muresan@usamvcluj.ro (V.M.);
simona.man@usamvcluj.ro (S.M.M.); iulian.rusu@usamvcluj.ro (I.E.R.); georgiana_fostoc@yahoo.com (G.F.);
sevastita.muste@usamvcluj.ro (S.M.)

[2] Department of Food Science, Faculty of Food Science and Technology, University of Agricultural Sciences
and Veterinary Medicine of Cluj-Napoca, 3–5 Mănăştur Street, 400372 Cluj-Napoca, Romania;
dan.vodnar@usamvcluj.ro

[3] Institute of Life Sciences, University of Agricultural Sciences and Veterinary Medicine Cluj-Napoca,
3–5 Calea Mănăştur, 400372 Cluj-Napoca, Romania

[4] Department of Microbiology and Immunology, University of Agricultural Sciences and Veterinary Medicine,
3–5 Mănăştur Street, 400372 Cluj-Napoca, Romania; adriana.urcan@usamvcluj.ro

* Correspondence: simona.chis@usamvcluj.ro; Tel.: +40-742-009731

Received: 19 October 2020; Accepted: 9 November 2020; Published: 10 November 2020

Abstract: In recent years, food products manufactured with buckwheat and sprouts flours have attracted widespread interest due to their high nutritional value with various health benefits, becoming more and more popular. The purpose of this study was to assesses the influence of buckwheat and sprouts flours on the nutritional, sensorial and textural characteristics on the final baked products. In order to achieve these goals, methods like HPLC-RID (High-Perfomance Liquid Chromatography with Refractive Index Detection), aluminum chloride colorimetric assay, Folin-Ciocalteu and 1,1-Diphenyl-2-picrylhydrazyl (DPPH) were used to determine fructose, glucose, sucrose, maltose; total flavonoids, total phenols and antioxidant activity. Sensorial analysis was realized by using hedonic test and texture profile was performed on a CT 3 Texture Analyzer. The results proved that wheat flour could be successfully replaced by 20% buckwheat and 10% sprouts flours, respectively, improving their nutritional value, without negative influence on texture parameters and sensorial features. The obtained buns were accepted by consumers with a total hedonic score of 9.1 and 8.7, respectively. Hardness, gumminess and adhesiveness were improved by using Magimix improver, meanwhile cohesiveness, springiness, gumminess and adhesiveness were improved by using guar gum.

Keywords: buckwheat flour; buckwheat sprouts; buns; quality and textural parameters

---

1. Introduction

The base of human food pyramid are cereal-based products due to their nutritional value rich in carbohydrates, proteins and lipids that provides 56% of the total energy world consumption and 50% of the protein consumed globally [1]. Wheat (*Triticum aestivum*) is one of the main cereal crops used worldwide in bakery manufacturing [2]. However, through wheat milling various bioactive compounds amount could be reduced or even lost [2–4].

During the last decades, consumer requirements regarding bakery products became more focused towards products obtained from unconventional flours or from composite ones. Pseudocereals such as buckwheat, quinoa and amaranth have gained the researches attention due to their precious chemical composition rich in fibers, proteins, phenolic acids, flavonoids, fatty acids, vitamins and minerals [5–9], which are known for their benefits on human health [5].

Buckwheat is a pseudocereal with a significant level of proteins in comparison to cereals (wheat, sorghum, corn, rice etc.) with a range between 7–21%, being easily digestible [8]. This pseudocereal is also rich in minerals, especially in copper, manganese, iron, potassium, sodium and zinc [5,10,11]. Likewise, buckwheat is known for its B complex vitamin content, vitamins E and PP [9,12] as well as for its high dietary fibers content [13]. Its bioactive compounds such as phenolic acids and flavonoids (rutin and quercitin mainly) contribute to a rich antioxidant capacity, with multiple positive implications on consumers health [14–16].

Recently, sprouts consumption has increased mainly because of their nutritional value an excellent source of proteins, fibers, vitamins, minerals and various bioactive compounds. Germination facilitates the improvement of digestibility by increasing the bioavailability of minerals and increases the phenolic content, which could enhance the antioxidant activity [17–20].

In light of this evidence, buckwheat and buckwheat sprouts help to prevent and treat various diseases such as oxidative anxiety, neurodegenerative disorders, cardiovascular diseases and even skin cancer. These functional properties increased their application in food and pharmaceutical products manufacturing [21].

Considering that both buckwheat and sprouts flours are gluten-free, their introduction in bread manufacturing process represent a new technology challenge in order to obtain products with desirable sensorial characteristics; therefore, the introduction of some improvers such as special enzymes or hydrocolloids is necessary to mimic gluten network, which is responsible for dough structure [22,23].

Guar gum is an inexpensive and renewable additive used in various industries; however, its main use is in the food industry [24,25]. In breadmaking, guar ensures uniform hydration of the flour, improving the kneading and shaping of the dough. The high-water absorption capacity leads to a desired consistency of the dough. As a result, final baked products have a higher level of moisture retention even after baking, and the storage time is considered to be longer [24,26,27].

Xylanase, alpha-amylase and glucose oxidase are enzymes used in the bakery industry due to their various benefits. There are several studies which have reported that alpha amylase optimizes the ability to retain gases in the dough, promotes the increase in volume of products and their elasticity, improves color and helps to extend the shelf life by decreasing the crumb hardness [28,29]. Regarding xylanase, several authors reported an improvement of sensory parameters such as finer texture; more uniform crumb structure; lower adhesiveness; better color, taste and aroma as the control sample [30,31].

It is considered that the addition of glucose oxidase improves the qualities of the dough due to the strengthening effect of the gluten network, an important function especially in the case of lean flours. The use of this enzyme helps to obtain baked goods with increased volume and improves crumb properties [32].

This research aimed to establish the optimum addition levels of buckwheat and sprouted buckwheat flours on the final baked buns and to assesses their influence from the nutritional, sensorial and textural parameters. Hydrocolloids like guar gum and complex baking improver were used in order to improve the buns' texture parameters.

2. Materials and Methods

2.1. Experimental Design

In the buns manufacturing, the buckwheat flour and sprouted buckwheat flour were mixed in different percentages with wheat flour (10%, 20% and 30% respectively). The flours mixes were further entitled as composite flours.

The buns formulation contained the following ingredients: composite flours consisting of wheat flour, buckwheat and sprouted buckwheat flours, in different percentages (10%, 20%, and 30% respectively), water, salt, yeast, guar gum and Magimix improver, as showed in Tables 1 and 2. The following codes were used for composite flours with 10, 20, 30% buckwheat flour (BW): 10BW, 20BW and 30BW, meanwhile for composite flour with sprouts flour (SF) the abbreviations were: 10SF, 20SF and 30SF, respectively.

Table 1. Buns recipe by adding buckwheat flour.

Sample	Ingredients [%]						
	Wheat Flour	BW Flour	Water	Salt	Yeast	G	M
WF	100	0	60.7	2	2		
10BW1	90	10	68	2	2	0.2	
20BW1	80	20	76.5	2	2	0.2	
30BW1	70	30	81.1	2	2	0.2	
10BW2	90	10	68	2	2		0.2
20BW2	80	20	76.5	2	2		0.2
30BW2	70	30	81.1	2	2		0.2

WF: wheat flour, control sample; BW: flour-buckwheat flour; G: guar gum; M: baking improver Magimix; 10BW1, 20BW1, 30BW1: buns with 10%, 20%, 30% buckwheat flour addition and 0.2% guar gum; 10BW2, 20BW2, 30BW2: buns with 10%, 20%, 30% buckwheat flour and 0.2% Magimix.

Table 2. Buns recipe by adding sprouted buckwheat flour.

Sample	Ingredients [%]						
	Wheat Flour	SF Flour	Water	Salt	Yeast	G	M
10SF1	90	10	60.7	2	2	0.2	
20SF1	80	20	60.7	2	2	0.2	
30SF1	70	30	60.7	2	2	0.2	
10SF2	90	10	60.7	2	2		0.2
20SF2	80	20	60.7	2	2		0.2
30SF2	70	30	60.7	2	2		0.2

SF flour: sprouted buckwheat flour, G: guar gum, M: baking improver Magimix; 10SF1, 20SF1, 30SF1: buns with 10%, 20%, 30% sprouted buckwheat flour and 0.2% guar gum; 10SF2, 20SF2, 30SF2: buns with 10%, 20%, 30% sprouted buckwheat flour and 0.2% Magimix.

Buns were obtained by mixing all ingredients for 3 min at medium speed and 4 min at high speed, by using a mixer (KitchenAid® Precise Heat Mixing Bowl., Greenville, OH, USA). The doughs were left to leaven for 23 h at 2–4 °C. After the cold fermentation, doughs were left for leavening for 1 h at 29 °C and 80% relative humidity. This stage aims to raise the temperature of the doughs as close as possible to the temperature of the final leavening; further, doughs were divided and shaped into buns weighing 120 g and left for final fermentation for 30 min at 30 °C and 75% relative humidity in a thermoclimatic chamber (Zanolli, Verona, Italy). The baking of buns samples was carried out in a professional oven for 30 min at 220 °C (Zanolli, Verona, Italy). After baking, buns were left to cool at room temperature for 1 h, before being packaged and stored until testing.

2.2. Materials

Wheat flour was produced by a local Romanian mill (Europan, Sarmasag, Romania) and sold as type 550 according to ash content by Romanian classification (moisture 14.3%, wet gluten 29.3% and Falling Number 340 s. Buckwheat flour (BW) and green seeds were purchased from the health Romanian food stores; compressed yeast Fala (Bonopan) and salt from the local market. Guar gum and the complex baking improver Magimix-Bonopan (which contains alpha-amylase, maltogenic alpha-amylase, amyloglucosidase, glucose oxidase, xylanase, lipase and ascorbic acid) were acquired from Romanian ingredients market specialized on bakery industry (Lessafre-Marcq-en-Baroeul, France) and Solina Group (Alba Iulia, Romania).

The sugars standards were purchased from Merck (Darmstadt, Germany), and the analytical reagents and chemicals were acquired from Sigma Aldrich (St. Louis, MO, USA), being analytical grade.

2.3. Buckwheat Sprouts (SF) Manufacturing

In order to obtain buckwheat sprouts, achenes were soaked and germinated using the EasyGreen Germinator (EasyGreen factory, Langlade, France) at the temperature of 23–25 °C for 5 days (first 24 h by maintaining the light and 96 h by alternating: 12 h of light and 12 h of darkness). During germinating, seeds were sprayed with a mixture of tap and distilled water 1:1 *v/v* for 15 min every 12 h. Sprouted buckwheat achenes were dried in a food drier Biovita-DEH450 (Cluj-Napoca, Romania) at 35 °C for 24 h; flour was obtained by using a grinding machine ZBPP (Sadkiewicz-Instruments, Bydgoszcz, Poland) and sieved until the 0.6 mm granulation was reached. Afterward was stored at 20 °C until further analysis.

2.4. Physico-Chemical Analyses

Physico-chemical parameters of the raw materials and of the final baked products like moisture, ash, protein (total nitrogen × 5.7), lipids were determined according to AACC approved methods 44–15.02, 08–01.01, 46–11 A and 30–25.01, respectively (AACC-American Association of Cereal Chemists [33]). TTA (total titrable acidity) was determined by using a WTW pH-meter (Hanna Instruments, Vöhringen, Germany) and expressed as mL of NaOH, according to Corsetti et al., [34].

2.5. Free Sugars Estimation

Free sugars were analyzed by HPLC (High-Performance Liquid Chromatography) and the samples extraction was realized as follows: 2.5 g of sample was dissolved in distilled water (40 mL) and transferred quantitatively into a 50 mL volumetric flask and filled up to the mark with water. The solution was filtered through a 0.45 μm membrane filter (Millipore, Merck 164 KGaA, Darmstadt, Germany), collected in sample vials and placed in an autosampler for analysis. The HPLC analysis of the free sugars was carried out on a modified Alltima Amino 100 stainless steel column (Hicrom, Berkshire, UK), with the following characteristics: 4.6 mm diameter, 250 mm length, particle size 5 μm, according to the method developed by Bonta et al. [35]. The High Performance Liquid Chromatograph (LC -10AD VP model, Shimadzu, Kyoto, Japan) was equipped with degasser, two pumps, autosampler, thermostat oven, controller and refractive index detector. The injection volume was 10 μL and the flow rate 1.3 mL/min. The mobile phase was a solution composed by of acetonitrile and ultrapure water (80/20 *v/v*). For the quantification of main sugars, different calibration curves in the range of 50–0.25 g/100 g (fructose 50–20 g/100 g; glucose 10–40 g/100 g; sucrose 0.3–15 g/100 g; turanose, maltose, isomaltose, erlose 0.25–5 g/100 g), with regression coefficients (R^2) higher of 0.998 were used. The results are expressed in g/100 g f.w. (fresh weight).

2.6. Methanolic Samples Extraction

In order to analyze total flavonoids, total phenols and antioxidant activity, samples were prepared according to Bunea et al. [36] and Păucean et al. [37], with some modifications. Briefly, 1 g of sample, in three replicated each, was mixed with 100 mL of acidified methanol (CH_3OH: HCl, 85:15 *v/v*). Extraction was carried out for 1 h at room temperature using a stirrer (Velp magnetic stirrer, Usmate (MB)-Italy). Afterwards, the samples were maintained for 23 h at 4–8 °C and the mixture was filtered under vacuum through Whatman filter paper no.1. The filtrates were combined in a total extracts and the solvent was removed by using a vacuum-evaporator (Laborota 4010 digital rotary evaporator, Heidolph Instruments GmbH & Co. KG, Schwabach, Germany) at 35 °C, until the extraction solvent became colorless. The residue was recovered with/in 10 mL of methanol and filtered through a nylon filter 0.45 μm (Millipore, Merck KGaA, Darmstadt, Germany); the obtained extracts were stored at −20 °C until use.

2.7. Determination of Total Flavonoids Content

Flavonoids were measured by the aluminum chloride colorimetric assay adapted for use on a 96-well microplate reader (Synergy™ HT BioTek Instruments, Winooski, VT, USA), using quercetin as reference standard [38,39]. A volume of 25 μL of sample was added into 100 μL distilled water and 10 μL of 5% sodium nitrate ($NaNO_2$) solution. After 5 min, 15 μL of 10% aluminum chloride (AlCl3) was added and the samples were maintained to rest for 5 min.

Afterward, 50 μL of 1 M sodium hydroxide was added to the mixture and 50 μL of distilled water, respectively. Absorbance of the mixture was measured at 510 nm. A standard curve of quercetin was used to establish the final amount of total flavonoids content (y = 0.0003 x + 0.0029, R^2 = 0.9916) and the results were expressed as mg quercetin equivalents (mg QE/100 g). The assays were run in triplicate.

2.8. Total Phenolic Estimation

The total phenolic content from flours and final products was expressed as gallic acid equivalents/100 g of fresh product and determined using Folin-Ciocalteu method, adapted by Đorđević et al. [40] and Chis et al. [41]. Briefly, in a 10 mL volumetric flask, 100 μL of methanolic extract was mixed with 1 mL of distilled water and 500 mL of Folin-Ciocalteu's reagent; after 10 s of shaking and 5 min of rest at room temperature, 5 mL of 7.5% sodium carbonate solution was added. Distilled water was used until the graduation marking of the flask was reached. The obtained mixtures were allowed to stand for 90 min in a dark place without shaking. The absorbance was measured at 760 nm in a spectrophotometer (Shimadzu Scientific Instruments, Kyoto, Japan), meanwhile methanol was used as blank.

2.9. Antioxidant Activity

Antioxidant capacity was determined by reacting 0.1 mL methanolic extract with 3.9 mL of 0.025 g/L 2,2-diphenyl-1-picryl-hydrazyl (DPPH) as described by Đorđević et al. [40] and Chiș et al. [42] with small modifications. The absorbance of samples was read in triplicate at 515 nm by using a spectrophotometer Shimadzu 1700 (Shimadzu Scientific Instruments, Kyoto, Japan). Results were expressed as a percentage against the absorbance of the standard DPPH solution according to the following equation: radical scavenging ability (RSA) (%) = $[(A_0 − A_1)/A_0] \times 100$, where A_0 was the absorbance of the control (blank, without extracts), A_1 was the absorbance in the presence of the samples of the extracts. Methanol was used as a blank.

2.10. Rheological Assessment

Texture profile analysis of the buns was performed by using a Brookfield CT3 Texture Analyzer (Brookfield Engineering Labs, Middleboro, MA, USA) equipped with 10 kg load cell and the TA11/1000 cylindrical probe (25.4 mm diameter AOAC Standard Clear Acrylic 21 g, 35 mm length; 40% target deformation, 1 mm s^{-1} test and post-test speed, 5g trigger load, and 5 s recovery time) having as reference the method described by Păucean et al. [36] and Man et al. [43].

Baked samples were left 2 h before the assessment to cool, after cut into cube with side of 25 mm, placed on the analyzer plate and compressed twice with cylindrical probe. Texture parameters like hardness 1 and 2, total work 1 and 2, cohesiveness, springiness, chewiness, gumminess and adhesiveness were calculated using Texture Pro CT V1.6 software (Brookfield Engineering Labs, Middleboro, MA, USA). Hardness 1 and hardness 2 are defined as the peak force of the first and second compression cycles, respectively, meanwhile total work 1 and total work 2 represent the energy required to deform the samples during the two compression cycles.

2.11. Sensorial Analysis

From the sensorial point of view characteristics like appearance, texture, color, taste, flavor and overall acceptability were analyzed by 25 semi-trained panelists, staff and students of the Faculty of Food Science and Technology (70% female and 30% male, range 19–63 years), from Cluj-Napoca, Romania. A nine hedonic test was used for the evaluation of these parameters, ranging for 9 as like extremely to 1 as dislike extremely, according to the method described by Chis et al. [41].

2.12. Statistical Analysis

Data were analyzed using one-way ANOVA followed by a post hoc Duncan's multiple comparison test, performed in GraphPad Prism8 Software (San Diego, CA, USA), with a confidence interval of 95%, $p < 0.05$. The results of three independent (n = 3) replicates were averaged and expressed as mean ± standard deviations.

3. Results and Discussions

3.1. Flours Characterization

3.1.1. Physico-Chemical Parameters of Flours

The physico-chemical parameters obtained for wheat, buckwheat and sprouted buckwheat flours are presented in Table 3. The results indicated that the levels of ash, total fat, protein and fibers were significantly higher in sprouted buckwheat flours compared to buckwheat and wheat flours, respectively.

Table 3. Physico-chemical parameters of flours.

Sample	Moisture Content [%]	Ash [%]	Total Fat [%]	Protein [%]	Fibers [%]	TTA
WF	10.55 ± 0.23 [d]	0.54 ± 0.11 [a]	1.36 ± 16 [a]	10.80 ± 0.25 [a]	1.14 ± 0.11 [a]	1.93 ± 0.10 [a]
BW	9.24 ± 0.27 [a]	1.64 ± 0.10 [d,e]	2.36 ± 0.11 [c]	12.23 ± 0.23 [c,d]	4.08 ± 0.24 [e]	5.40 ± 0.21 [f]
10BW	10.36 ± 0.31 [c,d]	0.71 ± 0.12 [a,b]	1.69 ± 0.15 [a,b]	11.14 ± 0.32 [a,b]	1.78 ± 0.10 [b]	2.43 ± 0.16 [b]
20BW	10.25 ± 0.34 [b,c]	0.88 ± 0.15 [b,c]	1.83 ± 0.17 [a,b,c]	11.42 ± 0.11 [a,b,c]	2.03 ± 0.25 [b]	2.97 ± 0.12 [c]
30BW	10.10 ± 0.25 [b]	1.05 ± 0.22 [c]	2.03 ± 0.19 [b,c]	11.67 ± 0.22 [a,b,c]	2.53 ± 0.21 [i,c,d]	3.50 ± 0.20 [d]
SF	9.16 ± 0.19 [a]	2.53 ± 0.34 [f]	5.54 ± 0.22 [f]	18.75 ± 0.56 [f]	4.67 ± 0.12 [f]	4.37 ± 0.16 [e]
10SF	10.36 ± 0.27 [c,d]	1.00 ± 0.17 [c]	2.23 ± 0.18 [b,c]	12.14 ± 0.31 [b,c,d]	2.02 ± 0.10 [b]	2.33 ± 0.12 [a,b]
20SF	10.26 ± 0.32 [b,c]	1.42 ± 0.23 [d]	3.06 ± 0.21 [d]	12.83 ± 0.32 [d,e]	2.48 ± 0.15 [c]	2.63 ± 0.25 [b,c]
30SF	10.18 ± 0.25 [b,c]	1.76 ± 0.27 [e]	3.90 ± 0.11 [e]	13.66 ± 0.25 [e]	2.96 ± 0.11 [d]	3.00 ± 0.11 [c]

WF: wheat flour; BW flour: buckwheat flour; 10BW, 20BW, 30BW: composite flour with 10%, 20%, 30% buckwheat flour addition; SF flour: sprouted buckwheat flour; 10SF, 20SF, 30SF: composite flour with 10%, 20%, 30% sprouted buckwheat flour. Different lowercase letters within a column indicate significant differences between the means of the same parameter analyzed at $p < 0.05$ and (n = 3) according to Duncan's multiple range test. All results are expressed to fresh weight (f.w.) product.

In the present study, the SF protein content (18.75%) is in line with the values presented by Lim Kim et al. [44] (19%) and Lee and Kim [45] (19.35 ± 2.14%), respectively. Kim et al. [44] reported also higher levels for ash (2.82%) and fibers (4.39%) of sprouts compared to seeds (2.80% ash and 3.82% fiber contents). In this light, Lee and Kim [45] reported a value of 3.51% ash for sprouts and 1.78% for ungerminated seeds. With respect to total lipid amount, a higher level of total lipids in sprouts compared to seeds (2.59% and 3.02%, respectively) were reported [44,45].

Regarding the composite flours, the partial addition of these unconventional flours to wheat flour, led to an enrichment of samples from a physico-chemical point of view (Table 3), where buckwheat sprouts flour had a greater impact on these parameters.

3.1.2. Free Sugars Content of Flours

The free sugars determined in the present study were: glucose, sucrose, maltose and fructose, as showed in Table 4. The main sugar from the analyzed flours was glucose (4.80%) which was identified in higher extent amount in SF flour, meanwhile WF and BW flours reached smaller values such as 0.07% and 0.03%, respectively. Fructose, was the second major sugar identified in a bigger amount in SF flour (2.5%), followed by sucrose which was identified mainly in BW flour (0.74%). The presence of glucose and fructose in SF flour could be justified by the germination process which lead to the degradation of disaccharides, trisaccharides and tetrasaccharides into simple sugars in order to provide the energy needed for the sprouts grow [34,46,47].

Table 4. Free sugars content of flour samples.

Sample	Glucose [%]	Sucrose [%]	Maltose [%]	Fructose [%]
RT (min)	11.20	17.80	22.70	9.10
WF	0.07 ± 0.01 [a]	0.21 ± 0.02 [a]	0.07 ± 0.02 [d]	0.05 ± 0.01 [a]
BW	0.03 ± 0.01 [a]	0.74 ± 0.05 [d]	0.11 ± 0.10 [f]	0.20 ± 0.05 [a]
10BW	0.08 ± 0.02 [a]	0.26 ± 0.01 [a,b]	0.07 ± 0.01 [d]	0.07 ± 0.01 [a]
20BW	0.08 ± 0.02 [a]	0.30 ± 0.01 [b,c]	0.07 ± 0.01 [d]	0.08 ± 0.01 [a]
30BW	0.07 ± 0.01 [a]	0.35 ± 0.02 [c]	0.08 ± 0.01 [e]	0.09 ± 0.02 [a]
SF	4.80 ± 0.12 [d]	0.27 ± 0.05 [a,b]	0.02 ± 0.01 [a]	2.50 ± 0.10 [c]
10SF	0.94 ± 0.08 [b]	0.21 ± 0.08 [a]	0.06 ± 0.01 [c]	0.21 ± 0.01 [a]
20SF	2.00 ± 0.10 [c]	0.21 ± 0.05 [a]	0.06 ± 0.01 [c]	0.40 ± 0.05 [a,b]
30SF	2.61 ± 0.09 [c]	0.23 ± 0.04 [a]	0.05 ± 0.01 [b]	0.63 ± 0.05 [b]

RT = retention time; WF: wheat flour; BW flour: buckwheat flour; 10BW, 20BW, 30BW: composite flour with 10%, 20%, 30% buckwheat flour addition; SF flour-sprouted buckwheat flour, 10SF, 20SF, 30SF: composite flour with 10%, 20%, 30% sprouted buckwheat flour. Different lowercase letters within a column indicate significant differences between the means of the same parameter analyzed at $p < 0.05$ and (n = 3) according to Duncan's multiple range test.

Sucrose was the major compound in buckwheat (0.74%) and belongs to the category of low-molecular-weight sugars; Mazza and Omah reported also that sucrose has the highest amount among the low-molecular-weight sugars in buckwheat flour [48].

With respect to WF sugars content, sucrose was the mainly simple sugar identified (0.21%), followed by glucose (0.07%), maltose (0.07%) and fructose (0.05%). The results are similar to those reported by Codina et al. [49] who identified for WF values of 0.05% for glucose, maltose and fructose and 0.2% for sucrose, respectively. Sahlström et al. [50] analyzed flours from six wheat cultivars and obtained values between 0.04–0.05% for glucose, 0.188–0.246% for sucrose, 0.44–0.063% for maltose, but a higher value for fructose 0.79–1.04%.

Therefore, composite flours with BW had higher values for sucrose and maltose, meanwhile flours with added SF obtained higher levels of glucose and fructose as shown in Table 4.

3.1.3. Total Flavonoid Content of Flours

The increased interest for the use of buckwheat and sprouts in food industry is also due to their flavonoid content, which are important compounds with an antioxidant role.

The total flavonoid of sprouted buckwheat flour was 659.92 mg QE/100 g f.w. Other studies reported values of 1500 mg/100 g f.w. [51] or values in the range of 100–1250 mg/100 g f.w. [52]. The differences between values could be explained by the buckwheat varieties but also by time and germination condition. An important germination condition could be the absence or presence of light and also its type (e.g., blue, red, fluorescent).

As illustrated in Table 3 buckwheat flour has a lower amount of flavonoids −275.58 mg QE/100 g compared with SF flour, which reached a value of 659.92 mg QE/100 g. The total flavonoids content of buckwheat flour was reported by the literature in a range of 67–225 mg/100 g by Qin et al. [47], meanwhile, Ren et al., [53] reported a value of 400 mg/100 g for buckwheat flour. The differences between the results could be explained by the buckwheat varieties and the cultivating conditions Qin et al. [47] such as the fertilizer soil treatments with *Azospirillum* and *Azobacter* [54].

As it was expected, the content of total flavonoids in wheat flour extract was the lowest: 10.10 mg QE/100 g compared to sprouted buckwheat and buckwheat flours, respectively.

Partial substitution of wheat flour resulted on a significant increase of total flavonoids amount in composite flours; 10BW, 20BW and 30BW reached values of 38.71 mg QE/100 g, 67.30 mg QE/100 g and 99.96 mg QE/100 g, respectively. Furthermore, 10SF, 20SF and 30SF samples had even higher values: 72.71 mg QE/100 g, 140.73 mg QE/100 g and 211.73 mg QE/100 g.

3.1.4. Total Phenolic Content of Flours

Sprouted buckwheat flour has a higher phenol content (678 mg GAE/100 g f.w.) compared to buckwheat and wheat flours which reached values of 365.67 mg GAE/100 g f.w. and 43.17 mg GAE/100 g f.w, respectively as illustrated in Table 5 These results are consistent with the values reported by Alvarez-Jubete et al. [7] who obtained values of 670 ± 12.3 mg GAE/100 g for buckwheat sprouts, 323 ± 14.1 mg GAE/100 g for buckwheat and 53.1 ± 2.8 mg GAE/100 g for wheat.

Table 5. Total flavonoids, total phenols and radical scavenging activity of flour samples.

Sample	Total Flavonoid Content [mg QE/100 g f.w.]	Total Polyphenol Content [mg GAE/100 g f.w.]	Radical Scavenging Activity [%]
WF	10.10 ± 0.21 [a]	43.17 ± 0.18 [a]	39.18 ± 0.15 [a]
BW	275.58 ± 0.56 [g]	365.67 ± 0.39 [f]	62.56 ± 0.31 [d]
10BW	38.71 ± 0.14 [b]	54.33 ± 0.22 [a]	45.15 ± 0.25 [b]
20BW	67.30 ± 0.22 [c]	77.00 ± 0.31 [b]	48.54 ± 0.51 [b,c]
30BW	99.96 ± 0.34 [d]	96.83 ± 0.25 [c]	50.83 ± 0.48 [b,c]
SF	659.92 ± 0.61 [h]	678.00 ± 0.70 [g]	87.31 ± 0.24 [e]
10SF	72.71 ± 0.24 [c]	78.83 ± 0.27 [b]	47.05 ± 0.31 [b,c]
20SF	140.73 ± 0.35 [e]	150.17 ± 0.29 [d]	52.54 ± 0.20 [c]
30SF	211.73 ± 0.49 [f]	218.83 ± 0.32 [e]	59.04 ± 0.14 [d]

f.w.: fresh weight; WF: wheat flour; BW flour: buckwheat flour, 10BW, 20BW, 30BW: composite flour with 10%, 20%, 30% buckwheat flour addition; SF flour: sprouted buckwheat flour; 10SF, 20SF, 30SF: composite flour with 10%, 20%, 30% sprouted buckwheat flour. Different lowercase letters within a column indicate significant differences between the means of the same parameter analyzed at $p < 0.05$ and (n = 3) according to Duncan's multiple range test.

Partial replacement of wheat flour in different percentages (10%, 20% and 30%) with buckwheat flour and sprouted buckwheat flour respectively, led to an increase of the phenolic content in composite flours (Table 5). This could be explained due to the higher total phenols content of buckwheat and sprouts flours compared to wheat flour.

3.1.5. Radical Scavenging Activity of Flours

The results for the raw materials radical scavenging activity (RSA) are presented in Table 5 Sprouted buckwheat flour was characterized with the higher radical scavenging activity 87.31%, followed by buckwheat flour with a score of 62.56%. Wheat flour had the lowest value for this parameter: 39.18%. As expected, the composite flours with the addition of sprouted buckwheat flour showed higher values than the composite flours with the addition of buckwheat flour as follows: 10SF:47.05%, 20SF:52.54%, 30SF:59.04%, while 10BW:45.54%, 20BW:48.54%, 30BW:50.83%.

Zhang et al. [21] reported a similar value for sprouted buckwheat flour. Some authors explained the high antioxidant activity of sprouted buckwheat flour by the fact that germination enhanced the accumulation of antioxidant compounds. There is also mentioned in the literature, that the enzymes contained in sprouts could be responsible for enhancing the radical scavenging activity [17,21].

Concerning RSA buckwheat flour, Inglett et al. [55] obtained values ranging between 65–71% RSA, after assessment of four buckwheat flour; and for wheat flour, Han et al. [56] presented a result of 28.0 ± 5.2%.

3.2. Final Products Characterization

3.2.1. Physico-Chemical Parameters of Buns

Table 6 presents the following physico-chemical buns parameters: moisture, ash, total fat, protein, fibers and TTA. The analyzes performed on the final baked products showed that buns moisture increases with the augmentation amount of buckwheat added compared to the control buns manufactured only with wheat flour. This could be explained due to the higher fiber content of BW compared to WF, increasing the water absorption capacity of buckwheat flour compared to the wheat. Several authors also reported higher moisture in various bakery products in which buckwheat was added compared to the control sample, attributing this to BW increased water absorption capacity [40–43].

Table 6. Physico-chemical parameters of buns.

Sample	Moisture Content [%]	Ash [%]	Total Fat [%]	Protein [%]	Fibers [%]	TTA
PM	39.94 ± 0.61 [b]	0.56 ± 0.11 [a]	1.17 ± 0.22 [a]	7.88 ± 0.28 [a]	0.69 ± 0.10 [a]	2.09 ± 0.15 [a]
10BW1	42.06 ± 0.65 [d]	0.72 ± 0.25 [b]	1.68 ± 0.11 [b]	8.51 ± 0.10 [b]	0.71 ± 0.14 [a,b]	2.53 ± 0.19 [b,c]
20BW1	43.76 ± 0.71 [e]	0.90 ± 0.23 [c]	1.83 ± 0.17 [b,c]	8.63 ± 0.18 [b]	0.85 ± 0.11 [d]	2.83 ± 0.28 [d]
30BW1	45.40 ± 0.52 [f]	1.05 ± 0.28 [d]	2.05 ± 0.15 [c]	8.82 ± 0.24 [c]	1.15 ± 0.21 [f]	3.23 ± 0.11 [j,f]
10BW2	41.14 ± 0.53 [c]	0.71 ± 0.14 [b]	1.74 ± 0.21 [b]	8.55 ± 0.31 [b]	0.71 ± 0.15 [a,b]	2.62 ± 0.18 [c]
20BW2	43.07 ± 0.59 [e]	0.89 ± 0.28 [c]	1.84 ± 0.11 [b,c]	8.64 ± 0.29 [b]	0.87 ± 0.06 [d]	2.87 ± 0.13 [d]
30BW2	44.83 ± 0.6 [f]	1.02 ± 0.30 [c,d]	2.04 ± 0.23 [c]	8.83 ± 0.25 [c]	1.13 ± 0.20 [f]	3.25 ± 0.38 [f]
10SF1	39.00 ± 0.54 [a]	1.10 ± 0.18 [d]	2.74 ± 0.28 [d]	9.03 ± 0.11 [d]	0.77 ± 0.13 [c]	2.45 ± 0.19 [b]
20SF1	38.96 ± 0.48 [a]	1.45 ± 0.17 [e]	2.98 ± 0.25 [d]	9.26 ± 0.32 [e]	0.96 ± 0.11 [e]	2.78 ± 0.17 [d]
30SF1	38.87 ± 0.39 [a]	1.80 ± 0.18 [f]	3.92 ± 0.30 [e]	9.38 ± 0.25 [e]	1.17 ± 0.14 [f]	3.12 ± 0.21 [e]
10SF2	39.00 ± 0.42 [a]	1.09 ± 0.21 [d]	2.79 ± 0.25 [d]	9.08 ± 0.24 [d]	0.78 ± 0.21 [c]	2.48 ± 0.27 [b]
20SF2	38.80 ± 0.37 [a]	1.40 ± 0.17 [e]	3.01 ± 0.11 [d]	9.26 ± 0.34 [e]	0.97 ± 0.13 [e]	2.87 ± 0.24 [d]
30SF2	38.72 ± 0.52 [a]	1.79 ± 0.21 [f]	3.84 ± 0.25 [e]	9.35 ± 0.31 [e]	1.15 ± 0.16 [f]	3.17 ± 0.28 [e,f]

PM: control sample; 10BW1, 20BW1, 30BW1: buns with 10%, 20%, 30% buckwheat flour addition and 0.2% guar gum; 10BW2, 20BW2, 30BW2: buns with 10%, 20%, 30% buckwheat flour and 0.2% Magimix; 10SF1, 20SF1, 30SF1—buns with 10%, 20%, 30% sprouted buckwheat flour and 0.2% guar gum; 10SF2, 20SF2, 30SF2: buns with 10%, 20%, 30% sprouted buckwheat flour and 0.2% Magimix. Different lowercase letters within a column indicate significant differences between the means of the same parameter analyzed at $p < 0.05$ and (n = 3) according to Duncan's multiple range test. All results are expressed to fresh weight (f.w.) product.

On the other side, the addition of sprouted buckwheat flour did not bring any modification of this parameter, the absorption capacity being the same: either 10% or 30% of this flour was added. This could be explained due to the higher fat content of SF compared to BW. The fat content of the raw material could have a directly influence on lowering the water absorption capacity, as previously demonstrated by [57].

The highest values for ash and total fat content were registered by buns supplemented with buckwheat sprouts flour followed by the products with addition of buckwheat flour. As we expected, wheat-based products had the lowest levels concerning the ash and total fat content.

The highest fiber content was identified in samples with the addition of SF, followed by samples manufactured with BW. The samples made with WF registered the lower fiber values content. This could be explained by the richest chemical composition of SF fiber (4.67%) compared to the BW content (4.08%), and WF content (1.14%), respectively Table 6. The use of improver and guar gum did not change in a statistically significant way the final fiber content.

Bhavsar et al. [58] reported for wheat bread 1.98% fiber content, 2.58% fiber content for the bread with 10% buckwheat flour and a value of 3.16% for the product with 20% buckwheat flour. These results are higher than those obtained in the present study, probably due to the fiber initial raw materials amount. As far as we know, no studies have been found in the literature regarding the content of fiber content in baked goods with the addition of sprouted buckwheat flour.

Regarding the protein content, it changed in all products. It is known that proteins undergo a series of changes during baking, of which the most important is the denaturation of gliadin and glutenins; gluten is transformed from an elastic structure into a semi-rigid one, thereby defining the final shape of the product [59].

The protein content in wheat-based buns (PM) was 7.88%, meanwhile, the final products supplemented with BW 10%, 20% and 30% registered values of 8.51%, 8.63% and 8.82%, respectively (Table 6). This could be justified by the higher protein content of BW compared to WF (Table 3). No significant differences were registered between sample with different improvers.

On the other side, the SF use in the manufacture of final baked buns determined an increment of the protein content, as showed in Table 6 This could be explained by the higher protein content of SF compared with BW and WF, respectively (Table 3).

Another parameter that has changed was the TTA, which increased as a result of alcoholic fermentation. The addition of both buckwheat and sprouted buckwheat flour led to the increment of this parameter. Drobot et al. [60] justified the increased TTA value due to the presence of a higher amount of organic acids in buckwheat compared to wheat flour. Selimović et al. [61] and Bojňanská et al. [62] indicated also an augmentation of TTA in bakery products supplemented with buckwheat flour, with positive effects on taste. According to Požrl et al. [46] the increase of TTA causes a decrease of phytic acid content, which is an antinutrient that binds minerals and proteins, changing their solubility, absorption and digestibility. Long fermentation is also associated with quantitative decrease of phytic acid.

3.2.2. Free Sugars Content of Final Products

Dough sugars are subjected to transformations during baking, so the temperature gradient between the core and the dough surface is responsible for the behavior of the starch as follows: in the center, being lower temperature, the starch becomes sticky and with a colloidal structure, thus forming the crumb. At the surface, the higher temperature causes dextrinization and caramelization of the available sugars. Volatile gases and compounds are already depleted, and the Maillard reaction between sugars and amino acids leads to the formation of new compounds, responsible for the organoleptic properties typical of the manufactured assortment [63].

The free sugars content of the final baked products increased as BW and SF flour percentages increased due to their richer chemical composition in simple sugars (Table 3) and also as a possible consequence of two reactions: enzymatic hydrolysis of starch and fructans during kneading and fermentation of dough and due to thermal degradation during baking process, which lead to the formation of lower sugars, especially maltose and sucrose [63]. Samples with BW addition registered an augmentation of glucose, sucrose, maltose and fructose content; while in buns supplemented with SF, glucose was reduced, but fructose increased (Table 7).

To the best of our knowledge, no studies have been found in the literature regarding the content of free sugars in baked goods with the addition of these two unconventional flours.

Table 7. Free sugars content from buns.

Sample	Glucose [%]	Sucrose [%]	Maltose [%]	Fructose [%]
RT (min)	11.20	17.80	22.70	9.10
PM	0.16 ± 0.08 [a]	0.95 ± 0.10 [a]	0.83 ± 0.10 [b]	0.06 ± 0.01 [a]
10BW1	0.16 ± 0.07 [a]	0.99 ± 0.10 [d]	0.85 ± 0.12 [b,c]	0.08 ± 0.01 [a]
20BW1	0.16 ± 0.08 [a]	1.04 ± 0.12 [e]	0.86 ± 0.09 [c]	0.10 ± 0.01 [a]
30BW1	0.19 ± 0.09 [c]	1.07 ± 0.08 [f]	0.89 ± 0.05 [d]	0.11 ± 0.05 [a]
10BW2	0.16 ± 0.05 [a]	0.99 ± 0.08 [d]	0.84 ± 0.07 [b,c]	0.08 ± 0.02 [a]
20BW2	0.17 ± 0.05 [b]	1.04 ± 0.09 [e]	0.86 ± 0.12 [c]	0.10 ± 0.03 [a]
30BW2	0.20 ± 0.04 [d]	1.07 ± 0.06 [f]	0.89 ± 0.08 [d]	0.12 ± 0.02 [a]
10SF1	0.16 ± 0.08 [a]	0.95 ± 0.04 [a]	0.75 ± 0.05 [a]	0.23 ± 0.05 [b]
20SF1	0.19 ± 0.05 [c]	0.95 ± 0.07 [a]	0.74 ± 0.10 [a]	0.43 ± 0.05 [c]
30SF1	0.21 ± 0.07 [e]	0.96 ± 0.06 [a,b,c]	0.73 ± 0.10 [a]	0.67 ± 0.07 [d]
10SF2	0.17 ± 0.08 [b]	0.95 ± 0.08 [a,b]	0.75 ± 0.07 [a]	0.23 ± 0.02 [b]
20SF2	0.19 ± 0.01 [c]	0.96 ± 0.07 [a,b,c]	0.75 ± 0.05 [a]	0.43 ± 0.08 [c]
30SF2	0.22 ± 0.08 [f]	0.97 ± 0.10 [c,d]	0.73 ± 0.02 [a]	0.67 ± 0.05 [d]

RT: Retention time; PM: control sample, 10BW1, 20BW1, 30BW1: buns with 10%, 20%, 30% buckwheat flour addition and 0.2% guar gum; 10BW2, 20BW2, 30BW2: buns with 10%, 20%, 30% buckwheat flour and 0.2% Magimix; 10SF1, 20SF1, 30SF1: buns with 10%, 20%, 30% sprouted buckwheat flour and 0.2% guar gum; 10SF2, 20SF2, 30SF2: buns with 10%, 20%, 30% sprouted buckwheat flour and 0.2% Magimix. Different lowercase letters within a column indicate significant differences between the means of the same parameter analyzed at $p < 0.05$ and (n = 3) according to Duncan's multiple range test.

3.2.3. Total Flavonoid Content of Buns

The values resulting from the quantitative determination of the buns total flavonoid are represented graphically in Table 8. The smallest value was obtained for control buns (PM), with a content of 3.99 mg QE/100 g, meanwhile, the highest value was represented by the buns with 30% sprouts flour 93.97 mg QE/100 g. This could be justified by the highest content in total flavonid of SF, compared to WF (Table 5).

Table 8. Buns' total flavonoids, total phenols and radical scavenging activity.

Sample	Total Flavonoid Content [mg QE/100 g FW]	Total Polyphenol Content [mg GAE/100 g FW]	Radical Scavenging Activity [%]
PM	3.99 ± 0.12 [a]	20.33 ± 0.12 [a]	35.21 ± 0.12 [a]
10BW1	13.99 ± 0.15 [a]	36.66 ± 0.10 [a,b]	37.58 ± 0.15 [b]
20BW1	30.23 ± 0.23 [b]	51.33 ± 0.26 [b,c]	42.87 ± 0.21 [c]
30BW1	40.31 ± 0.25 [b]	64.66 ± 0.31 [c]	46.28 ± 0.23 [d]
10BW2	13.98 ± 0.12 [a]	36.00 ± 0.12 [a,b]	37.63 ± 0.19 [b]
20BW2	30.23 ± 0.26 [b]	52.00 ± 0.18 [b,c]	42.90 ± 0.29 [c]
30BW2	40.32 ± 0.21 [b]	64.33 ± 0.31 [c]	46.32 ± 0.34 [d]
10SF1	30.98 ± 0.25 [b]	48.33 ± 0.35 [b,c]	41.87 ± 0.21 [c]
20SF1	63.52 ± 0.32 [c]	82.00 ± 0.51 [d]	48.99 ± 0.28 [e]
30SF1	93.93 ± 0.39 [d]	150.33 ± 0.52 [e]	54.83 ± 0.31 [f]
10SF2	30.97 ± 0.21 [b]	49.66 ± 0.04 [b,c]	41.88 ± 0.15 [c]
20SF2	63.52 ± 0.34 [c]	81.33 ± 0.56 [d]	49.09 ± 0.25 [e]
30SF2	93.94 ± 0.41 [d]	151.00 ± 0.48 [e]	54.96 ± 0.36 [f]

f.w.: fresh weight; PM: control sample; 10BW1, 20BW1, 30BW1: buns with 10%, 20%, 30% buckwheat flour addition and 0.2% guar gum; 10BW2, 20BW2, 30BW2: buns with 10%, 20%, 30% buckwheat flour and 0.2% Magimix; 10SF1, 20SF1, 30SF1: buns with 10%, 20%, 30% sprouted buckwheat flour and 0.2% guar gum; 10SF2, 20SF2, 30SF2: buns with 10%, 20%, 30% sprouted buckwheat flour and 0.2% Magimix. Different lowercase letters within a column indicate significant differences between the means of the same parameter analyzed at $p < 0.05$ and (n = 3) according to Duncan's multiple range test.

According to Chlopicka et al. [64], the flavonoid content decreases 2–4 times in the final bakery products compared to the flour. In their study, wheat-based buns had a 2.03 mg/100 g total flavonoid content; product with 15% buckwheat flour obtained 3.34 mg/100 g total flavonoids and, surprisingly, buns with 30% buckwheat had a score of 3.29 mg/100 g.

Constantini et al. [65] obtained similar, lower values for the final baked products correlated to the raw material, so for bread from 100% wheat flour total flavonoid content was 60 ± 20 mg RE/100 g, while for wheat flour this value was higher −80 ± 5 mg RE/100 g (RE: rutin).

3.2.4. Total Phenolic Content of Final Products

The content of polyphenols is closely related to the antioxidant capacity; baking causes partial loss of polyphenols and therefore of the antioxidant capacity [56,60]. According to Selimović et al. [61] and Chlopicka et al. [64] values of total polyphenols in baked products are lower than in flour and dough.

The lowest result obtained in this research was for the control sample—20.33 mg GAE/100 g (Table 8) which is in accordance with the value reported by Alvarez-Jubete et al. [7] (29.1 ± 0.6 mg GAE/100 g).

The addition of 10%, 20% and 30% SF increased the total phenolic content by 2.37 (48.33 mg GAE/100 g), 4.03 (82 mg GAE/100 g) and 7.39 (150.33 mg GAE/100 g) times compared to wheat-based products. Total phenols content of the samples supplemented with BW were higher compared to the product based on wheat flour, but lower than the SF final baked products (Table 8). This could be explained through the chemical composition of SF which was higher in total phenols content compared to BW (Table 5). No significant difference ($p < 0.05$) in total phenols content were found between samples with guar gum and improver, respectively.

The current state of art regarding the phenolic content in bakery products manufactured with BW, reported different values for total phenolic content. For instance, Selimović et al. [61] obtained a value of 12 ± 24 mg GAE/100 g for control, 25 ± 31 mg GAE/100 g for products with 15% buckwheat flour and 46 ± 62 mg GAE/100 g for bread with 30% buckwheat.

In the study carried out by Chlopicka et al. [64] the total phenolic compounds were 1.7 ± 0.07 mg GAE/g for the control sample, 2.1 ± 0.08 mg GAE/g for bread with 15% buckwheat and 2.65 ± 0.1 mg GAE/g product with 30% buckwheat.

To the best of our knowledge, regarding the determination of total polyphenolic content for the bakery products fortified with SF, no reports were found.

3.2.5. Radical Scavenging Activity of Samples

Buns with the addition of SF have a higher radical scavenging activity (10SF1-41.87%, 20SF1-48.99%, 30SF1-54.83%) than products manufactured with buckwheat flour (10BW1-37.58%, 20BW1-42.87%, 30BW1-46.28%), respectively those with wheat flour (35.21%). Between the samples with 0.2% guar and those with 0.2% improver not significant differences ($p < 0.05$) were noticed (Table 8).

There are many researches which show that baking process causes a decrease of antioxidant capacity baked goods [64,66–68]. The addition of different flours with a rich antioxidant activity might be able to enhance the antioxidant capacity of the final baked goods or to minimize its thermal losses [64,68,69].

According to the literature there are also studies that showed an increase in the radical scavenging capacity of wheat-based products compared to flour and dough as a result of the conjugated phenolic compounds release, especially in crust. Moreover, these works attribute this increase in antioxidant capacity to the Maillard reaction too [56,68].

3.2.6. Rheological Evaluation of Products

In the present study, the following textural parameters have been analyzed: total work, hardness, cohesiveness, springiness, chewiness, gumminess and adhesiveness, as listed in Table 9.

Table 9. Textural analysis of baked products.

Sample	Total Work Cycle 1 [mJ]	Hardness Cycle 1 [g]	Total Work Cycle 2 [mJ]	Hardness Cycle 2 [g]	Cohesiveness [n.a.]	Springiness [n.a.]	Chewiness [g]	Gumminess [g]	Adhesiveness [g]
PM	33.90 ± 0.65 [f]	482.00 ± 2.58 [c]	28.97 ± 0.21 [b,c,d]	463.00 ± 2.11 [b,c]	0.81 ± 0.11 [e,f]	1.22 ± 0.21 [g]	407.67 ± 2.33 [c]	398.67 ± 2.15 [b]	3.00 ± 0.12 [b,c]
10BW1	29.63 ± 0.34 [d]	353.00 ± 1.99 [b]	29.23 ± 0.19 [b,c,d]	477.33 ± 2.65 [c]	0.85 ± 0.10 [g]	0.95 ± 0.17 [f]	411.67 ± 2.57 [c]	430.67 ± 2.18 [b]	3.00 ± 0.10 [b,c]
20BW1	31.13 ± 0.35 [e]	393.00 ± 2.65 [b]	34.90 ± 0.25 [c,d]	588.00 ± 2.49 [d]	0.83 ± 0.21 [f,g]	0.95 ± 0.11 [f]	495.33 ± 2.56 [c]	505.00 ± 1.95 [c]	5.67 ± 0.15 [e]
30BW1	34.43 ± 0.28 [f]	526.00 ± 2.24 [d]	38.60 ± 0.31 [d]	623.33 ± 3.10 [d]	0.85 ± 0.18 [g]	0.94 ± 0.13 [f]	526 ± 3.11 [d]	537.00 ± 2.12 [c]	8.67 ± 0.21 [f]
10BW2	15.37 ± 0.21 [a]	508.60 ± 2.18 [c,d]	30.53 ± 0.29 [b,c,d]	486.33 ± 2.00 [c]	0.79 ± 0.25 [e]	0.86 ± 0.21 [b,c]	237.33 ± 2.01 [a]	400.67 ± 2.85 [b]	2.67 ± 0.18 [a,b]
20BW2	22.00 ± 0.19 [b]	346.67 ± 2.59 [b]	18.97 ± 0.17 [a,b]	298.33 ± 1.85 [a]	0.76 ± 0.17 [d]	0.83 ± 0.18 [a,b]	280.33 ± 2.15 [a,b]	527.33 ± 2.68 [c]	3.67 ± 0.17 [c,d]
30BW2	18.30 ± 0.20 [b]	251.65 ± 1.85 [a]	13.20 ± 0.15 [a]	243.00 ± 1.19 [a]	0.75 ± 0.31 [c,d]	0.80 ± 0.20 [a]	422.67 ± 3.08 [c]	576.33 ± 2.75 [c]	4.33 ± 0.14 [d]
10SF1	26.40 ± 0.25 [c]	385.00 ± 2.10 [b]	21.87 ± 0.22 [a,b]	367.67 ± 2.25 [b]	0.80 ± 0.22 [e]	0.93 ± 0.11 [e,f]	286.00 ± 2.54 [a,b]	306.67 ± 13.68 [a]	2.33 ± 0.10 [a]
20SF1	83.30 ± 1.02 [h]	1379.33 ± 5.25 [f]	67.50 ± 0.63 [e]	1176.00 ± 3.87 [e]	0.76 ± 0.15 [d]	0.91 ± 0.14 [d,e,f]	830.00 ± 2.98 [e]	949.33 ± 3.11 [d]	2.67 ± 0.18 [a,b]
30SF1	101.47 ± 1.29 [i]	2407.00 ± 6.29 [h]	115.90 ± 0.76 [g]	2408.00 ± 4.85 [f]	0.73 ± 0.18 [c]	0.89 ± 0.10 [c,d,e]	1317.33 ± 3.98 [f]	1632.67 ± 4.02 [f]	4.00 ± 0.22 [d]
10SF2	28.17 ± 0.25 [d]	404.00 ± 2.15 [b,c]	22.60 ± 0.23 [a,b,c]	376.33 ± 1.42 [b]	0.70 ± 0.12 [b]	0.87 ± 0.18 [b,c,d]	279.00 ± 2.11 [a,b]	301.33 ± 2.68 [a]	3.67 ± 0.17 [c,d]
20SF2	79.77 ± 1.00 [g]	950.67 ± 2.21 [e]	39.50 ± 0.35 [d]	658.00 ± 2.85 [d]	0.68 ± 0.13 [b]	0.85 ± 0.11 [b,c]	705.00 ± 2.54 [e]	839.00 ± 3.11 [d]	3.67 ± 0.11 [c,d]
30SF2	109.77 ± 1.19 [j]	1907.00 ± 4.25 [g]	82.60 ± 0.68 [f]	1167.33 ± 3.86 [e]	0.65 ± 0.14 [a]	0.79 ± 0.19 [a]	1235.00 ± 3.02 [f]	1418.00 ± 3.00 [e]	3.33 ± 0.26 [c,d]

PM: control sample; 10BW1, 20BW1, 30BW1: buns with 10%, 20%, 30% buckwheat flour addition and 0.2% guar gum; 10BW2, 20BW2, 30BW2: buns with 10%, 20%, 30% buckwheat flour and 0.2% Magimix; 10SF1, 20SF1, 30SF1: buns with 10%, 20%, 30% sprouted buckwheat flour and 0.2% guar gum; 10SF2, 20SF2, 30SF2: buns with 10%, 20%, 30% sprouted buckwheat flour and 0.2% Magimix. Different lowercase letters within a column indicate significant differences between the means of the same parameter analyzed at $p < 0.05$ and (n = 3) according to Duncan's multiple range test.

The principal mechanical characteristic of solid foods is represented by hardness [70]. This parameter is defined as the force needed to obtain a given deformation of the product [36]. In the present study, hardness 1 and 2 values increased as the percentage of buckwheat flour increased. The same pattern was observed on the final baked products manufactured with different percentages of SF flour. The explanation for the hardness 1 and 2 increased values could be the reduction of gluten content matrix, considering that BW and SF flours are gluten free. This is in line with Man et al. [43], who reported that the dough reduction in gluten matrix could lead to a harder bread texture. Furthermore, Moradi et al. [71] showed that gluten matrix is also involved in the elasticity of the dough and its capacity to retain gases during fermentation process, having a key role in bakery foods manufacture.

On the other side, the use of guar gum and Magimix improver, lead to significantly different values for hardness 1 and 2 (Table 9). For instance, for hardness 2, the samples manufactured with Magimix registered lower values, than those made with guar gum. This could be due to the chemical composition of Magimix, mainly to alpha-amylase content which could be involved in the improvement of dough properties. In this light, Savkina et al. [72] reported that improvers with alpha-amylase could enhanced bread texture and elasticity. Moreover, it is important to mentioned that Magimix contains also xylanase, which positively influenced the bread hardness, as reported by Shah et al., [73]. Likewise, Rodge et al. [27] proved that guar gum had no significant influence on the crust of the white bread, therefore, its hardness. The chewiness and gumminess parameters, showed proportional trends with hardness, samples with 0.2% guar registered higher values than samples with 0.2% improver.

On the other hand, regarding the cohesiveness, a parameter used to describe the internal cohesion of the material [36], the products manufactured with guar gum, showed better results. Cohesiveness value is desirable to reach a high value because it helps to form a bolus during mastication process, instead of disintegration of the product [74]. This is in concordance with Encina-Zelada et al. [74] who demonstrated that gum improved cohesiveness even in the gluten free products. In the same light, Turkut et al. [75] proved that the moisture content of the final baked goods could be involved in lowering cohesiveness and chewiness values. In the present study, the moisture content of the buns manufactured with different percentages of BW changed, reaching higher value when 30% BW was added, meanwhile in SF final baked goods no statistical differences were noticed (Table 6).

Springiness is a textural parameter linked to freshness and elasticity of the final baked products [75] and therefore is desirable to have high values. Samples with guar gum registered higher values that those with improver (Table 4). Rodge et al. [27] demonstrated that guar gum could improve dough stability and could enhance the elasticity of the final baked product.

Adhesiveness is defined as the necessary force needed to remove the product to the palate with tongue help [76]. In the present study, the adhesiveness of the final baked products made with guar gum were statistically different that those made with improver. This could be explained by the capacity of guar gum to have a positive influence on the dough stickiness, as reported by Ghodke, [77].

3.2.7. Sensorial Analysis

The final baked products manufactured with BW flour were more appreciated by panelists than those made with SF flour (Table 10). From the sensorial point of view, 20BW1and 10SF1 buns were the most appreciated by the panelists, reaching a final hedonic score of 9.1 and 8.7 respectively. This is in agreement with Starowicz et al. [78] who reported that the use of buckwheat flour in a range of 20–30% in bakery products manufacture reached the highest sensory acceptance. Bilgiçli & İbanoğlu [79] stated that breads produced with 10% and 20% buckwheat flour have been shown to have acceptable sensory qualities.

Table 10. Hedonic scores for buns.

Sample	Visual Appearance	Texture	Odor	Taste	Flavor	Overall Acceptability
PM	9.10 ± 0.15 [g]	9.00 ± 0.21 [f]	9.00 ± 0.15 [f]	9.00 ± 0.18 [g]	9.10 ± 0.10 [d]	9.20 ± 0.14 [g]
10BW1	9.00 ± 0.21 [f,g]	8.60 ± 0.32 [d,e,f]	9.00 ± 0.25 [f]	9.00 ± 0.14 [g]	9.00 ± 0.10 [d]	8.90 ± 0.12 [f,g]
20BW1	9.10 ± 0.10 [g]	8.90 ± 0.25 [f]	9.00 ± 0.15 [f]	9.10 ± 0.10 [g]	9.10 ± 0.12 [d]	9.30 ± 0.10 [g]
30BW1	7.10 ± 0.21 [b]	5.90 ± 0.17 [b]	6.70 ± 0.32 [b]	6.50 ± 0.35 [a,c]	6.20 ± 0.45 [a]	7.00 ± 0.42 [c]
10BW2	8.60 ± 0.17 [d,e,f]	8.50 ± 0.23 [d,e,f]	8.30 ± 0.25 [d,e]	8.20 ± 0.24 [e]	8.70 ± 0.25 [c,d]	8.90 ± 0.22 [f,g]
20BW2	8.50 ± 0.28 [d,e]	8.30 ± 0.24 [d,e]	8.20 ± 0.15 [d,e]	8.30 ± 0.23 [e,f]	8.50 ± 0.11 [c,d]	8.50 ± 0.25 [e,f]
30BW2	6.80 ± 0.30 [b]	5.60 ± 0.52 [a,b]	6.20 ± 0.29 [a]	6.20 ± 0.29 [a,b,c]	5.80 ± 0.19 [a]	6.00 ± 0.36 [a]
10SF1	8.70 ± 0.45 [f]	8.90 ± 0.20 [f]	8.90 ± 0.18 [f]	8.80 ± 0.11 [g]	8.90 ± 0.45 [d]	9.20 ± 0.32 [g]
20SF1	8.90 ± 0.25 [e,f,g]	8.70 ± 0.17 [e]	8.60 ± 0.18 [e,f]	8.70 ± 0.17 [f,g]	8.60 ± 0.11 [c,d]	8.60 ± 0.25 [e,f]
30SF1	6.70 ± 0.32 [b]	5.50 ± 0.36 [a,b]	6.30 ± 0.34 [a,b]	6.00 ± 0.35 [a]	6.00 ± 0.25 [a]	6.50 ± 0.11 [b]
10SF2	8.30 ± 0.38 [c,d]	8.00 ± 0.30 [c,d]	7.90 ± 0.25 [c,d]	8.00 ± 0.21 [d,e]	8.30 ± 0.16 [b,c]	8.10 ± 0.15 [d,e]
20SF2	7.90 ± 0.22 [c]	7.60 ± 0.17 [c]	7.60 ± 0.52 [c]	7.70 ± 0.28 [d]	7.90 ± 0.32 [b]	7.80 ± 0.46 [d]
30SF2	6.00 ± 0.15 [a]	5.00 ± 0.50 [a]	6.00 ± 0.21 [a]	6.00 ± 0.45 [a,b]	6.00 ± 0.50 [a]	5.70 ± 0.50 [a]

PM: control sample; 10BW1, 20BW1, 30BW1: buns with 10%, 20%, 30% buckwheat flour addition and 0.2% guar gum; 10BW2, 20BW2, 30BW2: buns with 10%, 20%, 30% buckwheat flour and 0.2% Magimix; 10SF1, 20SF1, 30SF1: buns with 10%, 20%, 30% sprouted buckwheat flour and 0.2% guar gum; 10SF2, 20SF2, 30SF2: buns with 10%, 20%, 30% sprouted buckwheat flour and 0.2% Magimix. Different lowercase letters within a column indicate significant differences between the means of the same parameter analyzed at $p < 0.05$ and (n = 3) according to Duncan's multiple range test.

The consumers' preference through the products with buckwheat flour could be justified by their better taste, structure and color, compared to those with SF. This is in line with Xu at al. 2014, who reported that 12% SF addition could produce remarkable negative impact on the textural and sensorial properties of bread. Besides, the color of the products made with BW flour became less darker (Figure 1), compared with the buns manufactured with SF flour (Figure 2).

On the other hand, the use of guar gum in the manufactured of baked products could have a positive influence on sensorial features such as taste and color [71,78]. The positive influence of the guar gum on the final color of the bakery products is confirmed also by Rodge et al. [27] who reported that the polysaccharides for the guar gum could be involved in the Maillard reaction and influencing the intensity of the brown color.

Due to the germination process, the amount of alpha-amylase increased favoring the hydrolysis of starch and enhancing the dextrin level during baking process. These enzymes in a higher amount could lead to sticky final baked products with a low satiety value, according to Xu et al. [80]. Moreover, buckwheat sprouts have a bitter and astringent taste due to their high-level content of flavonoids, causing at high level percentages the product's consumer unacceptability [80].

Figure 1. Buns with addition of buckwheat flour and 0.2% guar gum, 0.2% baking improver, respectively and sections of the obtained final baked buns. PM: control sample; 10BW1: buns with 10% buckwheat flour addition and 0.2% guar gum; 20BW1: buns with 20% buckwheat flour addition and 0.2% guar gum; 30BW1: buns with 30% buckwheat flour addition and 0.2% guar gum; 10BW2: buns with 10% buckwheat flour addition and 0.2% Magimix; 20BW2: buns with 20% buckwheat flour addition and 0.2% Magimix; 30BW2—buns with 30% buckwheat flour addition and 0.2% Magimix.

Figure 2. Buns with addition of sprouted buckwheat flour and 0.2% guar gum, 0.2% baking improver, respectively and sections of the obtained final baked buns. PM: control sample; 10SF1: buns with 10%, sprouted buckwheat flour and 0.2% guar gum; 20SF1: buns with 20%, sprouted buckwheat flour and 0.2% guar gum; 30SF1: buns with 30%, sprouted buckwheat flour and 0.2% guar gum; 10SF2: buns with 10%, sprouted buckwheat flour and 0.2% Magimix; 20SF2: buns with 20%, sprouted buckwheat flour and 0.2% Magimix; 30SF2: buns with 30%, sprouted buckwheat flour and 0.2% Magimix.

4. Conclusions

Buckwheat and sprouts buckwheat flours are valuable raw materials, which had a serious impact on the nutritional enrichment on the final baked products. For instance, the use of 10SF and 20BW composite flours increased the total flavonoids content of the final baked products with 26.24% and 26.99% respectively. The same trend was observed at total phenols where 10SF1 was 2.37 times higher than the control sample, meanwhile 20BW1 was 2.52 times higher. In the same light, radical scavenging activity of 10SF1 and 20BW1 increased compared to the control with 6.66% and 7.66% respectively.

Furthermore, there uses in buns manufacturing are limited due to their negative influence on the sensorial and rheological features. The sensorial analysis showed that samples manufactured with 20% buckwheat flour and guar gum and buns made with 10% sprouts buckwheat flour with guar gum were the most appreciated final baked goods regarding the following parameters: overall appearance, taste, flavor, odor, color and texture, reaching a final hedonic score of 9.3 and 9.2, respectively.

With respect to rheological features, it is necessary the assessment of further rheological parameters to have broader understanding how these additives could influence the texture of the final baked products.

To conclude, we can assess that buckwheat and sprouts buckwheat flours are precious raw materials that could be successfully used for the improvement of bakery products nutritional values.

Author Contributions: Conceptualization, A.S. and S.M.; methodology, D.C.V., S.M.M., M.S.C. and A.P.; software, A.C.U. and A.S.; validation, A.S., M.S.C., D.C.V. and A.P.; formal analysis, M.S.C., A.C.U., V.M., G.F. and I.E.R.; writing—original draft preparation, A.S., M.S.C. and A.P.; writing—review and editing, A.S. and M.S.C.; supervision, S.M.; project administration, A.S. and S.M. All authors have read and agreed to the published version of the manuscript.

Funding: This research publication was supported by funds from the Ministry of Education and Research (FDI 0252).

Acknowledgments: The authors acknowledge the technical support provided by the researcher Laura Stan and by the researcher Orsolya Borsai.

Conflicts of Interest: The authors declare no conflict of interest.

References

1. Shahidi, F.; Chandrasekara, A. Millet grain phenolics and their role in disease risk reduction and health promotion: A review. *J. Funct. Foods* **2013**, *5*, 570–581. [CrossRef]

2. Dapčević-Hadnađev, T.; Torbica, A.; Hadnađev, M. Rheological properties of wheat flour substitutes/alternative crops assessed by Mixolab. *Procedia Food Sci.* **2011**, *1*, 328–334. [CrossRef]

3. Giacco, R.; Clemente, G.; Cipriano, D.; Luongo, D.; Viscovo, D.; Patti, L.; Di Marino, L.; Giacco, A.; Naviglio, D.; Bianchi, M.; et al. Effects of the regular consumption of wholemeal wheat foods on cardiovascular risk factors in healthy people. *Nutr. Metab. Cardiovasc. Dis.* **2010**, *20*, 186–194. [CrossRef] [PubMed]

4. Dziki, D.; Różyło, R.; Gawlik-Dziki, U.; Świeca, M. Current trends in the enhancement of antioxidant activity of wheat bread by the addition of plant materials rich in phenolic compounds. *Trends Food Sci. Technol.* **2014**, *40*, 48–61. [CrossRef]

5. Stamatovska, V.; Nakov, G.; Uzunoska, Z.; Kalevska, T.; Menkinoska, M. Potential use of some pseudocereals in the food industry. *ARTTE* **2018**, *6*, 54–61. [CrossRef]

6. Alvarez-Jubete, L.; Arendt, E.K.; Gallagher, E. Nutritive value and chemical composition of pseudocereals as gluten-free ingredients. *Int. J. Food Sci. Nutr.* **2009**, *60*, 240–257. [CrossRef]

7. Alvarez-Jubete, L.; Wijngaard, H.; Arendt, E.K.; Gallagher, E. Polyphenol composition and in vitro antioxidant activity of amaranth, quinoa buckwheat and wheat as affected by sprouting and baking. *Food Chem.* **2010**, *119*, 770–778. [CrossRef]

8. Păucean, A. *Tendințe Moderne Privind Creșterea Valorii Nutritive a Făinii de Grâu Și a Produselor de Panificație;* Editura MEGA: Cluj-Napoca, Romania, 2017; p. 39.

9. Mir, N.A.; Riar, C.S.; Singh, S. Nutritional constituents of pseudo cereals and their potential use in food systems: A review. *Trends Food Sci. Technol.* **2018**, *75*, 170–180. [CrossRef]

10. Khan, F.; Arif, M.; Khan, T.U.; Khan, M.I.; Bangash, J.A. Nutritional evaluation of common buckwheat of four different villages of gilgit-baltisan. *ARPN J. Agric. Biol. Sci.* **2013**, *8*, 264–266.

11. Golob, A.; Stibilj, V.; Kreft, I.; Germ, M. The feasibility of using tartary buckwheat as a se-containing food material. *J. Chem.* **2015**, *2015*, 246042. [CrossRef]

12. Fabjan, N.; Rode, J.; Košir, I.J.; Wang, Z.; Zhang, Z.; Kreft, I. Tartary buckwheat (*Fagopyrum tataricum* Gaertn.) as a source of dietary rutin and quercitin. *J. Agric. Food Chem.* **2003**, *51*, 6452–6455. [CrossRef]

13. Brennan, C.S. Dietary fibre, glycaemic response, and diabetes. *Mol. Nutr. Food Res.* **2005**, *49*, 560–570. [CrossRef]

14. Békés, F.; Schoenlechner, R.; Tömösközi, S. Chapter 14—Ancient Wheats and Pseudocereals for Possible Use in Cereal-Grain Dietary Intolerances. In *Cereal Grains 2nd Edition Assessing and Managing Quality*; Wrigley, C., Batey, I., Miskelly, D., Eds.; Woodhead Publishing: Cambridge, UK, 2017; pp. 353–389.

15. Wijngaard, H.H.; Arendt, E.K. Buckwheat—Review. *Cereal Chem.* **2006**, *83*, 391–401. [CrossRef]

16. Frutos, M.J.; Rincón-Frutos, L.; Valero-Cases, E. *Chapter 2.14—Rutin, Nonvitamin and Nonmineral Nutritional Supplements*; Academic Press: London, UK, 2019; pp. 111–117.

17. Brajdes, C.; Vizireanu, C. Sprouted buckwheat an important vegetable source of antioxidants. *Ann. Univ. Dunarea Jos Galati Fascicle VI Food Technol.* **2012**, *36*, 53–60.

18. Márton, M.; Mándoki, Z.; Csapó-Kiss, Z.; Csapó, J. The role of sprouts in human nutrition. A review. *Aliment. Hung. Univ. Transylvania* **2010**, *3*, 81–117.

19. Schendel, R. Chapter 10—Phenol content in sprouted grains. In *Sprouted Grains Nutritional Value, Production and Applications*; Feng, H., Nemzer, B., DeVries, J.W., Eds.; Woodhead Publishing: Duxford, UK, 2019; pp. 247–316.

20. Nemzer, B.; Lin, Y.; Huang, D. Chapter 3—Antioxidants in sprouts of grains. In *Sprouted Grains Nutritional Value, Production and Applications*; Feng, H., Nemzer, B., DeVries, J.W., Eds.; Woodhead Publishing: Duxford, UK, 2019; pp. 55–68.

21. Zhang, G.; Xu, Z.; Gao, Y.; Huang, X.; Zou, Y.; Yang, T. Effects of germination on the nutritional properties, phenolic profiles, and antioxidant activities of buckwheat. *J. Food Sci.* **2015**, *80*, H1111–H1119. [CrossRef] [PubMed]

22. Mir, S.A.; Shah, M.A.; Naik, H.R.; Zargar, I.A. Influence of hydrocolloids on dough handling and technological properties of gluten-free breads. *Trends Food Sci. Technol.* **2016**, *51*, 49–57. [CrossRef]

23. Li, J.-M.; Nie, S.-P. The functional and nutritional aspects of hydrocolloids in foods. *Food Hydrocoll.* **2016**, *53*, 46–61. [CrossRef]

24. Gupta, S.; Variyar, P. Chapter 12—Guar Gum: A Versatile Polymer for the Food Industry. In *Biopolymers for Food Design*; Grumezescu, A.M., Holban, A.M., Eds.; Academic Press: London, UK, 2018; Volume 20, pp. 487–514.

25. Thombare, N.; Jha, U.; Mishara, S.; Siddiqui, M.Z. Guar gum as a promising starting material for diverse applications: A review. *Int. J. Biol. Macromol.* **2016**, *88*, 361–372. [CrossRef] [PubMed]

26. BeMiller, J.N. Chapter 9—Guar, Locust Bean, Tara and Cassia Gums. In *Carbohydrate Chemistry for Food Scientists*, 3rd ed.; Academic Press: West Lafayette, IN, USA, 2018; pp. 241–252.

27. Rodge, A.B.; Sonkamble, S.M.; Salve, R.V.; Hashmi, S.I. Effect of hydrocolloid (guar gum) incorporation on the quality characteristics of bread. *J. Food Process. Technol.* **2012**, *3*, 136. [CrossRef]

28. Tebben, L.; Shen, Y.; Li, Y. Improvers and functional ingredients in whole wheat bread: A review of their effects on dough properties and bread quality. *Trends Food Sci. Technol.* **2018**, *81*, 10–24. [CrossRef]

29. Matsushita, K.; Santiago, D.M.; Noda, T.; Tsuboi, K.; Kawakami, S.; Yamauchi, H. The bread making qualities of bread dough supplemented with whole wheat flour and treated with enzymes. *Food Sci. Technol. Res.* **2017**, *23*, 403–410. [CrossRef]

30. Driss, D.; Bhiri, F.; Siela, M.; Bessess, S.; Chaabouni, S.; Ghorbel, R. Improvement of breadmaking quality by xylanase GH11 from Penicillium occitanis Pol6. *J. Texture Stud.* **2012**, *44*, 75–84. [CrossRef]

31. Ghoshal, G.; Shivhare, U.; Banerjee, U. Effect of xylanase on quality attributes of whole-wheat bread. *J. Food Qual.* **2013**, *36*, 172–180. [CrossRef]

32. Bock, J.E. Chapter 9—Enzymes in breadmaking. In *Improving and Tailoring Enzymes for Food Quality and Functionality*, 1st ed.; Yada, R.Y., Ed.; Woodhead Publishing: Vancouver, BC, Canada, 2015.

33. Cai, Y.; Corke, H.; Wang, D.; Li, W. Buckwheat: Overview. *Encycl. Food Grains* **2016**, *1*, 307–315. [CrossRef]

34. Corsetti, A.; Gobbetti, M.; De Marco, B.; Balestrieri, F.; Paoletti, F.; Russi, L.; Rossi, J. Combined effect of sourdough lactic acid bacteria and additives on bread firmness and staling. *J. Agric. Food Chem.* **2000**, *48*, 3044–3051. [CrossRef]

35. Bonta, V.; Marghitas, L.A.; Stanciu, O.; Laslo, L.; Dezmirean, D.; Bobis, O. High-performance liquid chromatographic analysis of sugars in Transylvanian honeydew honey. *Bull. Univ. Agric. Sci. Vet. Med. Cluj-Napoca-Anim. Sci. Biotechnol.* **2008**, *65*, 229–232.

36. Bunea, A.; Rugina, D.O.; Pintea, A.M.; Sconta, Z.; Bunea, C.I.; Socaciu, C. Comparative polyphenolic content and antioxidant activities of some wild and cultivated blueberries from Romania. *Not. Bot. Horti Agrobot.* **2011**, *39*, 70–76. [CrossRef]

37. Păucean, A.; Man, S.M.; Chiş, M.S.; Mureşan, V.; Pop, C.R.; Socaci, S.A.; Mureşan, C.C.; Muste, S. Use of pseudocereals preferment made with aromatic yeast strains for enhancing wheat bread quality. *Foods* **2019**, *8*, 443. [CrossRef]

38. Urcan, A.D.; Criste, A.D.; Dezmirean, D.S.; Mărgăoan, R.; Caeiro, A.; Campos, M.G. Similarity of data frombee bread with the same taxa collected in India and Romania. *Molecules* **2018**, *23*, 2491. [CrossRef]

39. Chiş, M.S.; Păucean, A.; Man, S.M.; Vodnar, D.C.; Teleky, B.-E.; Pop, C.R.; Stan, L.; Borsai, O.; Kadar, C.B.; Urcan, A.C.; et al. Quinoa sourdough fermented with *Lactobacillus plantarum* ATCC 8014 designed for gluten-free muffins—A powerful tool to enhance bioactive compounds. *Appl. Sci.* **2020**, *10*, 7140. [CrossRef]

40. Đorđević, T.; Šiler-Marinković, S.S.; Dimitrijević-Branković, S.I. Effect of fermentation on antioxidant properties of some cereals and pseudo cereals. *Food Chem.* **2010**, *119*, 957–963. [CrossRef]

41. Chiş, M.S.; Păucean, A.; Man, S.M.; Mureşan, V.; Socaci, S.A.; Pop, A.; Stan, L.; Rusu, B.; Muste, S. Textural and sensory features changes of gluten free muffins based on rice sourdough fermented with *Lactobacillus spicheri* DSM 15429. *Foods* **2020**, *9*, 363. [CrossRef] [PubMed]

42. Chiş, M.S.; Păucean, A.; Stan, L.; Mureşan, V.; Vlaic, R.A.; Man, S.; Muste, S. *Lactobacillus plantarum* ATCC 8014 in quinoa sourdough adaptability and antioxidant potential. *Rom. Biotechnol. Lett.* **2018**, *23*, 13581–13591.

43. Man, S.M.; Păucean, A.; Călian, I.D.; Mureşan, V.; Chiş, M.S.; Pop, A.; Mureşan, A.; Bota, M.; Muste, S. Influence of fenugreek flour (*Trigonella foenum-graecum* L.) addition on the technofunctional properties of dark wheat flour. *J. Food Qual.* **2019**, *2019*, 8635806. [CrossRef]

44. Kim, S.L.; Son, Y.K.; Hwang, J.J.; Kim, S.K.; Hur, H.S.; Park, C.H. Development and utilization of buckwheat sprouts as functional vegetables. *Fagopyrum* **2001**, *18*, 49–54.

45. Lee, E.H.; Kim, C.J. Nutritional changes of buckwheat during germination. *Korea J. Food Cult.* **2008**, *23*, 121–129.

46. Požrl, T.; Kopjar, M.; Kurent, I.; Hribar, J.; Jänes, A.; Simčič, M. Phytate degradation during breadmaking: The influence of flour type and breadmaking procedures. *Czech J. Food Sci.* **2009**, *27*, 29–38. [CrossRef]

47. Qin, P.; Wang, Q.; Shan, F.; Hou, Z.; Ren, G. Nutritional composition and flavonoids content of flour from different buckwheat cultivars. *Int. J. Food Sci. Technol.* **2010**, *45*, 951–958. [CrossRef]

48. Mazza, G.; Oomah, B.D. Buckwheat. In *Encyclopedia of Food Sciences and Nutrition*; Caballero, B., Trugo, L.C., Finglas, P.M., Eds.; Academic Press: London, UK, 2003; pp. 692–699.

49. Codină, G.; Mironeasa, S.; Voica, D.; Mironeasa, C. Multivariate analysis of wheat flour dough sugars, gas production, and dough development at different fermentation times. *Czech J. Food Sci.* **2013**, *31*, 222–229. [CrossRef]

50. Sahlström, S.; Park, W.; Shelton, D.R. Factors influencing yeast fermentation and the effect of LMW sugars and yeast fermentation on hearth bread quality. *Cereal Chem. J.* **2004**, *81*, 328–335. [CrossRef]

51. Nam, T.G.; Lim, Y.J.; Eom, S.H. Flavonoid accumulation in common buckwheat (*Fagopyrum esculentum*) sprout tissues in response to light. *Hortic. Environ. Biotechnol.* **2018**, *59*, 19–27. [CrossRef]

52. Yiming, Z.; Hong, W.; Linlin, C.; Zhou, X.; Wen, T.; Xinli, S. Evolution of nutrient ingredients in tartary buckwheat seeds during germination. *Food Chem.* **2015**, *186*, 244–248. [CrossRef] [PubMed]

53. Ren, S.-C.; Sun, J.-T. Changes in phenolic content, phenylalanine ammonia-lyase (PAL) activity, and antioxidant capacity of two buckwheat sprouts in relation to germination. *J. Funct. Foods* **2014**, *7*, 298–304. [CrossRef]

54. Singh, R.; Babu, S.; Avasthe, R.K.; Yadav, G.S.; Chettri, T.K.; Phempunadi, C.D.; Chatterjee, T. Bacterial inoculation effect on soil biological properties, growth, grain yield, total phenolic and flavonoids contents of common buckwheat (*Fagopyrum esculentum* Moench) under hilly ecosystems of North-East India. *Afr. J. Microbiol. Res.* **2015**, *9*, 1110–1117.

55. Inglett, G.E.; Chen, D.; Berhow, M.; Lee, S. Antioxidant activity of commercial buckwheat flours and their free and bound phenolic compositions. *Food Chem.* **2011**, *125*, 923–929. [CrossRef]

56. Han, H.-M.; Koh, B.-K. Antioxidant activity of hard wheat flour, dough and bread prepared using various processes with the addition of different phenolic acids. *J. Sci. Food Agric.* **2010**, *91*, 604–608. [CrossRef]

57. Akinyede, A.; Amoo, I. Chemical and functional properties of full fat and defatted cassia fistula seed flours. *Pak. J. Nutr.* **2009**, *8*, 765–769. [CrossRef]

58. Bhavsar, G.; Sawate, A.R.; Kshirsagar, R.B.; Chappalwar, V.M. Studies on physico-chemical characteristics of buckwheat and its exploration in breads as functional food. *Int. J. Eng. Res. Technol.* **2013**, *2*, 3971–3980.

59. Shewry, P.R.; Hey, S.J. The contribution of wheat to human diet and health. *Food Energy Secur.* **2015**, *4*, 178–202. [CrossRef]

60. Drobot, V.; Semenova, A.; Smirnova, J.; Myhonic, L. Effect of buckwheat processing products on dough and bread quality made from whole-wheat flour. *Int. J. Food Stud.* **2014**, *3*, 1–12. [CrossRef]

61. Selimović, A.; Miličevic, D.; Jašić, M.; Selimović, A.; Ačkar, Đ.; Pešić, T. The effect of baking temperature and buckwheat flour addition on the selected properties of wheat bread. *Croat. J. Food Sci. Technol.* **2014**, *6*, 43–50.

62. Bojňanská, T.; Chlebo, P.; Gažar, R.; Horna, A. Buckwheat-enriched bread production and its nutritional benefits. *Eur. J. Plant Sci. Biotechnol.* **2009**, *3*, 49–55.

63. Hidalgo, A.; Brandolini, A. *Bread—Bread from Wheat Flour. Encyclopedia of Food Microbiology*; Batt, C.A., Tortorello, M.L., Eds.; Academic Press: London, UK, 2014; Volume 1, pp. 303–308.

64. Chlopicka, J.; Pasko, P.; Gorinstein, S.; Jedryas, A.; Zagrodzki, P. Total phenolic and total flavonoid content, antioxidant activity and sensory evaluation of pseudocereal breads. *LWT* **2012**, *46*, 548–555. [CrossRef]

65. Costantini, L.; Lukšič, L.; Molinari, R.; Kreft, I.; Bonafaccia, G.; Manzi, L.; Merendino, N. Development of gluten-free bread using tartary buckwheat and chia flour rich in flavonoids and omega-3 fatty acids as ingredients. *Food Chem.* **2014**, *165*, 232–240. [CrossRef] [PubMed]

66. Martinez-Villaluenga, C.; Michalska, A.; Frias, J.; Piskuła, M.K.; Vidal-Valverde, C.; Zieliński, H. Effect of flour extraction rate and baking on thiamine and riboflavin content and antioxidant capacity of traditional rye bread. *J. Food Sci.* **2009**, *74*, C49–C55. [CrossRef] [PubMed]

67. Moore, J.; Luther, M.; Cheng, Z.; Yu, L. (Lucy) Effects of baking conditions, dough fermentation, and bran particle size on antioxidant properties of whole-wheat pizza crusts. *J. Agric. Food Chem.* **2009**, *57*, 832–839. [CrossRef]

68. Karrar, E.M.A. A review on: Antioxidant and its impact during the bread making process. *Int. J. Nutr. Food Sci.* **2014**, *3*, 592. [CrossRef]

69. Yu, L.; Nanguet, A.-L.; Beta, T. Comparison of antioxidant properties of refined and whole wheat flour and bread. *Antioxidants* **2013**, *2*, 370–383. [CrossRef]

70. Amigo, J.M.; Alvarez, A.D.O.; Engelsen, M.M.; Lundkvist, H.; Engelsen, S.B. Staling of white wheat bread crumb and effect of maltogenic α-amylases. Part 1: Spatial distribution and kinetic modeling of hardness and resilience. *Food Chem.* **2016**, *208*, 318–325. [CrossRef]

71. Moradi, M.; Bolandi, M.; Karimi, M.; Nahidi, F.; Baghaei, H. Improvement of gluten-free taftoon bread properties during storage by the incorporation of potato powder (*Satrina V.*), guar gum, sodium caseinate and transglutaminase into the matrix. *J. Food Meas. Charact.* **2020**, *14*, 2282–2288. [CrossRef]

72. Savkina, O.; Kuznetsova, L.; Burykina, M.; Kostyuchenko, M.; Parakhina, O. The influence of the flour amylolytic enzymes activity, dosage of ingredients and bread making method on the sugar content and the bread quality. *Agron. Res.* **2020**, *18*, 1873–1887.

73. Shah, A.R.; Shah, R.; Madamwar, D. Improvement of the quality of whole wheat bread by supplementation of xylanase from *Aspergillus foetidus*. *Bioresour. Technol.* **2006**, *97*, 2047–2053. [CrossRef] [PubMed]

74. Encina-Zelada, C.R.; Cadavez, V.; Monteiro, F.; Teixeira, J.A.; Gonzales-Barron, U. Physicochemical and textural quality attributes of gluten-free bread formulated with guar gum. *Eur. Food Res. Technol.* **2019**, *245*, 443–458. [CrossRef]

75. Turkut, G.M.; Cakmak, H.; Kumcuoglu, S.; Tavman, S. Effect of quinoa flour on gluten-free bread batter rheology and bread quality. *J. Cereal Sci.* **2016**, *69*, 174–181. [CrossRef]

76. Morais, E.; Cruz, A.G.; Faria, J.; Bolini, H.M.A. Prebiotic gluten-free bread: Sensory profiling and drivers of liking. *LWT* **2014**, *55*, 248–254. [CrossRef]

77. Ghodke, S.K. Effect of guar gum on dough stickiness and staling in chapatti-an Indian unleavened flat bread. *Int. J. Food Eng.* **2009**, *5*, 1–21. [CrossRef]

78. Starowicz, M.; Koutsidis, G.; Zieliński, H. Sensory analysis and aroma compounds of buckwheat containing products—A review. *Crit. Rev. Food Sci. Nutr.* **2017**, *58*, 1767–1779. [CrossRef]
79. Bilgiçli, N.; Ibanoğlu, Ş. Effect of pseudo cereal flours on some physical, chemical and sensory properties of bread. *J. Food Sci. Technol.* **2015**, *52*, 7525–7529. [CrossRef]
80. Xu, F.-Y.; Gao, Q.-H.; Ma, Y.-J.; Guo, X.; Wang, M. Comparison of tartary buckwheat flour and sprouts steamed bread in quality and antioxidant property. *J. Food Qual.* **2014**, *37*, 318–328. [CrossRef]

Publisher's Note: MDPI stays neutral with regard to jurisdictional claims in published maps and institutional affiliations.

Article

Characterization of Quinoa Seeds Milling Fractions and Their Effect on the Rheological Properties of Wheat Flour Dough

Ionica Coțovanu, Ana Batariuc and Silvia Mironeasa *

Faculty of Food Engineering, Stefan cel Mare University of Suceava, 720229 Suceava, Romania;
ionica.cotovanu@usm.ro (I.C.); ana.batariuc@usm.ro (A.B.)
* Correspondence: silviam@fia.usv.ro; Tel.: +40-741-985-648

Received: 16 September 2020; Accepted: 13 October 2020; Published: 16 October 2020

Abstract: Replacement of refined wheat flour with milling fractions of quinoa seeds represents a useful way for the formulation of value-added baked products with beneficial characteristics to consumers. The aim of this study was to assess the chemical composition and physical properties of different particle sizes of quinoa flour on Falling number index (FN) and dough rheological properties determined by Mixolab in a planned research based on design of experiment by using full factorial design. The ash and protein contents were higher in medium particle size, whereas the carbohydrates presented a lower value, this fraction having also the highest water absorption and water retention capacity. The reduction of particles led to an increased swelling capacity and a decreased bulk density. The particle size significantly influenced the FN values in linear and quadratic terms ($p < 0.05$), showing a decrease with the particle size increasing. Particle size decrease significantly increased water absorption and the rate of protein weakening due to heat (C1–2), whereas starch gelatinization rate (C3–2), starch breakdown rate related to amylase activity (C3–4) and starch retrogradation speed (C5–4) decreased. By increasing the amount of quinoa flour (QF) in wheat flour, the dough stability and the torques C2, C3, C4 and C5 followed a decreased trend, whereas water absorption and dough development time rose. Optimization, determined by particle size and level of QF added in wheat flour based on which of the combination gives the best rheological properties, showed that the composite flour containing 8.98% quinoa flour of medium particle size was the most suitable.

Keywords: quinoa flour; particle size; rheological properties; optimization

---

1. Introduction

Quinoa (*Chenopodium quinoa Willd*) is a pseudocereal with large applications in the food industry and represent a current trend in the human diet because it has excellent nutritional and nutraceutical value and is a gluten-free grain [1,2]. It can be used for functional food development due to its functional and rheological properties, sensory characteristics, nutrient profile and stability [3]. This crop is considered a whole vegetable protein and whole grain carbohydrate [4] and contains high quality oil and high amounts of vitamins, minerals and a large diversity of bioactive compounds [5]. With respect to the distribution of nutritive compounds within quinoa seed tissues, carbohydrates can be found mostly in the perisperm, while proteins, fat and minerals are concentrated in the endosperm and germ or embryo, which represent approximately 30% of the whole seed [4,6]. The most important carbohydrate from quinoa seed is starch, which varies between 52.2–69.2% of the dry matter [4,7,8], and having as low as 11% amylose content [9]. Total fiber content in quinoa seeds varies from 7.0 to 26.5%, being mostly comprised of insoluble polysaccharides (78 of the total dietary fiber content) [10]. The protein content of quinoa varies from 8 to 22% depending on the varieties [11,12] and it is of very high quality compared to wheat. Proteins from quinoa seeds are rich in amino acids such as lysine, threonine and methionine, which are deficient in cereals [13], making them complete proteins [14]. Lipid levels from quinoa (4.0–7.6%) are much higher than those from wheat [15,16], unsaturated fatty acids being the predominant fatty acids (71–84.5% of total lipids) [16,17]. From the nutritional quality of the lipid fraction, quinoa oil presents a ω-6/ω-3 ratio of 4.7–19.6, with high quality and nutritious character [16,18]. Vitamins and minerals of quinoa seeds are among the features that make quinoa such an excellent nutrient source [19,20]. Vega-Galvez et al. (2010) [20] found that the total mineral content in quinoa seeds is about twice that of cereals, being particularly rich in calcium, magnesium, iron and zinc. Regarding bioactive compounds, the main phenolic compounds presented in quinoa seeds are gallic acid (167.2–308.3 mg/100 g dw) and flavonoids (rutin, quercitin and kaempferol) [21,22], and variation in levels exists between species, and depending on the growing conditions.

Nutritionally, quinoa is recognized as a super grain and can be used in the bakery industry to improve final product quality or to develop new products, and the properties of starch present in seeds is similar to those found in wheat [23]. Investigations on the substitution of wheat flour with quinoa flour at various levels has been made in many investigations [24–26], but no previous studies concerning how the variation in particle size can influence the composition, physical and rheological properties have been done.

Milling of quinoa seeds, a grinding process performed to produce flour or different flour fractions, is an important step that influences the product's quality. Significant differences in the nutritional composition, especially ash, protein and starch levels were reported for different particle sizes of milled grains, which may be related to cell-wall materials [27]. Therefore, the processing techniques and final product quality is greatly dependent on the particle size of the flour used and its constituents. Particle fractions can improve water or oil-holding capacity, the release of enzymes and physical properties by exposing their surface area during processing stages such as mixing, fermenting or baking. Nutritionally, particle size has a significant impact on food digestibility in the gastrointestinal system [28], digestibility lowering with increased particle size [29]. Size reduction is also applied extensively in the grain milling industry to separate the endosperm from the undesirable components of the whole grain, such as saponins in quinoa hull.

Some studies reported that the intermediate particle size (250–500 µm) gave best results in the supercritical dioxide extraction of quinoa oil [30]. In another study, it was found that the development of a gluten-free pasta product improved nutritionally required grits of particles with sizes between 200 to 600 µm [31]. Solaesa et al. (2020) [4] showed that different particle fractions of quinoa flour showed wide variation in chemical composition and functionality with high impact on rheological properties. Flour particle size influences hydration and, in consequence, dough rheology, affecting the quality of the final product [32]. Ahmed et al. (2019) [28] showed that separated particle fractions of quinoa flour presented wide variations in chemical composition and functionality, influencing rheological properties. The effects of a range of particle sizes on the rheological properties of various flour doughs has been investigated previously [33–36] concluding that particle size and composition remarkably affect rheological properties, but there was no definite trend.

Different rheological tests are used in the cereal industry to measure rheological properties, providing information on the flour quality and performance during processing. A Mixolab device is used to assess the processing quality of the flour dough when subjected to simultaneous mechanical shear stress and temperature limitation. This device is able to measure the torque of the dough in real time during an increase in temperature in a single test. So far, there is no evidence of the use of different quinoa seeds milling fractions at different levels in wheat flour breads and their effect on the dough rheological behavior assessed by using Mixolab. Thus, the objective of this study was to investigate the impact of different quinoa flour particle size additions at different levels in refined wheat flour on dough rheological properties, and to find the optimal formulation of quinoa particle size level added, in order to predict the bread-making behavior of this composite formulation. Furthermore, the influence of particle size on the proximate composition and some physical properties of quinoa flour fractions were investigated. The formulation can be use to obtain new bread products with improved nutritional profiles.

2. Materials and Methods

2.1. Materials

A commercial wheat flour of 650 type was purchased from S.C. Mopan S.A. (Suceava, Romania). Quinoa flours were obtained by grinding white quinoa seeds acquired from S.C. SANOVITA S.R.L (Suceava, Romania) with a laboratory grinder (Grain Mill, KitchenAid, Model 5KGM, Whirlpool Corporation, USA) and sieved through a Retsch Vibratory Sieve Shaker AS 200 basic (Haan, Germany) to obtain quinoa flour (QF) at three different particle sizes: large, L > 300 μm, medium, M > 180 μm, < 300 μm and small fractions, S < 180 μm.

2.2. Proximate Composition

The wheat flour used in this study contained 14.00% moisture (ICC 110/1), 12.60% protein (ICC 105/2), 1.40% fat (ICC 105/1), 0.65% ash (ICC 104/1), 30.00% wet gluten (ICC 106/1), 6.00 mm gluten deformation index (SR 90:2007) and 312 s Falling number index (ICC 107/1). These characteristics, determined according to the Romanian standard SR 877:1996, revealed that the wheat flour had a strong quality and was good flour for bread making.

White quinoa seeds (*Chenopodium quinoa*) presented the following analytical characteristics: moisture content 13.28% (SR EN ISO 665:2003), fat content 5.61% (SR EN ISO 659:2009), protein content 14.12% (SR EN ISO 20483:2007) and ash content 2.00% (SR ISO 2171:2009). The protein content was calculated with a general factor of 6.25. Carbohydrate content of the flour samples was determined by difference by applying the equation described by Alonso-Miravalles and O'Mahaony (2018) [37] (Equation (1)):

$$\% \text{ carbohydrates } = 100 - (protein + fat + ash + moisture\ content) \tag{1}$$

The analyses were made at least in duplicate, and the values of the parameters were expressed as the average ± standard deviation.

2.3. Physical Properties of Quinoa Flour Fractions

The water absorption capacity, water retention capacity, swelling capacity and the bulk density for the three particle sizes of quinoa flour (QL, QM, QS) were determined at least in duplicate for each sample.

2.3.1. Water Absorption Capacity

The water absorption capacity (WAC) was determined by the method described by Sosulski et al. (1976) [38]. The sample (1 g) was mixed with 10 mL distilled water, kept at ambient temperature for 30 min and centrifuged for 10 min at 2000 rpm in a centrifuge model MPW-223e (MPW Med. Instruments, Warszaw, Poland). Water absorption capacity was expressed as percent water bound per gram of the sample.

2.3.2. Water Retention Capacity

The water retention capacity (WRC) of quinoa flour fractions was measured following the method of Onipe et al. (2017) [39] with slight modifications. Two grams of each quinoa flour sample was weighed into test tubes and 20 mL of distilled water was added. Samples were vortexed on a water bath with continuous stirring at 25 °C (Memmert Waterbath, Germany) and allowed to stand for 1 h at

25 °C before centrifuging at 1600 rpm for 25 min. Water excess (supernatant) was decanted and samples were allowed to drain. The weights of water bound were determined by difference (Equation (2)).

$$\text{WRC (g/g)} = \frac{\text{Residue hydrated weight} - \text{Residue dry weight}}{\text{Residue dry weight}} \tag{2}$$

2.3.3. Swelling Capacity

The method of Olapade et al. (2011) [40], with some modifications, was used for determination of swelling capacity (SC). One gram (1 g) of quinoa flour sample from each particle size (QL, QM, QS) was mixed with 10 mL of water in a centrifuge tube. The weighed tube was heated in a water bath at 80 °C for 15 min. After this operation, the tube was centrifuged at 2000 rpm for 30 min and the supernatant was decanted and then weighed. Swelling capacity was expressed as percent swelled per gram flour (Equation (3)).

$$\text{SC (\%)} = \frac{\text{swelled gel (ml)}}{\text{sample weight (g)}} \tag{3}$$

2.3.4. Bulk Density

Bulk density (BD) of quinoa flour fractions was determined according to the method suggested by Ikegwu (2009) [41]. The sample (50 g) was put in a 100 mL graduated cylinder and tapped 20–30 times. The bulk density was calculated as weight of flour (g) divided by flour volume (mL).

2.4. α-Amylase Activity

The α-amylase activity in wheat-quinoa composite flour formulated according to the experimental design (Table 1) was evaluated using the Falling number (FN) method (ICC standard method 107/1), based on viscosity, and FN index value was determined at least in duplicate for each sample.

Table 1. Coded and real values of factors used in experimental design.

Run	Coded Values		Real Values	
	X_1	X_2	Particle Size (μm)	QF (%)
1	1.00	−1.00	380	0
2	1.00	−0.50	380	5
3	−1.00	0.00	180	10
4	1.00	0.00	380	10
5	−1.00	1.00	180	20
6	−1.00	0.50	180	15
7	0.00	0.50	280	15
8	−1.00	−1.00	180	0
9	0.00	1.00	280	20
10	0.00	−1.00	280	0
11	−1.00	−0.50	180	5
12	0.00	−0.50	280	5
13	1.00	1.00	380	20
14	1.00	0.50	380	15
15	0.00	0.00	280	10

2.5. Mixolab Measurements

A Mixolab rheological device was used to measure rheological properties of composite flour dough samples formulated as the experimental design from Table 1 shows. The protocol used to assess the rheological behavior of composite quinoa-wheat flour as a function of mixing and temperature by means of the Mixolab device (Chopin, Tripette et Renaud, Paris, France) is established by ICC-Standard Method No. 173 (ICC, 2010) [42]. To characterize the dough behavior in this work, the standard option Chopin+ protocol was applied. The setting of the test was similar to the one presented in a previous paper by Coțovanu et al. (2020) [43] for amaranth flour supplementation and consisted of a dough mixing duration time of 8 min at constant mixing rate of 80 rpm and temperature of 30 °C, a heating rate of 4 °C/min up to 90 °C, a cooling rate of 4 °C/min up to 50 °C, and a total analysis time of 45 min. A few preliminary mixing tests were performed to determine the optimum hydration level to reach the maximum consistency of dough corresponding to the C1 torque value of 1.1 N·m, and then the complete test was made in order to determine the dough properties of the composite flour formulated. The parameters from the registered Mixolab curves were reported in previously research by Coțovanu et al. (2020) [43]. The parameters determined during mixing were water absorption, WA (%) dough development time, DDT (min) and stability, ST (min); then, when dough temperature increased due to protein reduction, the minimum consistency C2 (N·m) torque was recorded. When dough temperature increased to certain values, starch gelatinization was given by the values of maximum consistency C3 torque (N·m), the stability of starch gel by minimum consistency C4 torque (N·m) and starch retrogradation when cooling through maximum consistency, C5 torque (N·m). In addition, from the Mixolab curve, the following secondary parameters were determined: the difference between torques C1 and C2 (C1–2), which is a measure of the protein weakening; the difference between torques C3 and C2 (C3–2), which estimates the starch gelatinization; the difference between torques C3 and C4 (C3–4), which is related to starch breakdown and the difference between torques C5 and C4 (C5–4) corresponding to starch recrystallization when cooling the dough.

2.6. Design of Experiments and Statistics

In this study, the simultaneous effects of two factors, particle size at three fractions (L, M and S) and addition level of quinoa flour (QF) in wheat flour at five levels (0, 5, 10, 15 and 20%) on the Falling number (FN) index were investigated; water absorption of quinoa-wheat flour dough and rheological properties, as responses, were also investigated. The main and interaction effects of the factors, particle size and addition level of QF in wheat flour, on the responses were evaluated by using response surface methodology (RSM) by means of a general factorial design. RSM has been widely used in the design and formulations of new products, and optimization in conjunction with desirability function to find, the optimal condition during processing [44] or to obtain the optimum factors, describing the relationships between responses and factors [45]. A numerical optimization technique was used in the optimization process as showed in some studies [43,46–48] allowing us to find the optimal value of the factors in relation with dough rheology. In Table 1 shows the experimental design with the coded and real values of the factors used in this study, consisting of 15 combinations. For all responses a polynomial quadratic regression model (Equation (4)) was proposed.

$$Y = b_0 + b_1 \cdot X_1 + b_2 \cdot X_2 + b_{11} \cdot X_1^2 + b_{22} \cdot X_2^2 + b_{12} \cdot X_1 \cdot X_2 \tag{4}$$

In the model, Y represents the response variable and b_0, b_1, b_2, b_{11}, b_{22} and b_{12} are the regression coefficients represented as intercept, linear, quadratic and interaction effects of X_1 (particle size of quinoa flour) and X_2 (level of quinoa flour added in wheat flour) factors.

The most adequately predictive model for each response was chosen through sequential *F*-test, coefficients of determination (R^2), adjusted coefficients of determination ($Adj.-R^2$) and significant probabilities. For this purpose, a multiple linear regression analysis was applied to fit the data obtained for response to linear, two-factor interactions, quadratic and cubic models. The statistical significance of the coefficients in each predictive model were found by analysis of variance (ANOVA) and the experimental and predictive values were compared by paired *t* test ($p < 0.05$) to determine the validity of the model for each response. The polynomial response surfaces were generated to show the dependence between factors, particle size and addition level, and responses. Design-Expert software, trial version 12 (Stat-Ease, Inc., Minneapolis, MN, USA) was used to carry out the experimental design, the model analysis, generation of the response surface graphs, test of model adequacy and the optimum level of formulation factors finding.

For numerical optimization, the desired goal for each response was selected as follows: dough stability (ST) and starch gelatinization (C3–2) at maximum value, protein weakening (C1–2) and starch recrystallization during paste cooling (C5–4) were minimized, while all remaining responses were kept within range. Through one-way analysis of variance (ANOVA) the significant difference within means was tested at a 5% significance level.

3. Results and Discussion

3.1. Proximate Composition

The proximate composition for the three fractions, large (QL), medium (QM) and small (QS) of quinoa seeds flour is shown in Table 2. The moisture content of the samples ranged from 9.75 to 10.46%, a decrease in the moisture content with decreasing particle size occurring. According to the obtained results, there were no significant differences ($p > 0.05$) in the moisture content between the particle sizes. The same results of moisture of different particle sizes of quinoa flours in different fractions was observed by Ahmed et al. (2019) [28] and for rye and barley flours [49]. The decrease of moisture when quinoa particle size decreases can be explained by higher possible heat generation during the milling of the finer particle size flours [50]. The low moisture observed for the quinoa flours was a good indicator of their potential for longer shelf life. In agreement with Offia-Olua (2014) [51] materials such as flour and starch containing up to 12% moisture have higher storage stability than those with higher moisture contents.

Table 2. Chemical composition and physical properties of the quinoa flours fractions: large particle size (QL), medium particle size (QM) and small particle size (QS).

Parameters	Particle Size		
	QL	QM	QS
Chemical composition			
Moisture (%)	10.46 ± 0.05 [a]	10.04 ± 0.01 [a]	9.75 ± 0.50 [a]
Ash (%)	1.75 ± 0.01 [a]	3.20 ± 0.04 [c]	2.53 ± 0.00 [b]
Protein (%)	11.96 ± 0.08 [a]	18.86 ± 0.05 [c]	15.69 ± 0.19 [b]
Fat (%)	6.44 ± 0.00 [a]	6.41 ± 0.01 [a]	6.36 ± 0.03a [a]
Carbohydrates (%)	69.39 ± 0.94 [c]	61.49 ± 0.09 [a]	65.66 ± 0.65 [b]
Physical properties			
WAC (%)	2.50 ± 0.02 [a]	2.92 ± 0.00 [c]	2.75 ± 0.04 [b]
WRC (g/g)	4.32 ± 0.04 [a]	4.91 ± 0.10 [b]	4.31 ± 0.04 [a]
SC (mL/g)	5.26 ± 0.08 [a]	5.55 ± 0.23 [a]	6.21 ± 0.17 [b]
BD (g/mL)	0.55 ± 0.00 [c]	0.52 ± 0.01 [b]	0.46 ± 0.00 [a]

WAC—water absorption capacity; WRC—water retention capacity; SC—swelling capacity; BD—bulk density. Lower-case letter ([a], [b] and [c]) refers to the comparison of the same compound between the different particle size quinoa flour samples; results followed by the lowercase letter are significantly different according to Tukey's HSD post hoc test ($p < 0.05$).

The ash contents ranged between 1.75 and 3.2% and the values increased significantly with the reduction of particle size, but the highest values were observed at medium particle size (QM) (3.2%), followed by small particle size (QS) (2.53%). There were significant differences ($p < 0.0001$) between the samples regarding the ash content correlated with mineral content [52]. The ash content is closely related to the chemical composition of the quinoa seed and also to the morphological structure. Some studies reported that mineral content of quinoa seems to vary dramatically due to soil type, mineral composition and fertilizer application [20]. Minerals, such as P, K, and Mg are located in the embryo, while Ca and P are in the pericarp associated with pectic compounds of the cell wall [20], which may explain the variation in ash from different quinoa fractions. The cell wall material from broken endosperm may contribute significantly to the ash content [53]. Increasing ash content with reducing quinoa particle size was observed earlier by Ahmed et al. (2019) [28], by Alonso-Miravalles and O'Mahony (2018) [37] for quinoa (1.8 to 3.60%), for whole amaranth flour (2.4 to 6.86%), for buckwheat (1.51 to 3.05%) and also for barley [50]. Many minerals in quinoa are found at concentrations greater than those reported for most grain crops [23]. High ash content, resulting from quinoa flour addition to wheat flour, could involve an increased amount of minerals in the composite flour samples.

Crude protein content of quinoa flours with different particle sizes (11.96–18.86%) is comparable to the values (12.0–16.0%) reported in other studies [11,14,20,28]. The crude protein content of the flours was significantly different ($p < 0.05$) from each other. It can be seen from the data presented in Table 2 that the protein content from quinoa flours has the same trend as the ash content. The highest content of crude protein was observed for particle size QM (18.86%). Other authors found a direct increase of protein content with a decrease of particle size [31], but Sulivan et al. (1962) [53] opined that the change of protein with particle size is not universal and it depends on the structure of endosperm and type of endosperm cells. The data obtained by Ando et al. (2002) [54] indicated that proteins are found especially in the embryo (23.5%) whereas in perisperm are only 7.2%. D'Amico et al. (2019) [55] found higher concentrations of up to 38% in the embryo and less than 5% in the perisperm. The low amount of proteins in the perisperm was also confirmed by Lindeboom et al. (2005) [56] and Chauhan et al. (1992) [57]. These data suggest that quinoa seeds store proteins in the embryo to provide nutrients for growing and development [58]. This result indicates that the substitution of wheat flour with medium particle size of quinoa flour (QM) would increase the protein content of final products, producing more shelf stable products due to their lower moisture content.

The fat content of the quinoa flours fractions ranged from 6.36 to 6.44%, but the particle sizes did not show significant differences in lipid content ($p > 0.05$). The content of lipids increased with particle size increase. Small quinoa particle size (QS) has been considered an alternative oilseed crop due to its lipidic fraction [59]. Similar data for fat content, which ranged between 1.8 and 9.5%, were reported earlier in other studies [59,60]. Prego et al. (1998) [61] showed that lipids are stored in the cells of the endosperm and embryo tissues, which explains this variation.

The carbohydrate content ranged from 61.49 to 69.39%. The carbohydrate contents for quinoa particle size flours were significantly different ($p < 0.05$) from each other. The largest particle (QL) showed a very high percentage of carbohydrate content of 69.38%. While the protein and ash content of samples increased, a decrease in carbohydrate content for the flour samples was observed (Table 2).

3.2. Physical Properties

The water absorption capacity (WAC) of flours measures the water absorption ability by the starch after swelling in excess water. The results showed that WAC for quinoa flours fractions ranged between 2.50 and 2.92% (Table 2). The results were in accordance with the reported values by Wang et al. (2016) [26] and Rodriguez-Sandoval et al. (2012) [62]. The least WAC was observed for large particle (QL), and all three particle sizes presented significant difference ($p < 0.0001$) between them, the highest WAC being presented by the medium particle size (QM), making it desirable for use in bakery products. The increase of WAC in QM can be explained by its higher level of protein compared to small particle size (QS), amino acid composition, protein conformation and surface polarity affecting water-binding properties of flours [63]. Solaesa et al. (2020) [4] reported similar results on three different particle sizes of quinoa flour. WAC may be related to starch characteristics which are reflected by amylose content, granular structure and molecular structure of amylose and amylopectin [64].

Water retention capacity (WRC) allows assessment of the flour's ability to retain water under a centrifugal gravity force, considering physically entrapped, capillary-bound and hydrodynamic water [65]. The results obtained showed that there were no statistically significant differences ($p > 0.05$) among the large and small particle sizes (Table 2), but significant differences between large and small particle fractions and medium particle size ($p < 0.05$) were found, which had the maximum WRC (4.91%). The increase of WRC in QM can be related to high protein content and ash, hydration properties being influenced by the particle fraction composition. The results were in range with those found by James (2009) [8] for quinoa flour. Approximate results observed between WRC and particle size have been reported in the literature [48] for rye and barley flours. The values were higher than the results reported by Ogungbenle et al. (2009) [66], who mentioned that quinoa flours were able to retain 1.47 g of weight of water. The relationship between WRC and the particle size was observed by other authors [67] and can be explained by the large surface area presented by finer flours. Nevertheless, other studies were either not able to establish a correlation between particle size and WRC, or no decrease in WRC was found with reductions in particle size [36]. These differences could be connected to different compositions of particles with different sizes and to the morphology and particle size distribution, which was not evaluated in other studies.

The results of the swelling capacity (SC) for each particle fraction are shown in Table 2. The values ranged between 5.26 and 6.21 mL/g. The SC values obtained for the small particles (QS) differed significantly ($p < 0.05$) from the other two particle sizes, QL and QM, because QS had the highest swelling capacity. These values can be explained by the extent of granular swelling during heating. Quinoa starch granules are approximately 1–3 μm and the small size of granules facilitates the penetration of water molecules into quinoa starch granules, thereby increasing the SC with particle size decrease. Similar results regarding SC of different quinoa particle size were found by Tang et al. (2015) [68] and Wang et al. (2016) [26].

Knowing the bulk density (BC) is essential when considering packaging and transport. The results of BC values obtained for quinoa flour fractions ranged from 0.46 for the small particle size, 0.52 for medium fraction to 0.55 for the large particle size (Table 2). Significant differences between samples ($p < 0.05$) were observed, bulk density decreasing with particle size decrease. This trend can be explained, probably, by the negative correlation between bulk density and lipids, and by the positive correlation between bulk density and carbohydrates. Similar findings were obtained by Ratnawati et al. (2019) [69]. Other authors have reported similar results for quinoa flour in different fractions [4]. The lowest moisture content (9.75%) of the small particles (QS) could contribute to their low bulk density due to the ability of water moisture to stick to flour particles, thus reducing their specific volume [69].

3.3. Fitting the Models

Based on the statistical processing of data regarding the effects of two factors, particle size and addition level, predictive models for FN index of quinoa-wheat composite flour and dough rheological properties in terms of Mixolab parameters were evaluated. The most fitting quadratic models were obtained for the following parameters: Falling number (FN), water absorption (WA), dough stability (ST), dough development time (DT), C2 torque as minimum consistency during protein weakening stage (C2), difference between torques C1 and C2 (C1–2), C3 torque as maximum consistency during starch gelatinization stage (C3), difference between torques C3 and C2 (C3–2), C4 torque as minimum consistency corresponding to hot starch stability gel (C4), C3–4—difference between torques C3 and C4, C5 torque as maximum consistency during starch retrogradation stage (C5) and difference between torques C5 and C4 (C5–4). The coefficients of determination (R^2) values (0.69–0.97), indicated the fitted models represented the experimental data well, elucidating the relationships between factors and responses.

3.4. α-Amylase Activity

The α-amylase activity of the formulated quinoa-wheat composite flours, defined through the Falling number (FN) index, showed that FN values varied from 277 to 347 s. The regression model obtained for FN (s) is represented by Equation (5):

$$\text{FN} = 307.99 - 15.65 \cdot X_1 - 4.53 \cdot X_2 - 11.95 \cdot X_1 \cdot X_2 + 9.25 \cdot X_1^2 - 0.38 \cdot X_2^2 \tag{5}$$

The results of the regression analysis for the FN index showed that the quadratic model was highly significant ($p < 0.0001$) and defined well the real α-amylase activity of the composite flour through FN index ($R^2 = 0.88$). The particle size of quinoa flour added had a significant negative effect ($p < 0.05$) on the FN, while the level added had a nonsignificant effect ($p > 0.05$). Probably, the phenolic acids present in the small particle size bind to α-amylase changing its conformation and reducing its hydrolytic activity, thus determining an increase in FN index. FN was significantly negatively correlated ($p < 0.05$) with the interaction effect between QF addition level in wheat flour and particle size, and also with quadratic term of QF addition level. The negative coefficient of the QF addition level indicated that the FN index of quinoa-wheat composite flour decreased with substitution of quinoa seed flour level increase. The effects of particle size and QF addition level indicated a decrease of FN index with particle size and QF addition increase (Figure 1a), suggesting an increase of the α-amylase activity. It is well known that the FN index is inversely proportional with α-amylase activity in flour [70]. The increase of α-amylase activity of the composite mixture is related to the fact that α-amylase is a metalloenzyme which depends on the presence of calcium ions in its molecule for activity [71]. It is known that quinoa seeds contain a high level of calcium [59] and, therefore, an increased level of quinoa flour in the quinoa-wheat composite flour would result in an increase of α-amylase activity mixture, improving final product quality.

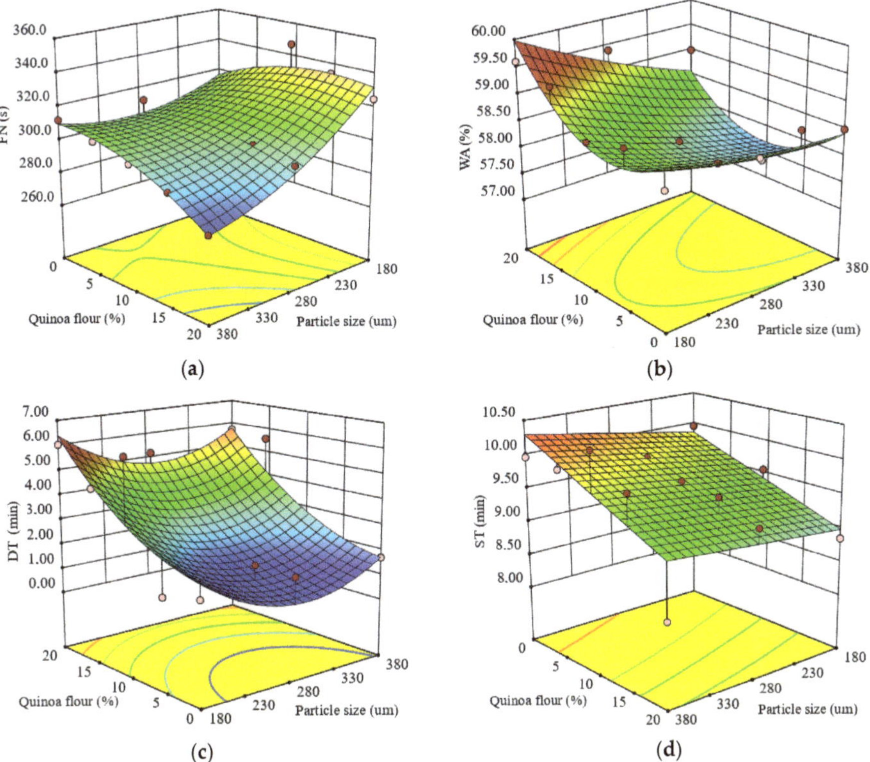

Figure 1. Response surface graph for combined effects of quinoa flour particle size and levels added in refined wheat flour on: (**a**) Falling number index and Mixolab parameters during dough development stage: (**b**) water absorption (WA), (**c**) development time (DT), (**d**) dough stability (ST).

3.5. Dough Rheological Properties during Mixing Stage of the Mixolab

3.5.1. Water Absorption

The quadratic polynomial model describing water absorption (WA) as a simultaneously function of particle size and level of QF added in wheat flour is presented as follows:

$$\text{WA} = 58.06 - 0.46 \cdot X_1 + 0.32 \cdot X_2 - 0.24 \cdot X_1 \cdot X_2 + 0.10 \cdot X_1^2 + 0.80 \cdot X_2^2 \tag{6}$$

This model is suitable to predict WA as a function of formulation factors ($R^2 = 0.78$) defining the real behavior of the dough among water absorption. WA of quinoa-wheat composite flour dough ranged from 57.3 to 59.6%. The results of regression analyses for WA indicated that the linear terms of particle size and QF level had significant ($p < 0.05$) effects on WA. The model showed that the linear terms of particle size, and the interaction between particle size and QF level added, had a negative influence on WA, which indicated that the water absorption of quinoa-wheat composite flour dough increased when levels of QF added in wheat flour increased and particle size decreased. When the particle sizes of flours are smaller, they have a higher surface area to absorb more water. The effect of different addition levels of QF in wheat flour on WA of quinoa-wheat composite flour dough is shown in Figure 1b.

The response surface graph indicates that with the increase of QF addition level in wheat flour, WA increased. These results are in agreement with the results obtained by Ahmed et al. (2019) [28] and Drakos et al. (2017) [49]. It seems that water absorption value depends on quinoa particle size, a lower size increasing the values of dough water absorption in accordance with the findings of Ahmed et al. (2016) [36]. The medium particle fractions, followed by small particles, absorbed more water compared to the large particle sizes of QF, a result that can be related to water absorption capacity values (Table 2). This trend can be influenced by various factors, in particular particle size composition. According to Rao et al. (2016) [72], damaged starch could contribute to the increase of the WA of quinoa-wheat composite flours due to the interaction between starch and nonstarch components like proteins and cell wall matrices. A decrease in WA with an increase of particle size can be explained by the chemical structure of quinoa seed, which is rich in dietary fibers made up mostly of insoluble polysaccharides and cellulose [10] and mainly located in endosperm [73], while starch is situated in perisperm. Similar result was reported by Coțovanu et al. (2020) [43], Iuga et al. (2019) [34] and Mironeasa et al. (2019) [33] which demonstrated that the small particle size of nongluten flours increases the values of WA.

3.5.2. Dough Development Time

The regression model for dough development time (DT) is represented by Equation (7):

$$\text{DT} = 1.93 - 0.2940 \cdot X_1 + 2.05 \cdot X_2 - 0.0020 \cdot X_1 \cdot X_2 + 1.33 \cdot X_1^2 + 0.7762 \cdot X_2^2 \tag{7}$$

The test for precision of the model indicated that the quadratic model was suitable to predict DT as a function of the formulation factors. The R^2 value of 0.70 was satisfactory to confirm the adequacy of the model. Dough DT was significantly affected ($p < 0.05$) by the level of QF added in wheat flour, while the particle size did not significantly ($p > 0.05$) influence DT which ranged from 1.28 to 6.13 min. DT increased remarkably in the composite flour as the QF level increased, showing that a long time was needed between the addition of water and the time when the dough reached the optimal rheological characteristics. The increase in DT can be explained by the addition of nongluten flour which diluted the gluten network. As DT represents a measure of dough strength, the higher it is the stronger the dough. The effect of particle size and QF level added in wheat flour on DT value represented by a response surface graph is shown in Figure 1c. The response surface graph revealed that DT increased as the level increased, depending on particle size. DT increased about 2.7–3.5 times in the composite flour with small particle size as the QF level increased above 10% compared to the control. This increase of

DT could be due to the increase of water when the level of quinoa flour is higher thereby requiring more mixing time. Larger particles, which are rich in dietary fibers, need more time for hydration and, thus, a higher DT.

3.5.3. Dough Stability

The regression model for dough stability (ST) is represented by Equation (8):

$$ST = 9.86 + 0.1660 \cdot X_1 - 0.4867 \cdot X_2 - 0.0490 \cdot X_1 \cdot X_2 - 0.1240 \cdot X_1^2 - 0.3848 \cdot X_2^2 \qquad (8)$$

Dough ST was significantly influenced ($p < 0.05$) by particle size and by the level of QF added in wheat flour. The quadratic model fitted to the experimental results of dough ST showed a good coefficient of determination ($R^2 = 0.69$). The statistical analysis of the model coefficients showed that the linear terms were highly significant only for the level of QF added ($p < 0.05$), while the linear term of particle size, interaction coefficient and quadratic terms were nonsignificant ($p > 0.05$).

The dough ST ranged from 8.40 to 10.35 min for the composite flours' formulations. An increase in QF addition level decreased significantly ($p < 0.05$) dough ST (Figure 1d). Compared to the control sample, it seemed that the weakening of dough caused by the addition of QF was not excessive, probably due to the percentage of lipids from the QF which are able to form lipoprotein complexes, leading to the stabilization of dough [74]. The decrease in dough ST with the increase of QF level may be related to the lower water availability in the dough system because it can be seen that the water absorption capacity of dough decreased with an increase of QF level. Gluten dilution, when nongluten flour was added, diminished the dough's viscoelastic properties [75] and, therefore, the ST decreased. At a high QF level the gluten networks may be disrupted, which can lead to dough weakening.

3.6. Dough Rheological Properties during Heating

3.6.1. Protein Weakening

The quadratic model describing C2 torque variation during protein weakening related to mechanical shear and temperature constraint is presented in Equation (9):

$$C2 = 0.4934 + 0.0145 \cdot X_1 - 0.0267 \cdot X_2 + 0.0071 \cdot X_1 \cdot X_2 - 0.0063 \cdot X_1^2 + 0.0152 \cdot X_2^2 \qquad (9)$$

The regression model was statistically significant at $p < 0.005$ and fitted the experimental data for C2, showing a high coefficient of determination ($R^2 = 0.82$). It was found that particle size and QF level had a highly significant linear effect ($p < 0.05$ and $p < 0.005$) on C2. The results showed an increase of C2 as the particle size increased and a decrease of C2 when QF addition increased (Figure 2a). The decrease of C2 can be explained by the lower water availability in the dough system due to the fact that the water absorption capacity of dough increases when the QF addition level increases, or it may be related to gluten dilution or to the activity intensification of proteolytic enzymes. C2 torque decreased with particle size reduction, especially in the case of small particle sizes, and may be linked either to a release of water molecules in dough or to the presence of a high content of proteins and the quinoa flour interfering with the protein unfolding. An increase of protein content in medium particle size, followed by small particle size, compared to large particles was observed (Table 2). Therefore, the balance between protein and starch from quinoa-wheat composite flour could play an essential role for dough consistency, particularly the amylose to amylopectin ratio.

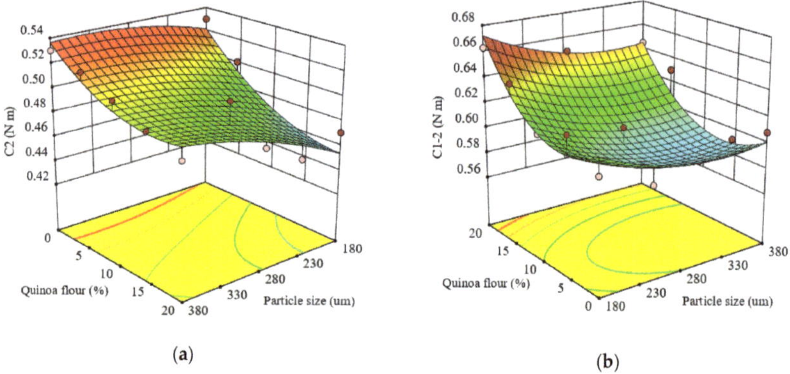

Figure 2. Response surface graph for combined effects of quinoa flour particle size and level added in refined wheat flour on Mixolab parameters during protein weakening stage: (**a**) C2 torque (C2) and (**b**) difference between torques C1 and C2 (C1–2).

The protein network strength under increased heating, represented by the difference between torques C1 and C2 (C1–2), indicated that the quadratic model (Equation (10) was statistically significant at $p < 0.01$, and fitted adequately the experimental results of C1–2 with a high coefficient of determination ($R^2 = 0.78$).

$$C1-2 = 0.4934 + 0.0145 \cdot X_1 - 0.0267 \cdot X_2 + 0.0071 \cdot X_1 \cdot X_2 - 0.0063 \cdot X_1^2 + 0.0152 \cdot X_2^2 \qquad (10)$$

The linear effect of particle size and level of QF addition on C1–2 was significant ($p < 0.05$ and $p < 0.005$), while the interaction and quadratic effects of particle size and QF addition in wheat flour were not significant ($p > 0.05$).

It can be seen from Figure 2b that increasing the concentration of QF increased the C1–2 torque but increasing the particle size led to a decrease in C1–2. This decrease in C1-2 could be due mainly to the particles' composition. As observed from Table 2, high carbohydrate content was found in the large particle size with a low level of proteins, compared to the medium and small fractions. With respect to the combined effect between particle size and QF addition level on protein network strength, it was observed that the protein network became weaker under increased heating. The addition of QF caused a decrease of protein weakening speed with temperature rise. This is probably due to the changes in protein network structure which favor the enzymatic attack resulting from less compact proteins and, therefore, increased speed of protein weakening at higher temperature. The protein network decrease may be related with increased dough proteolytic activity, the enzyme having an optimal activity within the temperature range of this Mixolab stage. This result could also be also associated with the area-specific surface of quinoa starch, that is larger than for wheat starches and makes them more sensitive to hydrolysis by α-amylase than wheat starch [76].

3.6.2. Starch Gelatinization

The regression model obtained for C3 torque, which represent maximum consistency during starch gelatinization process that occurs when dough is heated above 60 °C is presented in Equation (11):

$$C3 = 1.86 + 0.0397 \cdot X_1 - 0.1090 \cdot X_2 + 0.0236 \cdot X_1 \cdot X_2 - 0.0223 \cdot X_1^2 - 0.0085 \cdot X_2^2 \qquad (11)$$

The quadratic regression equation was fitted ($p < 0.0001$) for prediction of factors' effects on C3 torque and showed a high coefficient of determination ($R^2 = 0.96$).The linear coefficient of the QF addition level indicated a highly significant negative effect ($p < 0.0001$) on the torque C3, while particle size had a positive effect ($p < 0.005$). It was observed that interaction between PS and QF addition level had a significant effect ($p < 0.05$) on C3. Similar observations regarding particle size effect on the starch gelatinization process were reported by Ahmed et al. (2019) [28] for quinoa flour and may be due to low amylose content compared to the control. Collar et al. (2017) [77] showed that nonstarch components such as lipids, proteins and dietary fibers could restrict swelling and gelatinization during cooking, in addition to a diluting effect due to the interaction with starch polymers (lipids and proteins), and to the competition for water (proteins and dietary fiber), interfering with starch swelling. As observed in Figure 3a, C3 gradually decreased with the QF substitution increase, and a higher level of QF (20%) addition in wheat flour led to a decreased value for C3 torque (1.67 N · m) in the case of smaller particle size. This behavior could be related to the amylase activity of quinoa flour and the amylose-lipid complexes formed during heating of starch slurries. The results are in agreement with those obtained when amaranth was added in wheat flour [43]. The C3 torque decreased when the particle size of quinoa flour decreased. This was expected because the small particles fraction was a starch-rich particle and lead to a lower C3 torque compared to coarse particles that are fiber-rich particle fractions. Similar observations were reported by Martinez-Villaluenga et al. (2020) [1] and, Haros and Schoenlechner (2017) [76]. Additionally, the medium and small fractions had less carbohydrates content compared to large particle size fractions (Table 2). The small particle sizes presented large surfaces and competed with starch granules from wheat flour, contributing to lower dough viscosity. An opposite trend was reported by Ahmed et al. (2019) [28] for larger particles of QF which led to a higher pasting viscosity explained by the smaller surface area with lower swelling. The differences among studied particle fractions was affected by their composition and might be due to the particle microstructure of the quinoa-starch that govern water absorption and swelling.

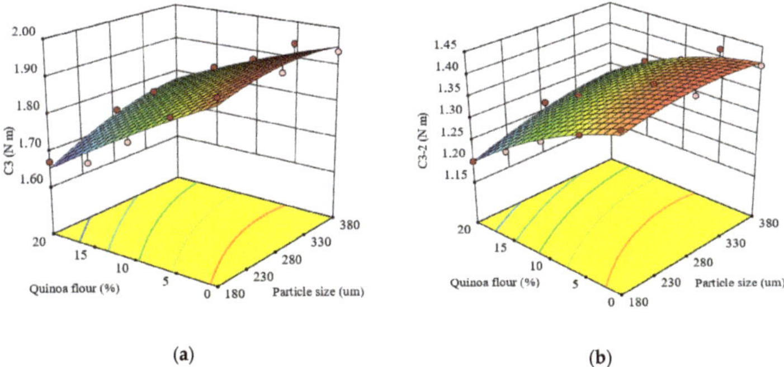

(a)   (b)

Figure 3. Response surface graph for combined effects of quinoa flour particle size and level added in refined wheat flour on Mixolab parameters during starch gelatinization stage: (**a**) C3 torque (C3) and (**b**) difference between torques C3 and C2 (C3–2).

The effects of particle size and QF addition level in wheat flour on the difference between C3 and C2 torques (C3–2) is described by Equation (12):

$$C3-2 = 1.36 + 0.0252 \cdot X_1 - 0.0827 \cdot X_2 + 0.0165 \cdot X_1 \cdot X_2 - 0.0160 \cdot X_1^2 - 0.0231 \cdot X_2^2 \qquad (12)$$

The quadratic model represented the experimental data of C3–2 very well with a high R^2 value (0.98). The linear terms of particle size and QF addition level in wheat flour on C3-2 were significant ($p < 0.0001$). C3-2 was significantly correlated ($p < 0.05$) with the quadratic effect of particle size and the interaction effect of particle size and QF addition level in wheat flour. The effect of particle size and QF dose added in wheat flour indicated an increase of C3–2 with particle size increase and an increase of this response with level decrease (Figure 3b). This behavior might be due to the combined effect of increased α-amylase activity in composite flour and to the compounds present in the particle fractions' composition of quinoa flour, according to the observations reported by Wang et al. (2016) [26].

3.6.3. Hot Starch Stability Gel

The stability of hot-formed starch gel or cooking stability is represented by C4 torque. A quadratic regression equation (Equation (13)) was fitted to predict the effect of particle size and added QF level in wheat flour on the variation of C4 (Figure 4a).

$$C4 = 1.67 + 0.0079 \cdot X_1 - 0.2385 \cdot X_2 + 0.0130 \cdot X_1 \cdot X_2 - 0.367 \cdot X_1^2 + 0.0074 \cdot X_2^2 \tag{13}$$

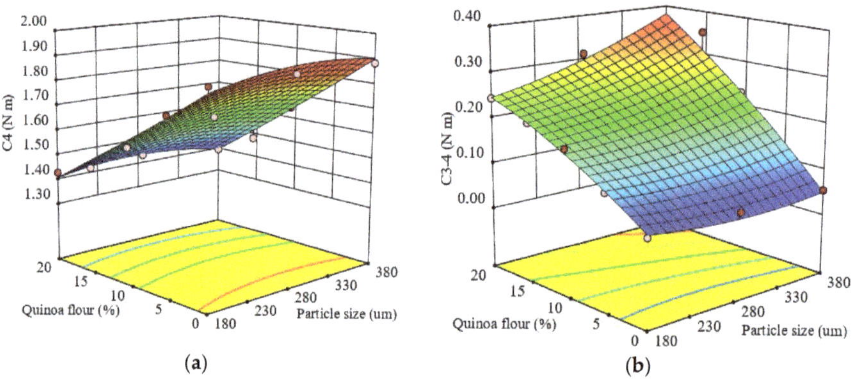

(a)   (b)

Figure 4. Response surface graph for combined effects of quinoa flour particle size and level added in refined wheat flour on Mixolab parameters during cooking stage: (**a**) C4 torque (C4) and (**b**) difference between torques C3 and C4 (C3–4).

The regression model obtained was very good because the value of the coefficient of determination (R^2) was high and close to 1 (0.98), which indicated that 98% of total variation was explained by the model and only 2% of the total variation was unexplained. The effects of the formulation factor of QF added in wheat flour on the C4 torque showed a highly significant ($p < 0.0001$) decrease with the QF level increase. The relationship with particle size decrease effect was not significant ($p > 0.05$). The stability of hot-formed starch gel can be related to the stability of the already broken starch granules under heating (Liu et al., 2016) [78]. The heating process at high temperature led to a decrease of viscosity associated with paste resistance to disintegration at high temperature, the lowest C4 value (1.39 N · m) being obtained for large particle sizes at a level of 20% QF added to wheat flour. For all formulations, the C4 torque had lower values compared to the control. The decrease of C4 torque with the amount of quinoa flour added can be explained by its high content of soluble fiber, mainly in small particle sizes, which can bind water by hydrogen bonds leading to a decrease of available water for the starch granules. On the other hand, Ahmed et al. (2019) [28] showed that the low amylose content of quinoa seeds can lead to a more fluid gel formation without such a defined three-dimensional structure, contributing to a decrease in C4 torque.

The difference of the points C3–4 was significantly affected by particle size and QF addition level in wheat flour at $p < 0.01$ and $p < 0.0001$, respectively. The regression model obtained (Equation (14)) fitted to the experimental results of the difference between torques C3–4 showed a high coefficient of determination ($R^2 = 0.97$), and indicated that the model explained 97% of the observed data variation.

$$C3 - 4 = 0.1860 + 0.0318 \cdot X_1 + 0.1295 \cdot X_2 + 0.0366 \cdot X_1 \cdot X_2 + 0.0144 \cdot X_1^2 - 0.015 \cdot X_2^2 \tag{14}$$

Figure 4b show the response surface graph for the combined effect of particle size and QF level addition in wheat flour on C3–4. It can be seen that as the particle size and QF level increased in samples, C3–4 increased.

All the samples with QF addition had higher values of C3–4 compared to the control sample, depending on the particle size, even if the quinoa flour did not add amylase to the dough system. An increase of C3-4 could be associated with reduced gel stability when hot and may be due to starch damage.

3.6.4. Starch Retrogradation

The regression equation obtained for C5 torque, which represents starch retrogradation during the cooling period of the Mixolab, is given by the Equation (15).

$$C5 = 2.54 + 0.0559 \cdot X_1 - 0.4681 \cdot X_2 + 0.0297 \cdot X_1 \cdot X_2 - 0.0791 \cdot X_1^2 + 0.0730 \cdot X_2^2 \tag{15}$$

The ANOVA results show that the quadratic model fitted for C5 torque was well adjusted to the experimental data ($R^2 = 0.98$).

The maximum consistency during starch retrogradation (C5) was significantly ($p < 0.0001$) affected by the particle sizes and different levels of QF added to wheat flour. It was found that there was a significant ($p < 0.05$) positive effect of increasing particle size and a significant ($p < 0.0001$) negative effect of QF level added to wheat flour on C5 torque. The response surface graph for the combined effect of particle size and level of QF added to wheat flour on C5 is given in Figure 5a. As the level of QF increased and the particle size decreased, the C5 torque decreased. This indicates that starch retrogradation depends on the starch proportions of amylose and amylopectin [79], amylose recrystallizing faster than amylopectin. The decreases of C5 torque suggested that particle size composition influenced α-amylase activity in quinoa-wheat composite flours, a high activity causing a lowering of C5 value.

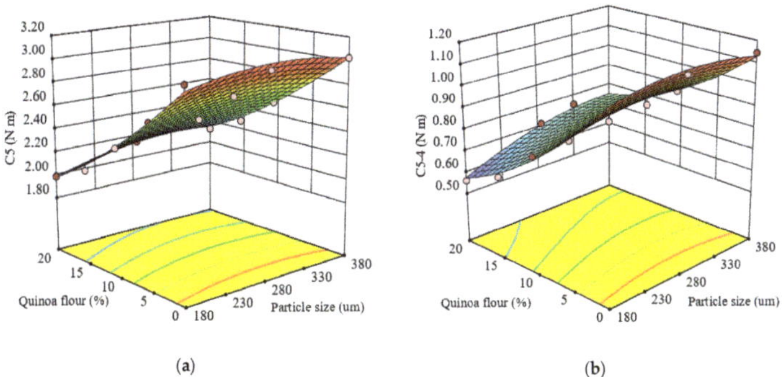

(a) (b)

Figure 5. Response surface graph for combined effects of quinoa flour particle size and level added in refined wheat flour on Mixolab parameters during starch retrogradation: (**a**) C5 torque (C5) and (**b**) difference between torques C5 and C4 (C5–4).

The quadratic regression model obtained from the ANOVA for the difference between C5 and C4 torques (C5–4) (Equation (16)) was found to be significant ($p < 0.0001$) and the R^2 value of 0.97 confirmed the adequacy of the model.

$$C5 - 4 = 0.8712 + 0.0480 \cdot X_1 - 0.2295 \cdot X_2 + 0.0427 \cdot X_1 \cdot X_2 - 0.0424 \cdot X_1^2 + 0.0656 \cdot X_2^2 \tag{16}$$

Based on the obtained results, it was found that there was a significant effect of factors, particle size and QF level added to wheat flour ($p < 0.005$ and $p < 0.0001$), and of their interaction ($p < 0.05$), on the C5-4. All the formulations had a lower value for C5-4 torque compared with the control sample.

The response surface graph for combined effect of factors on C5–4 is given in Figure 5b. As the particle size decreased and QF addition increased, the C5–4 decreased, the lowest value (0.56 N·m) being achieved at 20% QF of small particle size which substituted the wheat flour. This decrease may be attributed to the quinoa flour addition to wheat flour limiting starch retrogradation and preserving bread freshness depending on particle size.

3.7. Optimization

The numerical optimization procedure revealed that the most suitable composite flour would have 8.98% quinoa flour of 280 µm particle size. Predicted values for responses are shown in Table 3 compared to the values for control samples. As can be seen, optimal values for water absorption, dough stability and protein weakening in terms of C2 and C1–2 were very close to the values of control samples and were not statistically different at the 95% confidence level. It can be said that during the early stage of bread, the optimal dough formulation can retain gas similar to the control sample.

Table 3. Optimized quinoa-wheat flour dough compared to the wheat flour control sample.

Parameters	Values	
	Optimized Sample	**Control Sample**
FN (s)	308.37	313.13
WA (g/100 g)	58.03	58.43
DT (min)	1.72	1.43
ST (min)	9.91	9.78
C2 (N · m)	0.49	0.51
C1-2 (N · m)	0.59	0.60
C3 (N · m)	1.86	1.92
C3-2 (N · m)	1.37	1.41
C4 (N · m)	1.69	1.89
C3-4 (N · m)	0.17	0.03
C5 (N · m)	2.58	3.02
C5-4 (N · m)	0.89	1.13

FN—Falling number index; WA—water absorption; DT—development time; ST—stability; C2—torque as minimum consistency during protein weakening; C1–2—difference between torques C1 and C2; C3—torque as maximum consistency during starch gelatinization; C3–2—difference between torques C3 and C2; C4—torque as minimum consistency corresponding to hot starch stability gel; C3–4—difference between torques C3 and C4; C5—torque as maximum consistency during starch retrogradation; C5–4—difference between torques C5 and C4.

Dough development time increased compared to the control, suggesting an increase of gluten network strength indicating that optimal dough can sustain the mechanical treatment for a longer period during the bread making process. A similar finding was reported for optimized amaranth-wheat composite flour [43].

During the pasting stage, a slight decrease was obtained for the Mixolab parameters which revealed that the optimal quinoa-wheat flour formulation could be adequate for bread-making and the final products could have a longer shelf-life without problems of going stale. The medium particle size of added QF could delay baked goods going stale.

4. Conclusions

Milling fractions of quinoa seeds revealed variations in chemical composition and functionality depending on particle size. Medium particle sizes, followed by small particle sizes, have high content of protein and ash, whereas large particle sizes are enriched with carbohydrates. A higher water absorption capacity and water retention capacity was found for medium particle sizes, while the small particle sizes had a higher swelling capacity. Dough rheological properties were remarkably influenced by particle size and level of quinoa flour added to wheat flour, all the regression models obtained for responses being significant ($p < 0.05$) and with high coefficients of determination ($R^2 > 0.69$). A simultaneous optimization of multiple responses allowed us to obtain the optimal particle size and quinoa flour level added to wheat flour to achieve dough with the best rheological properties. For this purpose, the formulation with medium particle size, 8.98% quinoa flour and 91.02% wheat flour, was considered to be the most appropriate to develop new breadmaking products. It can be concluded that the processor should maintain a certain particle size to obtain a desired product consistency together with the appropriate rheological parameters. This optimal formulation does not alter the dough matrix, leads to the best technological parameters and can be used to enrich baked products due to their particle size composition.

Author Contributions: Conceptualization, I.C. and S.M.; methodology, I.C. and S.M.; software, I.C.; validation, I.C., A.B. and S.M.; formal analysis, I.C., A.B. and S.M.; investigation, I.C. and A.B.; resources, I.C., A.B. and S.M.; data curation, I.C., A.B. and S.M.; writing—original draft preparation, I.C.; writing—review and editing, I.C. and S.M.; visualization, I.C., A.B. and S.M.; supervision, S.M.; project administration, S.M.; funding acquisition, S.M. All authors have read and agreed to the published version of the manuscript.

Funding: This work was supported from contract no. 18PFE/16.10.2018 funded by Ministry of Research and Innovation within Program 1-Development of national research and development system, Subprogram 1.2-Institutional Performance-RDI excellence funding projects.

Acknowledgments: The authors thank Arcada Research laboratory, Arcada Mill, Romania for technical support.

Conflicts of Interest: The authors declare no conflict of interest.

References

1. Martínez-Villaluenga, C.; Peñas, E.; Hernández-Ledesma, B. Pseudocereal grains: Nutritional value, health benefits and current applications for the development of gluten-free foods. *Food Chem. Toxicol.* **2020**, *137*, 111178. [CrossRef] [PubMed]

2. Kahlon, T.S.; Avena-Bustillos, R.J.; Chiu, M.C.M. Sensory evaluation of gluten-free quinoa whole grain snacks. *Heliyon* **2016**, *2*, e00213. [CrossRef] [PubMed]

3. Burgos, V.E.; López, E.P.; Goldner, M.C.; Del Castillo, V.C. Physicochemical characterization and consumer response to new Andean ingredients-based fresh pasta: Gnocchi. *Int. J. Gastron. Food Sci.* **2019**, *16*, 100142. [CrossRef]

4. Solaesa, Á.G.; Villanueva, M.; Vela, A.J.; Ronda, F. Protein and lipid enrichment of quinoa (cv. *Titicaca*) by dry fractionation. Techno-functional, thermal and rheological properties of milling fractions. *Food Hydrocoll.* **2020**, *105*, 105770. [CrossRef]

5. Pereira, E.; Encina-Zelada, C.; Barros, L.; Gonzales-Barron, U.; Cadavez, V.; Ferreira, I.C. Chemical and nutritional characterization of *Chenopodium quinoa* Willd (*quinoa*) grains: A good alternative to nutritious food. *Food Chem.* **2019**, *280*, 110–114. [CrossRef] [PubMed]

6. Tanwar, B.; Goyal, A.; Irshaan, S.; Kumar, V.; Sihag, M.K.; Patel, A.; Kaur, I. Quinoa. In *Whole Grains and their Bioactives: Composition and Health*; Johnson, J., Walace, T., Eds.; John Wiley & Sons Ltd.: Hoboken, NJ, USA, 2019; pp. 269–305. [CrossRef]

7. Repo-Carrasco-Valencia, R.; Arana, J.V. Carbohydrates of Kernels. In *Pseudocereals: Chemistry and Technology*; Haros, M., Schoenlechner, R., Eds.; Wiley: Oxford, UK, 2017; pp. 28–48.

8. James, L.E.A. Quinoa (*Chenopodium quinoa* Willd.): Composition, chemistry, nutritional, and functional properties. *Adv. Food Nutr. Res.* **2009**, *58*, 1–31. [CrossRef]

9. Ahamed, N.T.; Singhal, R.S.; Kulkami, P.R.; Palb, M. Physicochemical and functional properties of Chenopodium quinoa starch. *Carbohydr. Polym.* **1996**, *31*, 99–103. [CrossRef]

10. Lamothe, L.M.; Srichuwong, S.; Reuhs, B.L.; Hamaker, B.R. Quinoa (*Chenopodium quinoa* W.) and amaranth (*Amaranthus caudatus* L.) provide dietary fibres high in pectic substances and xyloglucans. *Food Chem.* **2015**, *167*, 490–496. [CrossRef]

11. Nowak, V.; Du, J.; Charrondière, U.R. Assessment of the nutritional composition of quinoa (*Chenopodium quinoa* Willd.). *Food Chem.* **2016**, *193*, 47–54. [CrossRef]

12. Rosero, O.; Marounek, M.; Břeňová, N. Phytase activity and comparison of chemical composition, phytic acid P content of four varieties of quinoa grain (*Chenopodium quinoa* Willd.). *Acta Agronómica* **2013**, *62*, 13–20.

13. Rizzello, C.G.; Lorusso, A.; Montemurro, M.; Gobbetti, M. Use of sourdough made with quinoa (*Chenopodium quinoa*) flour and autochthonous selected lactic acid bacteria for enhancing the nutritional, textural and sensory features of white bread. *Food Microbiol.* **2016**, *56*, 1–13. [CrossRef] [PubMed]

14. Miranda, M.; Vega-Gálvez, A.; Quispe-Fuentes, I.; Rodríguez, M.J.; Maureira, H.; Martínez, E.A. Nutritional aspects of six quinoa (*Chenopodium quinoa* Willd.) ecotypes from three geographical areas of Chile. *Chil. J. Agric. Res.* **2012**, *72*, 175. [CrossRef]

15. Joshi, D.C.; Sood, S.; Hosahatti, R.; Kant, L.; Pattanayak, A.; Kumar, A.; Yadav, D.; Stetter, M.G. From zero to hero: The past, present and future of grain amaranth breeding. *Theor. Appl. Genet.* **2018**, *131*, 1807–1823. [CrossRef] [PubMed]

16. Tang, Y.; Li, X.; Chen, P.X.; Zhang, B.; Liu, R.; Hernandez, M.; Draves, J.; Marcone, M.F.; Tsao, R. Assessing the fatty acid, carotenoid, and tocopherol compositions of amaranth and quinoa seeds grown in Ontario and their overall contribution to nutritional quality. *J. Agric. Food Chem.* **2016**, *64*, 1103–1110. [CrossRef]

17. Vera, E.P.; Alca, J.J.; Saravia, G.R.; Campioni, N.C.; Alpuy, I.J. Comparison of the lipid profile and tocopherol content of four Peruvian quinoa (*Chenopodium quinoa* Willd.) cultivars ('Amarilla de Marangani', 'Blanca de Juli', INIA 415 'Roja Pasankalla', INIA 420 'Negra Collana') during germination. *J. Cereal Sci.* **2019**, *88*, 132–137. [CrossRef]

18. Paucar-Menacho, L.M.; Dueñas, M.; Peñas, E.; Frias, J.; Martínez-Villaluenga, C. Effect of dry heat puffing on nutritional composition, fatty acid, amino acid and phenolic profiles of pseudocereals grains. *Pol. J. Food Nutr. Sci.* **2018**, *68*, 289–297. [CrossRef]

19. Nascimento, A.C.; Mota, C.; Coelho, I.; Gueifão, S.; Santos, M.; Matos, A.S.; Gimenez, A.; Lobo, M.; Samman, N.; Castanheira, I. Characterisation of nutrient profile of quinoa (*Chenopodium quinoa*), amaranth (*Amaranthus caudatus*), and purple corn (*Zea mays* L.) consumed in the North of Argentina: Proximates, minerals and trace elements. *Food Chem.* **2014**, *148*, 420–426. [CrossRef]

20. Vega-Gálvez, A.; Miranda, M.; Vergara, J.; Uribe, E.; Puente, L.; Martínez, E.A. Nutrition facts and functional potential of quinoa (*Chenopodium quinoa* Willd.), an ancient Andean grain: A review. *J. Sci. Food Agric.* **2010**, *90*, 2541–2547. [CrossRef]

21. Han, Y.; Chi, J.; Zhang, M.; Zhang, R.; Fan, S.; Dong, L.; Huang, F.; Liu, L. Changes in saponins, phenolics and antioxidant activity of quinoa (*Chenopodium quinoa* Willd) during milling process. *LWT Food Sci. Technol.* **2019**, *114*, 108381. [CrossRef]

22. Rocchetti, G.; Lucini, L.; Rodriguez, J.M.; Barba, F.J.; Giuberti, G. Gluten-free flours from cereals, pseudocereals and legumes: Phenolic fingerprints and in vitro antioxidant properties. *Food Chem.* **2019**, *271*, 157–164. [CrossRef]

23. Gómez-Caravaca, A.M.; Segura-Carretero, A.; Fernández-Gutiérrez, A.; Caboni, M.F. Simultaneous determination of phenolic compounds and saponins in quinoa (*Chenopodium quinoa* Willd) by a liquid chromatography–diode array detection–electrospray ionization–time-of-flight mass spectrometry methodology. *J. Agric. Food Chem.* **2011**, *59*, 10815–10825. [CrossRef]

24. Tamba-Berehoiu, R.M.; Turtoi, M.O.; Popa, C.N. Assessment of quinoa flours effect on wheat flour doughs rheology and bread quality. *Ann. Univ. Dunarea Jos Galati. Fascicle Food Technol.* **2019**, *43*, 173–188. [CrossRef]

25. Codină, G.G.; Franciuc, S.G.; Todosi-Sănduleac, E. Studies on the influence of quinoa flour addition on bread quality. *Food Environ. Saf. J.* **2016**, *15*, 165–174.

26. Wang, S.; Zhu, F. Formulation and Quality Attributes of Quinoa Food Products. *Food Bioprocess Technol.* **2016**, *9*, 49–68. [CrossRef]

27. Sullivan, B.; Engebreston, W.E.; Anderson, M.L. The relation of particle size of certain flour characteristics. *Cereal Chem.* **1960**, *37*, 436–455.

28. Ahmed, J.; Thomas, L.; Arfat, Y.A. Functional, rheological, microstructural and antioxidant properties of quinoa flour in dispersions as influenced by particle size. *Food Res. Int.* **2019**, *116*, 302–311. [CrossRef] [PubMed]

29. Nguyen, A.; Pasquier, P.; Marcotte, D. Influence of groundwater flow in fractured aquifers on standing column wells performance. *Geothermics* **2015**, *58*, 39–48. [CrossRef]

30. Benito-Román, O.; Rodríguez-Perrino, M.; Sanz, M.T.; Melgosa, R.; Beltrán, S. Supercritical carbon dioxide extraction of quinoa oil: Study of the influence of process parameters on the extraction yield and oil quality. *J. Supercrit. Fluids* **2018**, *139*, 62–71. [CrossRef]

31. Joubert, M.; Morel, M.H.; Lullien-Pellerin, V. Pasta color and viscoelasticity: Revisiting the role of particle size, ash, and protein content. *Cereal Chem.* **2018**, *95*, 386–398. [CrossRef]

32. Tsatsaragkou, K.; Kara, T.; Ritzoulis, C.; Mandala, I.; Rosell, C.M. Improving carob flour performance for making gluten-free breads by particle size fractionation and jet milling. *Food Bioprocess Technol.* **2017**, *10*, 831–841. [CrossRef]

33. Mironeasa, S.; Iuga, M.; Zaharia, D.; Mironeasa, C. Rheological analysis of wheat flour dough as influenced by grape peels of different particle sizes and addition levels. *Food Bioprocess Technol.* **2019**, *12*, 228–245. [CrossRef]

34. Iuga, M.; Mironeasa, C.; Mironeasa, S. Oscillatory rheology and creep-recovery behaviour of grape seed-wheat flour dough: Effect of grape seed particle size, variety and addition level. *Bull. UASVM Food Sci. Technol.* **2019**, *76*, 40–51. [CrossRef]

35. Mironeasa, S.; Iuga, M.; Zaharia, D.; Dabija, A.; Mironeasa, C. Influence of particle sizes and addition level of grape seeds on wheat flour dough rheological properties. In Proceedings of the International Multidisciplinary Scientific GeoConference Surveying Geology & Mining Ecology Management, Vienna, Austria, 27–29 November 2017; Volume 17, pp. 265–272. [CrossRef]

36. Ahmed, J.; Al-Attar, H.; Arfat, Y.A. Effect of particle size on compositional, functional, pasting and rheological properties of commercial water chestnut flour. *Food Hydrocoll.* **2016**, *52*, 888–895. [CrossRef]

37. Alonso-Miravalles, L.; O'Mahony, J.A. Composition, protein profile and rheological properties of pseudocereal-based protein-rich ingredients. *Foods* **2018**, *7*, 73. [CrossRef] [PubMed]

38. Sosulski, F.; Humbert, E.S.; Bui, K.; Jones, J.D. Functional propreties of rapeseed flours, concentrates and isolate. *J. Food Sci.* **1976**, *41*, 49–1352. [CrossRef]

39. Onipe, O.O.; Beswa, D.; Jideani, A.I.O. Effect of size reduction on colour, hydration and rheological properties of wheat bran. *Food Sci. Technol.* **2017**, *37*, 389–396. [CrossRef]

40. Olapade, A.A.; Aworh, O.C.; Oluwole, O.B. Quality attributes of biscuit from acha (*Digitaria exilis*) flour supplemented with cowpea (*Vigna unguiculata*) flour. *J. Food Sci. Technol.* **2011**, *2*, 198–203. [CrossRef]

41. Ikegwu, O.J.; Nwobasi, V.N.; Odoh, M.O.; Oledinma, N.U. Evaluation of the pasting and some functional properties of starch isolated from some improved cassava varieties in Nigeria. *Afr. J. Biotechnol.* **2009**, *8*, 2310–2315.

42. ICC. *International Association for Cereal Sciences and Technology, ICC Standard No. 173*; ICC: Vienna, Austria, 2010.

43. Coțovanu, I.; Stoenescu, G.; Mironeasa, S. Amaranth influence on wheat flour dough rheology: Optimal particle size and amount of flour replacement. *J. Microbiol. Biotechnol. Food Sci.* **2020**, *10*, 1–10. Available online: https://www.jmbfs.org/in-next-issue/ (accessed on 16 September 2020).

44. Katina, K.; Heiniö, R.L.; Autio, K.; Poutanen, K. Optimization of sourdough process for improved sensory profile and texture of wheat bread. *LWT Food Sci. Technol.* **2006**, *39*, 1189–1202. [CrossRef]

45. Mironeasa, S.; Mironeasa, C. Dough bread from refined wheat flour partially replaced by grape peels: Optimizing the rheological properties. *J. Food Process Eng.* **2009**, *42*, e13207. [CrossRef]

46. Iuga, M.; Mironeasa, S. Grape seeds effect on refined wheat flour dough rheology: Optimal amount and particle size. *Ukr. Food J.* **2019**, *8*, 799–814. [CrossRef]

47. Mironeasa, S.; Iuga, M.; Zaharia, D.; Mironeasa, C. Optimization of grape peels particle size and flour substitution in white wheat flour dough. *Sci. Study Res. Chem. Eng. Biotechnol. Food Ind.* **2019**, *20*, 29–42.

48. Mironeasa, S.; Iuga, M.; Zaharia, D.; Mironeasa, C. Optimization of white wheat flour dough rheological properties with different levels of grape peels flour addition. *Bull. UASVM Food Sci. Technol.* **2019**, *76*, 27–39. [CrossRef]

49. Drakos, A.; Kyriakakis, G.; Evageliou, V.; Protonotariou, S.; Mandala, I.; Ritzoulis, C. Influence of jet milling and particle size on the composition, physicochemical and mechanical properties of barley and rye flours. *Food Chem.* **2017**, *215*, 326–332. [CrossRef]

50. Prasopsunwattana, N.; Omary, M.B.; Arndt, E.A.; Cooke, P.H.; Flores, R.A.; Yokoyama, W.; Toma, A.; Chongcham, S.; Lee, S.P. Particle size effects on the quality of flour tortillas enriched with whole grain waxy barley. *Cereal Chem.* **2009**, *86*, 439–451. [CrossRef]

51. Offia-Olua, B.I. Chemical, functional and pasting properties of wheat (*Triticum* spp)-Walnut (*Juglansregia*) flour. *Food Nutr. Sci.* **2014**, *5*, 1591–1604. [CrossRef]

52. Schuck, P.; Jeantet, R.; Dolivet, A. *Analytical Methods Food Dairy Powders*; Blackwell: Hoboken, NJ, USA, 2012.

53. Sullivan, J.T. Evaluation of Forage Crops by Chemical Analysis. A Critique. *Agron. J.* **1962**, *54*, 511–515. [CrossRef]

54. Ando, H.; Chen, Y.C.; Tang, H.; Shimizu, M.; Watanabe, K.; Mitsunaga, T. Food components in fractions of quinoa seed. *Food Sci. Technol. Res.* **2002**, *8*, 80–84. [CrossRef]

55. D'Amico, S.; Jungkunz, S.; Balasz, G.; Foeste, M.; Jekle, M.; Tömösközi, S.; Schoenlechner, R. Abrasive milling of quinoa: Study on the distribution of selected nutrients and proteins within the quinoa seed kernel. *J. Cereal Sci.* **2019**, *86*, 132–138. [CrossRef]

56. Lindeboom, N.; Chang, P.R.; Falk, K.C.; Tyler, R.T. Characteristics of starch from eight quinoa lines. *Cereal Chem.* **2005**, *82*, 216–222. [CrossRef]

57. Chauhan, G.S.; Eskin, N.A.M.; Tkachuk, R. Nutrients and antinutrients in quinoa seed. *Cereal Chem.* **1992**, *69*, 85–88.

58. Herman, E.M.; Larkins, B.A. Protein storage bodies and vacuoles. *Plant Cell* **1999**, *11*, 601–613. [CrossRef] [PubMed]

59. Koziol, M.J. Quinoa: A potential new oil crop. In *New Crops*; Janick, J., Simon, J.E., Eds.; Wiley: New York, NY, USA, 1993; pp. 328–336.

60. Oshodi, H.N.; Ogungbenle, M.O.; Oladimeji, A.A. Chemical composition, nutritionally valuable minerals and functional properties of benniseed (*Sesamum radiatum*), pearl millet (*Pennisetum typhoides*) and quinoa (*Chenopodium quinoa*) flours. *Int. J. Food Sci. Nutr.* **1999**, *50*, 325–331. [CrossRef]

61. Prego, I.; Maldonado, S.; Otegui, M. Seed structure and localization of reserves in *Chenopodium quinoa*. *Ann. Bot.* **1998**, *82*, 481–488. [CrossRef]

62. Rodriguez-Sandoval, E.; Sandoval, G.; Cortes-Rodríguez, M. Effect of quinoa and potato flours on the thermomechanical and breadmaking properties of wheat flour. *Braz. J. Chem. Eng.* **2012**, *29*, 503–510. [CrossRef]

63. Fekria, A.M.; Isam, A.M.A.; Suha, O.A.; Elfadil, E.B. Nutritional and functional characterization of defatted seed cake flour of two Sudanese groundnut (*Arachis hypogaea*) cultivars. *Int. Food Res. J.* **2012**, *19*, 629–637.

64. Pérez, S.; Bertoft, E. The molecular structures of starch components and their contribution to the architecture of starch granules: A comprehensive review. *Starch Stärke* **2010**, *62*, 389–420. [CrossRef]

65. Joardder, M.U.H.; Mourshed, M.; Masud, M.H. *State of Bound Water: Measurement and Significance in Food Processing*; Springer: Berlin/Heidelberg, Germany, 2019. [CrossRef]

66. Ogungbenle, H.N.; Oshodi, A.A.; Oladimeji, M.O. The proximate and effect of salt applications on some functional properties of quinoa (*Chenopodium quinoa*) flour. *Pak. J. Nutr.* **2009**, *8*, 49–52. [CrossRef]

67. Protonotariou, S.; Batzaki, C.; Yanniotis, S.; Mandala, I. Effect of jet milled whole wheat flour in biscuits properties. *LWT- Food Sci. Technol.* **2016**, *74*, 106–113. [CrossRef]

68. Tang, Y.; Li, X.; Chen, P.X.; Zhang, B.; Hernandez, M.; Zhang, H.; Marcone, M.F.; Liu, R.; Tsao, R. Characterisation of fatty acid, carotenoid, tocopherol/tocotrienol compositions and antioxidant activities in seeds of three *Chenopodium quinoa* Willd genotypes. *Food Chem.* **2015**, *174*, 502–508. [CrossRef] [PubMed]

69. Ratnawati, L.; Desnilasari, D.; Surahman, D.N.; Kumalasari, R. Evaluation of physicochemical, functional and pasting properties of soybean, mung bean and red kidney bean flour as ingredient in biscuit. In *IOP Conference Series: Earth and Environmental Science*; IOP Publishing: Bristol, UK, 2019; Volume 251, p. 012026. [CrossRef]

70. Struyf, N.; Verspreet, J.; Courtin, C.M. The effect of amylolytic activity and substrate availability on sugar release in non-yeasted dough. *J. Cereal Sci.* **2016**, *69*, 111–118. [CrossRef]

71. Sundarram, A.; Murthy, T.P.K. α-amylase production and applications: A review. *J. Appl. Environ. Microbiol.* **2014**, *2*, 166–175. [CrossRef]

72. Rao, B.D.; Anis, M.; Kalpana, K.; Sunooj, K.V.; Patiul, J.V.; Ganesh, T. Influence of milling methods and particle size hydration properties of sorghum flour and quality of sorghum biscuits. *LWT Food Sci. Technol.* **2016**, *67*, 8–13.

73. Haros, C.M.; Sanz-Penella, J.M. Food uses of whole pseudocereals. In *Pseudocereals: Chemistry and Technology*; Haros, M., Schoenlechner, R., Eds.; Wiley: Oxford, UK, 2017; pp. 163–192.

74. Carr, N.O.; Daniels, N.W.; Frazier, P.J. Lipid interactions in breadmaking. *Crit. Rev. Food Sci. Nutr.* **1992**, *31*, 237–258. [CrossRef]

75. Mohammed, I.; Ahmed, A.R.; Senge, B. Effects of chickpea flour on wheat pasting properties and bread making quality. *J. Food Sci. Technol.* **2014**, *51*, 1902–1910. [CrossRef]

76. Haros, C.M.; Schoenlechner, R. *Pseudocereals: Chemistry and Technology*; John Wiley & Sons: Hoboken, NJ, USA, 2017; pp. 163–192.

77. Collar, C. Significance of heat-moisture treatment conditions on the pasting and gelling behaviour of various starch-rich cereal and pseudocereal flours. *Food Sci. Technol. Int.* **2017**, *23*, 623–636. [CrossRef]
78. Liu, X.L.; Mu, T.H.; Sun, H.N.; Zhang, M.; Chen, J.W. Influence of potato flour on dough rheological properties and quality of steamed bread. *J. Integr. Agric.* **2016**, *15*, 2666–2676. [CrossRef]
79. BeMiller, J.N. Starches: Molecular and granular structures and properties. In *Carbohydrate Chemistry for Food Scientists*, 3rd ed.; BeMiller, J.N., Ed.; Elsevier: Amsterdam, The Netherlands, 2019; pp. 159–189.

Publisher's Note: MDPI stays neutral with regard to jurisdictional claims in published maps and institutional affiliations.

Article

Nutritional and Functional Properties of Gluten-Free Flours

Alina Culetu, Iulia Elena Susman, Denisa Eglantina Duta * and Nastasia Belc

National Institute of Research & Development for Food Bioresources, IBA Bucharest, 6 Dinu Vintila Street, 021102 Bucharest, Romania; alinaculetu@gmail.com (A.C.); iulia.susman@bioresurse.ro (I.E.S.); nastasia.belc@bioresurse.ro (N.B.)
* Correspondence: denisa.duta@bioresurse.ro

Abstract: This study characterized and compared 13 gluten-free (GF) flours (rice, brown rice, maize, oat, millet, teff, amaranth, buckwheat, quinoa, chickpea, gram, tiger nut, and plantain) for their nutritional and functional properties. For all GF flours investigated, starch was the major component, except for gram, chickpea, and tiger nut flours with lower starch content (<45%), but higher fiber content (8.8–35.4%). The higher amount of calcium, magnesium, zinc, potassium, phosphorus, similar values for iron and lower content of sodium in gram, makes this flour a good alternative to chickpea or other GF flour to develop healthier food products. Amaranth flour had a high protein digestibility, while tiger nut and millet flours were less digestible. Gram, chickpea, quinoa, buckwheat, and oat flours fulfilled amino acids recommendation for daily adult intake showing no limiting amino acid. Total polyphenolic content and antioxidant capacity showed higher values for buckwheat, followed by quinoa and maize flours. Gram, chickpea, maize, and quinoa flours are good candidates to improve health conditions due to lower saturated fatty acid content. The findings of this study provide useful insights into GF flours and may contribute to the development of novel gluten-free products like bread, cookies, or pasta.

Keywords: gluten-free flour; gram; plantain; chickpea; tiger nut; pseudo-cereal; oat; millet; teff; rice

Citation: Culetu, A.; Susman, I.E.; Duta, D.E.; Belc, N. Nutritional and Functional Properties of Gluten-Free Flours. *Appl. Sci.* **2021**, *11*, 6283. https://doi.org/10.3390/app11146283

Academic Editor: Georgiana Gabriela Codină

Received: 16 June 2021
Accepted: 5 July 2021
Published: 7 July 2021

Publisher's Note: MDPI stays neutral with regard to jurisdictional claims in published maps and institutional affiliations.

1. Introduction

Coeliac disease or gluten-related disorders including, wheat allergy and non-celiac gluten sensitivity cause major health problems for people when ingesting small amount of gluten [1]. In addition to people who are forced to consume gluten-free products, the demand for these products has also increased for people who want to follow a healthy diet. In this sense, it is necessary to expand and diversify the food industry both in terms of progress in ingredients and formulations, and in the production of functional foods [2]. Because coeliac people have various nutrient deficiencies, there are many challenges in the development of gluten-free products [3,4]. In addition to nutrient deficiencies, other issues make gluten products difficult to replace. For example, gluten-free dough is more difficult to handle due to a lack of cohesiveness, elasticity, and baking quality [5,6]. In general, gluten-free products are characterized by high starch content, low fiber content, short shelf life or texture issues, like increased bread crumb hardness [7]. Within this aim, more and more researches are being conducted to find formulations between different gluten-free flours and ingredients to obtain products that are similar to wheat-containing products.

Rice is the most used cereal flour for gluten-free products development [8]. Rice has unique nutritional, hypoallergenic, colorless, and bland taste properties and a low level of prolamin. An alternative to rice is brown rice (unpolished rice), which contains many nutritional and bioactive components, including fiber, amino acids, minerals, and phenolic compounds [9]. Despite the many qualities of rice flour, it was necessary to expand the raw materials used for the development of gluten-free products to diversify the product range and to enrich the nutrient content. Thus, a long list of raw materials was used in various studies to develop the gluten-free sector as much as possible. Among them are cereals (maize, sorghum), whole grains (brown rice, millet, teff, oatmeal), pseudo-cereals (amaranth, quinoa, buckwheat), legumes (pea, lentils, soybean, chickpea, gram), seeds (flax seeds, pumpkin seeds), nuts (almond, walnut, peanuts), tuberous rhizomes (tiger nut, jerusalem artichokes), and other types of raw materials (plantain, coconut) [3,10].

Hager et al. [11] showed that whole grains flours had higher fiber content compared to wheat flour. In addition, the protein content is higher in case of teff flour and lower for oat flour, but oat protein is superior to wheat protein due to its higher lysine content.

Pseudo-cereals are a good alternative to wheat flour too because they are an important source of minerals (calcium, iron, and zinc), vitamins, and phytochemicals (saponins, polyphenols, phytosterols, phytosteroids and betalains), which present a real potential health benefit [12].

Legumes flours are usually used in gluten-free products due to their nutritionally properties. All of them are an important source of nutrients such as proteins, complex carbohydrates, fibers, micronutrients, and antioxidant compounds [13]. Chickpea is a legume rich in protein and has good emulsifying properties bringing an improvement in the gluten-free bread volume [14]. Chickpea flour was used in combination with tiger nut flour as an alternative to emulsifier and shortening in gluten-free bread. The effect of reducing or eliminating the shortening or emulsifier was due to the interaction between chickpea protein and tiger nut fat [14]. Tiger nut is a tuber rich in carbohydrates, lipids, and fiber; therefore, tiger nut flour was used in bakery products as well as for gluten-free bread with good baking and nutritional characteristics [15]. Another legume category is represented by Bengal gram, which is a pulse crop that contains approximately 17–22% protein, 6.48% fat, 3.82% crude fiber, and 50% carbohydrates [16].

Over the last few years, the use of plantain and other varieties of banana has increased around the world. Plantain flour contains large amounts of starch, cellulose, hemicellulose, and lignin in the pulp and shows high resistance to hydrolysis by digestive enzymes [17]. Some studies have used plantain flour in gluten-free products [17–20].

The aim of this study was to have an overall view of different sources of gluten-free flours. Within this aim, 13 gluten-free flours (rice, brown rice, maize, oat, millet, teff, amaranth, buckwheat, quinoa, chickpea, gram, tiger nut, and plantain) were compared in terms of their nutritional and functional properties in order to identify the most suitable ones for gluten-free products' development.

2. Materials and Methods

2.1. Gluten-Free Flour Samples

Thirteen different commercially available gluten-free (GF) flours were evaluated. Maize, oat, buckwheat, chickpea, tiger nut, and plantain flours were from Biorganik (Budapest, Hungary), while rice, millet, and amaranth flours were provided by Biosviat (Sofia, Bulgaria). Brown rice, wholemeal teff, quinoa, and gram flours were acquired from different suppliers, namely: Biopont Ltd. (Ercsi, Hungary), 3Pauly (Detmold, Germany), Infinity Foods Co-operative Limited (Brighton, UK), and Doves Farm (Berkshire, UK), respectively. The flours were stored in airtight brown glass jars at room temperature.

2.2. Proximate Composition

Moisture, protein, fat, and ash content of GF flours were determined according to the AOAC methods 925.10, 920.152, 922.06, and 923.03. Briefly, moisture content was analyzed by the drying method, protein content through Kjeldahl method using a nitrogen-to-protein conversion factor of 6.25, fat content by Soxhlet extraction with petroleum ether, and ash content through the gravimetric method by sample burning at 550 °C in a furnace. Dietary fiber, starch, and amylose levels were determined using enzymatic kits (K-TDFR, K-TSTA, and K-AMYL) and following the procedures recommended by the supplier (Megazyme International Ltd., Bray, Ireland). Ridascreen® Gliadin Kit (R-Biopharm AG, Darmstad, Germany) was used for analyzing gluten content.

2.3. Minerals Analysis

One gram of flour was microwave digested (EthosEasy Advanced microwave system, Milestone, Italy) at 550 °C with 5 mL HNO_3 and 1 mL H_2O_2. After digestion, the sample were transferred to volumetric flasks (100 mL), filled with deionized water, and analyzed by inductively coupled plasma mass spectrometry (NexION 300Q ICP-MS, Perkin Elmer, Waltham, MA, USA). An external standard (multi-element solution for ICP, SPEX CertiPrep, Metuchen, NJ, USA) was used for calibration.

2.4. Determination of Fatty Acid Profile

Fatty acid profile was determined according to ISO 12966-2 [21]. A volume of 1 μL of the resulting fatty acid methyl esters samples was injected into a Perkin Elmer-Clarus 500 gas chromatograph with flame ionization detector (Perkin Elmer, Waltham, MA, USA) in a split mode (1:100). Operating parameters were as follows: detector temperature: 250 °C; injector temperature: 260 °C; oven temperature: 180 °C to 220 °C (5 °C/min). A BPX70 column (60 m–0.25 mm–0.25 μm; SGE Analytical Science, Victoria, Australia) was used and carrier gas was hydrogen 1 mL/min. Individual fatty acid methyl esters were identified by comparison to the standard mixture of Supelco 37 Component FAME Mix (Sigma-Aldrich, Bellefonte, PA, USA) and expressed as g of fatty acids/100 g fat.

2.5. Amino Acid Analysis

Amino acid composition was determined using Agilent 1260 Series HPLC (Agilent Technologies, Waldbronn, Germany) with diode array detector following a previous methodology [22]. The samples were hydrolyzed with 6 M HCl at 110 °C for 24 h and the protein hydrolysates were treated with phenyl isothiocyanate to form phenylthiocarbamyl derivatives of the amino acids.

The amino acid score (AAS) was calculated by dividing the amino acid content of the sample by its reference value established by FAO [23]. The reference value for adults expressed as g/100 g protein were: 1.5 for histidine, 3 for isoleucine, 5.9 for leucine, 4.5 for lysine, 2.2 for the sulphur amino acids (methionine and cysteine), 3.8 for the aromatic amino acids (phenylalanine and tyrosine), 2.3 for threonine, and 3.9 for valine [23]. The protein digestibility corrected amino acid score (PDCAAS) was obtained by multiplying the lowest value of the AAS by the percentage of the protein digestibility of the respective sample with the aim to assess how well dietary protein can match the demand for amino acids, and to allow the prediction of dietary protein utilization [23].

Other protein quality indexes such as Protein Score, Essential Amino Acid Index (EAAI), Biological Value (BV), Protein Efficiency Ratio (PER), and Nutritional Index (NI) were calculated from the content of essential amino acids, as described in a previous paper [24].

2.6. Determination of Protein Digestibility

The method proposed by Hsu et al. [25] was used for the analysis of the in vitro digestion of proteins. Briefly, flour samples were suspended in distilled water to achieve an amount of 6.25 mg protein/mL, followed by pH setting to 8.0. After the suspensions were incubated at 37 °C, trypsin (1.6 mg/mL; type IX-S, Sigma-Aldrich, Saint Louis, USA) was added and the decrease in pH was measured after 10 min. The percentage of the protein digestibility was determined as: % = 210.46 − 18.10 · pH [25].

2.7. Determination of Total Polyphenolic Content and Antioxidant Capacity

For extract preparation, 1 g of flour was extracted with 6 mL of 80% methanol for 3 h on a vortex. The extract obtained was centrifuged at 11,000 rpm for 30 min and the collected supernatant was used for determination of the total polyphenolic content (TPC) and the antioxidant capacity, following the methods proposed by Horszwald and Andlauer [26] with some modifications. For TPC, 500 μL extract was mixed with 5 mL freshly prepared Folin-Ciocalteu reagent (15-fold diluted), incubated for 10 min in dark, and then 500 μL of 20% Na_2CO_3 was added. After 20 min reacting in the dark, the absorbance of the mixture was measured at 755 nm (Specord 200 Spectrophotometer, Analytik Jena AG, Jena, Germany). Gallic acid was used for calibration. Results were expressed as mg of gallic acid equivalents (GAE) per g of gluten-free flour on dry matter basis. For the antioxidant capacity, 400 µL extract was mixed with 6 mL of 0.04 mg/mL DPPH (1,1-diphenyl-2-picrylhydrazyl) solution in methanol. The mixture was kept in the dark for 30 min and the absorbance of the solution was measured at 517 nm. Trolox (6-hydroxy-2,5,7,8-tetramethylchroman-2-carboxylic acid) was used as the standard. Results were expressed as mg of Trolox per g of gluten-free flour on a dry matter basis.

2.8. Thermal Properties

A differential scanning calorimeter (DSC8000, Perkin Elmer, Waltham, MA, USA) was used to evaluate the thermal properties of the GF flours. The instrument was calibrated with indium (Indium Calibration Standard, Perkin Elmer) with melting temperature and enthalpy of 156.6 °C and 28.5 J/g. Distilled water was added into the flours at a ratio of 3:1 directly into 60 µL DSC stainless steel pans that were sealed and equilibrated at room temperature for 24 h before determination. The scanning conditions were: heating temperature from 20 to 120 °C, heating rate of 10 °C/min, nitrogen flow rate of 20 mL/min, and an empty pan as the reference. After the first run (which reflects the gelatinization thermal properties), the pans were stored at 4 °C for 7 days to enable starch retrogradation and rescanned under the same heating conditions. Endothermic transitions consisting of T_p (peak temperature), ΔH_g (gelatinization enthalpy), and ΔH_r (retrogradation enthalpy) were obtained through Pyris Manager software (Perkin Elmer). The degree of retrogradation (DR) was calculated as: DR (%) = $(\Delta H_r / \Delta H_g) \times 100$. All the measurements were performed in triplicate.

2.9. Functional Properties

The functional properties of the GF flours were determined according to the methods described by Klunklin and Savage [27]. Briefly, water and oil absorption were determined using 1 g of flour and 10 mL of distilled water or soybean oil. For the swelling power, 1 g of flour was mixed with 10 mL of distilled water and heated to form a paste. To determine foam capacity, 2 g of flour were added to 50 mL water in a cylinder and homogenized by Ultra-Turrax to allow foam formation. For the bulk density, 50 g flour was weight in a graduated cylinder and its volume was measured.

2.10. Statistical Analysis

The results of the gluten-free flour analysis were presented as the mean values of three replicates together with the standard deviation. Data were analyzed by one-way analysis of variance (ANOVA) supported by Tukey's test ($p < 0.05$). Pearson correlations analysis was done to calculate the correlations among data. Furthermore, the data were subjected to principal component analysis (PCA). The statistical analysis was performed using Minitab®20 Statistical Software (Minitab Ltd., Coventry, UK).

3. Results

3.1. Proximate Composition of GF Flours

The gluten content (expressed in ppm) for the flours under investigation was as follows: 14.15 (rice), 8.63 (brown rice), 7.22 (maize), 6.14 (oat), 8.75 (millet), 8.42 (teff), 5.3 (amaranth), 3.94 (buckwheat), 8.04 (quinoa), 4.66 (chickpea), 2.02 (gram), 4.78 (tiger nut), and 3.68 (plantain). All the flours had a gluten content lower than 20 ppm, which confirms their claim as gluten-free.

The proximate composition of the analyzed gluten-free flour is presented in Table 1. Among the different GF flours, protein content was significantly higher ($p < 0.05$) in both chickpea samples and decreased in the order: gram > chickpea > amaranth > buckwheat > quinoa > teff > oat > millet > brown rice > rice > maize > tiger nut > plantain. Protein content of gram flour was about 1.27 times the chickpea level. Compared to the rice flour—the most used GF flour—the protein content in gram was 8.2-fold higher. The protein content varies between varieties. For example, millet recorded different protein values depending on the type: 11.9% for common millet flour, 8% for pearl millet flour, and 7.3% for finger millet flour [28,29]. In addition, the protein content varies from 5.7–14.2% for buckwheat, 9.1–16.7% for quinoa, and 13.1–21.5% for amaranth [30–32].

Table 1. Chemical composition of the gluten-free flours.

Flours	Moisture, %	Protein, % d.m.	Fat, % d.m.	Ash, % d.m.	Fiber, % d.m.	Starch, % d.m.	Amylose, % Starch d.m.
Rice	12.20 ± 0.05 [d]	8.15 ± 0.02 [j]	0.68 ± 0.02 [h]	0.63 ± 0.01 [i]	0.88 ± 0.03 [j]	82.58 ± 0.29 [a]	25.15 ± 0.92 [c]
Brown rice	14.30 ± 0.10 [b]	8.81 ± 0.04 [i]	2.63 ± 0.04 [e]	1.73 ± 0.03 [f]	4.31 ± 0.21 [i]	76.19 ± 0.14 [c]	15.45 ± 0.55 [e]
Maize	14.50 ± 0.10 [a]	7.25 ± 0.05 [k]	2.16 ± 0.01 [g]	0.65 ± 0.03 [i]	7.30 ± 0.09 [g]	72.52 ± 0.33 [e]	24.02 ± 0.99 [cd]
Oat	9.77 ± 0.02 [i]	10.53 ± 0.08 [g]	5.87 ± 0.05 [c]	1.35 ± 0.01 [h]	6.31 ± 0.11 [h]	72.53 ± 0.22 [e]	22.10 ± 0.94 [d]
Millet	10.80 ± 0.05 [f]	9.98 ± 0.09 [h]	3.95 ± 0.01 [d]	1.45 ± 0.02 [g]	7.79 ± 0.20 [fg]	73.61 ± 0.25 [d]	25.31 ± 0.63 [c]
Teff	11.90 ± 0.04 [e]	11.89 ± 0.05 [f]	2.48 ± 0.02 [f]	2.51 ± 0.03 [c]	8.31 ± 0.20 [ef]	70.06 ± 0.21 [f]	28.71 ± 0.58 [ab]
Amaranth	9.07 ± 0.03 [j]	16.09 ± 0.09 [c]	6.20 ± 0.03 [b]	2.45 ± 0.02 [c]	9.01 ± 0.12 [d]	59.39 ± 0.27 [i]	6.51 ± 0.08 [f]
Buckwheat	12.80 ± 0.06 [c]	14.91 ± 0.07 [d]	2.31 ± 0.03 [g]	2.09 ± 0.01 [e]	10.69 ± 0.18 [c]	68.33 ± 0.32 [g]	28.51 ± 0.35 [ab]
Quinoa	10.20 ± 0.03 [gh]	13.40 ± 0.03 [e]	6.08 ± 0.08 [b]	2.25 ± 0.02 [d]	9.39 ± 0.40 [d]	63.85 ± 0.09 [h]	8.37 ± 0.20 [f]
Chickpea	10.10 ± 0.03 [h]	18.60 ± 0.01 [b]	5.79 ± 0.03 [c]	3.55 ± 0.01 [a]	19.61 ± 0.41 [b]	39.52 ± 0.12 [k]	27.97 ± 0.49 [b]
Gram	10.30 ± 0.06 [g]	23.75 ± 0.01 [a]	6.11 ± 0.02 [b]	2.61 ± 0.02 [b]	8.81 ± 0.20 [de]	45.74 ± 0.22 [j]	23.08 ± 0.56 [d]
Tiger nut	6.91 ± 0.05 [k]	4.65 ± 0.01 [l]	25.15 ± 0.14 [a]	2.28 ± 0.03 [d]	35.42 ± 0.09 [a]	25.41 ± 0.18 [l]	30.13 ± 0.57 [a]
Plantain	10.80 ± 0.10 [f]	2.91 ± 0.03 [m]	0.34 ± 0.02 [i]	2.05 ± 0.02 [e]	4.38 ± 0.10 [i]	78.54 ± 0.39 [b]	22.80 ± 0.94 [d]

Values are expressed as mean ± standard deviation (n = 3). Means that do not share a letter in a column are significantly different ($p < 0.05$). d.m. dry matter basis.

The fat content was relatively high in tiger nut flour (25.1%) when compared to the other flours, which might have a negative impact on the shelf-life and quality characteristics of the foods. Further, it varied from 6.11% for gram flour to 0.34% for plantain flour. The higher fat content of the flours leads to their ability to absorb and retain oil, improves the structure and mouth feel, and helps to enhance the flavor retention, also reducing moisture and fat losses of food products [33].

The ash content varied from about 0.6% (rice and maize flours) to 3.5% (chickpea flour). The high ash content (>2%) for chickpea, teff, amaranth, quinoa, tiger nut, buckwheat, and plantain flours denoted that these types of GF flours are an important source of minerals.

The highest levels of dietary fiber ($p < 0.05$) were found for tiger nut, chickpea, and buckwheat flours, while the rice flour had the lowest content (0.88%). The fiber content of tiger nut flour (35.4%) was much higher compared to the values reported in other studies, which varied between 5.6–22.3% [34–36]. This variance comes from the difference in varieties along with the milling procedure or environmental factors.

Rice, plantain, and brown rice had higher starch content (76.2–82.6%) than those of amaranth, gram, chickpea, and tiger nut flours (<60%) ($p < 0.05$).

Amylose content ranged between 6.5–8.4% and 30.1% starch corresponding to amaranth and quinoa flour for the lowest content and tiger nut for the highest content, respectively. Brown rice had a lower amylose content than rice flour ($p < 0.05$). As shown in Table 1, the percentage of amylose from the total starch was relatively similar with no significant difference ($p > 0.05$) between several groups of GF flours, mainly: (1) teff, buckwheat, and chickpea (28–28.7%); (2) millet, rice, and maize (24–25.3%); and (3) maize, gram, plantain, and oat (22.1–24%). The ratio between amylose and amylopectin has an impact on the starch digestion and, accordingly, to the glycemic index. Di Cairano et al. [37] showed a slight negative correlation (r = −0.368) between amylose content and predicted glycemic index in gluten-free cereals, pseudo-cereal, and legumes flours. Regarding the bread products, a much higher negative correlation was obtained (r = −0.964) [38]. Thus, the lower amylose content in quinoa (5.3%) led to a high glycemic index of quinoa bread, while a lower glycemic index was obtained when flours with higher amylose content (20.5–22.8%), such as oat and teff were used in the bread manufacture.

Even if there was variability on the chemical composition values of the selected GF flours with those obtained by other researchers [11,39,40] mainly because of the different varieties or technology for flour processing, the GF flours order for a specific parameter was similar.

3.2. Mineral Composition of GF Flours

The mineral composition of the flours is presented in Table 2. In general, calcium, magnesium, and iron are scarce in the gluten-free diet [40]. The authors underlined the importance of calcium intake and bone metabolism in people with coeliac disease. From the studied flours, calcium levels were high in amaranth (189.7 mg/100 g), teff (166.7 mg/100 g), and gram (150.1 mg/100 g), compared with the other flours with much lower values ranging from 2.2 to 64.3 mg/100 g. The lowest content was for maize and rice flours. Moreover, calcium content in brown rice was six-fold higher than in rice because dehulling process decreases the calcium levels as stated previously [41]. Gram contained 2.3 times more calcium than chickpea.

Table 2. Mineral composition of gluten-free flours (mg/100 g d.m.).

Flours	Calcium	Magnesium	Iron	Zinc	Potassium	Sodium	Phosphorous
Rice	2.25 ± 0.12 [k]	47.66 ± 0.34 [j]	0.80 ± 0.11 [j]	2.30 ± 0.07 [f]	113.50 ± 1.21 [l]	1.31 ± 0.03 [f]	112.02 ± 0.40 [k]
Brown rice	13.42 ± 0.23 [h]	152.72 ± 0.93 [e]	0.86 ± 0.07 [j]	2.74 ± 0.20 [d]	272.35 ± 1.08 [i]	3.81 ± 0.24 [c]	342.20 ± 2.30 [f]
Maize	4.28 ± 0.06 [j]	47.33 ± 0.41 [j]	1.19 ± 0.09 [i]	1.01 ± 0.07 [h]	174.76 ± 0.96 [k]	0.52 ± 0.05 [h]	95.83 ± 0.19 [l]
Oat	35.50 ± 0.13 [e]	87.63 ± 0.17 [i]	2.77 ± 0.05 [g]	2.72 ± 0.06 [de]	333.22 ± 4.20 [h]	3.77 ± 0.11 [c]	290.89 ± 1.40 [i]
Millet	9.23 ± 0.38 [i]	142.88 ± 0.62 [f]	4.18 ± 0.07 [e]	2.99 ± 0.06 [c]	248.44 ± 0.64 [j]	1.35 ± 0.11 [f]	309.87 ± 1.45 [h]
Teff	166.70 ± 0.30 [b]	226.47 ± 0.68 [b]	8.67 ± 0.17 [a]	4.00 ± 0.03 [a]	407.22 ± 1.45 [g]	16.05 ± 0.17 [a]	402.30 ± 0.84 [e]
Amaranth	189.74 ± 1.70 [a]	270.79 ± 1.38 [a]	7.43 ± 0.06 [b]	3.02 ± 0.05 [c]	513.69 ± 1.05 [f]	2.09 ± 0.06 [e]	597.93 ± 1.33 [b]
Buckwheat	16.19 ± 0.09 [g]	221.79 ± 1.75 [c]	3.39 ± 0.06 [f]	2.50 ± 0.10 [ef]	512.26 ± 1.26 [f]	0.85 ± 0.07 [g]	328.44 ± 1.13 [g]
Quinoa	31.80 ± 0.23 [f]	229.48 ± 2.65 [b]	5.62 ± 0.07 [d]	3.39 ± 0.08 [b]	627.37 ± 1.57 [e]	1.99 ± 0.03 [e]	450.20 ± 2.34 [c]
Chickpea	64.32 ± 0.29 [d]	134.49 ± 0.95 [g]	5.98 ± 0.03 [c]	3.16 ± 0.03 [bc]	1127.67 ± 0.45 [c]	9.67 ± 0.04 [b]	420.94 ± 1.20 [d]
Gram	150.09 ± 1.33 [c]	181.12 ± 1.73 [d]	5.64 ± 0.06 [d]	3.94 ± 0.03 [a]	1144.11 ± 0.72 [b]	3.57 ± 0.11 [c]	643.52 ± 2.78 [a]
Tiger nut	16.79 ± 0.59 [g]	89.20 ± 0.35 [i]	0.74 ± 0.01 [j]	1.45 ± 0.05 [g]	717.59 ± 2.15 [d]	2.97 ± 0.06 [d]	204.78 ± 0.59 [j]
Plantain	8.11 ± 0.16 [i]	101.17 ± 0.55 [h]	2.19 ± 0.10 [h]	0.38 ± 0.01 [i]	1175.39 ± 1.71 [a]	1.27 ± 0.03 [f]	91.71 ± 0.30 [l]

Values are expressed as mean ± standard deviation (n = 3). Means that do not share a letter in a column are significantly different ($p < 0.05$). d.m. dry matter basis.

Similar to the calcium content, magnesium levels were high in amaranth m and teff (226.5 mg/100 g), as well as quinoa flours (229.5 mg/100 g), followed by buckwheat (221.8 mg/100 g) and gram (181.1 mg/100 g).

High content of iron was found in teff (8.7 mg/100 g) and amaranth (7.4 mg/100 g) followed by chickpea, gram, and quinoa (5.6–6 mg/100 g). On the other side, tiger nut, rice, and brown rice flours were deficient in iron (<1 mg/100 g). The teff and gram flours were a good source of zinc (around 4 mg/100 g). From a nutritional point of view, zinc and iron are essential elements in human nutrition, especially for diabetic patients, enhancing insulin production [42,43].

Potassium content ranged from 113.5 mg/100 g (rice flour) to 1175.4 mg/100 g (plantain flour), being the highest element. Other flours with high potassium content were the chickpea flours between 1127.7–1144.1 mg/100 g (with higher value for gram than chickpea) and tiger nut (717.6 mg/100 g). Sodium content was between 0.52 mg/100 g (maize flour) and 16.1 mg/100 g (teff flour). Gram had a lower sodium content than chickpea. It is widely known that a diet low in sodium and high in potassium helps in lowering the blood pressure and reducing the risk of cardiovascular disease [44].

The phosphorus content ($p < 0.05$) decreased in the following order: gram > amaranth > quinoa > chickpea > teff > brown rice > buckwheat > millet > oat > tiger nut > rice > maize > plantain.

According to Hager et al. [11], flours with high mineral composition (such as teff and quinoa) had also higher content of phytic acid that has the ability to bind minerals and hinder their absorption. To overcome this drawback, technologies like the sourdough fermentation can be applied to reduce the phytate content [45].

Based on the dietary reference intake established by the USDA [46], 100 g gram flour provides 92%, 43–57%, 36–49%, and 71–31% of the daily required phosphorous (male/female), magnesium (male–female), zinc (male–female), and iron (male–female), respectively. In addition, 100 g brown rice, millet, teff, amaranth, buckwheat quinoa, or chickpea flours provided more than 40% of the daily-recommended allowance of magnesium for female population. Hager et al. [11] noted the same observation for quinoa, teff and buckwheat flour. A good contribution to the daily-required amount of iron can be provided by 100 g of teff, amaranth, millet, quinoa, or chickpea flour. The calculation for the dietary intakes for minerals for each type of GF flour is detailed in Supplementary Materials Table S1.

Several studies have analyzed the mineral content of GF flour. For example, Rybicka et al. [47] considered buckwheat, corn, oat, rice, amaranth, chickpea, chestnut, millet, teff, and acorn flours, while Hager et al. [11] focused on rice, oat, quinoa, buckwheat, sorghum, maize, and teff. Their results were in line with the present study with some differences coming from the original raw material and its processing. The mineral content of tiger nut and plantain flours were in contrast with those presented by Adegunwa et al. [35]. The type of cultivar and soil characteristics (type, mineral content) contributed to this difference [48].

3.3. Fatty Acid Profile

The most abundant fatty acids were oleic, linoleic, and palmitic acid (Figure 1a), while stearic and α-linolenic acid were in lower amounts (<8% of the total fat) (Figure 1b). The content of palmitic acid in plantain flour (34.9%) was significantly higher ($p < 0.05$) compared to the other GF sources. In addition, the content of stearic acid was significantly higher ($p < 0.05$) for plantain and tiger nut flours.

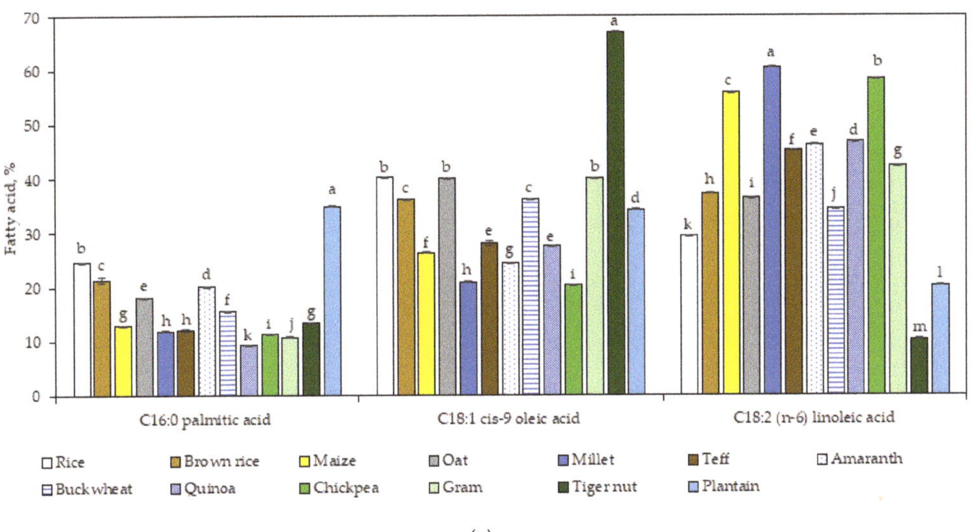

(a)

Figure 1. *Cont.*

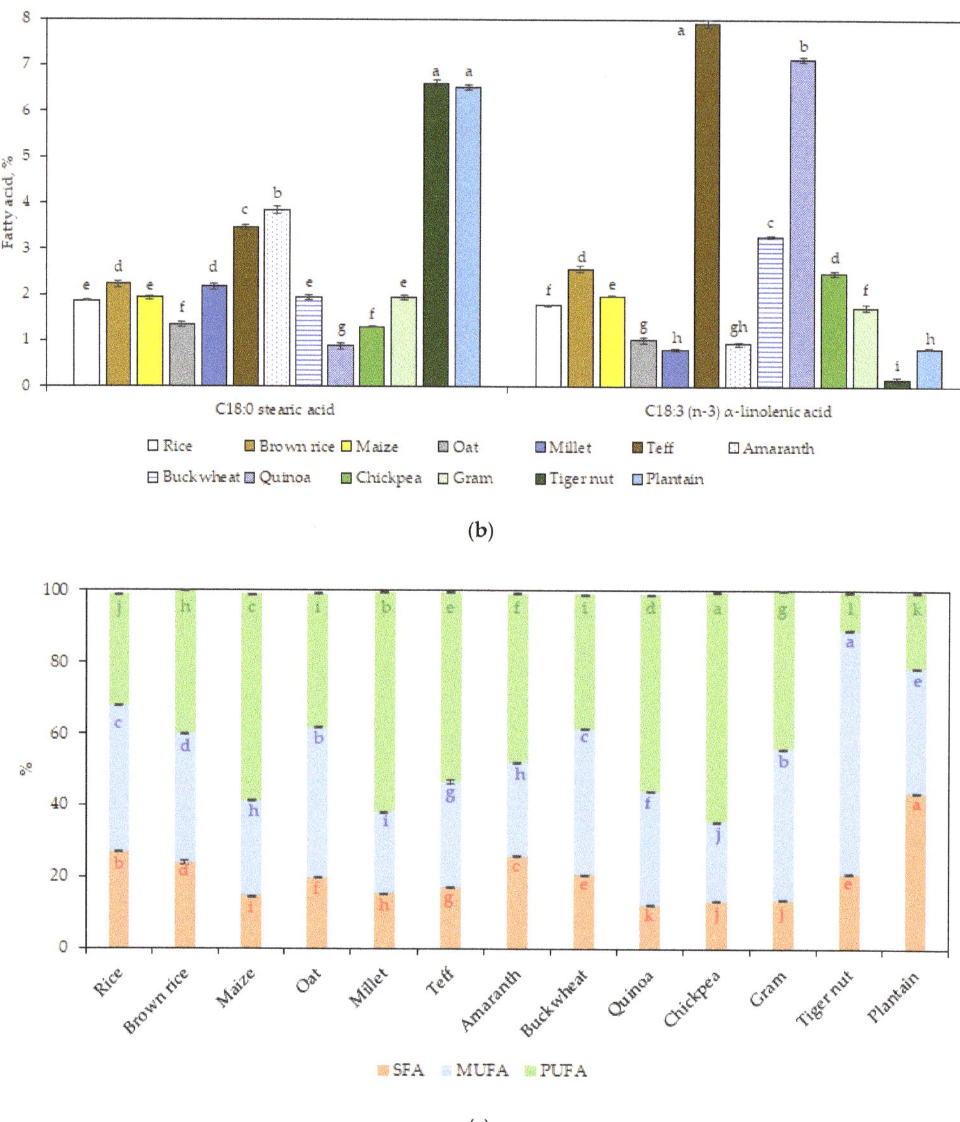

(b)

(c)

Figure 1. Fatty acid composition of investigated gluten-free flours: (**a**) palmitic acid, oleic acid, and linoleic acid; (**b**) stearic acid and α-linolenic acid; (**c**) Proportions of fatty acid groups (% of total fatty acids) of GF flours. The data are presented as mean values ± standard deviation (n = 3). Bars with different superscripted letters indicate significant differences ($p < 0.05$). SFA: saturated fatty acid; MUFA: monounsaturated fatty acid; PUFA: polyunsaturated fatty acid.

Tiger nut flour had the highest oleic acid content (67%), whereas the lowest was in the chickpea sample (20.4%) ($p < 0.05$). Gram, oat, and rice flours did not reveal significant differences in oleic acid content ($p > 0.05$); the values were around 40% (Figure 1a).

In contrast with the highest content in the oleic acid, tiger nut flour showed the lowest level in linoleic acid (10.2%) and α-linolenic acid (0.2%). Thus, tiger nut flour was the highest in MUFA and lowest in PUFA ($p < 0.05$) (Figure 1c). It was stated that tiger nut oil has a MUFA profile similar to olive oil [49]. Moreover, it was reported that tiger nut oil reduced low-density lipoprotein cholesterol and increased high-density lipoprotein cholesterol as well as stimulated the absorption of calcium in bones mainly attributed to the short and medium chain fatty acids, oleic acid, and essential fatty acids [50].

Teff flour showed a high amount of α-linolenic acid (7.9%; $p < 0.05$), followed by quinoa flour (7.1%). The high value found in teff flour is in line with a recent study [51], which stated also that α-linolenic acid level in teff was higher than other cereals: sorghum, millet, wheat rice, maize, or oat.

Summing up, the range for the two essential fatty acids varied as following: millet > chickpea > maize > quinoa > amaranth > teff > gram > brown rice > oat > buckwheat > rice > plantain > tiger nut (for linoleic acid) and teff > quinoa > buckwheat > brown rice $\approx$ chickpea > maize > gram $\approx$ rice > oat > amaranth $\approx$ plantain > tiger nut (for α-linolenic acid).

Comparing PUFA/SFA, the most favorable ratio was for chickpea (4.7) and quinoa (4.5), followed by millet and maize (3.9), gram (3.2), and teff (3.1) flours. The others flours had lower values (<2). The higher PUFA/SFA ratio, the more positive effect on cardiovascular health [52]. Regarding the ratio between omega 6 to omega 3, the values for millet (74.3), tiger nut (61.4), and amaranth (49) flours were higher than for the other GF sources analyzed i.e., chickpea, plantain, gram and maize flours with ratio in the range of 23.6 to 28.3 and rice, quinoa, buckwheat, and brown rice flours with the lowest ratios (6.1–14.5).

A recent study investigating the fatty acid profile of gluten-free bakery products showed that MUFA represent the majority fatty acid group, followed by SFA (30%) and PUFA (13%) [53]. Another study involving celiac children from Sweden remarked a high intake of SFA and low intake of PUFA in their diet [54].

The values of fatty acid composition were calculated also to g/100 g d.m. flour and are presented in Table S2.

3.4. Amino Acid Compositions

The amino acid composition of the GF flours studied is presented in Table 3 (expressed as g/100 g protein) and Table S3 (as g/100 g d.m. flour). There was a great variability in the amino acid content between the flours analyzed. Glutamic acid, aspartic acid, and arginine had the highest non-essential amino acids in GF flours, while leucine was the highest essential amino acid. Overall, tiger nut and plantain flours had the lower content in amino acid.

It is well known that lysine is a nutritionally limiting amino acid in cereals. Regarding its content, chickpea (8.5%) and gram (6.6%) flours, followed by the pseudo-cereals (quinoa, buckwheat, and amaranth, between 5.2–5.9%) and oat (5.2%) flours were found to be high in lysine. Srichuwong et al. [55] also confirmed that the pseudo-cereals amaranth and quinoa contained higher amount of lysine than cereals. It was showed that high levels of globulins and albumins was responsible for high lysine content [56]. Maize and millet flours showed a high content of leucine (12.3% and 11.8%, respectively), while the lowest content was for plantain and amaranth flours (5.9% and 5.6%, respectively). The percentage of methionine in the GF flours was significantly much higher in millet and teff flours (3.4–3.7%) than the others flours.

Glutamic acid was the most abundant amino acid, ranging from 12.9 to 23.3 g/100 g protein, except for plantain flour with significantly lower value (9.8%). In general, plantain flour had lower level of amino acids, except for threonine and histidine.

The higher arginine content (12%; $p < 0.05$) in chickpea and tiger nut will lead to a higher contribution of these flours to the Maillard reaction [57].

According to Table 4, gram, chickpea, quinoa, buckwheat, and oat flours did not contain any limiting amino acids as AAS value were higher than 1. Millet flour showed a lack of lysine. In addition, maize, teff, rice, brown rice, and plantain flours were lacking in lysine. Conversely, chickpea and gram flours recorded the highest scores for lysine. PDCAAS ranged between 0.27 and 0.97 (Table 4). A value of 1 is considered an optimum value. Accordingly, chickpea followed by oat, quinoa, gram, and buckwheat flours can be considered as GF sources with a good protein quality compared to the other flours investigated.

Table 3. Amino-acid compositions of the gluten-free flours (g/100 g protein).

Flours	Leu	Lys	Phe	Val	Ile	Thr	Met	His
Rice	8.10 ± 0.14 [cd]	3.31 ± 0.08 [fg]	4.61 ± 0.28 [de]	5.50 ± 0.21 [ab]	3.12 ± 0.21 [ef]	3.31 ± 0.21 [bc]	1.96 ± 0.14 [bcd]	2.28 ± 0.21 [def]
Brown rice	8.43 ± 0.20 [cd]	3.80 ± 0.20 [f]	5.12 ± 0.20 [bcd]	6.18 ± 0.20 [a]	4.46 ± 0.20 [b]	3.40 ± 0.20 [bc]	2.12 ± 0.13 [bc]	3.13 ± 0.20 [ab]
Maize	12.32 ± 0.41 [a]	2.90 ± 0.16 [g]	4.57 ± 0.25 [de]	4.95 ± 0.41 [bcd]	3.71 ± 0.16 [cd]	3.77 ± 0.25 [b]	2.26 ± 0.32 [bc]	3.12 ± 0.09 [ab]
Oat	8.95 ± 0.21 [bc]	5.26 ± 0.21 [d]	5.86 ± 0.16 [b]	6.25 ± 0.53 [a]	4.77 ± 0.34 [ab]	3.58 ± 0.21 [bc]	1.89 ± 0.11 [bcde]	2.11 ± 0.21 [ef]
Millet	11.79 ± 0.70 [a]	1.72 ± 0.06 [h]	5.46 ± 0.34 [bc]	4.98 ± 0.45 [bcd]	3.67 ± 0.19 [cde]	3.11 ± 0.28 [c]	3.67 ± 0.28 [a]	2.02 ± 0.22 [f]
Teff	7.83 ± 0.19 [d]	2.96 ± 0.10 [g]	4.93 ± 0.34 [cd]	4.65 ± 0.29 [cde]	3.06 ± 0.24 [f]	3.63 ± 0.19 [bc]	3.41 ± 0.11 [a]	2.23 ± 0.22 [def]
Amaranth	5.65 ± 0.21 [gh]	5.26 ± 0.14 [d]	3.90 ± 0.14 [e]	3.49 ± 0.21 [f]	2.83 ± 0.15 [fg]	3.08 ± 0.21 [c]	2.32 ± 0.14 [b]	2.28 ± 0.21 [def]
Buckwheat	6.28 ± 0.35 [fg]	5.95 ± 0.35 [c]	4.59 ± 0.39 [de]	4.74 ± 0.16 [bcde]	3.31 ± 0.21 [cdef]	3.33 ± 0.27 [bc]	1.80 ± 0.04 [cde]	2.18 ± 0.04 [ef]
Quinoa	6.76 ± 0.35 [ef]	5.34 ± 0.17 [cd]	3.99 ± 0.17 [e]	4.60 ± 0.13 [cde]	3.38 ± 0.30 [cdef]	3.57 ± 0.22 [bc]	2.24 ± 0.17 [bc]	2.69 ± 0.05 [bcd]
Chickpea	9.67 ± 0.12 [b]	8.47 ± 0.25 [a]	7.18 ± 0.36 [a]	5.32 ± 0.24 [bc]	5.08 ± 0.05 [a]	4.41 ± 0.27 [a]	1.93 ± 0.24 [bcd]	3.23 ± 0.06 [a]
Gram	7.54 ± 0.12 [de]	6.65 ± 0.31 [b]	5.60 ± 0.31 [bc]	4.23 ± 0.09 [def]	3.85 ± 0 [c]	3.38 ± 0.09 [bc]	1.60 ± 0.05 [de]	2.55 ± 0.07 [cde]
Tiger nut	5.24 ± 0.58 [h]	4.55 ± 0.35 [e]	2.77 ± 0.23 [f]	4.08 ± 0.27 [ef]	2.31 ± 0.23 [g]	3.47 ± 0.23 bc	1.90 ± 0 [bcde]	2.10 ± 0.16 [ef]
Plantain	5.91 ± 0.22 [fgh]	3.85 ± 0.01 [f]	3.85 ± 0.03 [e]	4.24 ± 0.05 [def]	3.23 ± 0.09 [def]	3.85 ± 0.07 [ab]	1.42 ± 0.04 [e]	3.02 ± 0.06 [abc]

Flours	Glu	Asp	Arg	Ala	Gly	Pro	Ser	Tyr	Cys
Rice	18.16 ± 1.40 [cd]	7.69 ± 0.42 [de]	8.10 ± 0.14 [d]	4.94 ± 0.21 [cd]	4.70 ± 0.21 [cde]	4.24 ± 0.21 [ef]	5.17 ± 0.28 [bc]	5.03 ± 0.28 [a]	1.54 ± 0.14 [def]
Brown rice	19.87 ± 0.66 [bc]	8.96 ± 0.33 [c]	8.43 ± 0.33 [cd]	5.56 ± 0.26 [c]	5.30 ± 0.26 [bc]	5.03 ± 0.13 [cd]	4.90 ± 0.26 [bc]	3.66 ± 0.08 [bc]	1.15 ± 0.08 [ghi]
Maize	17.81 ± 0.89 [cd]	6.19 ± 0.57 [g]	5.00 ± 0.65 [f]	7.42 ± 0.65 [b]	4.03 ± 0.32 [ef]	7.69 ± 0.41 [a]	4.36 ± 0.43 [cd]	4.03 ± 0.32 [b]	1.83 ± 0.09 [cd]
Oat	21.75 ± 1.61 [ab]	11.40 ± 0.80 [b]	6.60 ± 0.26 [e]	4.95 ± 0.21 [cd]	5.79 ± 0.11 [b]	3.82 ± 0.06 [fg]	5.68 ± 0.21 [ab]	4.00 ± 0.21 [b]	3.58 ± 0.11 [a]
Millet	21.11 ± 0.45 [ab]	4.94 ± 0.22 [h]	3.82 ± 0.22 [f]	10.33 ± 0.22 [a]	2.66 ± 0.17 [g]	7.45 ± 0.17 [a]	5.91 ± 0.28 [ab]	3.63 ± 0.17 [bc]	1.46 ± 0.22 [defg]
Teff	23.30 ± 0.38 [a]	4.14 ± 0.24 [h]	4.43 ± 0.06 [f]	5.67 ± 0.34 [c]	4.39 ± 0.38 [def]	6.11 ± 0.19 [b]	4.49 ± 0.19 [c]	3.44 ± 0.19 [c]	1.72 ± 0.17 [cd]
Amaranth	15.13 ± 0.31 [ef]	6.29 ± 0.27 [fg]	8.18 ± 0.31 [d]	3.53 ± 0.14 [f]	8.57 ± 0.31 [a]	4.49 ± 0.10 [de]	6.49 ± 0.34 [a]	2.78 ± 0.21 [de]	2.26 ± 0.21 [b]
Buckwheat	16.16 ± 0.77 [de]	7.57 ± 0.50 [de]	9.75 ± 0.59 [b]	3.80 ± 0.16 [ef]	5.36 ± 0.19 [bc]	3.46 ± 0.15 [g]	4.49 ± 0.12 [c]	2.49 ± 0.12 [e]	1.97 ± 0.09 [bc]
Quinoa	14.37 ± 0.33 [ef]	7.42 ± 0.21 [def]	9.19 ± 0.75 [bcd]	4.40 ± 0.17 [de]	5.40 ± 0.33 [bc]	3.99 ± 0.17 [efg]	4.38 ± 0.21 [c]	2.74 ± 0.08 [de]	1.22 ± 0.10 [fghi]
Chickpea	20.70 ± 0.48 [b]	13.00 ± 0.38 [a]	12.10 ± 0.45 [a]	5.04 ± 0.03 [cd]	4.93 ± 0.15 [cd]	5.54 ± 0.30 [bc]	6.38 ± 0.51 [a]	3.23 ± 0.12 [cd]	1.64 ± 0.15 [cde]
Gram	16.20 ± 0.23 [de]	10.39 ± 0.20 [b]	9.51 ± 0.24 [bc]	3.94 ± 0.05 [ef]	3.90 ± 0.09 [f]	4.41 ± 0.05 [def]	4.96 ± 0.24 [bc]	2.50 ± 0.03 [e]	1.31 ± 0.09 [efgh]
Tiger nut	12.94 ± 0.46 [f]	6.47 ± 0.46 [efg]	12.02 ± 0.46 [a]	5.24 ± 0.13 [c]	5.24 ± 0.35 [bc]	3.78 ± 0.35 [fg]	3.31 ± 0.13 [d]	1.26 ± 0.08 [f]	0.88 ± 0.01 [i]
Plantain	9.76 ± 0.22 [g]	8.22 ± 0.22 [cd]	4.36 ± 0.22 [f]	5.13 ± 0.22 [cd]	5.13 ± 0.22 [bcd]	4.36 ± 0.22 [ef]	4.49 ± 0.80 [c]	2.31 ± 0.01 [e]	1.05 ± 0.04 [hi]

Values are expressed as mean ± standard deviation (n = 3). Means that do not share a letter in a column are significantly different ($p < 0.05$).

Table 4. Amino acid scores (AAS) and protein digestibility corrected amino acid score (PDCAAS) of GF flours.

Flours	AAS							PDCAAS
	His	Ileeucine	Leu	Lys	Cys + Met	Phe + Tyr	Val	
Rice	1.52	1.04	1.37	0.73	1.59	2.54	1.41	0.56
Brown rice	2.09	1.49	1.43	0.84	1.49	2.31	1.58	0.61
Maize	2.08	1.24	2.09	0.65	1.86	2.26	1.27	0.47
Oat	1.40	1.59	1.52	1.17	2.49	2.59	1.60	0.89
Millet	1.35	1.22	2.00	0.38	2.33	2.39	1.28	0.27
Teff	1.49	1.02	1.33	0.66	2.33	2.20	1.19	0.49
Amaranth	1.52	0.94	0.96	1.17	2.08	1.76	0.89	0.70
Buckwheat	1.45	1.10	1.07	1.32	1.71	1.86	1.22	0.78
Quinoa	1.79	1.13	1.15	1.19	1.57	1.77	1.18	0.85
Chickpea	2.15	1.69	1.64	1.88	1.62	2.74	1.37	0.97
Gram	1.70	1.28	1.28	1.48	1.32	2.13	1.08	0.78
Tiger nut	1.40	0.77	0.89	1.01	1.26	1.06	1.05	0.54
Plantain	2.01	1.08	1.00	0.86	1.12	1.62	1.09	0.62

Values AAS were calculated based on the recommendation of amino acid requirements for daily adult intake [23]. Highlighted values represent the limiting value of AAS. When AAS < 1, the corresponding amino acid is considered a limiting one.

Other indexes which are used to estimate the protein quality are presented in Table 5. The protein score indicates the chemical score of the most limiting EAA that is present in the sample. The greater the EAAI, the more balanced amino acid composition and the higher quality and efficiency of the protein. EAAI, an indicator of the ratio of essential amino acids of the sample compared to the reference (hen's egg) and BV (which estimates the nitrogen potentially retained by the human body after consumption) were the highest for chickpea flour, followed by oat, brown rice, maize, and gram flours. The PER index, which describes the ability of a protein to support the body weight increase, was higher for maize and millet flours. Among all the indexes calculated, NI is the only one which considers both qualitative and quantitative factors and it is a global predictor of the quality of a protein source. Accordingly, gram and chickpea flours showed the higher nutritional index, followed by the pseudo-cereals flours, while plantain flour had the lowest NI.

Table 5. Other nutritional indexes of GF flours.

Flours	Protein Score, %	EAAI	BV	PER	NI
Rice	47.24	74.93	69.97	35.88	6.11
Brown rice	54.24	83.20	78.98	38.03	7.33
Maize	41.50	83.26	79.06	57.03	6.04
Oat	75.19	91.36	87.88	36.68	9.62
Millet	24.59	74.23	69.22	52.32	7.41
Teff	42.30	74.73	69.75	31.43	8.89
Amaranth	52.32	69.33	63.87	20.93	11.15
Buckwheat	61.27	74.69	69.71	25.76	11.14
Quinoa	60.73	75.59	70.70	28.83	10.13
Chickpea	62.62	99.97	97.27	44.01	18.59
Gram	51.07	78.29	73.64	33.99	18.59
Tiger nut	42.80	59.45	53.10	22.08	2.76
Plantain	43.42	68.23	62.67	26.15	1.99

EAAI: Essential Amino Acid Index; BV: Biological Value; PER: Protein Efficiency Ratio; NI: Nutritional Index.

3.5. Protein Digestibility

Figure 2 shows the values of the in vitro protein digestibility of the GF flours. Protein digestibility refers to the ability of the protein to be enzymatically hydrolyzed into amino acids. Amaranth flour had the highest digestibility (78.7%; $p < 0.05$) among all the flours studied. Tiger nut samples with the lowest protein and highest fiber content had the lowest protein digestibility (70.4%). A previous report stated that lower digestibility is correlated with increasing contents of fiber [58]. Plantain, brown rice, maize, and gram are a good source of protein as their protein digestibility (72.3–72.5%) were higher than chickpea, millet, and tiger nut (70.4–71.4%). In general, variations in protein digestibility take place after baking and Abdel-Aal [59] showed that the baking process resulted in significant improvement in protein digestion of spelt bread.

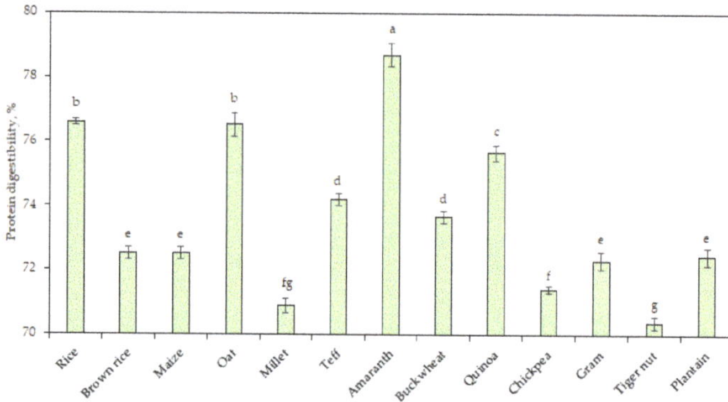

Figure 2. In vitro digestibility values. The data are presented as mean values ± standard deviation (n = 3). Bars with different superscripted letters indicate significant differences between corresponding digestibility values ($p < 0.05$).

3.6. Total Polyphenolic Content and Antioxidant Capacity

TPC and antioxidant capacity were highest in buckwheat (3.75 mg GAE/g d.m.) and the lowest in rice (0.20 mg GAE/g d.m.) flours, as shown in Figure 3. It was reported previously that buckwheat presents stronger antioxidative potential that other cereals due to the presence of flavonoids like as rutin, quercetin, epicatechin, and catechin [60].

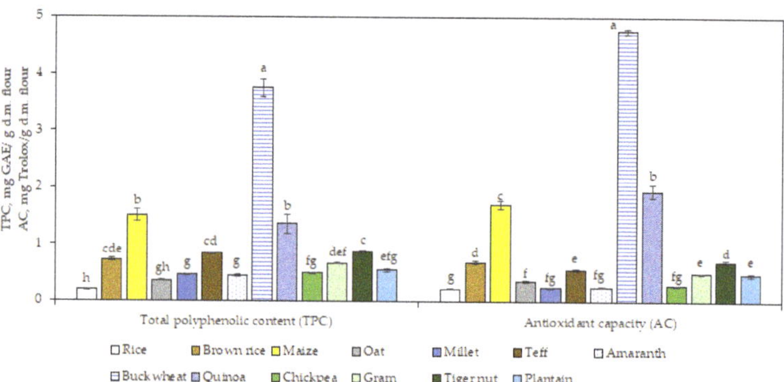

Figure 3. Total polyphenolic content and antioxidant capacity of different gluten-free flours. The data are presented as mean values ± standard deviation (n = 3). Bars that do not share a letter are significantly different ($p < 0.05$).

Comparable polyphenolic contents were obtained for maize and quinoa flours (1.36–1.51 mg GAE/g d.m.; $p > 0.05$), while brown rice, teff, and tiger nut flours showed similar value (0.72–0.88 mg GAE/g d.m.; $p > 0.05$). Lower values were obtained for the rest of the samples. The TPC in brown rice was 3.6-fold higher than white rice. No significant difference was found in TPC between chickpea and gram ($p > 0.05$). Significant differences in TPC ($p < 0.05$) were between the flours from pseudo-cereals, decreasing in the order:

buckwheat > quinoa > amaranth. Di Cairano et al. [37] reported also that buckwheat presented the highest TPC followed by quinoa, millet, and teff. Overall, it should be taken into consideration that TPC results can be overestimated because of the presence of some non-phenolic compounds, which can interfere with the Folin reagent [61].

In this study, a positive correlation was found between TPC and antioxidant capacity DPPH radical scavenging activities (r = 0.9892; $p < 0.05$). This result is in line with Rocchetti et al. [62], who found that the antioxidant capacity is strongly related to the phenolic profile in different gluten-free flours from cereals, pseudo-cereals, and legumes.

3.7. Gelatinization and Retrogradation Properties

Gelatinization temperatures and enthalpies associated with starch gelatinization varied among the GF flours (Table 6).

Table 6. Gelatinization and retrogradation thermal properties of GF flours.

Flours	Gelatinization of Starch		Retrogradation		DR, %
	T_p, °C	ΔH, J/g d.m.	T_p, °C	ΔH, J/g d.m.	
Rice	80.86 ± 0.12 [b]	11.44 ± 0.36 [b]	58.05 ± 0.27 [b]	5.74 ± 0.05 [b]	50.21 ± 1.98 [ab]
Brown rice	75.35 ± 0.09 [e]	11.69 ± 0.57 [b]	49.00 ± 0.25 [d]	0.12 ± 0.03 [hi]	1.05 ± 0.20 [g]
Maize	74.54 ± 0.26 [f]	10.29 ± 0.16 [c]	58.66 ± 0.64 [ab]	3.85 ± 0.05 [d]	37.41 ± 0.22 [d]
Oat	64.75 ± 0.08 [k]	6.80 ± 0.22 [f]	55.55 ± 0.68 [c]	0.54 ± 0.05 [g]	7.89 ± 0.54 [f]
Millet	78.04 ± 0.06 [d]	10.43 ± 0.27 [c]	58.54 ± 0.24 [ab]	4.38 ± 0.05 [c]	42.03 ± 1.40 [c]
Teff	71.62 ± 0.44 [i]	10.11 ± 0.15 [c]	58.44 ± 0.64 [ab]	0.24 ± 0.02 [h]	2.41 ± 0.22 [g]
Amaranth	79.10 ± 0.16 [c]	8.88 ± 0.46 [d]	49.02 ± 0.13 [d]	0.05 ± 0.01 [i]	0.60 ± 0.09 [g]
Buckwheat	72.27 ± 0.22 [h]	10.80 ± 0.36 [bc]	55.76 ± 0.39 [c]	1.48 ± 0.06 [f]	13.67 ± 0.30 [e]
Quinoa	66.83 ± 0.07 [j]	8.68 ± 0.15 [d]	58.66 ± 0.30 [ab]	0.06 ± 0.01 [i]	0.73 ± 0.08 [g]
Chickpea	75.63 ± 0.29 [e]	7.36 ± 0.13 [ef]	58.59 ± 0.17 [ab]	3.16 ± 0.09 [e]	42.88 ± 1.63 [c]
Gram	73.33 ± 0.09 [g]	7.73 ± 0.15 [e]	58.40 ± 0.02 [ab]	3.73 ± 0.06 [d]	48.27 ± 1.62 [b]
Tiger nut	81.87 ± 0.01 [a]	4.85 ± 0.34 [g]	58.44 ± 0.23 [ab]	0.14 ± 0.02 [hi]	2.96 ± 0.50 [g]
Plantain	72.38 ± 0.12 [h]	13.79 ± 0.26 [a]	59.22 ± 0.22 [a]	7.20 ± 0.07 [a]	52.21 ± 1.08 [a]

Values are expressed as mean ± standard deviation (n = 3). Means that do not share a letter in a column are significantly different ($p < 0.05$). T_p: peak temperature; ΔH: enthalpy; DR: degree of retrograded starch.

The variation in the thermal properties among the different flours are influenced by some factors such as: size of the starch granule, molecular structure of the amylopectin (branch, length and weight), starch, protein and dietary fiber content, as well as the presence of other compounds [63]. During starch gelatinization, the helix structure and crystallinity of the starch is lost and the granule is disrupted. Among all the samples under investigation, T_p was highest for the tiger nut flour ($p < 0.05$), which can be explained by its highest fat and fiber content. The swelling of the starch granule is disturbed by the presence of non-starch compounds such as fat, which leads to higher gelatinization temperatures [64]. Demirkesen et al. [36] also found higher gelatinization temperatures in tiger nut flour than in rice flour because of its higher amount of oil, fiber, and sugar. On the other hand, the lowest T_p was for the oat flour ($p < 0.05$). Lower gelatinization temperature is proof of shorter amylopectin chains as lower temperatures are needed for their completely dissociation [65]. The enthalpy of gelatinization reflects the amount of energy needed to break the molecular interactions in the starch during the gelatinization. Gelatinization enthalpy was significantly higher ($p < 0.05$) in plantain and both types of rice flours than in the other samples, because of the high crystallinity of the starch granules in rice and plantain [66,67]. In addition, the higher starch content in rice flours and plantain flour needs more energy to open the double helical structure of starch. A less thermal energy was needed for starch gelatinization in the tiger nut and oat flours, followed by chickpea and gram samples.

Retrogradation temperature and enthalpy were considerably lower than the gelatinization parameters, which denotes weaker starch crystallinity (Table 6). As described by Karim et al. [68], the lower values for the thermal transitions during retrogradation were attributed to improper realignment of amylose and amylopectin molecules that forms a less stable crystalline structure.

The peak corresponding to the retrograded starch presented similar values for most of the samples, except for brown rice, amaranth, oat, and buckwheat flours with significantly lower values ($p < 0.05$). Plantain, rice, millet, maize, gram, chickpea, and buckwheat presented a greater tendency to retrograde (higher value for the retrogradation enthalpy). The degree of retrogradation evaluated as the ratio of retrogradation to gelatinization enthalpy showed significantly lower values (DR < 10%; $p < 0.05$) for amaranth, quinoa, brown rice, teff, tiger nut, and oat compared to the other GF flours; thus, these samples present an advantage on retarding starch retrogradation. Srichuwong et al. [55] stated that the low retrogradation tendency of starch in quinoa and amaranth were the result of the short amylopectin branch chain and high content of the soluble dietary fiber. The use of GF flours with lower retrogradation enthalpy and DR are relevant for increasing the shelf-life of bakery products.

3.8. Functional Properties of GF Flours

Table 7 presents the functional properties of GF flours, which are influenced by the flour constituents and the relationships between them.

Table 7. Functional properties of the gluten-free flours.

Flours	Water Absorption Capacity, g/g	Oil Absorption Capacity, g/g	Swelling Power, g/g	Foaming Capacity, %	Bulk Density, g/mL
Rice	1.25 ± 0.01 [cd]	1.68 ± 0.01 [f]	7.32 ± 0.05 [a]	7.83 ± 0.02 [f]	0.77 ± 0.01 [a]
Brown rice	1.14 ± 0.04 [def]	1.80 ± 0.01 [bc]	6.21 ± 0.07 [b]	63.80 ± 3.08 [b]	0.71 ± 0.01 [b]
Maize	1.40 ± 0.02 [bc]	1.77 ± 0.01 [cd]	5.45 ± 0.02 [c]	57.58 ± 2.15 [c]	0.72 ± 0.03 [b]
Oat	1.26 ± 0.08 [cd]	1.74 ± 0.01 [de]	4.10 ± 0.06 [g]	7.76 ± 0.19 [f]	0.68 ± 0.01 [c]
Millet	1.03 ± 0.05 [ef]	1.66 ± 0.01 [f]	5.54 ± 0.02 [c]	66.02 ± 1.15 [b]	0.67 ± 0.01 [c]
Teff	0.95 ± 0.05 [f]	1.75 ± 0.03 [cde]	4.98 ± 0.04 [ef]	83.49 ± 1.39 [a]	0.77 ± 0.01 [a]
Amaranth	0.96 ± 0.01 [f]	1.88 ± 0.01 [a]	7.21 ± 0.13 [a]	34.05 ± 0.65 [d]	0.63 ± 0.01 [d]
Buckwheat	1.46 ± 0.15 [b]	1.66 ± 0.02 [f]	5.06 ± 0.03 [de]	58.07 ± 2.03 [c]	0.79 ± 0.01 [a]
Quinoa	1.15 ± 0.04 [de]	1.79 ± 0.04 [bcd]	5.41 ± 0.04 [c]	55.17 ± 1.54 [c]	0.71 ± 0.01 [b]
Chickpea	1.92 ± 0.09 [a]	1.84 ± 0.01 [ab]	4.88 ± 0.01 [f]	58.30 ± 1.04 [c]	0.67 ± 0 [c]
Gram	1.40 ± 0.01 [bc]	1.72 ± 0.01 [ef]	5.21 ± 0.02 [d]	65.20 ± 1.91 [b]	0.65 ± 0.01 [cd]
Tiger nut	1.36 ± 0.01 [bc]	1.88 ± 0.02 [a]	3.14 ± 0.02 [h]	26.92 ± 1.68 [e]	0.66 ± 0 [c]
Plantain	1.38 ± 0.05 [bc]	1.70 ± 0.02 [ef]	7.25 ± 0.08 [a]	9.84 ± 0.06 [f]	0.71 ± 0.01 [b]

Values are expressed as mean ± standard deviation (n = 3). Means that do not share a letter in a column are significantly different ($p < 0.05$).

Water absorption capacity reflects the amount of water that the flour can absorb and retain. The water absorption capacity was highest ($p < 0.05$) for chickpea flour (1.92 g/g) and lowest for teff and amaranth flours with values around 0.95 g/g. For an improved food texture of bread products, higher values for water absorption are desired. Higher water absorption values were attributed to the higher content of starch and fiber [27,69]. Patil and Arya [69] also stated that higher protein content tends to increased water absorption. However, in the present study, no good correlation was found between water absorption and protein content (r = 0.2099) or starch content (r = −0.4392).

Of all the GF flours studied, amaranth, tiger nut, and chickpea flours exhibited the highest oil absorption capacity (1.8–1.9 g/g), being suitable for retaining the flavor and enhance the mouthfeel when used in foods. For the other Gf flour, oil absorption capacity ranged between 1.66 and 1.80 g/g, with higher value for brown rice flour than rice. However, Di Cairano et al. [37] did not find any significant difference in the different GF studied regarding the oil absorption capacity. The water and oil absorption capacity depends on the type of protein, amino acid composition and protein polarity and hydrophobicity [70]. Moreover, variation in the amylose/amylopectin ratio contributes to differences in the water as well as oil absorption capacity of flour [71].

The degree of starch from the flour that absorbs water is expressed by the swelling power property. The swelling power of the GF flours varied from 3.14 to 7.32 g/g, where the lowest values were obtained for the tiger nut flour, while the highest were for rice, plantain, and amaranth flours. The high swelling power could be related with the higher content of amylopectin [27]. Tiger nut flour with the lowest amylopectin content (69.87%) showed the lowest swelling power (3.14%). However, only a small correlation was found between amylopectin content and swelling power (0.4694; $p > 0.05$). On the other side, fats inhibit the swelling property. The significantly higher fat content in the tiger nut flour (25.15%) explained the very low swelling capacity of this flour. Accordingly, the fat content was negatively correlated with the swelling power of the GF flours (-0.67211; $p < 0.05$).

The foaming capacity of the GF flours ranged from 7.8% in oat and rice flour to 83.5% in teff flour. A very strong difference was noted between rice and brown rice, the latter being 8.1-fold higher than normal rice. The foam capacity of a flour is dependent on the configuration of protein molecules and carbohydrates present in the flour [71,72]. Flour intended for use in bakery products should present good foam capacity.

Bulk density represents a measure of flour heaviness [34]. The highest bulk density ($p < 0.05$) was in case of buckwheat, teff, and rice flours, whereas amaranth flour had the lowest value ($p < 0.05$) among the flours analyzed. Bulk density and fat content of the flours were slightly negatively correlated (r = -0.4796; $p > 0.05$). This result was in contrast with the observation of Joshi et al. [72], who reported that full fat flours tended to exhibit higher bulk density as lipids might act as adhesives in the aggregation of the flour particles, leading to an increase in the bulk density. Di Cairano et al. [37] stated that the variable results in the bulk density of the different sources of GF analyzed are related to the nonhomogeneous particle size distributions of the flours.

3.9. Principal Component Analysis

Principal component analysis (PCA) was used to show the variation among the gluten-free flours and identify correlations between the parameters analyzed (Figure 4). Earlier studies also used PCA analysis to visualize variation between the characteristics of different flours [37,69].

The aim of PCA analysis is to reduce a big number of variables to a few variables, referred to as principal components (PCs) [73]. The results were projected onto the first principal component (PC1)/second principal component (PC2) plane. The PC1 and PC2 described 29.9% and 20.5% of variance, respectively. The plot shows the similarity between quinoa, buckwheat, and amaranth flours (the so-called pseudo-cereals sources) as these samples are located closely in the upper right part of the PCA plot. Other similarities were observed between several groups: oat and chickpea flours; rice and plantain flours; tiger nut, teff, brown rice, and maize flours as all these groups were clustered together on the plot. On the other hand, gram flour was very distinct compared to the other flours with a high negative score in PC2 (Figure 4a). Investigating different properties of GF flours, Di Cairano et al. [37] also found similarities on one hand between cereal flours (millet, sorghum and teff) and on the other between legumes flours (chickpea, red lentil, lentil, and pea). The same authors stated that buckwheat flour was differentiated by the other pseudo-cereals, amaranth and quinoa, mainly for the phenolic compounds and oil absorption capacity. This observation was opposite to the results from the present study, where the pseudo-cereal flours were clustered together.

Figure 4b shows that the parameters with curves in close proximity are positively correlated, while the curves in opposite senses are negatively correlated. PC1 had positive associations with fiber, fat, ash, and oil absorption capacity, while starch, gelatinization enthalpy, swelling power, amino acid score, retrogradation enthalpy, saturated fatty acid, and bulk density had negative associations with PC1. The second component is well-characterized by polyunsaturated fatty acid, foaming capacity, protein, total polyphenolic content, and antioxidant capacity. PC2 has large negative associations with monounsaturated fatty acid, degree of retrogradation, saturated fatty acid, fat, and fiber.

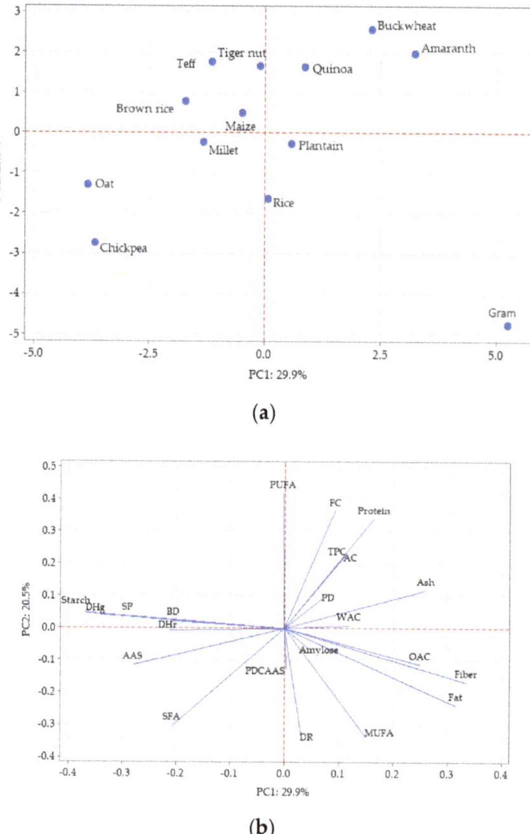

Figure 4. Principal component analysis: (**a**) score plot of first and second principal component (PC1 and PC2) describing the overall variability among the different gluten-free flours; (**b**) loading plot of PC1 and PC2 describing the variation among the parameters of the gluten-free flours. AAS: amino acid score; AC: antioxidant capacity; BD: bulk density; DHg: gelatinization enthalpy; DHr: retrogradation enthalpy; DR: degree of retrogradation; FC: foaming capacity; MUFA: monounsaturated fatty acid; OAC: oil absorption capacity; PD: protein digestibility; PDCAAS: protein digestibility corrected amino acid score; PUFA: polyunsaturated fatty acid; SFA: saturated fatty acid; SP: swelling power; TPC: total polyphenolic content; WAC: water absorption capacity.

4. Conclusions

To improve the quality of GF products, the trend is to use nutrient flours and find solutions for sensory and technology challenges.

The main idea of this study was to have a general overview and comparison of different gluten-free flours and to drag attention to the possibility to use other types of flours, such as gram or plantain flours in GF products. It was rather difficult to compare some of the results from this study with others from the literature because of the difference between the samples related to genotypes, environmental conditions, geographical region, and analysis methods.

Amino acid composition of the GF flours and the calculated value for PDCAAS indicated that chickpea, oat, quinoa, gram, and buckwheat flours were better protein quality sources compared to millet, maize, teff, tiger nut, rice, or plantain. To improve health conditions, the accent is to lower the consumption of SFA, increasing the PUFA and MUFA intake; thus, gram, chickpea, maize, and quinoa flours are good candidates.

The starch from amaranth, quinoa, brown rice, teff, tiger nut, and oat had improved cold-storage stability and from this point of view, are more appropriate for increasing the shelf life of the bakery products.

Amaranth, tiger nut, and chickpea flours having the highest oil absorption capacity could be better than other GF flours as a flavor retainer.

PCA analysis concluded that there were similarities between oat and chickpea flours; rice and plantain flours; tiger nut, teff, brown rice, and maize flours. Thus, combinations between them would be possible. Gram flour was more distinct from all the GF flours analyzed, highlighted by its position in the PCA plot. Gram flour was characterized by higher protein, fat and ash content, higher protein digestibility, lower starch content, lower SFA, and no limiting amino acids.

Information about the complete characterization of GF flours will allow not only to establish possible combinations between them with the aim to enhance the nutritional profile of the bakery products, but also to be used in other types of products, such as beverages, soups, sauces, or food adjuvants.

Supplementary Materials: The following are available online at https://www.mdpi.com/article/10.3390/app11146283/s1, Table S1: The contribution of 100 g flour to dietary reference intakes (RDA—Recommended Dietary Allowance or AI—Adequate Intake) for adult general population (male and female; life stage group 31 through 50 years old), expressed in %. Table S2: Fatty acid composition of investigated gluten-free flours (g/100 g d.m. flour). Table S3: Amino-acid compositions of gluten-free flours (g/100 g d.m. flour).

Author Contributions: Conceptualization, A.C., D.E.D., and N.B.; methodology, A.C. and D.E.D.; software, A.C.; formal analysis, A.C.; investigation, A.C. and I.E.S.; resources, A.C. and D.E.D.; data curation, A.C.; writing—original draft preparation, A.C. and I.E.S.; writing—review and editing, A.C. and D.E.D.; visualization, A.C. and D.E.D.; supervision, D.E.D. and N.B.; project administration, A.C.; funding acquisition, A.C., D.E.D., and N.B. All authors have read and agreed to the published version of the manuscript.

Funding: This work was achieved through the Core programme (PN 19 02), supported by the Ministry of Research, Innovation and Digitalization, project number 19 02 02 02.

Institutional Review Board Statement: Not applicable.

Informed Consent Statement: Not applicable.

Data Availability Statement: Data is contained within the article.

Acknowledgments: The authors acknowledge the technical support provided by researcher Marina Schimbator.

Conflicts of Interest: The authors declare no conflict of interest.

Abbreviations

Ala	Alanine
Arg	Arginine
Asp	Aspartic acid
Cys	Cystine
Glu	Glutamic acid
Gly	Glycine
His	Histidine
Ile	Isoleucine
Leu	Leucine
Lys	Lysine
Met	Methionine
Phe	Phenylalanine
Pro	Proline
Ser	Serine
Thr	Threonine

Tyr	Tyrosine
Val	Valine
AAS	amino acid score
AC	antioxidant capacity
BV	Biological Value
d.m.	dry matter
EAAI	Essential Amino Acid Index
MUFA	monounsaturated fatty acid
NI	Nutritional Index
PDCAAS	protein digestibility corrected amino acid score
PER	Protein Efficiency Ratio
PUFA	polyunsaturated fatty acid
SFA	saturated fatty acid
TPC	total polyphenolic content

References

1. Conte, P.; Fadda, C.; Drabińska, N.; Krupa-Kozak, U. Technological and nutritional challenges, and novelty in gluten-free breadmaking: A review. *Pol. J. Food Nutr. Sci.* **2019**, *69*, 5–21. [CrossRef]
2. Falguera, V.; Aliguer, N.; Falguera, M. An integrated approach to current trends in food consumption: Moving toward functional and organic products? *Food Control* **2012**, *26*, 274–281. [CrossRef]
3. Saturni, L.; Ferretti, G.; Bacchetti, T. The gluten-free diet: Safety and nutritional quality. *Nutrients* **2010**, *2*, 16–34. [CrossRef] [PubMed]
4. Vici, G.; Belli, L.; Biondi, M.; Polzonetti, V. Gluten free diet and nutrient deficiencies: A review. *Clin. Nutr.* **2016**, *35*, 1236–1241. [CrossRef]
5. Bender, D.; Schonlechner, R. Innovative approaches towards improved gluten-free bread properties. *J. Cereal Sci.* **2020**, *91*, 102904. [CrossRef]
6. Cappelli, A.; Oliva, N.; Cini, E. A systematic review of gluten-free dough and bread: Dough rheology, bread characteristics, and improvement strategies. *Appl. Sci.* **2020**, *10*, 6559. [CrossRef]
7. Demirkesen, I.; Ozkaya, B. Recent strategies for tackling the problems in gluten-free diet and products. *Crit. Rev. Food Sci. Nutr.* **2020**, 1–27. [CrossRef] [PubMed]
8. Gobbetti, M.; Pontonio, E.; Filanninob, P.; Rizzellob, C.G.; De Angelisb, M.; Di Cagnoa, R. How to improve the gluten-free diet: The state of the art from a food science perspective. *Food Res. Int.* **2018**, *110*, 22–32. [CrossRef] [PubMed]
9. Bolarinwa, I.F.; Lim, P.T.; Kharidah, M. Quality of gluten-free cookies from germinated brown rice flour. *Food Res.* **2019**, *3*, 199–207. [CrossRef]
10. Hosseini, S.M.; Soltanizadeh, N.; Mirmoghtadaee, P.; Banavand, P.; Mirmoghtadaie, L.; Shojaee-Aliabadi, S. Gluten-free products in celiac disease: Nutritional and technological challenges and solutions. *J. Res. Med. Sci.* **2018**, *23*, 109. [CrossRef] [PubMed]
11. Hager, A.-S.; Wolter, A.; Jacob, F.; Zannini, E.; Arendt, E.K. Nutritional properties and ultra-structure of commercial gluten free flours from different botanical sources compared to wheat flours. *J. Cereal Sci.* **2012**, *56*, 239–247. [CrossRef]
12. Martínez-Villaluenga, C.; Peñas, E.; Hernández-Ledesma, B. Pseudocereal grains: Nutritional value, health benefits and current applications for the development of gluten-free foods. *Food Chem. Toxicol.* **2020**, *137*, 111–178. [CrossRef]
13. Melini, F.; Melini, V.; Luziatelli, F.; Ruzzi, M. Current and forward-looking approaches to technological and nutritional improvements of gluten-free bread with legume flours: A critical review. *Compr. Rev. Food Sci. Food Saf.* **2017**, *16*, 1101–1122. [CrossRef] [PubMed]
14. Aguilar, N.; Albanell, E.; Miñarro, B.; Capellas, M. Chickpea and tiger nut flours as alternatives to emulsifier and shortening in gluten-free bread. *LWT Food Sci. Technol.* **2015**, *62*, 225–232. [CrossRef]
15. Aguilar, N.; Albanell, E.; Miñarro, B.; Guamis, B.; Capellas, M. Effect of tiger nut-derived products in gluten-free batter and bread. *Food Sci. Technol. Int.* **2015**, *21*, 323–331. [CrossRef]
16. Wani, I.A.; Farooq, G.; Qadir, N.; Wani, T.A. Physico-chemical and rheological properties of Bengal gram (*Cicer arietinum* L.) starch as affected by high temperature short time extrusion. *Int. J. Biol. Macromol.* **2019**, *131*, 850–857. [CrossRef] [PubMed]
17. Garcia-Valle, D.E.; Bello-Perez, L.A.; Flores-Silva, P.C.; Agama-Acevedo, E.; Tovar, J. Extruded unripe plantain flour as an indigestible carbohydrate-rich ingredient. *Front. Nutr.* **2019**, *6*, 2. [CrossRef]
18. Camelo-Méndez, G.A.; Silva, P.C.F.; Acevedo, E.Q.; Bello-Perez, L.A. Multivariable analysis of gluten-free pasta elaborated with non-conventional flours based on the phenolic profile, antioxidant capacity and color. *Plant. Foods Hum. Nutr.* **2017**, *72*, 411–417. [CrossRef] [PubMed]
19. Camelo-Méndez, G.A.; Tovar, J.; Pérez, L.A.B. Influence of blue maize flour on gluten-free pasta quality and antioxidant retention characteristics. *Food Sci. Technol.* **2018**, *55*, 2739–2748. [CrossRef]
20. Agama-Acevedo, E.; Bello-Perez, L.A.; Pacheco-Vargas, G.; Tovar, J.; Sáyago-Ayerdi, S.G. Unripe plantain flour as a dietary fiber source in gluten-free spaghetti with moderate glycemic index. *J. Food Process. Preserv.* **2019**, *43*, e14012. [CrossRef]

21. International Organization for Standardization (ISO). *Animal and Vegetable Fats and Oils–Gas Chromatography of Fatty Acid Methyl Esters–Part 2: Preparation of Methyl Esters of Fatty Acids*; EN ISO 12966-2:2017; International Organization for Standardization: Geneva, Switzerland, 2017.

22. Bidlingmeyer, B.A.; Cohen, S.A.; Tarvin, T.L. Rapid analysis of amino acids using pre-column derivatization. *J. Chromatogr.* **1984**, *336*, 93–104. [CrossRef]

23. Food and Agriculture Organization (FAO). Dietary protein quality evaluation in human nutrition: Report of an FAO expert consultation. *FAO Food Nutr. Pap.* **2013**, *92*, 1–66.

24. Curiel, J.A.; Coda, R.; Limitone, A.; Katina, K.; Raulio, M.; Giuliani, G.; Rizzello, C.G.; Gobbetti, M. Manufacture and characterization of pasta made with wheat flour rendered gluten-free using fungal proteases and selected sourdough lactic acid bacteria. *J. Cereal Sci.* **2014**, *59*, 79–87. [CrossRef]

25. Hsu, H.W.; Vavak, D.L.; Satterlee, L.D.; Miller, G.A. A multienzyme technique for estimating protein digestibility. *J. Food Sci.* **1977**, *42*, 1269–1273. [CrossRef]

26. Horszwald, A.; Andlauer, W. Characterisation of bioactive compounds in berry juices by traditional photometric and modern microplate methods. *J. Berry Res.* **2011**, *1*, 189–199. [CrossRef]

27. Klunklin, W.; Savage, G. Physicochemical, antioxidant properties and in vitro digestibility of wheat–purple rice flour mixtures. *Int. J. Food Sci. Technol.* **2018**, *53*, 1962–1971. [CrossRef]

28. Gull, A.; Prasad, K.; Kumar, P. Effect of millet flours and carrot pomace on cooking qualities, color and texture of developed pasta. *LWT Food Sci. Technol.* **2015**, *63*, 470–474. [CrossRef]

29. Burešová, I.; Tokár, M.; Mareček, J.; Hřivna, L.; Faměra, O.; Šottníková, V. The comparison of the effect of added amaranth, buckwheat, chickpea, corn, millet and quinoa flour on rice dough rheological characteristics, textural and sensory quality of bread. *J. Cereal Sci.* **2017**, *75*, 158–164. [CrossRef]

30. Joshi, D.C.; Chaudhari, G.V.; Sood, S.; Kant, L.; Pattanayak, A.; Zhang, K.; Fan, Y.; Janovská, D.; Meglič, V.; Zhou, M. Revisiting the versatile buckwheat: Reinvigorating genetic gains through integrated breeding and genomics approach. *Planta* **2019**, *250*, 783–801. [CrossRef]

31. Pereira, E.; Encina-Zelada, C.; Barros, L.; Gonzales-Barron, U.; Cadavez, V.; Ferreira, I.C.F.R. Chemical and nutritional characterization of *Chenopodium quinoa* Willd. (quinoa) grains: A good alternative to nutritious food. *Food Chem.* **2019**, *280*, 110–114. [CrossRef]

32. Thanh-Tien, N.N.; Le Ngoc, D.T.; Inoue, N.; Naofumi, M.; Van Hung, P. Nutritional composition, bioactive compounds, and diabetic enzyme inhibition capacity of three varieties of buckwheat in Japan. *Cereal Chem.* **2018**, *95*, 615–624. [CrossRef]

33. Sreerama, Y.N.; Sashikala, V.B.; Pratape, V.M.; Singh, V. Nutrients and antinutrients in cowpea and horse gram flours in comparison to chickpea flour: Evaluation of their flour functionality. *Food Chem.* **2012**, *131*, 462–468. [CrossRef]

34. Oladele, A.K.; Aina, J.O. Chemical composition and functional properties of flour produced from two varieties of tigernut (*Cyperus esculentus*). *Afr. J. Biotechnol.* **2007**, *6*, 2473–2476. [CrossRef]

35. Adegunwa, M.O.; Adelekan, E.O.; Adebowale, A.A.; Bakare, H.A.; Alamu, E.O. Evaluation of nutritional and functional properties of plantain (*Musa paradisiaca* L.) and tigernut (*Cyperus esculentus* L.) flour blends for food formulations. *Cogent Chem.* **2017**, *3*, 1383707. [CrossRef]

36. Demirkesen, I.; Sumnu, G.; Sahin, S. Quality of gluten-free bread formulations baked in different ovens. *Food Bioprocess. Technol.* **2013**, *6*, 746–753. [CrossRef]

37. Di Cairano, M.; Condelli, N.; Caruso, M.C.; Marti, A.; Cela, N.; Galgano, F. Functional properties and predicted glycemic index of gluten free cereal, pseudocereal and legume flours. *LWT Food Sci. Technol.* **2020**, *133*, 109860. [CrossRef]

38. Wolter, A.; Hager, A.-S.; Zannini, E.; Arendt, E.K. In vitro starch digestibility and predicted glycaemic indexes of buckwheat, oat, quinoa, sorghum, teff and commercial gluten-free bread. *J. Cereal Sci.* **2013**, *58*, 431–436. [CrossRef]

39. Banu, I.; Aprodu, I. Assessing the performance of different grains in gluten-free bread applications. *Appl. Sci.* **2020**, *10*, 8772. [CrossRef]

40. Alvarez-Jubete, L.; Arendt, E.K.; Gallagher, E. Nutritive value of pseudocereals and their increasing use as functional gluten-free ingredients. *Trends Food Sci. Technol.* **2010**, *21*, 106–113. [CrossRef]

41. Ghavidel, R.A.; Prakash, J. The impact of germination and dehulling on nutrients, antinutrients, in vitro iron and calcium bioavailability and in vitro starch and protein digestibility of some legume seeds. *LWT Food Sci. Technol.* **2007**, *40*, 1292–1299. [CrossRef]

42. Dubey, P.; Thakur, V.; Chattopadhyay, M. Role of minerals and trace elements in diabetes and insulin resistance. *Nutrients* **2020**, *12*, 1864. [CrossRef] [PubMed]

43. Zheng, Y.; Li, X.-K.; Wang, Y.; Cai, L. The role of zinc, copper and iron in the pathogenesis of diabetes and diabetic complications: Therapeutic effects by chelators. *Hemoglobin* **2008**, *32*, 135–145. [CrossRef] [PubMed]

44. Perez, V.; Chang, E.T. Sodium-to-potassium ratio and blood pressure, hypertension, and related factors. *Adv. Nutr.* **2014**, *5*, 712–741. [CrossRef] [PubMed]

45. Leenhardt, F.; Levrat-Verny, M.-A.; Chanliaud, E.; Rémésy, C. Moderate decrease of pH by sourdough fermentation is sufficient to reduce phytate content of whole wheat flour through endogenous phytase activity. *J. Agric. Food Chem.* **2005**, *53*, 98–102. [CrossRef] [PubMed]

46. USDA (United States Department of Agriculture). Dietary Reference Intakes. 2006. Available online: https://www.nal.usda.gov/sites/default/files/fnic_uploads/DRIEssentialGuideNutReq.pdf (accessed on 31 May 2021).

47. Rybicka, I.; Gliszczyńska-Świgło, A. Minerals in grain gluten-free products. The content of calcium, potassium, magnesium, sodium, copper, iron, manganese, and zinc. *J. Food Compos. Anal.* **2017**, *59*, 61–67. [CrossRef]

48. Zakpaa, H.D.; Mak-Mensah, E.E.; Adubofour, J. Production and characterization of flour produced from ripe "apem" plantain (*Musa sapientum* L. var. paradisiacal; *French horn*) grown in Ghana. *J. Agric. Biotech. Sustain. Dev.* **2010**, *2*, 92–99.

49. Sanchez-Zapata, E.; Fernandez-Lopez, J.; Angel Perez-Alvarez, J. Tiger nut (*Cyperus esculentus*) commercialization: Health aspects, composition, properties, and food applications. *Compr. Rev. Food Sci. Food Saf.* **2012**, *11*, 366–377. [CrossRef]

50. Ibitoye, O.B.; Aliyu, N.O.; Ajiboye, T.O. Tiger nut oil-based diet improves the lipid profile and antioxidant status of male Wistar rats. *J. Food Biochem.* **2018**, *42*, e12587. [CrossRef]

51. Amare, E.; Grigoletto, L.; Corich, V.; Giacomini, A.; Lante, A. Fatty acid profile, lipid quality and squalene content of teff (*Eragrostis teff* (Zucc.) Trotter) and amaranth (*Amaranthus caudatus* L.) varieties from Ethiopia. *Appl. Sci.* **2021**, *11*, 3590. [CrossRef]

52. Chen, J.; Liu, H. Nutritional indices for assessing fatty acids: A mini-review. *Int. J. Mol. Sci.* **2020**, *21*, 5695. [CrossRef]

53. Maggio, A.; Orecchio, S. Fatty acid composition of gluten-free food (bakery products) for celiac people. *Foods* **2018**, *7*, 95. [CrossRef]

54. Öhlund, K.; Olsson, C.; Hernell, O.; Öhlund, I. Dietary shortcomings in children on a gluten-free diet. *J. Hum. Nutr. Diet.* **2010**, *23*, 294–300. [CrossRef]

55. Srichuwong, S.; Curti, D.; Austin, S.; King, R.; Lamothe, L.; Gloria-Hernandez, H. Physicochemical properties and starch digestibility of whole grain sorghums, millet, quinoa and amaranth flours, as affected by starch and non-starch constituents. *Food Chem.* **2017**, *233*, 1–10. [CrossRef] [PubMed]

56. Gorinstein, S.; Pawelzik, E.; Delgado-Licon, E.; Haruenkit, R.; Weisz, M.; Trakhtenberg, S. Characterisation of pseudocereal and cereal proteins by protein and amino acid analysis. *J. Sci. Food Agric.* **2002**, *82*, 886–891. [CrossRef]

57. Pastoriza, S.; Rufián-Henares, J.Á.; García-Villanova, B.; Guerra-Hernández, E. Evolution of the Maillard reaction in glutamine or arginine-dextrinomaltose model systems. *Foods* **2016**, *5*, 86. [CrossRef]

58. Wu, T.; Taylor, C.; Nebl, T.; Ng, K.; Bennett, L.E. Effects of chemical composition and baking on in vitro digestibility of proteins in breads made from selected gluten-containing and gluten-free flours. *Food Chem.* **2017**, *233*, 514–524. [CrossRef] [PubMed]

59. Abdel-Aal, E.-S.M. Effects of baking on protein digestibility of organic spelt products determined by two in vitro digestion methods. *LWT Food Sci. Technol.* **2008**, *41*, 1282–1288. [CrossRef]

60. Holasova, M.; Fiedlerova, V.; Smrcinova, H.; Orsak, M.; Lachman, J.; Vavreinova, S. Buckwheat–the source of antioxidant activity in functional foods. *Food Res. Int.* **2002**, *35*, 207–211. [CrossRef]

61. Fu, L.; Xu, B.T.; Xu, X.R.; Qin, X.U.; Gan, R.; Li, H.B. Antioxidant capacities and total phenolic contents of 56 wild fruits from South China. *Molecules* **2010**, *15*, 8602–8617. [CrossRef]

62. Rocchetti, G.; Lucini, L.; Rodriguez, J.M.L.; Barba, F.J.; Giuberti, G. Gluten-free flours from cereals, pseudocereals and legumes: Phenolic fingerprints and in vitro antioxidant properties. *Food Chem.* **2019**, *271*, 157–164. [CrossRef]

63. Schirmer, M.; Jekle, M.; Becker, T. Starch gelatinization and its complexity for analysis. *Starch* **2015**, *67*, 30–41. [CrossRef]

64. Jhan, F.; Gani, A.; Shah, A.; Ashwar, B.A.; Bhat, N.A.; Ganaie, T.A. Gluten-free minor cereals of Himalayan origin: Characterization, nutraceutical potential and utilization as possible anti-diabetic food for growing diabetic population of the world. *Food Hydrocoll.* **2021**, *113*, 106402. [CrossRef]

65. Jane, J.; Chen, Y.Y.; Lee, L.F.; McPherson, A.E.; Wong, K.S.; Radosavljevic, M.; Kasemsuwan, T. Effects of amylopectin branch chain length and amylose content on the gelatinization and pasting properties of starch. *Cereal Chem.* **1999**, *76*, 629–637. [CrossRef]

66. Hu, W.-X.; Chen, J.; Xu, F.; Chen, L.; Zhao, J.-W. Study on crystalline, gelatinization and rheological properties of japonica rice flour as affected by starch fine structure. *Int. J. Biol. Macromol.* **2020**, *148*, 1232–1241. [CrossRef] [PubMed]

67. Soares, C.A.; Gonçalves Peroni-Okita, F.H.; Cardoso, M.B.; Shitakubo, R.; Lajolo, F.M.; Cordenunsi, B.R. Plantain and banana starches: Granule structural characteristics explain the differences in their starch degradation patterns. *J. Agric. Food Chem.* **2011**, *59*, 6672–6681. [CrossRef] [PubMed]

68. Karim, A.A.; Norziah, M.H.; Seow, C.C. Methods for the study of starch retrogradation. *Food Chem.* **2000**, *71*, 9–36. [CrossRef]

69. Patil, S.P.; Arya, S.S. Nutritional, functional, phytochemical and structural characterization of gluten-free flours. *J. Food Meas. Charact.* **2017**, *11*, 1284–1294. [CrossRef]

70. Chandra, S. Assessment of functional properties of different flours. *Afr. J. Agric. Res.* **2013**, *8*, 4849–4852.

71. Chandra, S.; Singh, S.; Kumari, D. Evaluation of functional properties of composite flours and sensorial attributes of composite flour biscuits. *J. Food Sci. Technol.* **2015**, *52*, 3681–3688. [CrossRef]

72. Joshi, A.U.; Liu, C.; Shridhar, K.S. Functional properties of select seed flours. *LWT Food Sci. Technol.* **2015**, *60*, 325–331. [CrossRef]

73. Jolliffe, I.T.; Cadima, J. Principal component analysis: A review and recent developments. *Phil. Trans. R. Soc. A* **2016**, *374*, 20150202. [CrossRef]

 *applied sciences*

Article

Quinoa Sourdough Fermented with *Lactobacillus plantarum* ATCC 8014 Designed for Gluten-Free Muffins—A Powerful Tool to Enhance Bioactive Compounds

Maria Simona Chiş [1], Adriana Păucean [1,*], Simona Maria Man [1], Dan Cristian Vodnar [2,3], Bernadette-Emoke Teleky [3], Carmen Rodica Pop [2], Laura Stan [2], Orsolya Borsai [4], Csaba Balasz Kadar [1], Adriana Cristina Urcan [5] and Sevastiţa Muste [1]

[1] Department of Food Engineering, Faculty of Food Science and Technology, University of Agricultural Sciences and Veterinary Medicine of Cluj-Napoca, 3-5 Mănăştur Street, 400372 Cluj-Napoca, Romania; simona.chis@usamvcluj.ro (M.S.C.); simona.man@usamvcluj.ro (S.M.M.); balasz-csaba@usamvcluj.ro (C.B.K.); sevastita.muste@usamvcluj.ro (S.M.)

[2] Department of Food Science, Faculty of Food Science and Technology, University of Agricultural Sciences and Veterinary Medicine of Cluj-Napoca, 3-5 Mănăştur Street, 400372 Cluj-Napoca, Romania; dan.vodnar@usamvcluj.ro (D.C.V.); carmen-rodica.pop@usamvcluj.ro (C.R.P.); laurastan@usamvcluj.ro (L.S.)

[3] Institute of Life Sciences, University of Agricultural Sciences and Veterinary Medicine Cluj-Napoca, 3-5 Calea Mănăştur, 400372 Cluj-Napoca, Romania; bernadette.teleky@usamvcluj.ro

[4] Agro Transilvania Cluster, Dezmir, Crişeni Street, FN, 407039 Dezmir, Romania; orsolya.borsai@usamvcluj.ro

[5] Department of Microbiology and Immunology, University of Agricultural Sciences and Veterinary Medicine, 3-5 Mănăştur Street, 400372 Cluj-Napoca, Romania; adriana.urcan@usamvcluj.ro

* Correspondence: adriana.paucean@usamvcluj.ro; Tel.: +40-745-264041

Received: 10 September 2020; Accepted: 9 October 2020; Published: 14 October 2020

Abstract: *Lactobacillus plantarum* ATCC 8014 was used to ferment quinoa flour, in order to evaluate its influence on the nutritional and rheological characteristics of both the sourdough and muffins. The quantification of carbohydrates and organic acids was carried out on a HPLC-RID system (high-performance liquid chromatography coupled with with refractive index detector), meanwhile HPLC-UV-VIS (high-performance liquid chromatography coupled with UV-VIS detector), AAS (Atomic absorption spectrophotometry), aluminum chloride colorimetric assay, Folin–Ciocalteu, and 1,1-Diphenyl-2-picrylhydrazyl radical scavenging activity (DPPH) methods were used to determine folic acid, minerals, flavonoids, total phenols, and radical scavenging activity, respectively. Two types of sourdough were used in this study: quinoa sourdough fermented with *L. plantarum* ATCC 8014 and quinoa sourdough spontaneous fermented. The first one influenced the chemical composition of muffins in terms of decreased content of carbohydrates, higher amounts of both organic acids and folic acid. Furthermore, higher amounts of flavonoids, total phenols and increased radical scavenging activity were recorded due to the use of *Lactobacillus plantarum* ATCC 8014 strain. These results indicate the positive effect of quinoa flour fermentation with the above strain and supports the use of controlled fermentation with lactic acid bacteria for the manufacturing of gluten free baked products.

Keywords: *Lactobacillus plantarum* ATCC 8014; nutritional effects; gluten-free muffins

---

1. Introduction

Quinoa (*Chenopodium quinoa* Willd) pseudo-cereal comes from the Andean region and it is considered an ancient grain [1]. Nowadays quinoa is receiving increased attention due to its various nutrients and bioactive compounds and is particularly used to design gluten free products. Celiac disease is defined as a chronic inflammatory autoimmune disorder of the small intestinal mucosa caused by the ingestion of gluten proteins [2] found mainly in wheat, barley, and rye [3]. A gluten free diet could lead to nutritional deficiencies in minerals, vitamins, folate [4], and the use of non-conventional raw materials, such as amaranth and quinoa, could represent new approaches to prevent the aforementioned nutritional deficiencies.

The valuable nutritional quality of quinoa seeds is mainly due to its high concentrations of proteins, minerals, and vitamins [5–7]. Quinoa proteins contains essential amino acids (lysine, threonine, and methionine), which are well-balanced and with a highly bioavailability, being superior to that of the common cereals. Its lipids contain unsaturated fatty acids (linoleic and linolenic acids) which are considered healthy [8,9]. Quinoa is also known as a high source of vitamins (folate and tocopherols), minerals (iron, calcium, copper, manganese, and potassium) and other phytochemicals (ecdysteroids, phenolic acids, and flavonoids such as kaempferol and quercetin) [8–17].

Although the nutritional composition of quinoa was characterized in many research articles, some issues concerning its applications in baking technology have received less attention [8]. Overall, the lack of gluten lead to a low baking quality of quinoa, while flavor, texture and appearance of baking products were generally considered with low or moderate acceptability by consumers [18,19]. Therefore, quinoa flour is considered a promising raw material for gluten-free products, but also a technological challenge for bakers who have to improve the textural and sensorial quality of these products [20].

It is generally recognized that fermentation could be a strategy to overcome these technological challenges [21]. The foremost fermentation used for baking purposes is with sourdough. Even more, it has been proven ideal to obtain bakery products with improved texture, taste, aroma, shelf life and nutritional value [22]. Thus, due to the microbial metabolic dynamics, the use of sourdough in gluten-free products (GF) manufacture may be considered a "tailored made" solution for improving their quality, safety, and acceptability [23].

Some of the beneficial aspects of use of sourdough fermentation are: the decrement of the glycemic response of bread, the improvement of minerals, phytochemicals, and vitamins uptake [24] along with the capacity of the lactobacilli metabolism to produce new functional compounds such as peptides, amino acid derivatives, and exo-polysaccharides [22]. The lactic acid bacteria from sourdough could influence the allergy and intolerance responses of cereal sensitive individuals due to their proteolytic activity [22].

Lactic acid bacteria (LAB) might be considered as cell factories able to deliver bioactive compounds and food ingredients producing improved quality GF products [25,26]. Raw matrix carbohydrates represent a key role in the adaptability of LAB in a new environment. Carbohydrates are the main fraction of flour from which starch, simple sugars, and fiber represent the main components. During milling, starch granules are broken, and the amylose and amylopectin glucose polymers could be hydrolyzed to simple molecules, such as maltose and glucose. During fermentation with LAB, the most important process is the utilization of carbohydrates, as a source of carbon [27], starch degradation being crucial for LAB growth and development [28]. Furthermore, LAB are able to metabolize starch, leading to the drop of the pH in the raw matrix and enhancing the production of organic acids such as lactic and acetic one [29].

It is generally accepted that *Lactobacillus plantarum*, like all facultative hetero fermentative lactobacilli, ferments hexoses to lactic acid via Embden-Meyerhof-Parnas pathway (EMP) and degrades pentoses and gluconate via the pentose phosphate (PP) pathway, producing acetic acid, ethanol, and formic acid [30,31]. *Lactobacillus plantarum* ATCC 8014 (*Lp*) was able to grow in quinoa sourdough, having a good adaptability and being able to improve the amino acids, volatile compounds, and sensory features of the final baked muffins through starch and protein degradation [3].

As supported by a large body of literature, dough fermented by LAB confer high sensory quality to bakery products [32], and they can improve the nutritional value of gluten-free flours [3,24,33]. On the other hand, spontaneous fermentation could lead to variations in the quality of the final baked goods; therefore, the use of selected starters is recommended [34]. Rizzello et al. [1] reported that the use of quinoa sourdough in bread improved the sensorial and textural quality, in terms of lower hardness, higher porosity specific volume, and protein content than the control (wheat bread). In the research published by Di Cagno et al. [35], *Lactobacillus sanfranciscensis* LS40 and LS41, and *Lactobacillus plantarum* CF1 were selected and used as sourdough starters for the manufacture of GF bread following a two-step fermentation process. By this approach, the improvement of the nutritional, textural and sensorial characteristics were targeted. The facultative heterofermentative *Lactobacillus plantarum* was among the dominant lactic acid biota in gluten-free sourdoughs from rice, amaranth, quinoa [36], or buckwheat and teff flour [20,37].

In the present study, quinoa flour fermented with *Lactobacillus plantarum* ATCC 8014 (*Lp*) was used as ingredient in gluten-free muffins manufacturing to improve their nutritional characteristics. In order to achieve this goal, the impact of *Lp* on carbohydrate metabolic conversion to organic acids and rheological features of quinoa sourdough was assessed.

2. Materials and Methods

2.1. Materials

Quinoa wholemeal flour (QWF), rice wholemeal flour (RWF), buckwheat flour, inulin, oatmeal, corn starch, baking powder, maple syrup, and coconut butter were acquired from Romanian specialized stores. *Lactobacillus plantarum* ATCC 8014 (*Lp*) was acquired from Microbiologics (St. Cloud, MN, USA). Maltose, glucose, fructose standards were purchased from Sigma Aldrich (Darmstadt, Germany), citric acid from Merck (Darmstadt, Germany) and lactic and acetic acid from Fluka (Saint Louis, MO, USA). All standard compounds were 99.5% pure. Before analysis, all the samples were filtered using a MF-MilliporeTM Membrane Filter (0.45 μm) from Merck (Darmstadt, Germany).

2.2. Microbial Starter Culture Preparation, Sourdough Preparation, and Muffins Formulation

Lactobacillus plantarum ATCC 8014 was purchased in lyophilized form and cultivated during 20 h in MRS (Man Rogosa Sharpe) broth at 37 °C, as reported in our previous study [38]. After that, the obtained biomass was centrifugated (Eppendorf R 5804 centrifuge, Hamburg, Germany) at 2300× *g*, 10 min, 4 °C and washed three times with sterile water. The SP initial cell concentration was 3.2 cfu/g sourdough. A spontaneous quinoa flour fermentation was used in the present study, as a control. Both sourdoughs were prepared by mixing quinoa flour with tap water in a ratio 1:1 (*v:v*), until a final dough yield of 200 was reached. The sourdoughs for controled fermentation were collected for prior analysis at 0, 4, 8, 12, and 24 h and coded with SP 0 h, SP 4 h, SP 8 h, SP 12 h, and SP 24 h. For the spontenous fermentation, the following abbreviations were used: OR 0 h, OR 4 h, OR 8 h, OR 12 h, and OR 24 h, respectively.

Muffins recipe was based on the following raw materials (*w/w*): mix dry raw materials (8% inulin, 10% oatmeal, 7% corn starch, 1.5% baking powder, and 8% buckwheat flour, 32.5% treated RWF), 15% sourdough with *Lp* strain (SP) or 15% sourdough without *Lp* strain (OR), eggs (8%), coconut butter (5%), and maple syrup (5%), as illustrated in Figure 1.

RWF was hydrothermally treated in order to improve its textural characteristics, as reported by Chiş et al. [39]. Different fermentation times (0, 12, and 24 h) were used for the muffins preparation with SP sourdough or with OR sourdough. The following codes were used for the final baked muffins manufactured with SP at 0, 12, and 24 h fermentation times: SP PF 0 h, SP PF 12 h, SP PF 24 h, and OR PF 0 h, OR PF 12 h, and OR PF 24 h for muffins with OR, respectively.

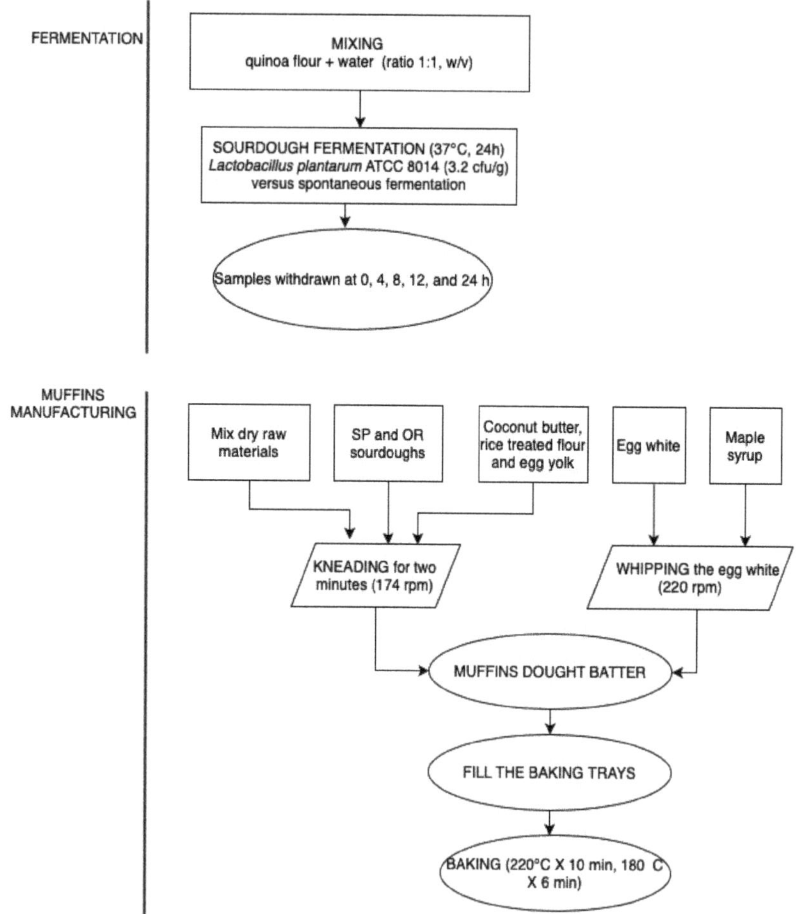

Figure 1. Flow diagram of sourdough (SP—sourdough fermented with *Lactobacillus plantarum*; OR—spontaneous fermented sourdough) and muffins' production.

2.3. Organic Acids and Glucose, Maltose, and Fructose Determination by HPLC-RID

High-performance liquid chromatography (HPLC-Agilent 1200 series, Santa Clara, CA, USA) equipped with solvent degasser, manual injector coupled with with refractive index detector (RID) (Agilent Techologies, Santa Clara, CA, USA) was used in order to analyze the organic acids and carbohydrates amount. Briefly, 1 g of sample was mixed with 5 mL of ultrapure water, vortexed for 1 min and sonicated for 2 h at 50 °C in a heated ultrasonic bath Elmasonis E 15H (Elma Schmidbauer GmbH, Singen, Germany). After that, the samples were centrifuged at $2300\times g$ for 10 min, in an Eppendorf 5804 centrifuge (Hamburg, Germany), filtered through Chromafil Xtra PA-45/13 nylon filter and 20 µL were injected in the HPLC-RID system.

The compounds were separated on a Polaris Hi-Plex H, 300×7.7 mm column (Agilent Techologies, Santa Clara, CA, USA) using the 5 mM H_2SO_4 mobile phase with a flow rate of 0.6 mL/min, column temperature T = 80 °C and RID temperature T = 35 °C. Elution of the compounds was made for 25 min. Data acquisition and results interpretation was performed using OpenLab software-ChemStation (Agilent Techologies, Santa Clara, CA, USA). The retention times for maltose, glucose, and fructose were 8.87 min, 10.24 min, and 10.88 min, respectively; meanwhile, the retention times for citric, lactic, and acetic acids were 9.39 min, 13.46 min, and 15.92 min, respectively.

2.4. Folic Acid Determination

Determination of folic acid was fulfilled in concordance with [40]. Briefly, an amount of 0.5 g of each sample was diluted with 5 mL of phosphate buffer (Ph = 7). After homogenization in a vortex, the mixtures were sonicated and centrifuged for 30 min at 3000 rpm (Eppendorf 5804, Hamburg, Germany). The supernatant was filtrated with nylon filter (0.45 μm) and 20 μL was injected in HPLC-UV detection (Agilent Technologies 1200 Series, Santa Clara, CA, USA).

A folic acid standard curve (y = 154.79x − 8.1463, R2 = 0.9954) having as minimum and maximum concentration 2 μg/mL and 25 μg/mL, respectively, was used in order to establish the concentration of folic acid, expressed in mg/l supernatant.

The operational parameters of the method were Lichrosphere 100 RP-18, (250 × 4.6 mm, 5 μm) column, mobile phase: ACN/AA 1% pH = 2.8, (20/80, *v/v*), flow rate: 0.6 mL/min, column temperature: 25 °C.

2.5. Minerals Content

Analysis of Macro and Microelements

Atomic absorption spectrophotometry (AAS) was used in order to determine the amounts of micro and macro elements, as described by [39,40]. Briefly, 3 g of each sample was burned for 10 h in a furnace (Nabertherm B150, Lilienthal, Germany) at a temperature of 550 °C. Afterwards, the ash was recovered in HCl 20% (*w/v*) in a volumetric flask, in order to achieve a final volume of 20 mL. The resulted samples were analyzed by AAS (Varian 220 FAA equipment, Germany). The results were calculated considering the samples fresh weight basis and expressed as mean value (n = 3) of three independent assays.

2.6. Total Flavonoids

Total flavonoids were determined according to the aluminum chloride colorimetric assay described by [41], adapted for the 96 well microplate reader (Synergy™ HT BioTek Instruments, Winooski, VT, USA). Quercetin was used as reference standard. Briefly, 25 μL of each sample methanolic extracts was mixed for 5 min with 100 μL distilled water and 10 μL of 5% sodium nitrate ($NaNO_2$) solution. Afterwards, 15 μL of 10% aluminum chloride ($AlCl_3$), 50 μL of 1 M sodium hydroxide (NaOH) and 50 μL of distilled water were added. The detection of total flavonoids was set at λ = 510 nm. A standard curve of quercetin was used to establish the final amount of total flavonoids content (y = 0.0003x + 0.0029, R^2 = 0.9916). The results were calculated as mg of Qe (quercetin equivalent) per g of extract.

2.7. Total Phenols Assay by Folin–Ciocalteau Reagent

In order to analyze the total phenols amount, 1 g of sample was homogenized with 100 mL acidified methanol (85:15 *v:v*, MeOH:HCl). Atfer that, the sample was dried at 40 °C by using a vacuum rotary evaporator (Laborota 4010 digital rotary evaporator, Heidolph Instruments GmbH & Co. KG, Schwabach, Germany), according to the method described by [42,43]. Folin–Ciocalteu colorimetric method was used to evaluate the total phenols amount, as follows: 100 μL of methanolic extract was mixed with 500 μL Folin–Ciocalteu reagent, 6 mL of distilled water, and 2 mL of 15% Na_2CO_3, as described by [34,44]. The solution was brought up to 10 mL by adding distilled water, kept in the dark for two hours at room temperature the absorbance was read at λ = 760 nm with a UV/visible spectrophotometer Schimadzu 1700 (Shimadzu Corporation, Kyoto, Japan). A standard calibration curve of galic acid was used to establish the final amount of total phenols (y = 1.022958x + 0.08740, R^2 = 0.99614) and the results were expressed as milligrams of gallic acid equivalent (GAE) per 100 g product.

2.8. Radical Scavenging Activity by DPPH Assay

The radical scavenging activity (RSA) was analyzed by DPPH method (1,1-Diphenyl-2-picrylhydrazyl) method as descriebed previously by [38]. Briefly, 0.1 mL of each methanolic extract was mixed with DPPH solution (3.9 mL), kept in the dark at room temperature for 30 min and recorded at $\lambda = 515$ nm, using an UV/visible spectrophotometer Schimadzu 1700 (Shimadzu Corporation, Kyoto, Japan). The following equation was used to calculate the radical scavening activity:

$$\text{RSA } [\%] \; = \; \frac{Abs\ DPPH - Abs\ Sample}{Abs\ DPPH} * 100 \tag{1}$$

where Abs_{DPPH} = absorbance of DPPH solution; Abs_{Sample} = absorbance of the sample.

2.9. Rheological Measurements

The SP and OR sourdoughs dynamic rheological characteristics were determined by using an Anton Paar MCR 72 rheometer (Anton Paar, Graz, Austria), supplied with a Peltier plate-plate system (P-PTD 200/Air) with temperature control and a 50 mm diameter smooth parallel plate geometry (PP-50-67300), according to the method described by [45]. The sourdough samples were analyzed before and after freezing (one week at −20 °C, and then defrosted at room temperature). Shortly, 3 g of sample was applied on the lower plate and the upper one was lowered to a plate distance set at a gap of 1 mm. Silicone oil was used in order to prevent sample moisture loss through testing. The storage modulus (G′) and loss modulus (G″) of each sourdough at an angular frequency of 0.628–628 rad/s^{-1} were tested, and the shear strain was set at a constant value of 0.1%, with 35 total measuring points, at a constant temperature of 30 °C.

2.10. Statistical Analysis

Duncan multiple comparison test by SPSS version 19 software (IBM Corp., Armonk, NY, USA) was used in order to analyze the results. All samples were analyzed in triplicates and the results were expressed as means ± standard deviations.

3. Results and Discussion

3.1. Carbohydrates, Organic Acids, Folic Acid, Minerals, Flavonoids, Total Phenols Content, and Radical Scavenging Activity of Quinoa Flour (QF)

3.1.1. Carbohydrates and Organic Acids from Quinoa Wholemeal Flour (QWF)

The values of simple carbohydrates (maltose, glucose, and fructose) and organic acids (lactic, acetic, and citric) of QWF are reported in Table 1. The main carbohydrates from QWF was glucose, followed by fructose and maltose.

The carbohydrate content in quinoa is higher than in the common cereals and might be associated with frost tolerance, as confirmed by [46] who reported that quinoa from mountain region had higher sugar content than those from the valleys region. Likewise, ref. [47] reported that also the method of extraction, the origin of quinoa seeds, cultivation, and environmental stress could also influence the sugars content.

With respect to fructose content, it is known that its consumption could induce oxidative stress, leading to different type of diseases such as obesity, hypertriglyceridemia, and cardiovascular diseases. Pasko et al. [48] showed that although quinoa seeds contain fructose, the seeds are able to reduce the oxidative stress due to its ability to increase MDA (malondialdehyde) level, indicating an intensive lipid peroxidation and protecting plasma against peroxidation. Moreover, reducing the oxidative stress could reduce the free radicals during some pathological states.

Table 1. Carbohydrates and organic acids content of quinoa flour.

Parameters	QWF	Retention Times (min)
Carbohydrates (mg/g f.w.)		
Maltose	6.64 ± 0.20	8.87
Glucose	89.45 ± 0.13	10.24
Fructose	12.12 ± 0.15	10.88
Organic acids (mg/g f.w.)		
Citric acid	8.59 ± 0.30	9.39
Lactic acid	n.d.	13.76
Acetic acid	n.d.	15.92

Values of three different determinations followed by standard deviation; QWF: quinoa wholemeal flour; f.w.: fresh weight; n.d.: not detected.

In our previous study, we determined a value of 68.2% total carbohydrates [26], close to the value reported by [1] of 67.9% and [49]. The total carbohydrates content in quinoa could vary between 48.6 and 68.1% of dry matter weight and 45.16–59.78%, respectively, as reported by [50,51]. Starch, the major carbohydrate of quinoa, ranges between 32 and 69% of total carbohydrates [52] and is followed by total dietary fiber (7–9.7%) [50] and fermentable sugars (2% from the total carbohydrate amount), as reported by [53].

With respect to organic acids, lactic, and acetic acids could not be found in QWF, but citric acid had a total amount of 8.59 ± 0.30 mg/g f.w. (fresh weight). Citric acid was also identified in quinoa flour by other researchers, in the range of 210–317 mg/100 g d.w., and from 0.40 to 0.71 g/100 g f.w., respectively [47,49].

3.1.2. Quinoa Wholemeal Flour (QWF) Content in Folic Acid

In the present study, the amount of folic acid determined in quinoa flour was 183 ± 0.03 µg/100 g f.w. (fresh weight). The folic acid amount from QF is higher than the folic acid content of wheat flour (10.62 µg/100 g f.w.) and green lentil flour (168.36 µg/100 g f.w.), respectively, as previously reported by Păucean et al. [40]. In general, the folic acid content in pseudo-cereals such as quinoa, is higher than the amount in cereal grains which could range from 29 to 143 µg/100 g f.w. [54]. For example, the raw pearl-millet was reported to have a folic acid content of only 26.2 µg/100 g f.w. Therefore, the researchers' attention is mainly focused on using raw materials with higher contents in folic acid in food manufacturing.

Many research data show that quinoa flour is as an important nutritional natural food source due to its valuable bioactive compounds such as B vitamin group (especially folic acid, named B9 vitamin), minerals, and essential amino acids [9,50,52,55]. Total folate content of quinoa is reported to be by about ten times higher than in wheat [56].

3.1.3. QWF Minerals Content

Quinoa is very rich in minerals like potassium, calcium, magnesium, zinc, and iron [57,58]. Table 2 displays the mineral content of quinoa flour and its composition in the following macro and microelements: calcium (Ca), magnesium (Mg), potassium (K), iron (Fe), copper (Cu), Zinc (Zn), manganese (Mn), and chromium (Cr). The ash content of quinoa flour could range from 2.4 to 4.8% [50]. In the present study, the ash content of quinoa flour was 2.3% (results previously published [3]). From the quantitative point of view, the main mineral is K (813.92 mg/100 g f.w.), followed by Mg, Ca, Fe, Mn, Zn, and Cu. The K and Mg content amounts are close to the values reported by Silva et al. [50], 926 mg/100 g, and 249.6 mg/100 g, respectively. Variation in different quinoa flour mineral content might be due to the environmental conditions (especially soil mineral availability) [59], fertilizer soil application [50], by the plant genotype [60,61], and also by the removal of the husk during milling process [51].

Table 2. Chemical composition and mineral content of quinoa flour.

Parameters	QWF
Minerals, mg/100 g f.w.	
Calcium (Ca)	18.09 ± 0.30
Magnesium (Mg)	303.43 ± 0.17
Potassium (K)	813.92 ± 0.11
Iron (Fe)	3.02 ± 0.20
Copper (Cu)	0.96 ± 0.03
Zinc (Zn)	1.82 ± 0.02
Manganese (Mn)	2.50 ± 0.01
Chromium (Cr)	n.d.

Values of three different determinations followed by standard deviation; f.w.: fresh weight. QWF: quinoa wholemeal flour; n.d.: not detected.

3.1.4. QWF Flavonoids

In the present study, the flavonoids content of QWF was 997 ± 0.52 mg Qe/100 g fresh weight. Quinoa flour has high flavonoid content that could range from 36.2 to 144.3 mg/100 g dry weight basis [10] but the total amount could be influenced by the extraction temperature, the solvent type used and non-application of ultrasounds [62]. The main flavonoids from quinoa are glycosides of the flavonols, kaempferol, and quercetin [47]. De Carvalho et al. [63] proved that besides quercetin and kaempferol glycosides, protocatechuic acid and a vanillic acid glucoside were also determined in QWF. Isoflavones, particularly Daidzein and Genistein, were found in different amounts in quinoa flour from different origins [64].

3.2. Carbohydrates, Organic Acids, Folic Acid, Minerals, Flavonoids, Total Phenols Content, Radical Scavenging Activity, and Rheological Features from Quinoa Sourdoughs with Lactobacillus plantarum (Lp) ATCC 8014 (SP) and without Lactobacillus plantarum ATCC 8014 (OR)

3.2.1. OR, SP Carbohydrates, and Organic Acids Contents

The OR and SP carbohydrates and organic acids amounts are displayed in Table 3. The glucose content of SP increased in the first 8 h of fermentation due to the starch degradation and conversion into glucose but decreased afterwards. There are two possible processes that could influence the glucose decrement. Firstly, glucose consumption during *Lp* cellular development and secondly, glucose conversion into lactic acid through via Embden-Meyerhof-Parnas pathway (EMP). On the other side, acetic acid increased during SP fermentation, probably due to the degradation of pentose and gluconate through via the pentose phosphate (PP) pathway [30,31] With respect to SP maltose content, the amount decreased after 24 h of fermentation, as reported in Table 3, reaching a final value of 1.05 mg/g f.w. after 24 h of fermentation.

On the other side, SP fructose content decreased from 4.1 mg/g f.w. to 2.69 mg/g f.w. after 24 h of fermentation. This could be due to the use of fructose by *Lp* as alternative external electron acceptor and its conversion into mannitol by mannitol dehydrogenase [65].

The SP content of lactic and acetic acids after 24 h of fermentation had a total value of 8.5 mg/g f.w. and 1.40 mg/g f.w., respectively, and the *Lp* growth during 24 h of fermentation reached a final value of log 6.7 cfu/g [38], thus emphasizing a good adaptability of the strain in the quinoa sourdough. The SP ratio between lactic and acetic acid, named FQ (fermentation quotient) was 6.07, indicating a good ratio between the two organic acids. With respect to OR fermentation, the FQ values was 1.27. This is in line with Montemurro et al. [66] who reported a FQ fermentation of 6.1 at quinoa flour fermented with *Lactobacillus plantarum* 1A7 strain.

Table 3. OR and SP carbohydrates and organic acids content during 24 h of fermentation.

Samples	Maltose mg/g f.w.	Glucose mg/g f.w.	Fructose mg/g f.w.	Citric Acid mg/g f.w.	Lactic Acid mg/g f.w.	Acetic Acid mg/g f.w.
OR 0 h	3.012 ± 0.02 [Aa]	42.662 ± 0.34 [Ac]	4.19 ± 0.19 [Aabc]	5.25 ± 0.33 [Abc]	n.d.	n.d.
SP 0 h	3.074 ± 0.21 [Aa]	42.034 ± 0.54 [Ac]	4.369 ± 0.28 [Babc]	5.06 ± 0.22 [Acd]	n.d.	n.d.
OR 4 h	2.979 ± 0.31 [Aa]	46.576 ± 0.32 [Abc]	4.21 ± 0.29 [Babc]	4.26 ± 0.11 [Ae]	n.d.	n.d.
SP 4 h	3.053 ± 0.11 [Aa]	46.085 ± 0.53 [Abc]	3.15 ± 0.22 [Ad]	4.92 ± 0.45 [Bcd]	n.d.	n.d.
OR 8 h	2.848 ± 0.22 [Aa]	51.126 ± 0.23 [Abc]	4.76 ± 0.39 [Ba]	4.74 ± 0.34 [Ad]	n.d.	n.d.
SP 8 h	2.916 ± 0.15 [Aa]	67.672 ± 0.61 [Ba]	3.75 ± 0.2 [Abcd]	5.9 ± 0.36 [Ba]	n.d.	n.d.
OR 12 h	2.85 ± 0.25 [Aa]	55.01 ±0.22 [Bb]	4.48 ± 0.5 [Bab]	5.06 ± 0.33 [Acd]	2.42 ± 0.02 [Aa]	0.50 ± 0.01 [Aa]
SP 12 h	1.814 ± 0.14 [Bb]	53.00 ± 0.45 [Abc]	3.08 ± 0.4 [Ad]	5.54 ± 0.12 [Bab]	4.60 ± 0.05 [Bb]	0.84 ± 0.02 [Bab]
OR 24 h	2.09 ± 0.03 [Ab]	46.02 ± 0.39 [Bbc]	3.66 ± 0.6 [Bcd]	5.15 ± 0.44 [Bbcd]	3.81 ± 0.31 [Ab]	2.98 ± 0.21 [Bc]
SP 24 h	1.05 ± 0.30 [Bc]	22.00 ± 0.27 [Ad]	1.69 ± 0.4 [Ae]	4.89 ± 0.33 [Acd]	8.50 ± 0.5 [Bc]	1.40 ± 0.21 [Ab]

Small letters in common indicate no significant differences between OR and SP samples withdrawn at different moments; Big different letters indicate significant differences between SP and OR samples at the same moment; f.w.: fresh weight; n.d.: not detected.

In the case of spontaneous sourdough, in the first 8 h of fermentation the carbohydrates conversion was almost similar to SP. After 8 h of spontaneous fermentation, the carbohydrates conversion lags behind, probably due to the low capacity of wild microbiota to multiply and to produce organic acids. This behavior is reflected by the FQ values compared to the same values for SP and indicate low amounts of lactic and acetic acids.

With respect to the citric acid amount, after 24 h of fermentation the SP's citric acid content decreased, but there was no significant difference between OR and SP ($p < 0.05$), as displayed in Table 3. The decrease of citric acid amount could be explained by the ability of *Lp* strain to use citric acid as an energy supply [45]. However, the *Lp* preference for energy source is reflected mainly in the use of fructose, decreasing its amount in SP sample from 4.36 to 1.69 mg/g f.w.

Salminen et al. [67] reported that in a raw matrix that contains glucose and fructose, the heterofermentative LABs will mainly use glucose as an energy source to grow and fructose as an electron acceptor. This is in line with [29] who proved that the same strain of *Lactobacillus plantarum* ATCC 8014 was able to growth in different MRS media supplemented with concentrations of glucose, fructose, sucrose, and maltose, and consumed all types of carbohydrates, although glucose being the easiest fermentable sugar for this strain.

On the other hand, [30] reported an increase amount of glucose and fructose during fermentation of wheat flour with LAB strains such as *Lactobacillus reuteri* and *Lactobacillus brevis*, respectively. Furthermore, *Lb. reuteri* R29 was able to metabolize maltose during sourdough fermentation, resulting in a low amount in the final sample.

Overall, it can be stated that during sourdough fermentation, the utilization of carbohydrates depends on the type of LAB strain which could have preferences towards a certain type of carbohydrates and on the chemical composition of the raw matrix. The fine link between bacterial strain and its favorite substrate is defined by the relationship between LAB and the raw material [68].

3.2.2. SP and OR folic Acid Content

In the present study, SP folic acid content improved during 24 h of fermentation, having a final value of 648.39 µg/100 g f.w. sourdough compared with OR sourdough, where the total folic acid amount was 169.12 µg/g dough f.w., as illustrated in Figure 2. The SP folic acid content is 3.8 times higher compared with OR sourdough and this is in agreement with [69] who reported that *L. plantarum* CRL 2107 + *L. plantarum* CRL 1964 strains are able to improve folic acid content during quinoa flour fermentation.

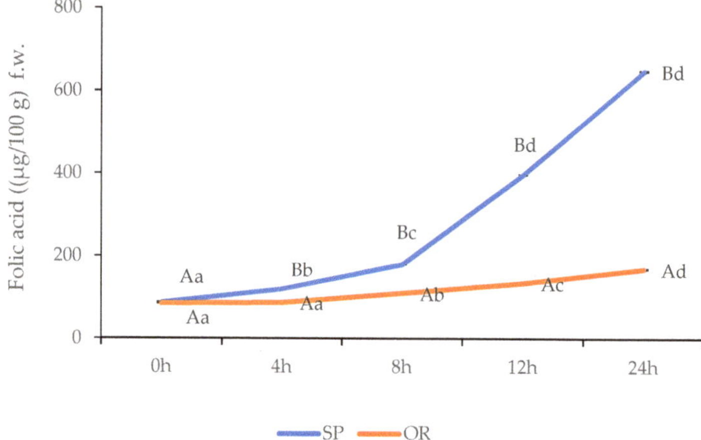

Figure 2. SP and OR folic acid content. Small letters in common indicate no significant differences between OR and SP samples withdrawn at different moments; Big different letters indicate significant differences between SP and OR samples withdrawn at the same moment.

The capacity of *Lactobacillus* strains to produce folic acid during fermentation is supported by a large body of literature [70–74]. Likewise, [55] reported that through fermentation of cereal based foods, the folate content could increase up to 700%. Furthermore, [74] showed that *Lactobacillus plantarum* CRL 1973 strain was able to produce folate during quinoa sourdough fermentation, due to its capability to synthesize B-group vitamins. Other strains of LAB isolated from cereals and seeds from Argen, like *Lactobacillus pentosus* ES124 and *Lactobacillus plantarum* ES137 were also reported to be able to produce high amounts of folate [75].

3.2.3. QP and QQ Macro and Microelements Content

During SP fermentation a significant increase of minerals was recorded (Table 4). Potassium content increased by 161.43 mg/100 g, magnesium content reached a maximum value of 294.59 mg/100 g, representing 1.5 times higher than the maximum value of the spontaneous fermented sourdough. Similar trends were recorded for Ca, Fe, and Mn contents for both SP 24 h and OR 24 h with ratios SP 24 h/OR 24 h between 1.5 and 1.6.

Table 4. Mineral content of OR and SP sourdoughs at different fermentation times.

Samples	Ca	Mg	K	Fe	Cu	Zn	Mn
OR 0 h	8.06 ± 0.13 Aa	156.09 ± 0.34 Aa	330.9 ± 0.89 Aa	0.61 ± 0.02 Aa	0.11 ± 0.01 Aa	0.46 ± 0.02 Aa	0.55 ± 0.02 Aa
SP 0 h	8.12 ± 0.11 Ba	151.00 ± 0.23 Aa	325.01 ± 0.99 Aa	0.63 ± 0.03 Aab	0.12 ± 0.02 Aa	0.49 ± 0.01 Aa	0.59 ± 0.06 Aab
OR 4 h	8.23± 0.17 Aa	159.03 ± 0.03 Aa	341.08 ± 0.77 Aa	0.69 ± 0.05 Aab	0.13 ± 0.04 Aa	0.50 ± 0.03 Aa	0.57 ± 0.05 Aa
SP 4 h	8.55 ± 0.33 Ba	163.09 ± 0.03 Bab	378.04 ± 0.88 Bbc	0.75 ± 0.34 Aabc	0.17 ± 0.01 Aab	0.53 ± 0.04 Aa	0.65 ± 0.07 Aab
OR 8 h	8.9 ± 0.22 Aa	172.04 ± 0.23 Aabc	353.56 ± 0.89 Aab	0.71 ± 0.11 Aab	0.23 ± 0.01 Aab	0.60 ± 0.01 Aa	0.60 ± 0.06 Aab
SP 8 h	9.6 ± 0.34 Bab	176.89 ± 0.56 Babc	456.67 ± 0.67 Bef	0.89 ± 0.33 Bd	0.29 ± 0.02 Ab	0.72 ± 0.03 Aa	0.72 ± 0.08 Abc
OR 12 h	9.53 ± 0.11 Aab	171.23 ± 0.45 Aabc	386.4 ± 0.69 Abc	0.77 ± 0.11 Abcd	0.16 ± 0.03 Aab	0.63 ± 0.02 Aa	0.58 ± 0.06 Aa
SP 12 h	11.67 ± 0.22 Bc	198.98 ± 0.64 Bc	407.61 ± 0.88 Bcd	1.11 ± 0.02 Be	0.51 ± 0.02 Bc	0.99 ± 0.01 Bb	0.89 ± 0.05 Bd
OR 24 h	10.86 ± 0.45 Abc	189.21 ± 0.59 Abc	427.53 ± 0.79 Ade	0.87 ± 0.02 Acd	0.21 ± 0.04 Aab	0.62 ± 0.02 Aa	0.78 ± 0.03 Acd
SP 24 h	17.06 ± 0.32 Bd	294.59 ± 0.89 Bd	486.44 ± 0.98 Bf	1.42 ± 0.04 Bf	0.79 ± 0.05 Bd	1.68 ± 0.00 Bc	1.23 ± 0.07 Be

The results are expressed in mg/100 g fresh weight. Small letters in common indicate no significant differences between OR and SP samples withdrawn at different moments; Big different letters indicate significant differences between SP and OR samples withdrawn at the same moment.

Quinoa contains about 1% phytic acid which reduces the bioavailability of magnesium, zinc, iron, and calcium due to the strong connection between phytate and these multivalent metal ions, acting as an excellent chelator of cations. As reported by the literature, phytate chelation of mineral cations could have a negative influence on the bioavailability of essential minerals like zinc, iron, calcium, and magnesium [76].

Several studies demonstrated that fermentation with LAB led to a significant reduction of the phytic acid amount, increasing the concentration of Ca and Mg [77,78].

The increment of mineral content in quinoa sourdough is caused by the diminution of phytic acid due to the acidic pH value, which activates flour endogenous phytase and due to the phytase activity of LAB. In the conditions of our study, after 24 h of fermentation the pH value of the SP was 4.2, compared to pH value of 5.8 for OR [38]. The drop of the pH to 4.2 for SP 24 h after 24 h of fermentation, enhanced lactic acid content, and decreased the phytic acid amount leading to higher amount of minerals such as calcium, potassium, iron, zinc, magnesium, manganese, and chromium. This is in agreement with [79] who proved that during fermentation of quinoa with *Lactobacillus plantarum* the phytate was tremendously reduced (82–98%) and iron amount increased three to fivefold.

Highlighting this idea, [69] reported that using *Lactobacillus plantarum* strains for fermenting quinoa sourdough could be considered as a bio-enrichment of it, due to the ability of these strains to increase mineral bioavailability such as Ca, Fe, and Mg, through phytate degradation. This is supported also by [1] who previously demonstrated that quinoa LAB sourdough had a phytase activity 2.75 times higher that raw quinoa flour.

3.2.4. Total Flavonoids Content of OR and SP Sourdoughs

Total flavonoids content of OR and SP sourdoughs are illustrated in Figure 3. The SP flavonoids content increased during 24 h of fermentation, having a total value of 1551 mg Qe/100 g f.w., meanwhile OR 24 h reached a final amount of 757 mg Qe/100 g f.w. This finding is consistent with [80] who reported that during fermentation with *Lactobacillus* genus the total flavonoid content could improve. Through enzymatic reactions, *Lactobacillus* strains were able to release from glycosides flavonoids and isoflavone aglycones, respectively [81].

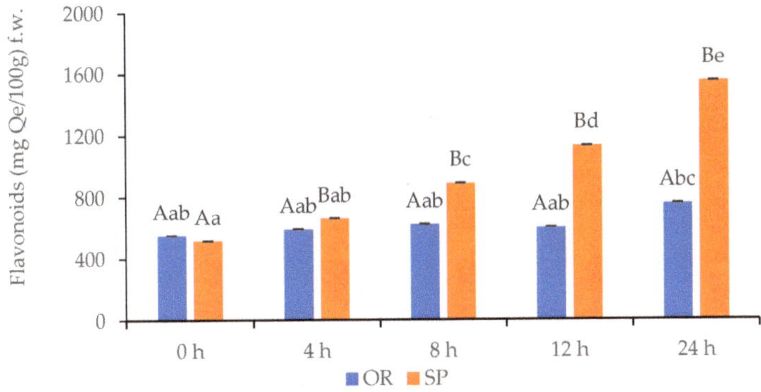

Figure 3. SP and OR total flavonoids content during 24 h of fermentation. Small letters in common indicate no significant differences between OR and SP samples withdrawn at different moments; Big different letters indicate significant differences between SP and OR samples withdrawn at the same moment.

3.2.5. Rheological Measurements

The rheological alterations of OR and SP sourdoughs during fermentation and freezing are presented in Figures 4–7. G′ represent the capability of materials to store the elastic deformation energy and G″ modul represent the viscous portion of the materials [45]. In general, the G″ was lower than the G′ in SP and OR fresh and frozen samples, indicating that the viscous properties of the sourdoughs increased while elastic behaviour decrease with the increasing hours of fermentation. This could be justified due to the possible *Lp* exopolysaccharides production through sourdough fermentation, which could act as viscosifiers and texturizers, having pseudoplastic rheological behavior and being involved in the water-binding capacity of sourdoughs [27,54,82] reported that strains from *Lactobacillus* genus could produce exopolysaccharides through sourdough fermentation. This is in agreement with [83] who confirmed that *Lactobacillus plantarum* is able to produce exopolysaccharides in sourdough through fermentation. On the other side, [84] reported that LAB might produce 12β-glucan during the growth process and metabolism, which could positively influence the viscosity and the water holding capacity of quinoa flour.

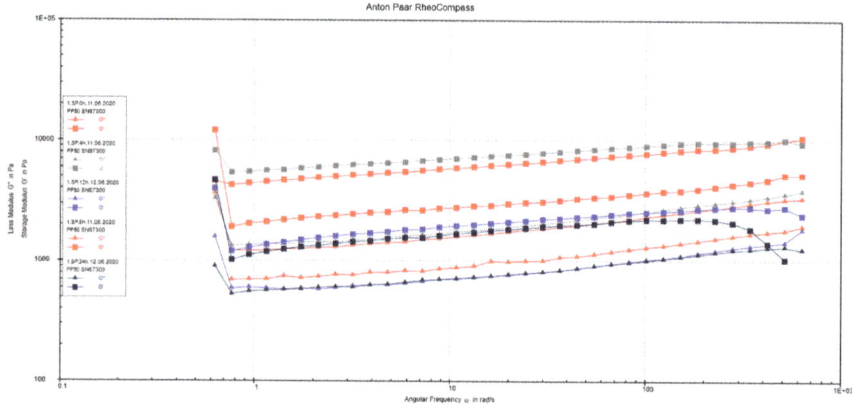

Figure 4. The storage (G′) and loss (G″) shear moduli for SP sourdough at different fermentation times: 0, 4, 8, 12, and 24 h.

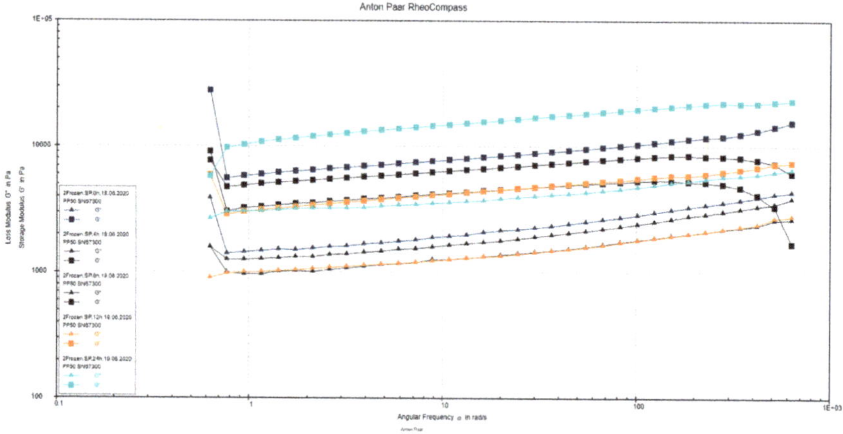

Figure 5. The storage (G′) and loss (G″) shear moduli for SP forzen sourdough at different fermentation times: 0, 4, 8, 12, and 24 h.

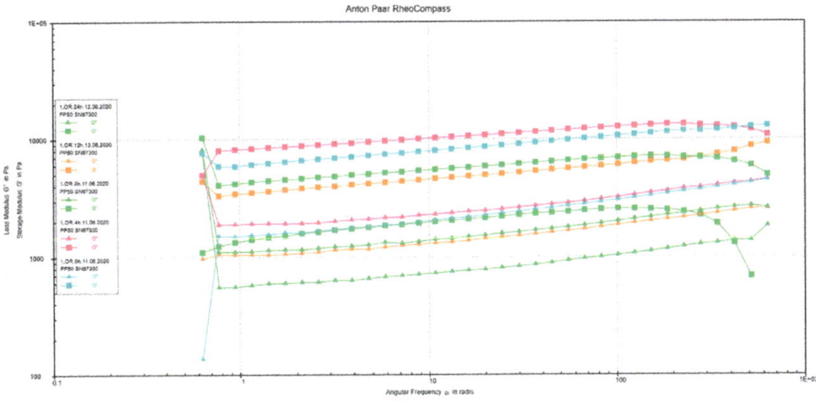

Figure 6. The storage (G′) and loss (G″) shear moduli for OR sourdough at different fermentation times: 0, 4, 8, 12, and 24 h.

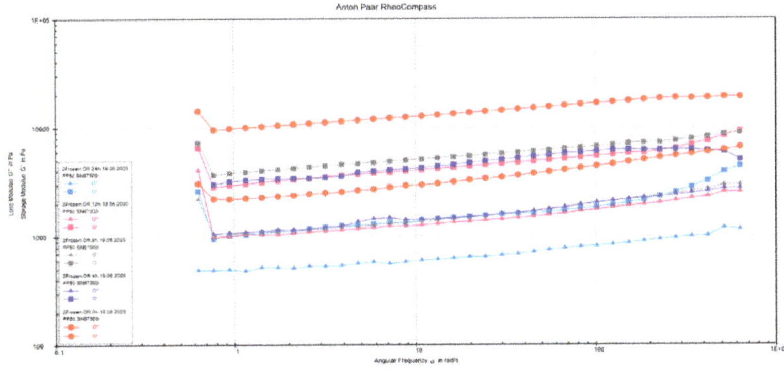

Figure 7. The storage (G′) and loss (G″) shear moduli for OR forzen sourdough at different fermentation times: 0, 4, 8, 12, and 24 h.

The low elastic properties of OR and SP fresh sourdoughs could be attributed to the starch degradation which might occur during fermentation [34]. Bolívar-Monsalve et al. [82] confirmed the amylolytic activity of *Lactobacillus plantarum*, activity which could influence the microstructure of quinoa starch and change the pasting properties of quinoa flour. SP 24 h frozen sample had the highest storage modulus (G′) (23127.00 Pa) and loss modulus (G″) (6574.7 Pa) at a a final angular frequency of 628 rad s^{-1}, while SP 24 h fresh sample had G′ value of 0.06 Pa and G″ value of 1852.2 Pa, respectively, indicating that through freezing the elastic behaviour of quinoa sourdough improved due to the freeze-thaw stability of quinoa starch [50]. Another possible explanation for the G″ improvement through freezing could be the reorganization of hydrogen bonds of amylose and amylopectin during the cooling period [82]. In addition, OR 24 h frozen sample registered values of 4463.80 Pa and 1161.40 Pa for G′ and G″, respectively, while OR 24 h fresh sample had values for G′ and G″ of 0.09 Pa and 1252.4 Pa supporting the idea that through freezing the viscous features of sourdough improved.

Gelation, water-holding, and foaming capacity represent the main technological applications of quinoa flour [52]. Water absorption capacity of quinoa flour is one of the most important physicochemical properties of carbohydrates content being influenced by the intermolecular association between starchy polymers [51]. Furthermore, this characteritics could influence the water loss in pastry or bakery final baked products [52]. Quinoa main carbohydrate is represented by starch having small granules (less than 3 μm in diameter) and higher maximum viscosity and significant swelling power compared with the barley and wheat starches [51]. Compared to the wheat flour, which has a gluten network, the gluten free flours network is mainly influenced by starch properties [84].

Due to its freeze-thaw stability, quinoa starch could be successfully used as an thickener in food manufacturing where resistance to retro degradation is desired. Likewise, quinoa starch is recomended for frozen baby food manufacturing proving good freeze-thawing stability. In addition to starch, dietary fiber (7–9%), such as pectin and xyloglucans, represent another carbohydrate group with importance on the viscosity and stability of the starch paste [50,57]. Moreover, due to its high content in soluble dietary fiber (1.41–2.3 % dry weight), quinoa flour could be used to improve the texture of highly viscous food products such as dough and final baked products [85].

3.3. Carbohydrates, Organic Acids, Folic Acid, Minerals, Flavonoids, Total Phenols Content, Radical Scavenging Activity, of OR PF and SP PF Muffins

3.3.1. Carbohydrates and organic acids content of gluten free muffins (GFM)

The GFM content in carbohydrates and organic acids are presented in Table 5. Glucose, fructose, and maltose amounts decreased in the gluten free muffins manufactured with SP 24 h sourdoughs, due to its lower carbohydrates content (Table 5). This is in line with [1] who reported that the starch utilization by the heterofermentative Labs such as *Lactobacillus plantarum* during quinoa sourdough fermentation improved the rate of starch hydrolysis and decreased the glycemic index in bread final baked product.

Table 5. Carbohydrates and organic acids of gluten free muffins manufactured with OR and SP sourdoughs at different fermentation times.

Sample	Maltose	Glucose	Fructose	Citric Acid	Lactic Acid	Acetic Acid
	mg/g f.w.	mg/g f.w.	mg/g f.w.	mg/g f.w.	mg/g f.w.	mg/g f.w.
OR PF 0 h	4.43 ± 0.03 [Ade]	20.06 ± 0.31 [Ad]	14.72 ± 0.56 [Ade]	1.72 ± 0.05 [Aa]	n.d.	n.d.
SP PF 0 h	4.15 ± 0.23 [Acd]	20.18 ± 0.22 [Ad]	14.17 ± 0.45 [Acd]	1.73 ± 0.03 [Aab]	n.d.	n.d.
OR PF 12 h	4.84 ± 0.23 [Be]	19.58 ± 0.34 [Bd]	15.39 ± 0.39 [Be]	1.72 ± 0.11 [Aa]	0.44 ± 0.02 [Aa]	0.80 ± 0.02 [Ba]
SP PF 12 h	3.14 ± 0.11 [Ab]	17.67 ± 0.67 [Ab]	12.92 ± 0.56 [Ab]	1.86 ± 0.43 [Aac]	0.93 ± 0.05 [Bc]	0.62 ± 0.04 [Aa]
OR PF 24 h	3.82 ± 0.03 [Bc]	18.65 ± 0.55 [Bc]	13.55 ± 0.88 [Bbc]	2.27 ± 0.02 [Ad]	0.82 ± 0.07 [Ab]	1.69 ± 0.09 [Bb]
SP PF 24 h	2.58 ± 0.34 [Aa]	16.37 ± 0.77 [Aa]	10.57 ± 0.65 [Aa]	2.30 ± 0.27 [Ad]	1.52 ± 0.08 [Bd]	1.02 ± 0.09 [Aa]

The results are expressed in mg/100 g fresh weight (f.w.). Small letters in common indicate no significant differences between OR and SP samples withdrawn at different moments; Big different letters indicate significant differences between SP and OR samples withdrawn at the same moment; n.d.: not detected.

The glucose and fructose content of the final baked muffins could be influenced also by maple syrup, which is considered a superior natural sweetener from the chemical point of view, being rich in minerals, flavor compounds, and antioxidant capacity [86]. Sucrose, glucose, and fructose are the main carbohydrates detected in maple syrup range between 61.2 and 65.8%, 0.13 and 0.39%, and 0.07 and 0.27%, respectively. Furthermore, the consumption of maple syrup could produce lower glucose and insulin responses, being considered a successful replacement of refined sugars in human diet [87].

The presence of the organic acids in GFM could be explained by controlled and spontaneous fermentation of SP and OR sourdoughs. The lactic content of SP PF 24 h is higher (statistically different $p < 0.05$) than the amount of OR PF 24 h. This was expected since the initial concentration of lactic acid was statistically different ($p < 0.05$) in SP 24 h sourdough than in OR 24 h sourdough (8.50 mg/100 g f.w., and 5.81 mg/100 g f.w., respectively).

The same trend was observed with respect to the acetic acid content of GFM as the initial content of SP 24 h and OR 24 h were statistically different ($p < 0.05$). Acetic acid is the most promising organic acid involved in the bio-preservation of bakery products [88], having an antifungal effect; meanwhile, lactic acid plays an important role in the storage and safety of the final baked goods [89]. Moreover, lactic acid could positively influence the aroma and also the texture of the final baked goods [53,90] and might also degrade the rate of starch digestion in bakery products [26]. Furthermore, the presence of lactic and acetic acids in the final baked good, formed during sourdough fermentation, has been proved to reduce insulinemic and acute glycemic responses [26].

Citric acid possesses antimicrobial activity and could be produced during fermentation of sourdough with *Lactobacillus plantarum* [45]. Furthermore, the addition of citric acid in the manufacture of baked leaved goods could improve their sensory characteristics, including flavor [91]. In the present study, even if *Lp* was able to increase the amount of citric acid, the differences were not significant ($p < 0.05$). The same trend was noticed in the muffins made with SP 24 h, compared with OR PF 24 h, whose values were 2.20 and 2.27 mg/g f.w., respectively (Table 5).

3.3.2. Folic Acid of Gluten Free Muffins (GFM)

The folic content of GFM manufactured with OR and SP sourdoughs at 0, 12, and 24 h of fermentation are illustrated in Figure 8. The folic content of GFM was influenced by the addition of the OR and SP, GFM produced with SP being statistically different ($p < 0.05$) from GFM produced with sourdough from spontaneous fermentation. This could be explained by the higher content in folic acid in sample SP 24 h (10.48 µg/g f.w) compared with the amount found in OR 24 h sample (3.2 µg/g f.w). This finding is consistent with [79] who proved that the use of sourdough fermented with Lactobacillus strains in bread manufacturing could counteract the thermal loss of bioactive compounds through baking process.

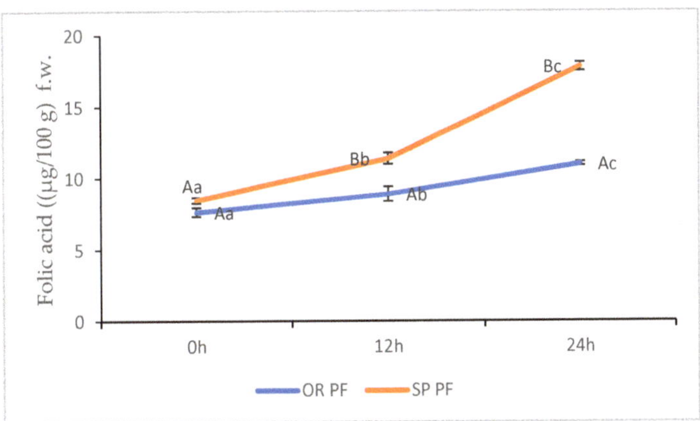

Figure 8. Folic acid content of gluten free muffins. Small letters in common indicate no significant differences between OR and SP samples withdrawn at different moments; Big different letters indicate significant differences between SP and OR samples withdrawn at the same moment.

It is noteworthy to mention that the content of folic acid in final products like bread, noodles, and cookies was improved when QWF was used [48]. Even if, folate is a temperature sensitive vitamin and baking process could diminish its amount, [48] reported a total folate of 17–98 µg/100 g d.m. (dry matter) in noodles, 18–62 µg/100 g d.m. in cookies, and 26–41 µg/100 g d.m. in breads, respectively.

Folate is the generic descriptor for folic acid, used to describe the folic acid and its derivatives, and it is involved in cell essential metabolism function. Folic acid is defined as a chemical form of folate and it is often used in the food fortification process [92]. The lack of folates in human body could lead to the development of different diseases, such as neural tube defects, malformations, megaloblastic anemia, cardiovascular diseases, and could play an important role in lung carcinogenesis [40,92]. Therefore, in the US, the fortification of food with folic acid is mandatory and the daily intake recommendation between 200 and 400 µg [40,92]. Moreover, recently [4] reported a lack of folate content in the diet of children diagnosed with celiac disease highlighting the necessity for mandated gluten free folate food fortification policy.

3.3.3. Minerals Content of Gluten Free Muffins (GFM)

As presented in Table 6, the mineral content of GFM manufactured with SP 24 h are significantly higher than OR 24 h. K, Mg, and Ca were the mainly minerals identified in the SP 24 h GFM; although, Mn, Zn, Fe, and Cu could be identified in smaller amounts.

Table 6. Mineral content of SP PF and OR PF muffins.

Samples	Ca	Mg	K	Fe	Cu	Zn	Mn
OR PF 0 h	8.21 ± 0.03 [Aa]	147.16 ± 0.07 [Aa]	411.67 ± 0.03 [Aa]	0.99 ± 0.02 [Aa]	0.40 ± 0.05 [Aa]	0.90 ± 0.23 [Aa]	0.89 ± 0.02 [Aa]
SP PF 0 h	8.34 ± 0.05 [Aa]	148.32 ± 0.09 [Aa]	410.36 ± 0.29 [Aab]	1.00 ± 0.03 [Aa]	0.41 ± 0.04 [Aa]	0.93 ± 0.02 [Aa]	0.90 ± 0.05 [Aa]
OR PF 12 h	10.30 ± 0.02 [Aab]	151.00 ± 0.07 [Aa]	420.69 ± 0.34 [Aab]	1.03 ± 0.05 [Aa]	0.41 ± 0.01 [Aa]	1.09 ± 0.05 [Ab]	0.99 ± 0.04 [Ab]
SP PF 12 h	12.06 ± 0.01 [Bd]	172.12 ± 0.04 [Bb]	465.24 ± 0.69 [Bc]	1.30 ± 0.06 [Ab]	0.40 ± 0.34 [Ab]	1.34 ± 0.07 [Ac]	1.22 ± 0.45 [Ac]
OR PF 24 h	11.20 ± 0.30 [Ac]	159.81 ± 0.02 [Aab]	430.95 ± 0.72 [Ab]	1.09 ± 0.03 [Aa]	0.45 ± 0.21 [Aab]	1.20 ± 0.11 [Ab]	1.12 ± 0.34 [Ad]
SP PF 24 h	14.25 ± 0.65 [Be]	195.99 ± 0.02 [Bc]	490.20 ± 0.89 [Bd]	1.55 ± 0.01 [Bc]	0.89 ± 0.11 [Bc]	1.75 ± 0.10 [Bd]	1.51 ± 0.22 [Be]

The results are expressed in mg/100 g fresh weight (f.w.). Small letters in common indicate no significant differences between OR and SP samples withdrawn at different moments; Big different letters indicate significant differences between SP and OR samples withdrawn at the same moment.

The final results for macro/micro minerals content demonstrated that through the lactic acid fermentation of a rich source of minerals, the final content of these nutrients was enhanced in the final baked product. These results are in agreement with previous findings such as [1,20].

Quinoa fermentation with Lp provided in the medium a value of pH optimal for enzymatic degradation of phytic acid, leading to final baked muffins enriched in minerals such as K, Ca, Mg, Mn, Fe, Zn, and Cu (Table 6). The successefully use of sourdough aiming to improve the minerals content of the gluten free products was prevously reported by [24,93]

The elimination of gluten involves sometimes the decrease of vitamins, minerals, fibers, and folate. Gluten free products had lower minerals content compared with the conventional ones and their bioavailability could range between 10 and 70% [94]. Furthermore, people who are undergoing a gluten free diet are exposed to mineral and vitamins deficiencies mainly because of their lower content in the final products [95] and due to the presence of phytic acid, an anti-nutritional factor which decreases the bioavailability of minerals such as calcium, magnesium, iron, or zinc [74].

3.3.4. Flavonoids, Total Phenols Content, Radical Scavenging Activity of Gluten Free Muffins (GFM)

The total flavonoids content of GFM is illustrated in Figure 9. In the final baked muffins manufactured with SP 24 h, a total content of 1561 mg Qe/100 g f.w. was determined, compared with OR PF 24 h, where flavonoids had a total amount of 1317 mg Qe/100 g f.w. The difference between the samples could be explained by the presence of SP or OR 24 h sourdoughs. Nonetheless, it is noteworthy to mention that [96] identified in buckwheat flour a total of 188 flavonoid metabolites that could positively influence the content of flavonoids in the muffins. On the other side, the thermal treatment of the final baked goods could have a negative influence on the total flavonoids content [97].

The total phenols (TF) and radical scavenging activity (RSA%) of gluten free muffins made with SP sourdough were significantly different to those made with OR sourdough (Figure 10) due to the capacity of Lp to produce higher extent of lactic acid amount in SP sample, that could influence through acidification the extractability of total phenols and the antioxidant potential. Moreover, the interaction of Folin–Ciolcateu reagent with other non-phenolic compounds like vitamins, amino acids, and proteins could also have an influence on the number of polyphenolic compounds. On the other side, the lower OR 24 h total phenols and radical scavenging activity suggested that spontaneous fermentation and quinoa endogenous enzymes were not able to decrease the pH and release antioxidant compounds [98].

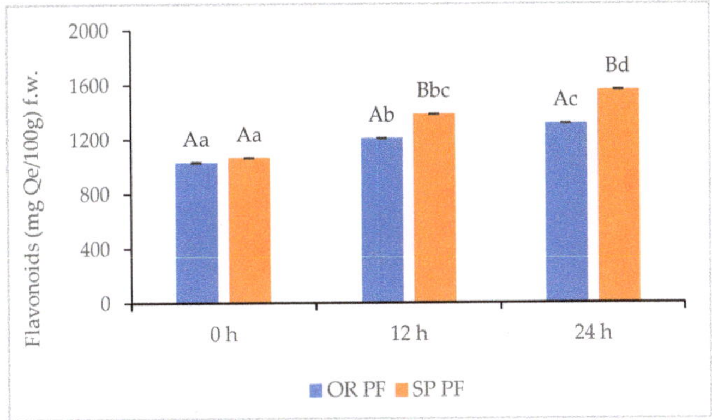

Figure 9. Total flavonoids of gluten free muffins. Small letters in common indicate no significant differences between OR and SP samples withdrawn at different moments; Big different letters indicate significant differences between SP and OR samples withdrawn at the same moments.

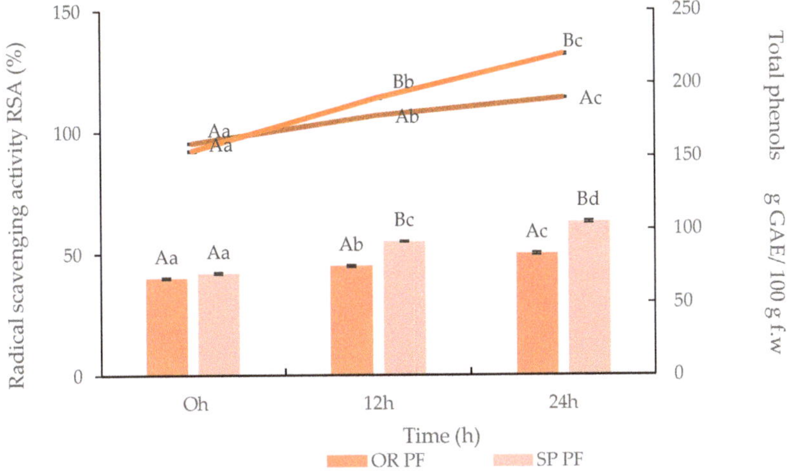

Figure 10. Total phenols and radical scavenging activity of gluten free muffins. Small letters in common indicate no significant differences between OR and SP samples withdrawn at different moments; Big different letters indicate significant differences between SP and OR samples withdrawn at the same moment.

It is important to mention that the TF amount of quinoa flour and RSA were 451 ± 0.3 mg GAE/100 g f.w. and 92 ± 0.5%, respectively [38]. The total phenols amount could vary between samples as reported by [50] who proved that QF from Chile had significant higher polyphenols content compared with the flour from Mexico (319 mg GAE/100 g and 180.4 mg GAE/100 g, respectively) and could be justified by the differences in the polyphenol extraction process [47].

The TF value is close to the value reported by [99] as 464 mg GAE/100 g. With respect to RSA, other studies reported a value of it up to 71.8% [1]. The differences between the results could be due to the different extraction conditions such as the extraction solvent and due to the duration of the extraction that could influence the total phenols content [100].

Through the quinoa fermentation with Lp, the TF and RSA content improved up to 350 mg GAE/100 g f.w. and 94%, respectively [38]. This idea is supported by [98] who indicated that fermentation of quinoa flour with LAB could lead to an improvement of the antioxidant activity. Briefly,

Rizello et al. indicated that Lb plantarum T6A10 strain was able to increase the radical scavenging activity from 32.7 to 84.8% during 24 h of fermentation, due to its ability to release peptides with antioxidant activity during controlled fermentation through proteolysis.

Overall, it can be stated that SP and OR 24 h sourdoughs had an important influence on the final amounts of flavonoids, TF, and RSA of SP PF 24 h and OR PF 24 h, increasing their amounts and highlighting the idea that fermentation of quinoa with Lp is a valuable source for exploiting its properties.

4. Conclusions

Fermentation of quinoa flour with *Lactobacillus plantarum* ATCC (*Lp*) 8014 leads to a sourdough nutritional enrichment and to an improvement of its rheological features. Briefly, glucose, maltose, and fructose were metabolized by *Lp*, enhancing lactic acid content during 24 h of fermentation; meanwhile, acetic acid was produced through the pentose phosphate pathway. The production of acids and the drop of the pH lead further to a bigger bioavailability of minerals, increasing their values at least twice. From the rheological point of view, the viscous properties of the *Lp* sourdough was improved, probably due to the production of exopolysaccharides and 12 β-glucan by *Lp*. Furthermore, the use of *Lp* sourdough in the final baked muffins improved their nutritional features such as folic acid, minerals, flavonoids, total phenols, and radical scavenging activity. The decrease of carbohydrates such as maltose, glucose, and fructose enhanced the presence of organic acids in the final leavened goods.

Lactobacillus plantarum ATCC 8014 and quinoa flour represent an optimum combination that needs to be explored further for the manufacturing of gluten free products.

Author Contributions: Conceptualization, M.S.C. and A.P.; methodology, L.S., S.M.M., S.M. and A.P.; software, M.S.C., O.B.; validation, M.S.C., L.S. and A.P.; formal analysis, M.S.C., B.-E.T., C.R.P., A.C.U. and C.B.K.; writing—original draft preparation, M.S.C. and A.P.; writing—review and editing, A.P. and L.S.; supervision, S.M.; project administration, M.S.C. and S.M.; funding acquisition, D.C.V. and A.P. All authors have read and agreed to the published version of the manuscript.

Funding: This research was funded by POC-A1-A1.1.1.-B-2015 and the publication was supported by National Research Development Projects to finance excellence (PFE)-37/2018–2020 granted by the Romanian Ministry of Research and Innovation.

Conflicts of Interest: The authors declare no conflict of interest.

References

1. Rizzello, C.G.; Lorusso, A.; Montemurro, M.; Gobbetti, M. Use of sourdough made with quinoa (*Chenopodium quinoa*) flour and autochthonous selected lactic acid bacteria for enhancing the nutritional, textural and sensory features of white bread. *Food Microbiol.* **2016**, *56*, 1–13. [CrossRef] [PubMed]

2. Kårlund, A.; Gómez-Gallego, C.; Korhonen, J.; Palo-Oja, O.M.; El-Nezami, H.; Kolehmainen, M. Harnessing microbes for sustainable development: Food fermentation as a tool for improving the nutritional quality of alternative protein sources. *Nutrients* **2020**, *12*, 1020. [CrossRef]

3. Chiş, M.S.; Păucean, A.; Stan, L.; Suharoschi, R.; Socaci, S.A.; Man, S.M.; Pop, C.R.; Muste, S. Protein metabolic conversion of nutritional features during quinoa sourdough fermentation and its impact on baked goods. *CyTA J. Food* **2018**, *280*, 744–753.

4. Cyrkot, M.; Anders, S.; Kamprath, C.; Liu, A.; Mileski, H.; Dowhaniuk, J.; Nasser, R.; Marcon, M.; Brill, H.; Turner, J.M.; et al. Folate content of gluten-free food purchases and dietary intake are low in children with coeliac disease. *Int. J. Food Sci. Nutr.* **2020**, *7486*, 1–13. [CrossRef] [PubMed]

5. El-Sohaimy, S.A.; Shehata, M.G.; Mehany, T.; Zeitoun, M.A. Nutritional, physicochemical, and sensorial evaluation of flat bread supplemented with quinoa flour. *Hindawi Int. J. Food Sci.* **2019**, *2019*, 1–15. [CrossRef]

6. Pappier, U.; Pinto, V.F.; Larumbe, G.; Vaamonde, G. Effect of processing for saponin removal on fungal contamination of quinoa seeds (*Chenopodium quinoa* Willd.). *Int. J. Food Microbiol.* **2008**, *125*, 153–157. [CrossRef] [PubMed]

7. Ceballos-Gonzalez, C.; Bolívar-Monsalve, J.; Ramírez-Toro, C.; Bolívar, G.A. Effect of lactic acid fermentation on quinoa dough to prepare gluten-free breads with high nutritional and sensory quality. *J. Food Process. Preserv.* **2017**, *42*, e13551. [CrossRef]

8. Vega-Galvez, A.; Miranda, M.; Vergara, J.; Uribe, E.; Puente, L.; Martinez, E.A. Nutrition facts and functional potential of quinoa (*Chenopodium quinoa* Willd.), an ancient Andean grain: A review. *J. Sci. Food Agric.* **2010**, *90*, 2541–2547. [CrossRef]

9. Nascimento, A.C.; Mota, C.; Coelho, I.; Gueifao, S.; Santos, M.; Matos, A.S.; Gimenez, A.; Lobo, M.; Samman, N.; Castanheira, I. Characterisation of nutrient profile of quinoa (*Chenopodium quinoa*), amaranth (*Amaranthus caudatus*), and purple corn (*Zea mays* L.) consumed in the North of Argentina: Proximates, minerals and trace elements. *Food Chem.* **2014**, *148*, 420–426. [CrossRef] [PubMed]

10. Repo-Carrasco-Valencia, R.; Hellström, J.K.; Pihlava, J.M.; Mattila, P.H. Flavonoids and other phenolic compounds in Andean indigenous grains: Quinoa (*Chenopodium quinoa*), kañiwa (*Chenopodium pallidicaule*) and kiwicha (*Amaranthus caudatus*). *Food Chem.* **2010**, *120*, 128–133. [CrossRef]

11. Alvarez-Jubete, L.; Wijngaard, H.; Arendt, E.K.; Gallagher, E. Polyphenol composition and in vitro antioxidant activity of amaranth, quinoa buckwheat and wheat as affected by sprouting and baking. *Food Chem.* **2010**, *119*, 770–778. [CrossRef]

12. Gómez-Caravaca, A.M.; Segura-Carretero, A.; Fernández-Gutiérrez, A.; Caboni, M.F. Simultaneous determination of phenolic compounds and saponins in quinoa (*Chenopodium quinoa* Willd.) by a liquid chromatography-diode array detection-electrospray ionization-time-of-flight mass spectrometry methodology. *J. Agric. Food Chem.* **2011**, *59*, 10815–10825. [CrossRef]

13. Kumpun, S.; Maria, A.; Crouzet, S.; Evrard-Todeschi, N.; Girault, J.P.; Lafont, R. Ecdysteroids from *Chenopodium quinoa* Willd., an ancient Andean crop of high nutritional value. *Food Chem.* **2011**, *125*, 1226–1234. [CrossRef]

14. Miranda, M.; Vega-Gálvez, A.; Uribe, E.; Lópeza, J.; Martínez, E.; Rodrígueza, M.J.; Quispea, I.; Di Scalac, K. Physicochemical analysis, antioxidant capacity and vitamins of six ecotypes of Chilean quinoa (*Chenopodium quinoa* Willd). *Procedia Food Sci.* **2011**, *1*, 1439–1446. [CrossRef]

15. Wang, S.; Zhu, F. Formulation and quality attributes of quinoa food products. *Food Bioprocess Technol.* **2015**, *9*, 1–20. [CrossRef]

16. Chillo, S.; Laverse, J.; Falcone, P.M.; Del Nobile, M.A. Quality of spaghetti in base amaranthus wholemeal flour added with quinoa, broad bean and chick pea. *J. Food Eng.* **2008**, *84*, 101–107. [CrossRef]

17. Lorusso, A.; Verni, M.; Montemurro, M.; Coda, R.; Gobbetti, M.; Rizzello, C.G. Use of fermented quinoa flour for pasta making and evaluation of the technological and nutritional features. *LWT Food Sci. Technol.* **2017**, *78*, 215–221. [CrossRef]

18. Stikic, R.; Glamoclija, D.; Demin, M.; Vucelic-Radovic, B.; Jovanovic, Z.; Milojkovic-Opsenica, D.; Jacobsen, S.E.; Milovanovic, M. Agronomical and nutritional evaluation of quinoa seeds (*Chenopodium quinoa* Willd.) as an ingredient in bread formulations. *J. Cereal Sci.* **2012**, *55*, 132–138. [CrossRef]

19. Rizzello, C.G.; Tagliazucchi, D.; Babini, E.; Sefora Rutella, G.; Taneyo Saa, D.L.; Gianotti, A. Bioactive peptides from vegetable food matrices: Research trends and novel biotechnologies for synthesis and recovery. *J. Funct. Foods* **2016**, *27*, 549–569. [CrossRef]

20. Wolter, A.; Hager, A.S.; Zannini, E.; Czerny, M.; Arendt, E.K. Impact of sourdough fermented with *Lactobacillus plantarum* FST 1.7 on baking and sensory properties of gluten-free breads. *Eur. Food Res. Technol.* **2014**, *239*, 1–12. [CrossRef]

21. Gallagher, E.; Gormley, T.R.; Arendt, E.K. Crust and crumb characteristics of gluten-free breads. *J. Food Eng.* **2003**, *56*, 153–161. [CrossRef]

22. Gobbetti, M.; Rizzello, C.G.; Di Cagno, R.; De Angelis, M. How the sourdough may affect the functional features of leavened baked goods. *Food Microbiol.* **2014**, *37*, 30–40. [CrossRef] [PubMed]

23. Coda, R.; Cagno, R.D.; Gobbetti, M.; Rizzello, C.G. Sourdough lactic acid bacteria: Exploration of non-wheat cereal-based fermentation. *Food Microbiol.* **2014**, *37*, 51–58. [CrossRef] [PubMed]

24. Gobbetti, M.; De Angelis, M.; Corsetti, A.; Di Cagno, R.; Calasso, R.; Archetti, G.; Rizzello, C.G. Novel insights on the functional/nutritional features of the sourdough fermentation. *Int. J. Food Microbiol.* **2019**, *302*, 1–11. [CrossRef] [PubMed]

25. Arendt, E.K.; Moroni, A.; Zannini, E. Medical nutrition therapy: Use of sourdough lactic acid bacteria as a cell factory for delivering functional biomolecules and food ingredients in gluten free bread. *Microb. Cell Fact.* **2011**, *10* (Suppl. 1), 1–9. [CrossRef]

26. Zannin, E.; Pontonio, E.; Waters, D.M.; Arendt, E.K. Applications of microbial fermentations for production of gluten-free products and perspectives. *Appl. Microbiol. Biotechnol.* **2012**, *93*, 473–485. [CrossRef]

27. Arena, M.P.; Russo, P.; Spano, G.; Capozzi, V. From microbial ecology to innovative applications in food quality improvements: The case of sourdough as a model matrix. *J. Multidiscip. Sci. J.* **2020**, *3*, 3. [CrossRef]

28. Axel, C.; Brosnan, B.; Zannini, E.; Peyer, L.C.; Furey, A.; Coffey, A.; Arendt, E.K. Antifungal activities of three different *Lactobacillus* species and their production of antifungal carboxylic acids in wheat sourdough. *Appl. Microbiol. Biotechnol.* **2016**, *100*, 1701–1711. [CrossRef]

29. Păucean, A.; Vodnar, D.C.; Socaci, S.A.; Socaciu, C. Carbohydrate metabolic conversions to lactic acid and volatile derivatives, as influenced by *Lactobacillus plantarum* ATCC 8014 and Lactobacillus casei ATCC 393 efficiency during in vitro and sourdough fermentation. *Eur. Food Res. Technol.* **2013**, *237*, 679–689. [CrossRef]

30. Salvetti, A.; Torriani, S.; Felis, G.E. The genus Lactobacillus: A taxonomic update. *Probiotics Antimicrob. Proteins* **2012**, *4*, 217–226. [CrossRef]

31. Hammes, W.P.; Vogel, R.F. The Genus Lactobacillus. In *The Genera of Lactic Acid Bacteria*; Wood, B.J.B., Holzapfel, W.H., Eds.; Blackie Academic & Professional: London, UK, 1995; pp. 19–54.

32. Corsetti, A.; Settanni, L. Lactobacilli in sourdough fermentation. *Food Res. Int.* **2007**, *40*, 539–558. [CrossRef]

33. De Vuyst, L.; Neysens, P. The sourdough microflora: Biodiversity and metabolic interactions. *Trends Food Sci. Technol.* **2005**, *16*, 43–56. [CrossRef]

34. Olojede, A.O.; Sanni, A.I.; Banwo, K. Rheological, textural and nutritional properties of gluten-free sourdough made with functionally important lactic acid bacteria and yeast from Nigerian sorghum. *LWT* **2020**, *120*, 1–8. [CrossRef]

35. Di Cagno, R.; Rizzello, C.G.; De Angelis, M.; Cassone, A.; Giuliani, G.; Benedusi, A.; Limitone, A.; Surico, M.F.; Gobbetti, M. Use of selected sourdough strains of Lactobacillus for removing gluten and enhancing the nutritional properties of gluten-free bread. *J. Food Prot.* **2008**, *71*, 1491–1495. [CrossRef] [PubMed]

36. Vogelmann, S.A.; Seitter, M.; Singer, U.; Brandt, M.J.; Hertel, C. Adaptability of lactic acid bacteria and yeasts to sourdoughs prepared from cereals, pseudocereals and cassava and use of competitive strains as starters. *Int. J. Food Microbiol.* **2009**, *130*, 205–212. [CrossRef]

37. Moroni, A.V.; Arendt, E.K.; Bello, F.D. Biodiversity of lactic acid bacteria and yeasts in spontaneously-fermented buckwheat and teff sourdoughs. *Food Microbiol.* **2011**, *28*, 497–502. [CrossRef]

38. Chiş, M.S.; Păucean, A.; Stan, L.; Mureşan, V.; Vlaic, R.A.; Man, S.; Biriş-Dorhoi, S.E.; Muste, S. Lactobacillus plantarum ATCC 8014 in quinoa sourdough adaptability and antioxidant potential. *Rom. Biotechnol. Lett.* **2018**, *23*, 13581–13591.

39. Chiş, M.S.; Păucean, A.; Man, S.M.; Bonta, V.; Pop, A.M.; Stan, L.; Beldean, B.V.; Pop, C.R.; Mureşan, V.; Muste, S. Effect of rice flour fermentation with *Lactobacillus spicheri* DSM 1549 on the nutritional features of gluten free muffins. *Foods* **2020**, *9*, 822. [CrossRef]

40. Păucean, A.; Moldovan, O.P.; Mureşan, V.; Socaci, S.A.; Dulf, F.; Man, M.S.; Mureşan, A.E.; Muste, S. Folic acid, minerals, amino-acids, fatty acids and volatile compounds of green and red lentils. Folic acid content optimization in wheat-lentils composite flours. *Chem. Cent. J.* **2018**, *12*, 1–9.

41. Urcan, A.D.; Criste, A.D.; Dezmirean, D.S.; Mărgăoan, R.; Caeiro, A.; Campos, M.G. Similarity of data from bee bread with the same taxa collected in India and Romania. *Molecules* **2018**, *23*, 2491. [CrossRef]

42. Bunea, A.; Rugină, D.O.; Pintea, A.M.; Sconţa, Z.; Bunea, C.I.; Socaciu, C. Comparative polyphenolic content and antioxidant activities of some wild and cultivated blueberries from Romania. *Not. Bot. Horti Agrobot.* **2011**, *39*, 70–76. [CrossRef]

43. Păucean, A.; Man, S.M.; Chis, M.S.; Mureşan, V.; Pop, C.R.; Socaci, S.A.; Mureşan, C.C.; Muste, S. Use of pseudocereals preferment made with aromatic yeast strains for enhancing wheat bread quality. *Foods* **2019**, *8*, 443. [CrossRef] [PubMed]

44. Dordević, T.M.; Šiler-Marinković, S.S.; Dimitrijević-Branković, S.I. Effect of fermentation on antioxidant properties of some cereals and pseudo cereals. *Food Chem.* **2010**, *119*, 957–963. [CrossRef]

45. Teleky, B.E.; Martău, A.G.; Ranga, F.; Cheţan, F.; Vodnar, D.C. Exploitation of lactic acid bacteria and Baker's yeast as single or multiple starter cultures of wheat flour dough enriched with soy flour. *Biomolecules* **2020**, *10*, 778. [CrossRef] [PubMed]

46. Schoenlechner, R. Quinoa: Its Unique Nutritional and Health-Promoting Attributes. In *Gluten-Free Ancient Grains. Cereal, Pseudocereals and Legumes: Sustainaible, Nutritious and Health-Promoting Foods for the 21st Century*, 1st ed.; Taylor, J., Awika, J.M., Eds.; Woodhead Publishing: Duxford, UK, 2017; pp. 105–129.

47. Pellegrini, M.; Lucas-Gonzales, R.; Ricci, A.; Fontecha, J.; Fernandez-Lopez, J.; Perez-Alvarez, J.A.; Viuda-Martos, M. Chemical, fatty acid, polyphenolic profile, techno-functional and antioxidant properties of flours obtained from quinoa (*Chenopodium quinoa* Willd) seeds. *Ind. Crop. Prod.* **2018**, *111*, 38–46. [CrossRef]

48. Pasko, P.; Barton, H.; Zagrodzki, P.; Izewska, A.; Krosniak, M.; Gawlik, M.; Gawlik, M.; Gorinstein, S. Effect of diet supplemented with quinoa seeds on oxidative status in plasma and selected tissues of high fructose-fed rats. *Plant Foods Hum. Nutr.* **2010**, *65*, 146–151. [CrossRef] [PubMed]

49. Pereira, E.; Encina-Zelada, C.; Barros, L.; Gonzales-Barron, U.; Cadavez, V.; Ferreira, C.F.R. Chemical and nutritional characterization of *Chenopodium quinoa* Willd (quinoa) grains: A good alternative to nutritious food. *Food Chem.* **2019**, *280*, 110–114. [CrossRef] [PubMed]

50. Angeli, M.; Silva, P.M.; Massuela, C.D.; Whan, M.W.; Hamar, A.; Khajehei, F.; Graeff-Hönninge, S.; Piatti, C. Quinoa (*Chenopodium quinoa* Willd.): An overview of the potentials of the "Golden Grain" and socio-economic and environmental aspects of its cultivation and marketization. *Foods* **2020**, *9*, 216. [CrossRef]

51. Vázquez-Luna, A.; Fuentes, F.; Rivadeneyra, E.; Hernández, C.; Díaz-Sobac, R. Nutrimental content and functional properties of Quinoa flour from Chile and Mexico. *Cienc. E Investig. Agrar.* **2019**, *46*, 144–153. [CrossRef]

52. Romano, N.; Ureta, M.M.; Guerrero-Sánchez, M.; Gómez-Zavaglia, A. Nutritional and technological properties of a quinoa (*Chenopodium quinoa* Willd.) spray-dried powdered extract. *Food Res. Int.* **2020**, *129*, 1–11. [CrossRef]

53. Franco, W.; Pérez-Díaz, I.M.; Connelly, L.; Diaz, J.T. Isolation of exopolysaccharide-producing yeast and lactic acid bacteria from quinoa (*Chenopodium quinoa*) sourdough fermentation. *Foods* **2020**, *9*, 337. [CrossRef] [PubMed]

54. Saubade, F.; Hemery, Y.M.; Rochette, I.; Guyot, J.P.; Humblot, C. Influence of fermentation and other processing steps on the folate content of a traditional African cereal-based fermented food. *Int. J. Food Microbiol.* **2018**, *266*, 79–86. [CrossRef]

55. Bilgiçl, N.; İbanoğlu, Ş. Effect of pseudo cereal flours on some physical, chemical and sensory properties of bread. *J. Food Sci. Technol.* **2015**, *52*, 7525–7529. [CrossRef]

56. Schoenlechner, R.; Wendner, M.; Siebenhandl-Ehn, S.; Berghofer, E. Pseudocereals as alternative sources for high folate content in staple foods. *J. Cereal Sci.* **2010**, *52*, 475–479. [CrossRef]

57. Li, G.; Zhu, F. Physicochemical properties of quinoa flour as affected by starch interactions. *Food Chem.* **2017**, *221*, 1560–1568. [CrossRef] [PubMed]

58. Iglesias-Puig, E.; Monedero, V.; Haros, M. Bread with whole quinoa flour and bifidobacterial phytases increases dietary mineral intake and bioavailability. *LWT Food Sci. Technol.* **2015**, *60*, 71–77. [CrossRef]

59. Valcárcel-Yamani, B.; Caetano da Silva Lannes, S. Applications of Quinoa (*Chenopodium quinoa* Willd) and amaranth (*Amaranthus* spp.) and their influence in the nutritional value of cereal based foods. *Food Public Health* **2012**, *2*, 265–275.

60. Prado, F.E.; Fernández-Turiel, J.L.; Tsarouchi, M.; Psaras, G.K.; González, J.A. Variation of seed mineral concentrations in seven quinoa cultivars grown in two agroecological sites. *Cereal Chem.* **2014**, *91*, 453–459. [CrossRef]

61. Gonzalez, J.A.; Konishi, Y.; Bruno, M.; Valoy, M.; Prado, F.E. Interrelationships among seed yield, total protein and amino acid composition of ten quinoa (*Chenopodium quinoa*) cultivars from two different agroecological regions. *J. Sci. Food Agric.* **2012**, *92*, 1222–1229. [CrossRef]

62. Carciochi, R.A.; Dimitrov, K. Optimization of antioxidant phenolic compounds extraction from quinoa (*Chenopodium quinoa*) seeds. *J. Food Sci. Technol.* **2014**, *52*, 4396–4404. [CrossRef]

63. De Carvalho, F.G.; Ovídio, P.P.; Padovan, G.J.; Jordão Junior, A.A.; Marchini, J.S.; Navarro, A.M. Metabolic parameters of postmenopausal women after quinoa or corn flakes intake-a prospective and double-blind study. *Int. J. Food Sci. Nutr.* **2014**, *65*, 380–385. [CrossRef] [PubMed]

64. Arneja, I.; Tanwar, B.; Chauhan, A. Nutritional composition and health benefits of golden grain of 21st century, quinoa (*Chenopodium quinoa* willd.): A review. *Pak. J. Nutr.* **2015**, *14*, 1034–1040. [CrossRef]

65. De Vuyst, L.; Vrancken, G.; Ravyts, F.; Rimaux, T.; Weckx, S. Biodiversity, ecological determinants, and metabolic exploitation of sourdough microbiota. *Food Microbiol.* **2009**, *26*, 666–675. [CrossRef] [PubMed]

66. Montemurro, M.; Pontonio, E.; Gobbetti, M.; Rizzello, C.G. Investigation of the nutritional, functional and technological effects of the sourdough fermentation of sprouted flour. *Int. J. Food Microbiol.* **2018**, *302*, 1–7. [CrossRef] [PubMed]

67. Von Wright, A.; Axerlsson, L. Lactic Acid Bacteria: An Introduction. In *Lactic Acid Bacteria: Microbiological and Functional Aspects*, 4th ed.; Lahtinen, S., Ouwehand, A.C., Salminen, S., Von Wright, A., Eds.; CRC Press, Taylor & Francis Group: Boca Raton, FL, USA; London, UK, 2011; pp. 1–17.

68. Petrova, P.; Petrov, K. Lactic acid fermentation of cereals and pseudocereals: Ancient nutritional biotechnologies with modern applications. *Nutrients* **2020**, *12*, 1118. [CrossRef] [PubMed]

69. Carrizo, S.L.; de Moreno de LeBlanc, A.; LeBlanc, J.G.; Rollán, G.C. Quinoa pasta fermented with lactic acid bacteria prevents nutritional deficiencies in mice. *Food Res. Int.* **2020**, *127*, 1–12. [CrossRef]

70. Florou-paneri, P.; Christaki, E.; Bonos, E. Lactic Acid Bacteria as Source of Functional Ingredients. In *Lactic Acid Bacteria*, 1st ed.; Kongo, M., Ed.; Intech Open Limited: Rijeka, Croatia, 2013; pp. 1–26.

71. Leblanc, J.G.; Savoy de Giori, G.; Smid, E.J.; Hugenholtz, J.; Sesma, F. Folate Production by Lactic Acid Bacteria and Other Food-Grade Microorganisms. In *Bioactive Foods as Dietary Interventions for Liver and Gastrointestinal Disease*, 1st ed.; Watson, R.R., Preedy, V., Eds.; Academis Press: Cambridge, MA, USA, 2007; pp. 329–339.

72. Carrizo, S.L.; Montes de Oca, C.E.; Laiño, J.E.; Suarez, N.E.; Vignolo, G.; LeBlanc, J.G.; Rollán, G. Ancestral Andean grain quinoa as source of lactic acid bacteria capable to degrade phytate and produce B-group vitamins. *Food Res. Int.* **2016**, *89*, 488–494. [CrossRef]

73. Motta, C.; Delgado, I.; Matos, A.S.; Gonzales, G.B.; Torres, D.; Santos, M.; Chandra-Hioe, M.V.; Arcot, J.; Castanheira, I. Folates in quinoa (*Chenopodium quinoa*), amaranth (*Amaranthus* sp.) and buckwheat (*Fagopyrum esculentum*): Influence of cooking and malting. *J. Food Compos. Anal.* **2017**, *64*, 181–187. [CrossRef]

74. LeBlanc, J.G.; Chain, F.; Martín, R.; Bermúdez-Humarán, L.G.; Courau, S.; Langella, P. Beneficial effects on host energy metabolism of short-chain fatty acids and vitamins produced by commensal and probiotic bacteria. *Microb. Cell Fact.* **2017**, *16*, 1–10. [CrossRef] [PubMed]

75. Salvucci, E.; Leblanc, J.G.; Perez, G. Technological properties of Lactic Acid Bacteria isolated from raw cereal material. *LWT Food Sci. Technol.* **2016**, *70*, 1–26. [CrossRef]

76. Taylor, J.R.N.; Belton, P.S.; Beta, T.; Duodu, K.G. Increasing the utilisation of sorghum, millets and pseudocereals: Developments in the science of their phenolic phytochemicals, biofortification and protein functionality. *J. Cereal Sci.* **2014**, *59*, 257–275. [CrossRef]

77. Magala, M.; Kohajdová, Z.; Karovičová, J. Degradation of phytic acid during fermentation of cereal substrates. *J. Cereal Sci.* **2015**, *61*, 94–96. [CrossRef]

78. Lopez, H.W.; Duclos, V.; Coudray, C.; Krespine, V.; Feillet-Coudray, C.; Messager, A.; Demigne', C.; Remesy, C. Making bread with sourdough improves mineral bioavailability from reconstituted whole wheat flour in rats. *Nutrition* **2003**, *19*, 524–530. [CrossRef]

79. Taylor, J.R.N.; Parker, M.L. Quinoa. In *Pseudocereals and Less Common Cereals, Grain Properties and Utilization Potential*, 1st ed.; Belton, S., Taylor, J.R.N., Eds.; Springer: Berlin/Heidelberg, Germany, 2002; pp. 93–122.

80. Saa, D.T.; Di Silvestro, R.; Dinelli, G.; Gianotti, A. Effect of sourdough fermentation and baking process severity on dietary fibre and phenolic compounds of immature wheat flour bread. *LWT Food Sci. Technol.* **2017**, *83*, 26–32. [CrossRef]

81. Montemurro, M.; Coda, R.; Rizzello, C.G. Recent advances in the use of Sourdough. *Foods* **2019**, *8*, 129. [CrossRef] [PubMed]

82. Bolívar-Monsalve, J.; Ceballos-González, C.; Ramírez-Toro, C.; Bolívar, G.A. Reduction in saponin content and production of gluten-free cream soup base using quinoa fermented with *Lactobacillus plantarum*. *J. Food Process. Preserv.* **2018**, *42*, 1–10. [CrossRef]

83. Sun, L.; Li, X.; Zhang, Y.; Yang, W.; Ma, G.; Ma, N.; Hu, Q.; Pei, F. A novel lactic acid bacterium for improving the quality and shelf life of whole wheat bread. *Food Control* **2020**, *109*, 1–9. [CrossRef]

84. Wolter, A.; Hager, A.S.; Zannini, E.; Czerny, M.; Arendt, E.K. Influence of dextran-producing *Weissella cibaria* on baking properties and sensory profile of gluten-free and wheat breads. *Int. J. Food Microbiol.* **2014**, *172*, 83–91. [CrossRef]

85. Turkut, G.M.; Cakmak, H.; Kumcuoglu, S.; Tavman, S. Effect of quinoa flour on gluten-free bread batter rheology and bread quality. *J. Cereal Sci.* **2016**, *69*, 174–181. [CrossRef]

86. Mellado-Mojica, E.; Seeram, N.P.; López, M.G. Comparative analysis of maple syrups and natural sweeteners: Carbohydrates composition and classification (differentiation) by HPAEC-PAD and FTIR spectroscopy-chemometrics. *J. Food Compos. Anal.* **2016**, *52*, 1–8. [CrossRef]

87. Nimalaratne, C.; Blackburn, J.; Lada, R.R. A comparative physicochemical analysis of maple (*Acer saccharum* Marsh.) syrup produced in North America with special emphasis on seasonal changes in Nova Scotia maple syrup composition. *J. Food Compos. Anal.* **2020**, *92*, 1–31. [CrossRef]

88. Debonne, E.; Vermeulen, A.; Bouboutiefskia, N.; Ruyssen, T.; Van Bockstaeled, F.; Eeckhout, M.; Devlieghere, F. Modelling and validation of the antifungal activity of DL-3-phenyllactic acid and acetic acid on bread spoilage moulds. *Food Microbiol.* **2020**, *88*, 1–10. [CrossRef] [PubMed]

89. Papadimitriou, K.; Zoumpopoulou, G.; Georgalaki, M.; Alexandraki, V.; Kazou, M.; Anastasiou, R.; Tsakalidou, E. *Sourdough Bread, Innovations in Traditional Foods*; Galanakis, C.M., Ed.; Woodhead Publishing: Cambridge, UK, 2019; pp. 127–158.

90. Reddy, G.; Altaf, M.; Naveena, B.J.; Venkateshwar, M.; Kumar, E.V. Amylolytic bacterial lactic acid fermentation—A review. *Biotechnol. Adv.* **2008**, *26*, 22–34. [CrossRef] [PubMed]

91. Kefalas, P.; Kotzamanidis, S.; Sabanis, D.; Yupsani, A.; Kefala, L.A.; Kokkalis, A.; Yupsanis, T. Bread making of durum wheat with chickpea sourdough or compressed baker's yeast. *J. Food. Qual.* **2009**, *32*, 644–668. [CrossRef]

92. Takata, Y.; Shu, X.O.; Buchowski, M.S.; Munro, H.M.; Wen, W.; Steinwandel, M.D.; Hargreaves, M.K.; Blot, W.J.; Cai, Q. Food intake of folate, folic acid and other B vitamins with lung cancer risk in a low-income population in the Southeastern United States. *Eur. J. Nutr.* **2020**, *59*, 671–683. [CrossRef]

93. Chis, M.S.; Păucean, A.; Man, S.M.; Muresan, V.; Socaci, S.A.; Pop, A.; Stan, L.; Rusu, B.; Muste, S. Textural and sensory features changes of gluten free muffins based on rice sourdough fermented with *Lactobacillus spicheri* DSM 15429. *Foods* **2020**, *9*, 363. [CrossRef]

94. Naqash, F.; Gani, A.; Gani, A.; Masoodi, F.A. Gluten-Free baking: Combating the challenges—A review. *Trends Food Sci. Technol.* **2017**, *66*, 98–107. [CrossRef]

95. Capriles, V.D.; dos Santos, F.G.; Arêas, J.A.G. Gluten-Free breadmaking: Improving nutritional and bioactive compounds. *J. Cereal Sci.* **2016**, *67*, 83–91. [CrossRef]

96. Li, L.; Shakhawat Hossain, M.D.; Ma, H.; Yang, Q.; Gong, X.; Yang, P.; Feng, B. Comparative metabolomics reveals diFferences in flavonoid metabolites among different coloured buckwheat flowers. *J. Food Compos. Anal.* **2019**, *85*, 1–8. [CrossRef]

97. Chlopicka, J.; Pasko, P.; Gorinstein, S.; Jedryas, A.; Zagrodzki, P. Total phenolic and total flavonoid content, antioxidant activity and sensory evaluation of pseudocereal breads. *LWT Food Sci. Technol.* **2012**, *46*, 548–555. [CrossRef]

98. Rizzello, C.G.; Lorusso, A.; Russo, V.; Pinto, D.; Marzani, B.; Gobbetti, M. Improving the antioxidant properties of quinoa flour through fermentation with selected autochthonous lactic acid bacteria. *Int. J. Food Microbiol.* **2017**, *241*, 252–261. [CrossRef] [PubMed]

99. Cannas, M.; Pulina, S.; Conte, P.; del Caro, A.; Urgeghe, P.P.; Piga, A.; Fadda, C. Effect of substitution of rice flour with quinoa flour on the chemical-physical, nutritional, volatile and sensory parameters of gluten-free ladyfinger biscuits. *Foods* **2020**, *9*, 808. [CrossRef] [PubMed]

100. Banu, I.; Vasilean, I.; Aprodu, I. Effect of lactic fermentation on antioxidant capacity of rye sourdough and bread. *Food Sci. Technol. Res.* **2010**, *16*, 571–576. [CrossRef]

Publisher's Note: MDPI stays neutral with regard to jurisdictional claims in published maps and institutional affiliations.

 *applied sciences*

Article

Use of Response Surface Methodology to Investigate the Effects of Sodium Chloride Substitution with Potassium Chloride on Dough's Rheological Properties

Andreea Voinea, Silviu-Gabriel Stroe * and Georgiana Gabriela Codină *

Faculty of Food Engineering, Stefan cel Mare University of Suceava, 720229 Suceava, Romania;
andy_v93@yahoo.com
* Correspondence: silvius@fia.usv.ro (S.-G.S.); codina@fia.usv.ro (G.G.C.);
  Tel.: +40-744-550-072 (S.-G.S.); +40-745-460-727 (G.G.C.)

Received: 22 May 2020; Accepted: 8 June 2020; Published: 11 June 2020

Abstract: Bakery products are one of the main sources of dietary sodium intake of the world's population. During the last decade, sodium intake has increased worldwide and nowadays the World Health Organization recommends reducing sodium intake by up to 2 g Na/day. KCl is the leading substitute for reducing sodium in bakery products. Therefore, the main purpose of our study was to investigate the impact of sodium reduction on dough's rheological properties by reformulating the dough recipe using two types of salts, namely NaCl and KCl, with different amounts added to wheat flour. In order to establish their combination for obtaining the optimum rheological properties of dough, the response surface methodology (RSM) by the Design Expert software was used. The effect of combined NaCl and KCl salts were made on mixing, viscometric and fermentation process by using Farinograph, Extensograph, Amylograph and Rheofermentometer devices. On dough's rheological properties, KCl and NaCl presented a significant effect ($p < 0.01$) on water absorption, stability, energy, dough resistance to extension, falling number and all Rheofermentometer-analyzed values. Mathematical models were achieved between independent variables, the KCl and NaCl amounts, and the dependent ones, dough rheological values. The optimal values obtained through RSM for the KCl and NaCl salts were of 0.37 g KCl/100 g and 1.31 g NaCl/100 g wheat flour, which leads to a 22% replacement of NaCl in the dough recipe.

Keywords: KCl; NaCl; rheological properties; multiple criteria optimization; desirability functions

---

1. Introduction

A high dietary sodium intake may lead to cardiovascular, bone demineralization and cancer diseases [1,2]. According to the American Heart Association, the cardiovascular diseases are the leading cause of mortality globally [3]. The close association between hypertension values and sodium intake is an important issue from a public health perspective. Nowadays, the World Health Organization (WHO) recommends to not exceed a sodium consumption of 5 g per day [4] and wants to reduce sodium intake up to 2 g Na/day [5]. In addition to the beneficial effects on health, sodium reduction also contributes to an annual decrease of medical expenditures [6]. Bread is considered worldwide to be an essential food for human nutrition. However, it might represent an important source of sodium intake. The increased consumption of bakery products increases the risk of diseases associated with sodium consumption [7,8]. The sodium sources in bakery products are provided by ingredients such as sodium bicarbonate, a baking agent, widely used in baking and sodium chloride (NaCl) which is one of the main ingredients used in the bakery manufacturing process [9]. The sodium chloride additions in bakery recipe are very important from a technological point of view [10]. Its reduction may lead to negative effects on technological process of bakery products and the quality of the finished products [7,11–13]. From the technological point of view, the sodium chloride addition increases dough strength and stability, its capacity to retain gases and, at low levels, yeast activity [7,10]. To bakery products, NaCl increases the shelf-life, due to the inhibition effect on microbial growth, and it improves bread texture and its sensory properties [12,13]. Due to the effect of sodium chloride on the technological process and the quality of the bakery products, its substitution in order to reduce the sodium content from the bakery products it is a problematic issue. Previous studies have shown that the potassium chloride (KCl) is the leading substitute for reducing sodium in bakery products [9,10,14–16]. Potassium chloride is a natural ingredient obtained from rock and sea salts with extraction methods similar to those of sodium chloride. The effect of potassium chloride consumption in the human diet is associated with a low risk of high blood pressure and other diseases associated with it, the effect being contrary to the intake of sodium chloride [17,18]. It has a salty taste but with metallic and bitter after tastes when high levels are incorporated in bakery recipes [19]. Therefore, the complete replacement of sodium chloride in bakery products it is not recommended. Its use in food products may only be in combination with sodium chloride in order to obtain products of a high quality [10,15]. Reducing sodium by replacing it with potassium chloride has to be done gradually because of its influence on technological process and quality of the bakery products [4]. The Response Surface Methodology (RSM) has been used in several food-related papers and applications. In the literature, some applications of RSM to flour, dough and bread, demonstrate its effectiveness. In particular, Cappelli et al. (2020) developed optimization charts regarding the milling process of wheat and for flour characterization [20]. Moreover, Cappelli et al. (2018) published predictive models of the rheological properties of doughs specifically developed with RSM [21]. The aim of this study was to analyze the effect of partial sodium chloride substitution with potassium chloride on the technological process of the bakery products. For this purpose, we used KCl and NaCl in different combinations by using response surface methodology (RSM) in order to analyze their effect on dough's rheological properties and to obtain their optimum formulation from the technological point of view.

2. Materials and Methods

2.1. Materials

Refined wheat flour (harvest 2019) was providing by the S.C Mopan S.A. (Suceava, Romania). The NaCl and KCl were purchased from the Romania market. A high-quality wheat flour was used. This is confirmed by the characteristics analyzed by the Romanian and international standard methods: 0.65 g/100 g ash (ICC 104/1), 14.0 g/100 g moisture (ICC 110/1), gluten deformation index 6 mm (SR 90:2007), 12.67 g/100 g protein (ICC 105/2), wet gluten 30 g/100 g (ICC106/1), falling number 442 s (ICC 107/1) [22].

2.2. Dough's Rheological Properties during Mixing and Extension

In order to analyze dough rheological properties during mixing a Farinograph device (Brabender, Duigsburg, Germany, 300 g capacity) was used. The dough's rheological properties during extension were analyzed using the Extensograph device (Brabender, Duigsburg, Germany). The Farinograph values analyzed through the ICC method 115/1 were water absorption (WA), dough stability (ST), dough development time (DDT) and degree of softening at 10 min (DS). The Extensograph values analyzed through the ICC method 114/1 were resistance to extension (R_{50}), maximum resistance to extension (R_{max}), energy (E) and ratio number (R/E) at a proving time of 135.

2.3. Dough Viscometric Rheological Properties

In order to analyze the dough viscometric rheological properties, Amylograph (Brabender OGH, Duisburg, Germany) and Falling Number (Perten Instruments AB, Sweden) devices were used. Amylograph trials were performed according to the ICC method 126/1: gelatinization temperature

(T_g), temperature at peak viscosity (T_{max}) and peak viscosity (PV_{max}). With respect to falling number trials, the ICC method 107/1 was applied.

2.4. Dough'sRheological Properties during Fermentation

The dough rheological properties during fermentation were determined by using a Rheofermentometer device (Chopin Rheo, type F3, Villeneuve-La-Garenne CEDEX, France). The fermentation parameters analyzed according to the AACC method 89-01.01. were maximum height of gaseous production (H'm), volume of the gas retained in the dough at the end of the test (VR), total CO_2 volume production (VT) and retention coefficient (CR).

2.5. Experimental Design and Statistical Analysis

In order to analyze the simultaneous effects of the KCl and NaCl amounts on the rheological properties of the wheat flour dough, the response surfaces methodology (RSM) was used. RSM has important application in the design, development and formulation of new products, to optimize the formulations factors [23–25] or to determine the optimum conditions for the process [26], showing the effect of the factors on the responses. Results optimization by the RSM method involved three main steps: the statistical design of the experiment, determination of the mathematical model coefficients and finally, prediction of the responses and checking the adequacy of the mathematical model within the design of the experiment (DOE) using the Design Expert software, trial version 12 (Stat-Ease, Inc., Minneapolis, MN, USA). For this study, two independent variables were chosen as follows: the influence of the variations of the potassium chloride amount ($A = X_1$) and sodium chloride ($B = X_2$) on the rheological parameters (dependent variables) of wheat flour dough. The experimental designs, with the real and coded values of the independent variables, are shown in Table 1.

Table 1. Real and coded values of independent variables used in the experimental design.

Run	Real Value		Coded Value	
	KCl [1] (g/100 g)	NaCl [1] (g/100 g)	X_1	X_2
1	0.3	0.3	−1	−1
2	1.5	0.3	1	−1
3	0.3	1.5	−1	1
4	0.9	0.9	0	0
5	0.9	0.9	0	0
6	0.9	0.9	0	0
7	1.5	0.9	1	0
8	0.9	0.9	0	0
9	1.5	1.5	1	1
10	0.9	1.5	0	1
11	0.9	0.9	0	0
12	0.9	0.3	0	−1
13	0.3	0.9	−1	0

[1] KCl—potassium chloride; NaCl—sodium chloride.

The rheological parameters determined by the Farinograph were WA—water absorption (Y_1); DT—development time (Y_2); ST—stability of dough (Y_3); DS—degree of softening (Y_4). The rheological parameters determined to Extensograph were: E—Energy (Y_5); R_{50}—resistance to extension up to 50 mm (Y_6); R_{max}—maximum resistance (Y_7); R/E—(Y_8). Moreover, the Falling Number index values (Y_9) have been determined. The rheological parameters determined by the Amylograph were T_g—gelatinization temperature (Y_{10}); PV_{max}—peak viscosity (Y_{11}); T_{max}—temperature at peak viscosity (Y_{12}), H'm—height under constraint of dough at maximum development time (Y_{13}), VT—total volume of CO_2 produced during fermentation (Y_{14}), VR—volume of the gas retained in the dough at the end of the test (Y_{15}) and CR—retention coefficient (Y_{16}). In order to minimize the measurement errors of the experimental data, the rheological values obtained for the wheat flour samples with different levels of KCl and NaCl addition according to our experimental design were carried out twice. In the statistical processing, their average values were used.

The predicted responses of the system ($Y_1 \ldots {}_n$) (Equation (1)) in factorial screening experiments have been defined by a mathematical model:

$$Y = f(X_1, X_2) = \beta_o + \sum_{i=1}^{n} \beta_i \cdot X_i + \sum_{i=1}^{n-1} \sum_{j=i+1}^{n} \beta_{ij} \cdot X_i \cdot X_j + \sum_{i=1}^{n} \beta_{ii} \cdot X_i^2 + \varepsilon \tag{1}$$

where β_o is the constant coefficient; β_i is the linear coefficient; β_{ij} is the interaction coefficient; β_{ii} is the quadratic coefficient; n is the number of factors studied and optimized in the experiment; X_i and X_j are the coded values of the independent variables and ε is the residual associated with the experiment. The residuals associated with the experiment were used to calculate the standard deviation values for each dependent variable. The significance of the model terms is evaluated by ANOVA, which performs a comparison of the variation in the response with the variation due to random error, at the probability value (p-value) of 95%. The suitability of the mathematical models has been checked by the F-tests and for the accuracy of the fitted polynomial equation was determined by adjusted coefficient of determination (Adjusted R^2). The non-significant coefficients were eliminated from the polynomial equations. In order to illustrate the dependence between the dependent and the independent variables, the three-dimensional graphical representation of the response surfaces was made.

3. Results and Discussion

3.1. Fitting Models

Following the statistical processing of the experimental data regarding the effects of independent variables on the predictive models for dough's rheological properties during the mixing of KCl–NaCl mixtures, the most fitting models (quadratic models)were obtained for the following parameters: water absorption (WA), dough development time (DT), dough stability (ST), degree of softening at 10 min (DS), the Falling Number value (FN), peak viscosity (PV_{max}), temperature at peak viscosity (T_{max}), height under constraint of dough at maximum development time (H'm), total volume of CO_2 produced during fermentation (VT), volume of the gas retained in the dough at the end of the test (VR) and retention coefficient (CR).

3.2. The Mixing and Extension Rheological Properties for the Mixes Samples

Applying the ANOVA method to the mixing and extension values, it was observed that KCl has a significant effect ($p < 0.01$) on the rheological parameters as E, R_{50}, R_{max}, R/E, T_g, H'm, VT, VR, CR, while NaCl has a significant effect ($p < 0.01$) on WA, ST, E, R_{50}, R_{max}, R/E, PV_{max}, H'm, VT, VR.

As seen Figure 1a, both types of salt led to a significantly decreased ($p < 0.01$) WA value. This may be due to the fact that, in the presence of salt ions, the electrostatic repulsions between gluten molecules are reduced as a consequence of their ability to partially shield the present charges between molecules. Thus, the gluten proteins aggregate in a higher extent due to the increased level of hydrophobic interactions between molecules fact that leads to a decrease of the water uptake ability [2].

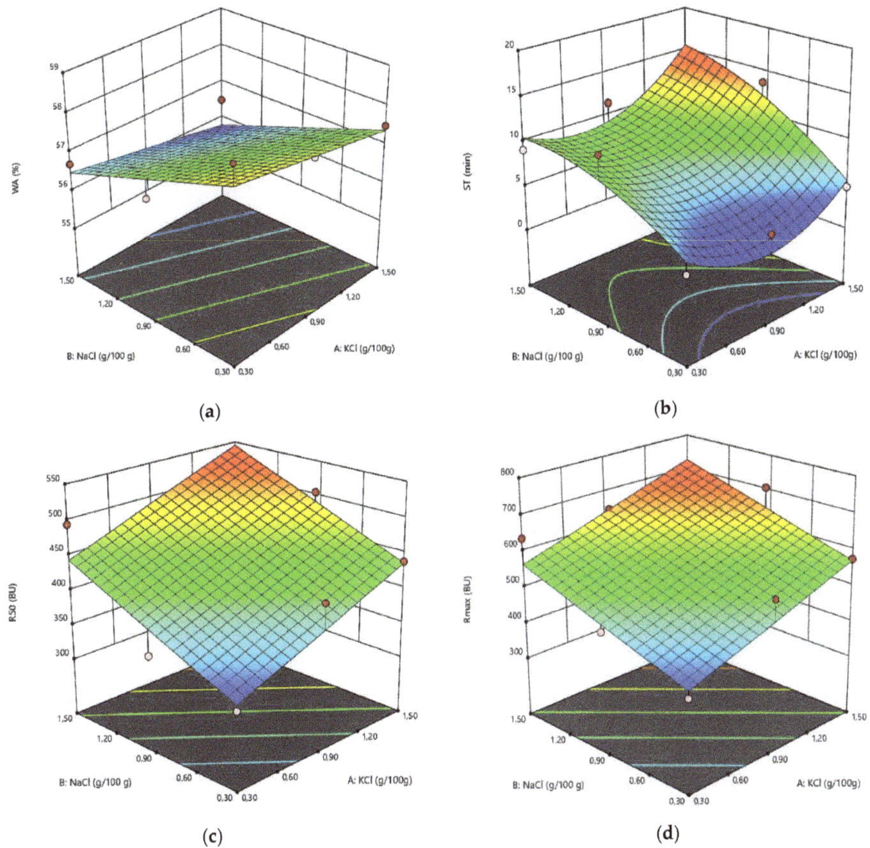

Figure 1. The graphical representations of the Farinograph and Extenograph parameters: (**a**) water absorption (WA); (**b**) stability of dough (ST); (**c**) resistance to extension up to 50 mm (R_{50}); (**d**) maximum resistance to extension (R_{max}).

From the two types of chloride salts it seems that NaCl presented a higher significant effect ($p < 0.01$) on WA value than KCl salt ($p < 0.01$). These results were similar with those reported by Tuhumury et al. [27] and Jeckle et al. [2], which concluded that the intensity of chloride salts on WA value depends on cation position in Hofmeister series K^+ being positioned before Na^+. A decrease of WA value with the increase of the salt level addition has also been reported by different researchers [5,9,28–30].

The graphical representation of the dough stability (ST) value in relation with the level of KCl and NaCl addition is shown in Figure 1b. This figure shows that there was a significant increase ($p < 0.01$) of this value with the addition of both independent variables. This indicates a strengthening effect of chloride salts on wheat dough. Gluten proteins present a surface hydrophobicity and contain almost 35% hydrophobic amino acids, which promotes a protein aggregation in a more extensive way when chloride salts are incorporated [9], leading to a higher dough stability. A significant increase of dough stability with the addition of chloride salts in wheat flour has also been reported by different researchers [2,11,31,32].

The effects of chloride salts on the Extenograph parameters curve are similar. According to Tuhumury et al. [27], this may be due to the fact that Na^+ and K^+ are situated nearby in the Hofmeister series and, that way, they exhibit similar effects on wheat dough properties. All the models for the

Extenograph values were linear. Both independent variables presented a significant positive effect ($p < 0.01$) on energy (E), resistance to extension (R_{50}), maximum resistance to extension (R_{max}) and ratio number (R/E). The effects of chloride salts on Extensograph values are related to their effect on gluten proteins. Their strengthen effect on dough due to the increase amount of hydrophobic interactions between molecules conducted to an increase value of E, R_{max}, R_{50} and R/E. These results were similar with those reported by McCann and Day [28], Miller and Hoseney [9], Tuhumury et al. [27], and Ortolan et al. [33], which concluded that chloride salts increased the resistance to extension, as seen in Figure 1c,d.

3.3. The Viscometricrheological Properties of the SampleMixes

The effect of NaCl and KCl addition in wheat flour on dough's viscometric properties, expressed as their corresponding regression coefficients and models, are shown in Table 2. From model analysis, the most significant models were those for quadratic model (*Adjusted* R^2 = 0.82), Falling Number value (FN), followed by those for quadratic model (*Adjusted* R^2 = 0.63), peak viscosity (PV_{max}) and 2FI model (*Adjusted* R^2 = 0.63) for gelatinization temperature (T_g) which was less significant (Table 3).

Table 2. Effects of independent variables, expressed as their corresponding coefficients on the predictive models for dough rheological properties during the mixing of KCl–NaCl mixtures.

Factors [b]	Parameters							
	Farinograph				Extensograph (Proving Time 135 min)			
	WA (%)	DT (min)	ST (min)	DS (UB)	E (cm^2)	R_{50} (BU)	R_{max} (BU)	R/E
Constant	56.75	1.55	6.85	54.86	106.54	439.62	566.85	4.11
A	−0.35 **	0.0167	1.55 *	0.666	14.33 ***	51.00 ***	81.83 ***	0.57 ***
B	−0.80 ***	−0.05	4.07 ***	−5.67 **	15.00 ***	54.67 ***	78.67 ***	0.50 ***
A × B	0.22	0.025	1.00	4.00	-	-	-	-
A^2	0.019	0.319 ***	3.78 **	−10.52 **	-	-	-	-
B^2	0.37	0.019	−1.67	11.48 **	-	-	-	-
Adjusted R^2	0.76	0.70	0.75	0.60	0.74	0.82	0.79	0.79
p-value [a]	0.0072 ***	0.031 **	0.0079 ***	0.0355 **	0.0005 ***	<0.0001 ***	0.0002 ***	0.0005 ***

[a] Significant at $p < 0.01$ ***, at $p < 0.05$ **, at $p < 0.1$ *. [b] A—KCl (g/100 g); B—NaCl (g/100 g); Adj. R^2 is measure of fit of the model. WA—water absorption; DT—development time; ST—dough stability; DS—degree of softening; E—Energy; R_{50}—resistance to extension up to 50 mm; R_{max}—maximum resistance, R/E—ratio number.

Table 3. Effects of independent variables, expressed as their corresponding coefficients on the predictive models for dough rheological properties during fermentation, gelatinization properties and α-amylase activity of KCl-NaCl mixtures.

Factors [b]	Parameters							
	FN (s)	T_g (°C)	PV_{max} (BU)	T_{max} (°C)	H'm (mm)	VT (mL)	VR (mL)	CR (%)
Constant	378.52	64.56	1221.66	89.00	61.95	1251.93	1117.07	89.28
A	1.50	0.62 ***	3.33	0.1167	−7.35 ***	−159.50 ***	−124.00 ***	2.12 ***
B	2.83	0.25 *	52.67 ***	0.25	−5.58 ***	−115.33 ***	−88.00 ***	1.62 **
A × B	−10.25 **	−0.1	−21.25	−0.05	−0.6750	−15.25	−24.00	−0.45
A^2	12.19 **	-	14.21	0.1879	−4.38 **	−95.26 **	−63.24 **	2.22 **
B^2	−23.81 ***	-	29.21	0.2879	−1.78	−61.76	−40.24	1.42
Adjusted R^2	0.82	0.63	0.633	0.40	0.90	0.87	0.91	0.75
p-value [a]	0.0028 ***	0.0071 ***	0.0143 **	0.1193	0.0004 ***	0.0008 ***	0.0003 ***	0.0073 ***

[a] Significant at $p < 0.01$ ***, at $p < 0.05$ **, at $p < 0.1$ *. [b] A—KCl (g/100 g); B—NaCl (g/100 g); Adj. R^2 is measure of fit of the model. FN—Falling Number; T_g—gelatinization temperature; PV_{max}—peak viscosity; T_{max}—temperature at peak viscosity; H'm—height under constraint of dough at maximum development time; VT—total volume of CO_2 produced during fermentation; VR—volume of the gas retained in the dough at the end of the test; CR—retention coefficient.

No significant model was obtained for T_{max}. Similar results were reported by Samutsri and Suphantharika [34], who concluded that different types of chloride salts did not presented any significant effect on pasting temperature on starch from rice. A positive effect on all viscometric properties was provided by the linear regression coefficients, suggesting that the increase in levels of

NaCl and KCl addition in wheat flour will lead to an increase in the viscometric values, as seen in Figure 2.

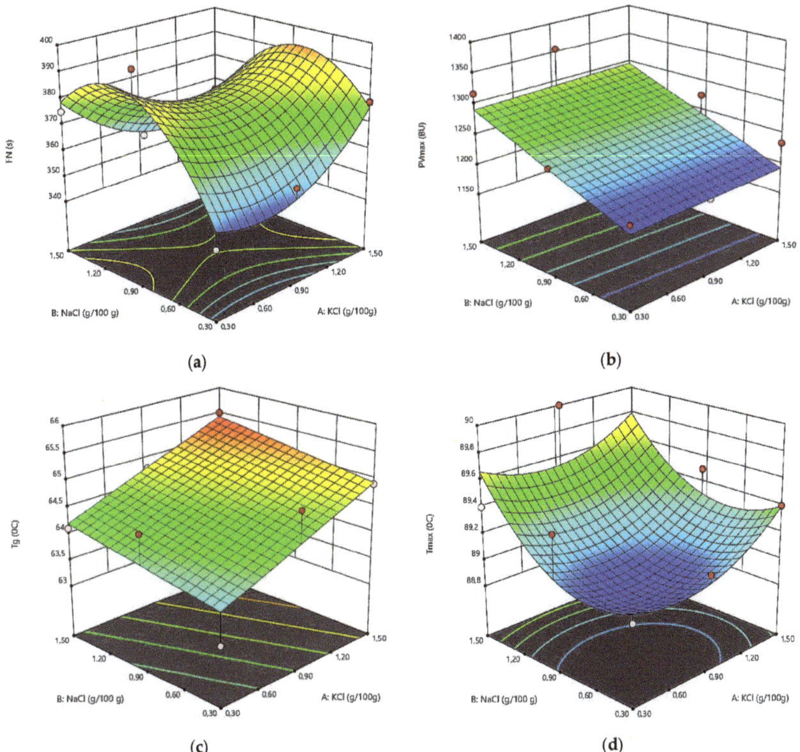

Figure 2. The graphical representations of the Falling Number and Amylograph parameters: (**a**) Falling Number value (FN); (**b**) peak viscosity (PV_{max}); (**c**) gelatinization temperature (T_g); (**d**) temperature at peak viscosity (T_{max}).

The increase of the FN and PV_{max} values with the increase of the NaCl and KCl addition (Figure 2a,b) may be due to the action of chloride salts on the protein part from the amylases structure. This fact reduces the activity of these enzymes with influence on the dough's rheological properties.

Previous studies have shown that, in the optimal range of pH activity of amylases, the chloride salts favor their activity in dough system, whereas outside of the pH range, it reduced its activity due to the shielding effect of the reactive groups of enzymes by the ions from the system as H^+, Na^+, K^+, Cl^- [35]. In general, wheat flour has a pH between 6.0–6.8, making its lightly acidic and even close to the neutral pH value. Chloride salts presents an alkaline pH. Therefore, a mix between chloride salts and wheat flour will lead to higher pH values. In Amylograph and Falling Number methods, wheat flour is mixed with distilled water with a pH value around 7.00 and different levels of chloride salts. Therefore, the mixes formed of wheat flour, distilled water and chloride salts will present pH values outside the optimal range of pH amylases activity which is around a value of $5.2 \div 5.4$ [36]. Due to the alkaline pH of chloride salts, the pH mixes analyzed to Amylograph and Falling Number will be even more outside of the optimal range of amylase activity with the increase level of chloride salts addition. Therefore, the amylases activity from the mixes from wheat flour, distilled water and chloride salts will decrease with the increase level of chloride salts addition. This fact leads to an increase in PV_{max} to the Amylograph device and to FN value to the Falling Number device which expresses α-amylase activity [37,38].

For the T_g value, a positive effect was provided by KCl and NaCl, as seen in Figure 2c, these data being similar with those reported by different researchers [39–41]. This behavior is due to the fact that these types of chloride salts decreased solubility of hydrophobic chains and enhanced water structure. When NaCl and KCl are incorporated in wheat flour dough it decreases water activity and increases the energy for physical and chemical reactions which involves water, delaying the starch gelatinization process [5,40].

3.4. The Fermentation Rheological Properties of the Mixes Samples

All the dependent variables analyzed through Rheofermentometer parameters were significantly affected ($p < 0.01$) by the levels of NaCl and KCl addition in wheat flour. Quadratic models (Table 3) for Rheofermentometer values showed a significant effect of the linear terms of NaCl and KCl, with a highly coefficient of determination (R^2) which varies mostly between 0.70 to 0.91.

The contour plots from Figure 3 for the Rheofermentometer values showed that the maximum height of gaseous production (H′m), total CO_2 volume production (VT) and volume of the gas retained in the dough at the end of the test (VR) significantly decreased ($p < 0.01$) with an increase in KCl and NaCl levels in wheat flour. The retention coefficient value (CR) increased with the level of KCl and NaCl addition in wheat flour.

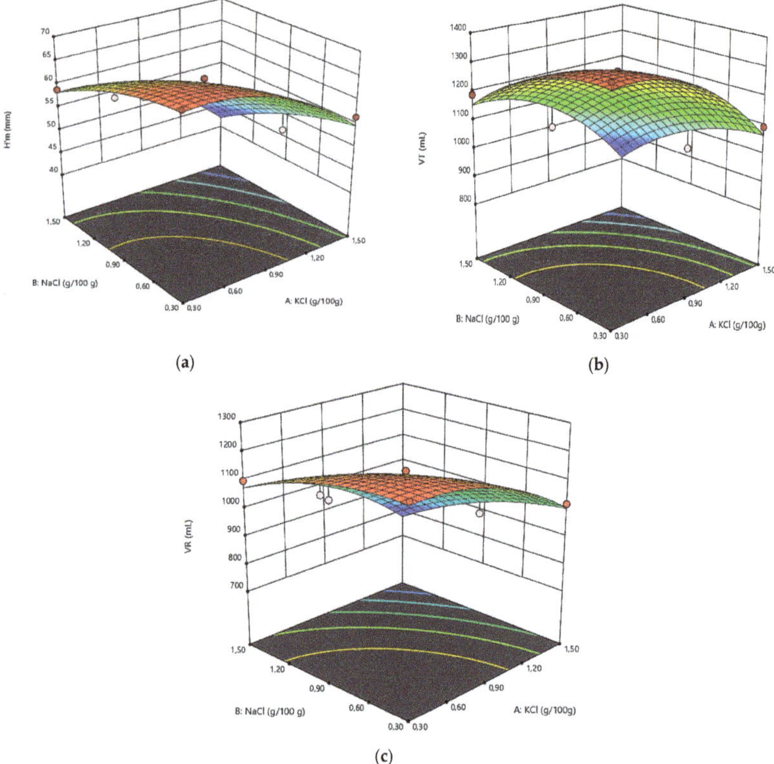

Figure 3. The graphical representations of the Rheofermentometer parameters: (**a**) maximum height of gaseous production (H′m); (**b**) total CO_2 volume production (VT); (**c**) volume of the gas retained in the dough at the end of the test (VR).

H′m is strongly affected by the yeast fermentation and also by the dough structure [2,42]. Thus, by chloride salts' addition, the gluten network becomes stronger and less extensible, leading to a lower dough expansion during fermentation. Additionally, chloride salts repress yeast activity by its osmotic pressure effect [12], leading to less CO_2 production and lower H′m values, as seen in Figure 3a. The decrease of the H′m, together with the increased level of KCl and NaCl addition, is in agreement with many previously made studies which reported that the addition of any type of chloride salts decreased the values of Rheofermentometer parameters [2,9,12,29,43].

The repressing effect of salt on yeast leads also to lower VT values, as seen in Figure 3b and, as a consequence, to lower VR values [43]. However, contrary to the negative effect of NaCl and KCl on H′m, VT and VR values, it presented a significantly ($p < 0.01$) positive effect on CR value. This behavior is due to the gluten network improvement which becomes stronger by chloride salt addition and with a higher ability to retain the gas released by fermentation [43].

3.5. Optimization of the KCl and NaClFormulation

An important objective of this research was to calculate the optimal values of the rheological parameters of the dough. For this purpose, the Derringer desirability function (Equation (2)), a multi-criterion decision-making method, was used [44,45]. The optimization process using the numerical method by the Design-Expert was performed.

$$D = \left(d_1^{r_1} \cdot d_2^{r_2} \cdot \ldots \cdot d_n^{r_n} \right)^{\frac{1}{\sum r_i}} \tag{2}$$

where $d_1, d_2, \ldots, d_n$ are the desirability indices for each dependent variables and $r_1, r_2, \ldots, r_i$ are the relative importance of the dependent variables. A non-zero value of D from zero implies that all responses are in desirable range and, for a D value close to 1, the response values are close to the desirable values (Figure 4). By applying the desirable function methodology, the optimal values of the independent variables were obtained.

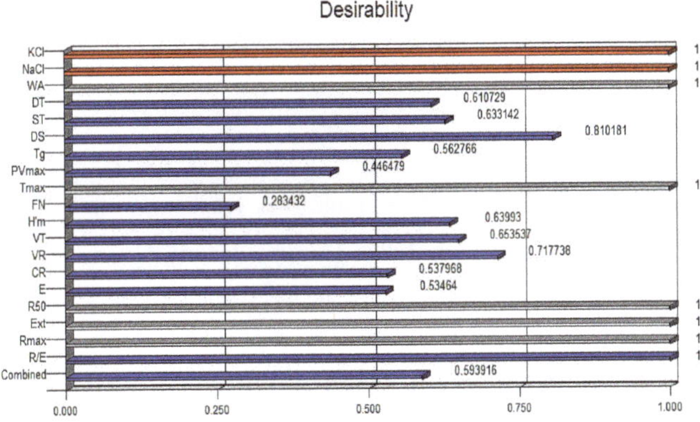

Figure 4. Desirability function scores for the independent variables and the studied dependent variables.

The optimum values of the amount of KCl are 0.37 g/100 g wheat flour, and the optimal amount of NaCl is 1.31 g/100 g wheat flour. For these optimal solutions, the optimal values for the dependent variables were obtained: WA—56.626%, DT—1.805 min, ST—10.471 min, DS—43.454 UB, T_g—64.149 °C, PV_{max}—1281.438 BU, T_{max}—89.513 OC, FN—383.978 s, H′m—59.714 mm, VT—1192.988 mL, VR—1098.555 mL, CR—92.027%, E—104.475 cm², R_{50}—433.33 BU, Ext—138.923 mm, R_{max}—549.359 BU and R/E—3.95 with a desirable function score of 0.594.

3.6. Strategy Approach for Bakery Products' Reformulation for Sodium Amount Reduction Related to Our KCl–NaCl Optimum Values

Bakery products are one of the main dietary sources of salt in most European countries. It seems that the highest consumption of bakery products occurred in the Eastern and Central Europe. In Northern Europe, other products, such as those of animal origin, are the main dietary sources of salt intake. Nowadays the daily salt intake in most EU countries ranges from 7 to 13 g per day (with the lowest intake values in Northern Europe countries and the highest ones in Central and Eastern countries), exceeding the World Health Organization's recommendation data [46]. Due to this, many European countries recommended food reformulation in order to reduce the salt content from its food, running many nutrition action plans for this purpose. For example, the Ministry of Health from some EU countries recommended reducing salt in bakery products by 15% up to 2015 in Austria, with 10% up to 2012 in Italy, with 20% up to 2014 in Spain, etc. [15]. The maximum target for the salt level that want to be achieved varies from one country to another. For example, of 2.35% for bread products from Hungary from 2019and of 1.4% for bakery products from Spain. These high differences between EU countries' targets are related to the usual levels that normally exist in these countries' bread products. For example, the level of salt from popular Hungarian bakery products is around of 3% [47], whereas in Spain the mean salt content from the bakery products is around 2% [48]. Besides the fact that salt reduction in bakery products affect its technological process, a fact developed in quite a large extent during this study, it also affects bakery products' quality. According to our study, a 22% replacement of NaCl in a dough recipe through KCl is optimum in order to obtain bakery products of a very good quality. According to the data from the international literature, a 20% sodium reduction in bakery products did not affected bread quality from the sensory (including taste) point of view [46]. Regarding the use of KCl as a NaCl substitute, previous studies have shown that its addition up to 20–30% did not affected the taste of the bakery products. However, levels higher than 30% of KCl addition in wheat flour led to metallic and bitter aftertastes, a fact that is not recommended it in bread making [10]. Therefore, our results are favorable for obtaining very good bakery products from a technological and sensory point of view. As mentioned above, nowadays different countries are trying gradual reductions of sodium levels from different foodstuffs. But this reduction is limited due to the consumers' acceptance, and those who are not willing to give up to their eating habits, especially from the sensory point of view. The use of KCl as an ingredient to reduce sodium in foodstuffs is expected to increase in the coming years [48]. This is in accordance with the many years of recommendations from the WHO, stating that people must reduce their Na intake and increase their K intake. Despite WHO's rigorous recommendation very little progress is being made worldwide in this direction [49]. Our optimum values obtained through RSM methodology reduces 22% of the sodium content from the bread recipe and increases the K level through KCl addition to around 200 mg/100 g bread. Our study proposes a formulation which leads to bakery products of a good quality in accordance with WHO recommendations of sodium reduction intake from foodstuffs. Additionally, the proposed sodium reduction is by NaCl substitution with KCl, a natural ingredient which has also been agreed by the World Health Organization.

4. Conclusions

According to the obtained data, it seems that both chloride salts have a similar effect on dough rheological properties. With respect to mixing properties, both types of salts presented a positive effect on dough stability, to the energy, and to dough extensibility values. During heating, the chloride salts increased dough viscosity, reflecting in an increase of PV_{max} and FN values. During fermentation, both of the chloride salts decreased the H'm, VT, and VR Rheofermentometer values and increased the CR value. The mathematical models obtained for the response variables were significant with high values of Adjusted $R^2 > 0.70$ (except for DS, PV_{max} and T_{max}), p-value < 0.05 (except for T_{max}) showing for most dependent variables no lack of fit. The optimum values, obtained with the numerical method, were for KCl—0.37 g/100 g wheat flour and for NaCl—1.31 g/100 g wheat flour. The use of the potassium chloride as a substitute of sodium chloride in bakery products has a double advantage, namely the reduction of sodium content as well the increase of potassium amount from the final products. Our optimum values obtained through RSM methodology lead to the best technological parameters and also reduced the amount of sodium from the bakery products by22%, a decreased level that, according to the data in the international literature, did not affect the sensory characteristics of the food products.

Author Contributions: A.V., S.-G.S. and G.G.C. contributed equally to the study design, collection of data, development of the sampling, analyses, interpretation of results, and preparation of the paper. All authors have read and agreed to the published version of the manuscript.

Funding: This research was funded by Ministry of Research and Innovation grant number [18PFE/16.10.2018].

Acknowledgments: This work was supported from contract no. 18PFE/16.10.2018 funded by Ministry of Research and Innovation within Program 1—Development of national research and development system, Subprogram 1.2—Institutional Performance—RDI excellence funding projects.

Conflicts of Interest: The authors declare no conflict of interest.

References

1. de Wardener, H.E.; MacGregor, G.A. Harmful effects of dietary salt in addition to hypertension. *J. Hum. Hypertens.* **2002**, *16*, 213–223. [CrossRef] [PubMed]
2. Jekle, M.; Necula, A.; Jekle, M.; Becker, T. Concentration dependent rate constants of sodium substitute functionalities during wheat dough development. *Food Res. Int.* **2019**, *116*, 346–353. [CrossRef] [PubMed]
3. Mozaffarian, D.; Benjamin, E.J.; Go, A.S.; Arnett, D.K.; Blaha, M.J.; Cushman, M.; de Ferranti, S.; Després, J.-P.; Fullerton, H.J.; Howard, V.J.; et al. Executive Summary: Heart Disease and Stroke Statistics—2015 Update. *Circulation* **2015**, *131*, 434–441. [CrossRef]
4. Zandstra, E.H.; Lion, R.; Newson, R.S. Salt reduction: Moving from consumer awareness to action. *Food Qual. Prefer.* **2016**, *48*, 376–381. [CrossRef]
5. Lopes, M.; Cavaleiro, C.; Ramos, F. Sodium Reduction in Bread: A Role for Glasswort (*Salicornia ramosissima* J. Woods). *Compr. Rev. Food Sci. Food Saf.* **2017**, *16*, 1056–1071. [CrossRef]
6. Bibbins-Domingo, K.; Chertow, G.M.; Coxson, P.G.; Moran, A.; Lightwood, J.M.; Pletcher, M.J.; Goldman, L. Projected effect of dietary salt reductions on future cardiovascular disease. *N. Engl. J. Med.* **2010**, *362*, 590–599. [CrossRef]
7. Raffo, A.; Carcea, M.; Moneta, E.; Narducci, V.; Nicoli, S.; Peparaio, M.; Sinesio, F.; Turfani, V. Influence of different levels of sodium chloride and of a reduced-sodium salt substitute on volatiles formation and sensory quality of wheat bread. *J. Cereal Sci.* **2018**, *79*, 518–526. [CrossRef]
8. Valerio, F.; Conte, A.; di Base, M.; Lattanzio, V.M.T.; Lonigro, S.L.; Padalino, L.; Pontonio, E.; Lavermicocca, P. Formulation of yeast-leavened bread with reduced salt content by using a Lactobacillus plantarum fermentation product. *Food Chem.* **2017**, *221*, 582–589. [CrossRef]
9. Miller, M.W. Role of Salt in Baking. *Cereal Foods World* **2008**, *53*, 4–6. [CrossRef]
10. Silow, C.; Axel, C.; Zannini, E.; Arendt, E.K. Current status of salt reduction in bread and bakery products—A review. *J. Cereal Sci.* **2016**, *72*, 135–145. [CrossRef]
11. Diler, G.; leBail, A.; Chevallier, S. Salt reduction in sheeted dough: A successful technological approach. *Food Res. Int.* **2016**, *88*, 10–15. [CrossRef] [PubMed]
12. Lynch, E.; Bello, F.D.; Sheehan, E.; Cashman, K.; Arendt, E.K. Fundamental studies on the reduction of salt on dough and bread characteristics. *Food Res. Int.* **2009**, *42*, 885–891. [CrossRef]
13. Moreau, L.; Lagrange, J.; Bindzus, W.; Hill, S. Influence of sodium chloride on colour, residual volatiles and acrylamide formation in model systems and breakfast cereals. *Int. J. Food Sci. Technol.* **2009**, *44*, 2407–2416. [CrossRef]
14. Belz, M.C.; Ryan, L.A.; Arendt, E.K. The Impact of Salt Reduction in Bread: A Review. *Crit. Rev. Food Sci. Nutr.* **2012**, *52*, 514–524. [CrossRef]
15. Belc, N.; Smeu, I.; Macri, A.; Vallauri, D.; Flynn, K. Reformulating foods to meet current scientific knowledge about salt, sugar and fats. *Trends Food Sci. Technol.* **2019**, *84*, 25–28. [CrossRef]

16. van Buren, L.; Dötsch-Klerk, M.; Seewi, G.; Newson, R.S. Dietary Impact of Adding Potassium Chloride to Foods as a Sodium Reduction Technique. *Nutrients* **2016**, *8*, 235. [CrossRef]

17. Binia, A.; Jaeger, J.; Hu, Y.; Singh, A.; Zimmermann, D. Daily potassium intake and sodium-to-potassium ratio in the reduction of blood pressure: A meta-analysis of randomized controlled trials. *J. Hypertens.* **2015**, *33*, 1509–1520. [CrossRef]

18. Aburto, N.J.; Hanson, S.; Gutierrez, H.; Hooper, L.; Elliott, P.; Cappuccio, F.P. Effect of increased potassium intake on cardiovascular risk factors and disease: Systematic review and meta-analyses. *BMJ* **2013**, *346*. [CrossRef]

19. Braschi, A.; Gill, L.; Naismith, D.J. Partial substitution of sodium with potassium in white bread: Feasibility and bioavailability. *Int. J. Food Sci. Nutr.* **2009**, *60*, 507–521. [CrossRef]

20. Cappelli, A.; Guerrini, L.; Parenti, A.; Palladino, G.; Cini, E. Effects of wheat tempering and stone rotational speed on particle size, dough rheology and bread characteristics for a stone-milled weak flour. *J. Cereal Sci.* **2020**, *91*, 102879. [CrossRef]

21. Cappelli, A.; Cini, E.; Guerrini, L.; Masella, P.; Angeloni, G.; Parenti, A. Predictive models of the rheological properties and optimal water content in doughs: An application to ancient grain flours with different degrees of refining. *J. Cereal Sci.* **2018**, *83*, 229–235. [CrossRef]

22. Popa, N.C.; Tamba-Berehoiu, R.; Popescu, S.; Varga, M.; Codină, G.G. Predictive model of the alveographic parameters in flours obtained from Romanian grains. *Rom. Biotechnol. Lett.* **2008**, *14*, 4234–4242.

23. Mironeasa, S.; Iuga, M.; Zaharia, D.; Mironeasa, C. Optimization of grape peels particle size and flour substitution in white wheat flour dough. Scientific Study & Research. *J. Food Proc. Eng.* **2019**, *20*, 29–42.

24. Codină, G.G.; Mironeasa, S. Use of response surface methodology to investigate the effects of brown and golden flaxseed on wheat flour dough microstructure and rheological properties. *J. Food Sci. Technol.* **2016**, *53*, 4149–4158. [CrossRef]

25. Mironeasa, S.; Iuga, M.; Zaharia, D.; Mironeasa, C. Optimization of White Wheat Flour Dough Rheological Properties with Different Levels of Grape Peels Flour Addition. *Bull. Univ. Agric. Sci. Vet. Med. Cluj-Napoca Food Sci. Technol.* **2019**, *76*, 27–39. [CrossRef]

26. Wang, Y.; Gao, Y.; Ding, H.; Liu, S.; Han, X.; Gui, J.Z.; Liu, D. Subcritical ethanol extraction of flavonoids from Moringaoleifera leaf and evaluation of antioxidant activity. *Food Chem.* **2017**, *218*, 152–158. [CrossRef]

27. Tuhumury, H.; Small, D.; Day, L. Effects of Hofmeister salt series on gluten network formation: Part I. Cation series. *Food Chem.* **2016**, *212*, 789–797. [CrossRef]

28. McCann, T.; Day, L. Effect of sodium chloride on gluten network formation, dough microstructure and rheology in relation to breadmaking. *J. Cereal Sci.* **2013**, *57*, 444–452. [CrossRef]

29. Beck, M.; Jekle, M.; Becker, T. Impact of sodium chloride on wheat flour dough for yeast-leavened products. I. Rheological attributes. *J. Food Sci. Agric.* **2012**, *92*, 585–592. [CrossRef]

30. Uthayakumaran, S.; Batey, I.L.; Day, L.; Wrigley, C.W. Salt reduction in wheat-based foods-technical challenges and opportunities. *Food Aust.* **2011**, *63*, 137–140.

31. Voinea, A.; Stroe, S.-G.; Codină, G.G. The Effect of Sodium Reduction by Sea Salt and Dry Sourdough Addition on the Wheat Flour Dough Rheological Properties. *Foods* **2020**, *9*, 610. [CrossRef]

32. Bernklau, I.; Neußfer, C.; Moroni, A.V.; Gysler, C.; Spagnolello, A.; Chung, W.; Jekle, M.; Becker, T. Structural, textual and sensory impactof sodium reduction on long fermented pizza. *Food Chem.* **2017**, *234*, 398–407. [CrossRef]

33. Ortolan, F.; Corrêa, G.P.; Cunha, R.L.; Steel, C.J. Rheological properties of vital wheat glutens with water or sodium chloride. *LWT* **2017**, *79*, 647–654. [CrossRef]

34. Samutsri, W.; Suphantharika, M. Effect of salts on pasting, thermal, and rheological properties of rice starch in the presence of non-ionic and ionic hydrocolloids. *Carbohydr. Polym.* **2012**, *87*, 1559–1568. [CrossRef]

35. Cordeiro, C.A.M.; Martins, M.L.L.; Luciano, A.B. Production and properties of α-amylase from termophilicBacillussp. *Braz. J. Microbiol.* **2002**, *33*, 57–61. [CrossRef]

36. Sinani, A.; Sana, M.; Seferi, E.; Sheahaj, M. The effect of α-amylase in rheology features of some wheat cultivars and their harmonization for producing baking according to customer requirements. *Glob. J. Biol. Agric. Health Sci.* **2014**, *4*, 58–64.

37. Struyf, N.; Verspreet, J.; Courtin, C. The effect of amylolytic activity and substrate availability on sugar release in non-yeasted dough. *J. Cereal Sci.* **2016**, *69*, 111–118. [CrossRef]

38. Codină, G.G.; Dabija, A.; Oroian, M. Prediction of Pasting Properties of Dough from Mixolab Measurements Using Artificial Neuronal Networks. *Foods* **2019**, *8*, 447. [CrossRef]

39. Nicol, T.W.; Isobe, N.; Clark, J.H.; Matubayasi, N.; Shimizu, S. The mechanism of salt effects on starch gelatinization from a statistical thermodynamic perspective. *Food Hydrocoll.* **2019**, *87*, 593–601. [CrossRef]

40. Moreira, R.; Chenlo, F.; Torres, M. Effect of sodium chloride, sucrose and chestnut starch on rheological properties of chestnut flour doughs. *Food Hydrocoll.* **2011**, *25*, 1041–1050. [CrossRef]

41. Salvador, A.; Sanz, T.; Fiszman, S.M. Dynamic rheological characteristics of wheat flour-water doughs. Effect of adding NaCl, sucrose and yeast. *Food Hydrocoll.* **2006**, *20*, 780–786. [CrossRef]

42. Codină, G.G.; Voica, D. The influence of different forms of bakery yeast Saccharomyces cerevisie type strain on the concentration of individual sugars and their utilization during fermentation. *Rom. Biotechnol. Lett.* **2010**, *15*, 5417–5422.

43. Pasqualone, A.; Caponio, F.; Pagani, M.A.; Paradiso, V.M.; Paradiso, V.M. Effect of salt reduction on quality and acceptability of durum wheat bread. *Food Chem.* **2019**, *289*, 575–581. [CrossRef]

44. Badwaik, L.S.; Prasad, K.; Seth, D. Optimization of ingredient levels for the development of peanut based fiber rich pasta. *J. Food Sci. Technol.* **2012**, *51*, 2713–2719. [CrossRef]

45. Candioti, L.V.; de Zan, M.M.; Cámara, M.S.; Goicoechea, H.C. Experimental design and multiple response optimization. Using the desirability function in analytical methods development. *Talanta* **2014**, *124*, 123–138. [CrossRef]

46. Kloss, L.; Meyer, J.D.; Graeve, L.; Vetter, W. Sodium intake and its reduction by food reformulation in the European Union—A review. *NFS J.* **2015**, *1*, 9–19. [CrossRef]

47. Lásztity, R. *The Chemistry of Cereal Proteins*; CRC Press: Boca Raton, FL, USA, 1995.

48. Farinós, N.P. Salt content in bread in Spain, 2014. *Nutr. Hospital.* **2018**, *35*, 650–654. [CrossRef]

49. Iwahori, T.; Miura, K.; Ueshima, H. Time to Consider Use of the Sodium-to-Potassium Ratio for Practical Sodium Reduction and Potassium Increase. *Nutrients* **2017**, *9*, 700. [CrossRef]

Article

Karl Fischer Water Titration—Principal Component Analysis Approach on Bread Products

Gabriela Popescu [1], Isidora Radulov [2], Olimpia A. Iordănescu [3], Manuela D. Orboi [4], Laura Rădulescu [5], Mărioara Drugă [6], Gabriel S. Bujancă [5], Ioan David [6], Daniel I. Hădărugă [7], Christine A. Lucan (Banciu) [6], Nicoleta G. Hădărugă [6,*] and Mircea Riviş [8]

[1] Department of Management and Rural Development, Banat's University of Agricultural Sciences and Veterinary Medicine "King Michael I of Romania" from Timişoara, Calea Aradului 119, 300645 Timişoara, Romania; gabrielapopescu@usab-tm.ro

[2] Department of Soil Sciences, Banat's University of Agricultural Sciences and Veterinary Medicine "King Michael I of Romania" from Timişoara, Calea Aradului 119, 300645 Timişoara, Romania; isidora_radulov@usab-tm.ro

[3] Department of Horticulture, Banat's University of Agricultural Sciences and Veterinary Medicine "King Michael I of Romania" from Timişoara, Calea Aradului 119, 300645 Timişoara, Romania; olimpiaiordanescu@usab-tm.ro

[4] Department of Economics and Company Financing, Banat's University of Agricultural Sciences and Veterinary Medicine "King Michael I of Romania" from Timişoara, Calea Aradului 119, 300645 Timişoara, Romania; orboi@usab-tm.ro

[5] Department of Food Control and Expertise, Banat's University of Agricultural Sciences and Veterinary Medicine "King Michael I of Romania" from Timişoara, Calea Aradului 119, 300645 Timişoara, Romania; laura.radulescu@usab-tm.ro (L.R.); gabrielbujanca@usab-tm.ro (G.S.B.)

[6] Department of Food Science, Banat's University of Agricultural Sciences and Veterinary Medicine "King Michael I of Romania" from Timişoara, Calea Aradului 119, 300645 Timişoara, Romania; marioaradruga@usab-tm.ro (M.D.); ioandavid@usab-tm.ro (I.D.); alexalukan@gmail.com (C.A.L.)

[7] Department of Applied Chemistry, Organic and Natural Compounds Engineering, Polytechnic University of Timişoara, Carol Telbisz 6, 300001 Timişoara, Romania; daniel.hadaruga@upt.ro

[8] Department of Dental Medicine, "Victor Babeş" University of Medicine and Pharmacy Timişoara, Eftimie Murgu Sq. 2, 300041 Timişoara, Romania; rivis.mircea@umft.ro

* Correspondence: nicoletahadaruga@usab-tm.ro; Tel.: +40-256-277-423

Received: 18 August 2020; Accepted: 14 September 2020; Published: 18 September 2020

Featured Application: Classification of bread products as a specific type of food product according to a standard database, using coupled Karl Fischer titration and multivariate analysis

Abstract: (1) Background: The water content and the way of bonding in the food matrices, including bread, can be easily and simply evaluated by Karl Fischer titration (KFT). The goal was to identify the main KFT parameters that influence the similarity/dissimilarity of commercial bread products, using multivariate statistical analysis. (2) Methods: Various commercial bread samples were analyzed by volumetric KFT and the water content, parameters from titration process and KFT kinetics were used as input for principal component analysis (PCA). (3) Results: The KFT water content was in the range of 35.1–44.2% for core samples and 19.4–22.9% for shell samples. The storage and transportation conditions consistently influence the water content of bread. The type of water molecules can be evaluated by means of KFT water reaction rates. The mean water reaction rates up to 2 min are consistently higher for bread core samples, which indicates a high fraction of "surface" water. PCA reveals the similarity of core samples and various bread types, as well as dissimilarity between bread parts, mainly based on KFT kinetic parameters. (4) Conclusions: KFT kinetics can be a useful tool for a rapid and simple differentiations between various types of bread products.

Keywords: bread; water content; Karl Fischer titration; KFT kinetics; principal component analysis

---

1. Introduction

Bread is one of the most consumed food products all over the world. There are various types of bread from the main ingredients and shapes points of view. Wheat bread is largely commercialized, made even by wheat flour or mixed with other cereal flour types such as rye [1,2]. On the other hand, rye bread has become more popular, as has whole grain bread [3,4]. Other bread types are "special" or "functional" food products [2,5,6]. Generally, white bread is made by common wheat flour (*Triticum aestivum* L. var. *aestivum*), water, yeast, shortening agents, sugar and salt. Nutritional and organoleptic characteristics can be enhanced by addition of oils and fats, oilseeds, almonds or legumes. On the other hand, gluten-free or low carbohydrate products can be obtained using rice, corn or cassava flour. Depending on the desired characteristics of the final products, additives such as enzymes, emulsifiers, gums modified starch or cellulose, as well as protein-rich ingredients (egg and soy), can be used [1,7–18].

The composition of bread strongly influences its final properties, such as nutritional value, taste and flavor, rheological properties and shelf-life [4,19]. In this regard, water plays an important role, in both dough and bread processing steps [20,21]. Water molecules can migrate between starch and gluten [22]. During these steps, starch gelatinizes and water can partially evaporate from the bread surface [23–25]. Water molecules migrate from the core to the surface, but the diffusive flow of liquid water is slower than the evaporation process. On the other hand, the partial water vaporization in the bread provide a vapor pressure gradient. Water vapor migrates to the core and condensates (mechanism of "evaporation–condensation") [26]. Consequently, the outer zone of the bread starts drying and forms the crust. These variations on water content and distribution influence the final properties of bread such as softness of the crumb or the crispness of the crust [26]. Among starch and gluten, which can range 45–58% in bread, proteins (~6%) also influence the water content and mobility [6,20].

It is a challenge to determine the water content, water molecule types and their mobility in such complex food matrices. There are many methods of water determinations including physical or chemical methods [27]. Many factors influence the results, such as limit of detection, type of water, compounds that bias the measurement method or sample solubility and diffusion of all water molecules. Physical methods are based on weight loss, such as oven, infrared and microwave drying or thermogravimetry, as well as spectral and other physical properties of the sample (e.g., conductimetry, refractive index or capacity) [28–33]. The water determination in bread is generally based on such methods. Other methods are based on chemical properties of samples. It is the case of Karl Fischer water titration (KFT, volumetric or coulometric techniques) and other types of titration methods (acyl halide, anhydride and calcium carbide). Moreover, water distribution between dough components, water hydration capacity or other properties can be evaluated by farinograph method [22,34].

KFT is a useful method for water determination in various materials, including less soluble or insoluble samples. The water content can be easily and selectively determined in complex food matrices such as dairy and meat products, honey, sugars, chocolate, fruits and vegetables, spices and cereal products [3,35–39]. There are many advantages for water determination by KFT: cover all water content ranges (ppm to 100%, by selecting coulometric, single- or bi-component volumetric techniques), determination of both free and "strongly retained" water, a wide range of solvent polarity and temperature of analysis, which allows protecting less stable/degradable samples, rapid and simple technique and equipment, increased sensitivity and accuracy and possibility of coupling with other equipment such as oven or distillation apparatus [38–40]. On the other hand, some compounds can interfere with the KFT reagents (such as aldehydes and ketones, which react with the alcohol used as solvent, or compounds that react with iodine—reducing agents) [27,31,33].

The KFT working parameters can be selected according to the composition and characteristics of the food sample. If the food product has a high content of oil and/or fat and lower content of water, the coulometric KFT technique can be used. On the other hand, these types of food products having higher content of water, such as in the case of butter, require volumetric KFT working with more hydrophobic solvent mixtures (e.g., methanol–decanol). The electrode coating for lecithin-based samples can be prevented by adding chloroform and formamide. Soybean, rapeseed, sunflower, sesame and pumpkin seed oils have been analyzed by coulometric KFT. The water content was in the range of 0.098–0.689% by automated coulometric KFT and 0.079–0.684% for manual KFT [38]. On the contrary, butter oil had 0.061–0.108% and 0.079–0.099% water by volumetric KFT working in methanol/chloroform and methanol/hexanol solvent mixtures, respectively [41].

Cereals, cereal-based products, sugars and dairy food ingredients and products have been widely studied from the KFT water content point of view. They generally have higher water content. On the other hand, the highly hydrophilic components (carbohydrates such as starch, mono- and disaccharides such as sucrose and lactose and proteins such as gluten) strongly interact with water molecules. Such components are less soluble in the KFT working medium and the extraction of water from the sample particles are enhanced by increasing the temperature analysis, using the homogenization or milling, and increasing the solvent polarity by adding polar solvents (e.g., formamide). Water content of wheat and rye flour mixtures have been determined using volumetric KFT at elevated temperature [3,12], while for sorghum the KFT water content at 65 °C slowly decreased from 12.9% to 12.25% by increasing the grinding time from 10 to 90 s [39]. Other studies used starch, chick-pea flour, dry alimentary small paste or toasted ground barley for the determination of water content by volumetric KFT. For a better accuracy, the determination temperature was increased to 50 °C, where the water content was up to 0.7% higher than for the analysis temperature of 25 °C [42]. Invert sugar and fructose syrups have been compared for their water content and the corresponding mass loss at elevated temperature. They have a water content of 26.67% and 29.46% for invert sugar and fructose syrups, respectively. On the other hand, the mass loss was in the range of 26.4–26.7% for inverted sugar syrup dried at 105 °C, while for fructose syrup the mass loss had no significant variation by halogen drying at 100–115 °C (mass loss of 29.06–29.46%) [32]. Similar observations were accounted for maltose (5.1–5.64% KFT water content or halogen drying mass loss), glucose monohydrate (8.34% by KFT and 7.72% mass loss), sucrose (0.024% by KFT and 0.055–0.062% mass loss) or isomalt- and sorbitol-based candies [30]. All determinations required elevated temperature (50 °C), formamide addition and internal homogenization. Sucrose I and II and fructose had the KFT water content of 0.0552–0.0556%, 0.0464–0.0469% and 0.0800–0.0803%, respectively, while for glucose monohydrate the determined water content was consistently higher (8.84%). α-Lactose (lactose monohydrate), anhydrous β-lactose and the amorphous form have different water content, which is difficult to determine. Volumetric KFT performed in fifteen laboratories at 40 °C provide relatively close results for water content of various commercial lactose ranging 4.48–5.19% [43]. A comprehensive study on the use of vaporization coulometric KFT technique have been performed by Kestens and co-workers [44]. The robustness and good repeatability of this coupled technique was demonstrated for various difficult food matrices, including toasted bread.

Dairy products and protein-containing food products have high content of water. Generally, proteins are more hydrophilic and formamide can be added in order to enhance the water extraction during volumetric KFT analysis [27]. Moreover, homogenization and enhancing analysis temperature can be applied (for example, for meat products) [35,45,46]. On the other hand, high fat dairy products need the addition of more hydrophobic solvents to the KFT working medium (such as chloroform or decanol). Isengard and collaborators [29] also used 1-propanol, 1-butanol or *tert*-butyl-methyl-ether in order to enhance the water extraction in the volumetric KFT analysis at 50 °C for water determination in dried milk products. When the gas extraction was coupled with an automated KFT technique, using methanol–octanol solvent mixture at 40 °C, valuable results for the water content of various commercial butter samples have been obtained [40]. Products having high content of fats and sugars need homogenization and cooling, as well as hydrophilic or more hydrophobic solvent mixture, depending of the characteristic of the main components—sugars and fats, respectively. Instant powders such as coffee substitutes, coffee and cocoa-based instant powders have been analyzed by volumetric KFT by addition of formamide [47].

The goal of the present study was to identify the main KFT parameters that influence the similarity/dissimilarity between commercial bread products, using multivariate statistical analysis—principal component analysis (PCA)—as a useful combined tool for a rapid and simple differentiations between various types of bread products.

2. Materials and Methods

2.1. Materials

Bread samples were obtained from the local market (the flour used for obtaining the bread products by various local producers was from the same manufacturer, Timişoara, Timiş county, Romania). Three major groups of bread samples were used for KFT and multivariate PCA analyses: (1) White wheat breads consisted of five sample types, which varied only in the shape and presence or absence of different seeds on the top of the bread (these extra-materials were not included in samples for analysis): "HC1", white wheat bread, non-packed (core); "HC2", white wheat bread, sliced and packed in plastic bag (core); "HC3", white wheat bread roll, non-packed (core); "HC4", white wheat bread roll, topped with various seeds, non-packed (core); and "HC5", homemade white wheat bread (core). A Moulinex OW 3022 Home Bread, having kneading (5, 20, 0.2 and 0.25 min, alternately with rests or growing), rest (5 min after the first kneading), growing (39, 25.5 and 49.45 min after the second to fourth kneading), baking (48 min, followed by 2.51 min), and maintaining at warm condition (60 min) cycles, was used. The homemade white wheat bread was obtained from 455 g of white wheat flour (from the same manufacturer as above), 250 mL of water, 30 g of sugar, 8 g of salt, 26 g of milk powder, 25 g of sunflower oil and 9 g of yeast. (2) Whole meal or Graham breads included codes "MC" and "GC" (cores). (3) Brown wheat breads included "BC1" (non-packed, from the market) and "BC2" (from the manufacturer). For comparison of KFT kinetic data, the shell (up to 1.5 mm from the outer surface of the bread) and pre-dried bread samples were included in the study. The shell samples were coded as "MS" or "BS" (whole meal and brown wheat bread shell samples, respectively), while the pre-dried samples were coded as "Cd" and "Sd" for core and shell, respectively. Pre-drying was performed at 37 °C in the oven with forced convection for 30 min, in order to evaporate the "surface" water and to compare the KFT kinetic parameters. All bread samples were kept in the dark at room temperature, in sealed bags prior analysis. Representative core and shell samples were prepared right before analysis by cutting in cubic portions with a side of 1–1.5 mm. Weighted samples were immediately sealed in the KF sample transfer unit until analysis.

2.2. Volumetric Karl Fischer Water Titration (KFT)

The water content of bread samples as well as the KFT kinetics were obtained using the volumetric bi-component KFT method. A Karl Fischer 701 Titrando apparatus equipped with a 10-mL dosing system and a 703 Ti Stand mixing system were used (Metrohm AG, Herisau, Switzerland). Component 1 was Titrant 5 apura® and Component 2 Solvent apura® (Merck&Co., Inc., Darmstadt, Germany). The titer of 4.4286 mg/mL for Component 1 was obtained using Water standard 1% apura®, standard for volumetric Karl Fisher titration (Merck&Co., Inc.). The sample amount was in the range of 0.057–0.419 g (Table 1). The following KFT parameters were used: I(pol) 50 μA, end point 250 mV, maximum rate 5 mL/min, stop criterion of drift, stop drift 15 μL/min and temperature of 25 ± 1 °C. Measurements were done in triplicate (unless otherwise stated).

The KFT kinetics was performed for pseudolinear ranges from the titration curves. The mean water reaction rate depends on the diffusion of water molecules to the sample surface during extraction and analysis. The water reaction rate can be determined by knowing the variation of the titration volume of Component 1 (iodine solution with known titer) on time, as well as the reaction volume (the volume of the working medium was 60 mL in all cases) (Figure 1). This variation of the iodine concentration on time for three pseudolinear ranges was obtained from the *Volume (normalized to the sample mass)* versus *Time* curve. This was the mean iodine or water reaction rate on the specified time range (expressed as mmol/L/s or mM/s).

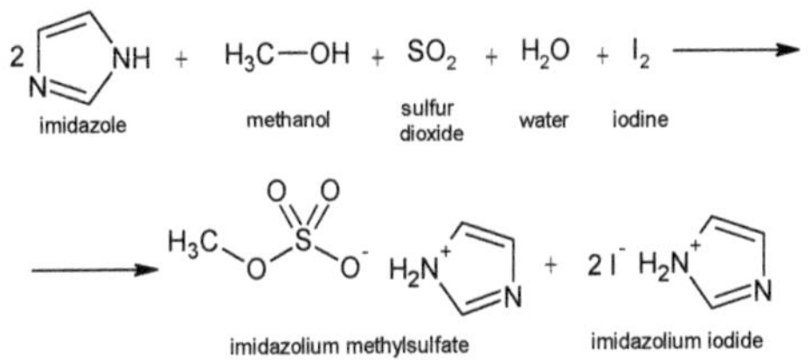

$$2 \text{ imidazole } (NH) + H_3C{-}OH \text{ (methanol)} + SO_2 \text{ (sulfur dioxide)} + H_2O \text{ (water)} + I_2 \text{ (iodine)} \longrightarrow$$

$$\longrightarrow H_3C{-}O{-}SO_2{-}O^- \; H_2N^+ \text{ (imidazolium methylsulfate)} + 2\,I^- \; H_2N^+ \text{ (imidazolium iodide)}$$

Figure 1. The overall KFT redox reaction (in the first step, methanol, sulfur dioxide and imidazole from Component 2 react and provide imidazolium methyl sulfite; in the second step, the latter reacts with the water from the sample and iodine from Component 1, providing imidazolium methyl sulfate and imidazolium iodide).

2.3. Statistics and Principal Component Analysis (PCA)

The KFT water content and values for KFT kinetic parameters were provided as mean (±standard deviation, SD). Moreover, the results from KFT analysis and kinetics for bread core, pre-dried bread core, raw and pre-dried bread shell samples were compared for significant difference using Tukey's HSD (honestly significant difference) test from the one-way ANOVA module in Statistica 7.1 software (StatSoft, Inc., Tulsa, OK, USA). KFT volume/sample mass ratio (V/m (mL/g)), water content (W) (%), variation of the KFT volume/sample mass on time ($\Delta(V/m)/\Delta t$) (mL/g/s) and the mean KFT reaction rate (v) (mM/s) were used as dependent variables. The categorical predictor (factor) was the bread sample code (*HC* (white wheat bread core), *MC* (whole meal bread core), *GC* (Graham bread core_, *BC* (brown wheat bread core), *MS* (whole meal bread shell, *BS* (brown wheat bread shell), *MCd* (pre-dried whole meal bread core), *BCd* (pre-dried brown wheat bread core), *MSd* (pre-dried whole meal bread shell) and *BCd* (pre-dried brown wheat bread shell)). The parameterization was sigma-restricted, while the confidence limit and significance level were set at 0.95 and 0.05, respectively. PCA is a multivariate statistical tool providing information on similarity/dissimilarity between bread samples, as well as the influence of independent variables for grouping. Both titration and kinetic parameters from the KFT analysis were used for PCA. This technique approximates the data matrix as a product of two reduced matrices that retain only the useful information. PC_1 (Factor 1, Principal Component 1) is that direction from the property space characterized by a maximum variance of the data. PC_2 (Factor 2, Principal Component 2) also has a maximum variance of the data but restricted by orthogonality to the first component. Other *PCs* can be obtained in the same manner, but only few *PCs* will retain the main information from all KFT data. The scores matrix vectors will provide information about similarity/dissimilarity of the samples (grouping of the cases). The loadings matrix vectors will provide the similarity/dissimilarity of properties, as well as the influence of the properties to the model. PC&CA module from Statistica 7.1 package (StatSoft), with centered data and cross-validation method, were used for PCA analysis of the KFT data.

3. Results

3.1. Karl Fischer Water Titration and Kinetics for Bread Samples

The water content of bread core samples varies in the range of 39–44.2% (Table 1). The highest water content of white wheat bread samples was obtained for commercial non-packed samples (43.1%), but with a higher standard deviation (probably due to a reduced uniformity of samples). This aspect was observed also for white bread roll samples, which were non-packed commercial products (39–39.4%

water content, with relatively high SD). The lowest SD in the case of white bread was obtained for the homemade samples; the KFT water content in this case was 41% (Table 1). The Graham samples have the lowest water content (36.5%), while the whole meal and brown bread water content depends on the sample source: the commercial whole meal and brown bread samples have higher water content (~44%), in comparison with the samples obtained directly from the manufacturer (35.1%); this difference can be explained by the possible use of other recipes and/or processing technique for the product from the market (i.e., colorant additives are indicated on the label).

Completely different results were obtained after pre-drying of the white bread core samples (Table 1). The water content after pre-drying is approximately the same for whole meal bread and brown bread core samples (~31%), in comparison with the total water content of 44% without pre-drying.

A similar approach was applied for some shell samples, without or with pre-drying. KFT water content of these samples was obtained only for comparison of the KFT kinetics and evaluating the corresponding parameters for similarity/dissimilarity of samples by PCA. The water content of the bread shell samples decreased to half in comparison with the core samples (22.9% and 19.4% for whole meal and brown bread shell samples, respectively, in comparison with ~44% for the corresponding core samples). On the other hand, the pre-drying of shell samples consistently decreased the water content (from 22.9% to 1.1% for whole meal bread shell and from 19.4% to 4.7% for brown bread shell samples, Table 1).

Table 1. KFT results for bread core samples, pre-dried bread core samples, as well as raw and pre-dried bread shell samples. Values are expressed as mean (±standard deviation, SD) of triplicate analysis (excepting * for duplicate or single analysis).

No.	Code	Sample Mass (g)	KFT Volume (mL)	KFT Volume/Sample Mass Ratio (mL/g) [1]	Water Content (%) [1]
1	HC1 *	0.0768 (±0.0149)	3.223 (±0.498)	42.16 (±1.71) [a]	43.12 (±6.66) [a]
2	HC2	0.0572 (±0.0070)	2.554 (±0.291)	44.66 (±0.84) [a]	39.56 (±0.74) [a]
3	HC3	0.0656 (±0.0178)	2.902 (±0.860)	43.16 (±0.18) [a]	39.04 (±1.00) [a]
4	HC4	0.4192 (±0.5131)	3.002 (±0.799)	44.05 (±1.85) [a]	39.37 (±1.47) [a]
5	HC5 *	0.0690 (±0.0141)	2.894 (±0.566)	41.69 (±0.71) [a]	41.06 (±0.34) [a]
6	GC	0.1288 (±0.0600)	5.328 (±2.491)	40.57 (±1.49) [a]	36.50 (±1.32) [b]
7	MC	0.0850 (±0.0032)	3.884 (±0.187)	46.45 (±0.97) [a]	44.23 (±1.14) [a]
8	BC1	0.0837 (±0.0018)	3.777 (±0.135)	44.06 (±0.59) [a]	44.11 (±0.61) [a]
9	BC2	0.0631 (±0.0376)	2.297 (±1.370)	37.86 (±1.70) [ab]	35.15 (±0.35) [b]
10	MCd *	0.0851	2.731	32.43 [b]	31.06 [c]
11	BCd	0.0851	2.752	32.74 [b]	31.30 [c]
12	MS	0.0909 (±0.0014)	2.154 (±0.081)	22.95 (±1.01) [c]	22.94 (±1.05) [d]
13	MSd *	0.0915	0.099	23.61 [c]	1.05 [e]
14	BS	0.0811 (±0.0084)	1.656 (±0.572)	17.56 (±4.04) [d]	19.44 (±4.72) [d]
15	BSd *	0.0926	0.452	5.35 [e]	4.72 [e]

[1] For a given parameter (KFT volume/sample mass ratio (V/m) and water content (W)), values with different letters are significantly different, according to Tukey's HSD (honestly significant difference) test ($p < 0.05$). All p-level values are presented in Supplementary Materials (Tables S2 and S3).

The aspect of *Volume/sample mass (V/m) versus Time* titration curves from the KFT analysis are very different for bread samples (Figures 2 and 3; see also Figures S1 and S2). Only two pseudolinear ranges were observed for all white bread samples and Graham bread samples, while whole meal and brown bread samples provide three pseudolinear ranges. The water reaction rates for white bread core samples, corresponding to "surface" water molecules, are in the range of 2.4–3.2 mM/s, the highest being determined for packed samples (code "HC2", Table 2). The Graham bread samples have similar behavior, but the water reaction rate for the first interval is lower (1.67 mM/s), while that for the second interval is in the same range (0.06 mM/s). Different results were obtained in the case of brown bread samples: the samples obtained from the market (containing colorants as additives) have similar behavior such as white bread samples (code "BC1", 2.17 mM/s for v_1 and 0.097 mM/s for v_2). It can be due to different recipes used, in comparison with the similar sample obtained from the manufacturer. Higher water reaction rates for both first and second interval (3.9–4.2 and 0.2–0.5 mM/s, respectively) were obtained in the case of whole meal bread and brown bread samples obtained from the manufacturer (code "*BC2*"). Even the core samples were pre-dried in the oven (the "surface" water was partially removed), these water reaction rates remain approximately the same (3.8–4 mM/s for the first interval and 0.3–0.5 mM/s for the second one). Some KFT analyses were performed on bread shell samples, but the results have high SDs in most cases (probably due to the higher hygroscopicity of the crust). The water reaction rates corresponding to the "surface" water molecules are relatively low (0.8–2.1 mM/s), while those corresponding to "strongly retained" water molecules vary in a wide range (0.32–0.44 and 0.06–0.3 mM/s for whole meal and brown bread samples, respectively, Table 2; v_3 values are presented in Table S1).

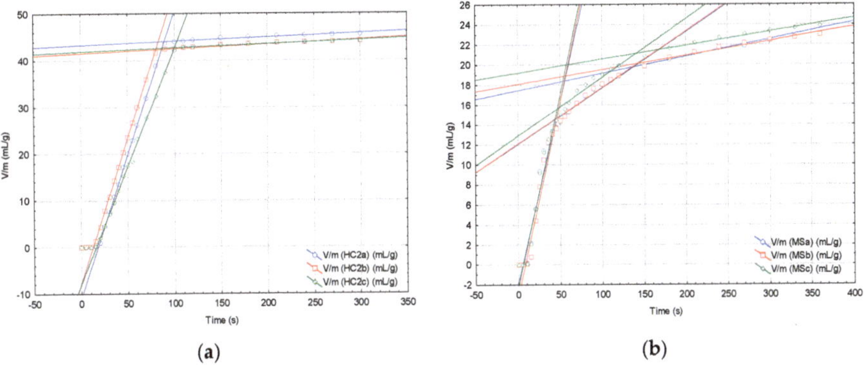

Figure 2. The *Volume/sample mass (V/m)* versus *Time* titration curves from KFT analysis of the selected bread samples (triplicates): (**a**) white wheat bread, sliced and packed in plastic bag (core), code "*HC2*"; and (**b**) whole meal wheat bread (shell), code "*MS*".

Table 2. The mean KFT water reaction rates (based on the variation of the normalized KFT volume on time, $\Delta(V/m)/\Delta t$) on the first two time ranges (pseudolinear) for bread core samples, pre-dried bread core samples and raw and pre-dried bread shell samples Values are expressed as mean (±standard deviation, SD) of triplicate analysis (except * for duplicate or single analysis).

No.	Code	Variation of the KFT Volume/Sample Mass on Time [$\Delta(V/m)_1/\Delta t_1$, mL/g/s] [1]	Variation of the KFT Volume/Sample Mass on Time [$\Delta V/m)_2/\Delta t_2$, mL/g/s] [1]	Mean KFT Reaction Rate v_2 (mM/s) [1]	Mean KFT Reaction Rate v_2 (mM/s) [1]
1	*HC1* *	0.434 (±0.078) [a]	0.0039 (±0.0027) [a]	2.42 (±0.43) [a]	0.022 (±0.015) [a]
2	*HC2*	0.581 (±0.066) [a]	0.0090 (±0.0008) [a]	3.23 (±0.37) [a]	0.050 (±0.004) [a]
3	*HC3*	0.503 (±0.124) [a]	0.0192 (±0.0093) [a]	2.80 (±0.69) [a]	0.107 (±0.052) [a]
4	*HC4*	0.495 (±0.112) [a]	0.0127 (±0.0067) [a]	2.75 (±0.62) [a]	0.071 (±0.037) [a]
5	*HC5* *	0.483 (±0.107) [a]	0.0139 (±0.0015) [a]	2.68 (±0.59) [a]	0.077 (±0.008) [a]
6	*GC*	0.301 (±0.156) [ab]	0.0111 (±0.0093) [a]	1.67 (±0.87) [ab]	0.062 (±0.052) [a]
7	*MC*	0.762 (±0.024) [c]	0.0380 (±0.0056) [b]	4.24 (±0.13) [c]	0.211 (±0.031) [b]
8	*BC1*	0.389 (±0.004) [a]	0.0175 (±0.0034) [a]	2.17 (±0.02) [a]	0.097 (±0.019) [a]
9	*BC2*	0.708 (±0.083) [c]	0.0848 (±0.0086) [c]	3.94 (±0.46) [c]	0.472 (±0.048) [c]
10	*MCd* *	0.717 [c]	0.0557 [d]	3.99 [c]	0.310 [d]
11	*BCd* *	0.682 [c]	0.0857 [cd]	3.79 [c]	0.476 [cd]
12	*MS*	0.380 (±0.015) [a]	0.0571 (±0.0013) [d]	2.12 (±0.09) [a]	0.318 (±0.007) [d]
13	*MSd* *	0.136 [d]	0.0785 [c]	0.76 [d]	0.436 [c]
14	*BS* *	0.180 [d]	0.0533 [d]	1.00 [d]	0.296 [d]
15	*BSd* *	0.205 [d]	0.0108 [a]	1.14 [d]	0.060 [a]

[1] For a given parameter (variation of the KFT volume/sample mass on time ($\Delta(V/m)/\Delta t$) and mean KFT reaction rate (v)), values with different letters are significantly different, according to Tukey's HSD (honestly significant difference) test ($p < 0.05$). All p-level values are presented in the Supplementary Materials (Tables S4 and S5).

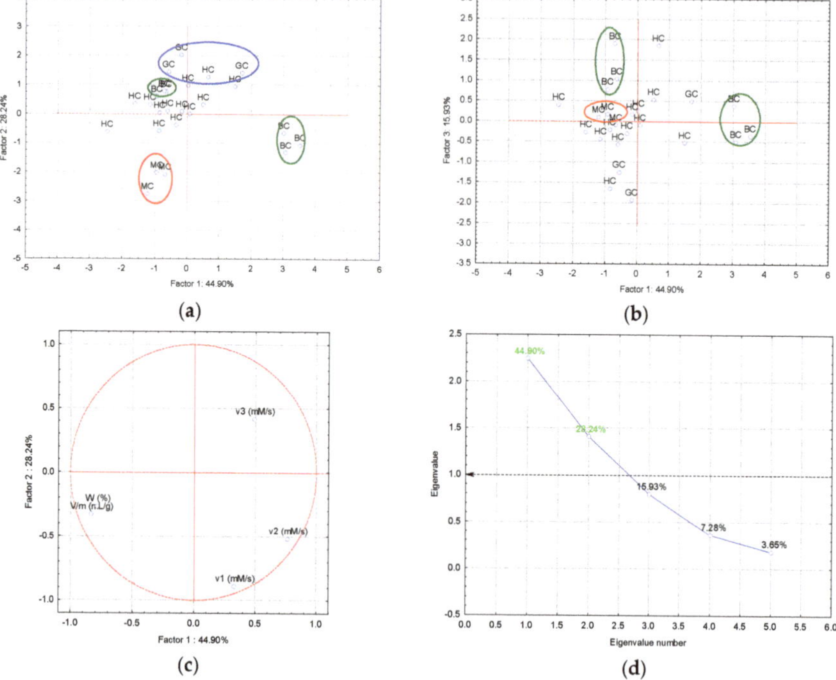

Figure 3. PCA results for the KFT data of bread core samples: (**a**) PC_2 versus PC_1 scores plot; (**b**) PC_3 versus PC_1 scores plot; (**c**) PC_2 versus PC_1 loadings plot; and (**d**) eigenvalues of the correlation matrix. Meaning of color lines in (**a,b**): Green—brown wheat bread core (code "BC"); red—whole meal bread core (code "MC"); blue—Graham bread core (code "GC"). Meaning of color lines in (**d**): Green numbers – relevant eigenvalues.

3.2. Principal Component Analysis on KFT Data for Bread Samples

Valuable classifications were obtained in the case of bread samples using PCA technique. All white bread core samples were located in the center of the scores plot, but less grouped (codes "HC", Figure 3a,b). The whole meal and brown bread core samples are well grouped, according to both PC_2 versus PC_1 and PC_3 versus PC_1 scores plots (codes "BC" (two groups) and "MC"). Some brown bread samples are more similar with white bread samples (samples from the market). Graham bread samples are partially similar with white bread; they are grouped in the upper side of the PC_2 versus PC_1 scores plot (code "GC").

Evident grouping was observed from PCA analysis of the KFT water reaction rate data for core, "C" and shell "S" bread samples. The core samples are especially located in the center of the scores plot, while shell samples were classified in the left of this plots (both PC_2 versus PC_1 and PC_3 versus PC_1, Figure 4a,b). These classifications are due to the KFT water content and normalized volume (W and V/m), as well as KFT reaction rates corresponding to "strongly retained" water for PC_1 (v_2 and v_3) and to KFT reaction rates corresponding to both "surface" and "strongly retained" water for PC_2 (v_1 and v_2, Figures 3c and 4c). The water reaction rate corresponding to the "normal" drift in the KFT process has weak influence in the multivariate analysis. The first two PCs are enough to retain the important information from the KFT data, according to eigenvalues of the correlation matrix from the PCA analysis (Figures 3d and 4d). These PCs explain 73.14% and 80.43% from the variance of the KFT data corresponding to bread core samples and all bread core and shell samples, respectively. Other PCA results are presented in the Supplementary Materials (Figures S3 and S4).

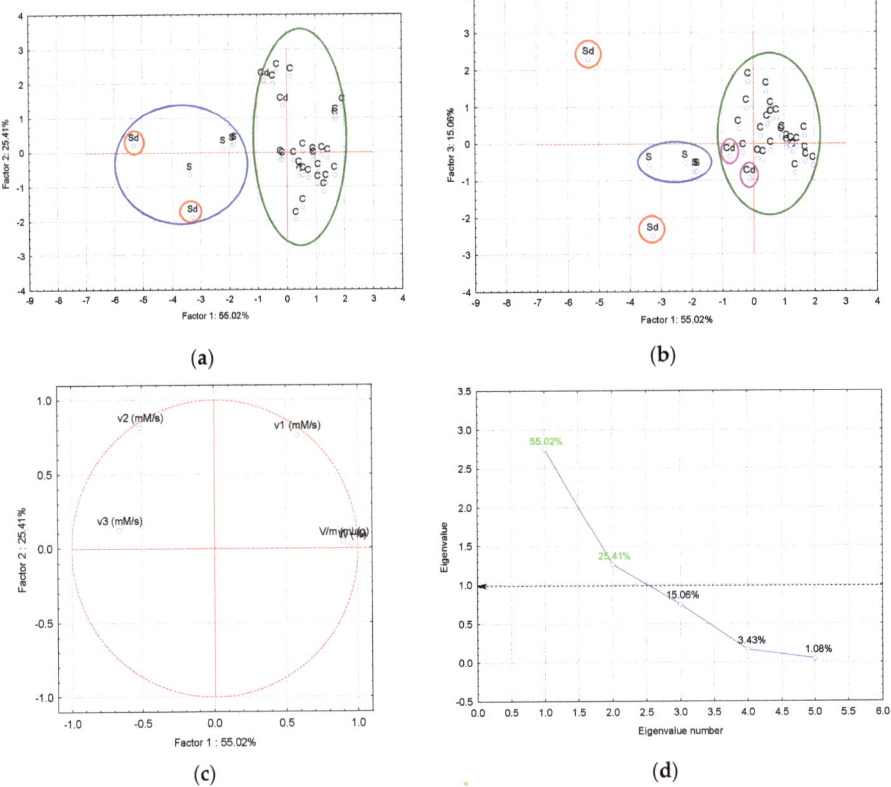

Figure 4. PCA results for the KFT data of all bread core and shell samples: (**a**) PC_2 versus PC_1 scores plot; (**b**) PC_3 versus PC_1 scores plot; (**c**) PC_2 versus PC_1 loadings plot; and (**d**) eigenvalues of the correlation matrix. Meaning of color lines in (**a**,**b**): Green—raw bread core (code "*C*"); pink—pre-dried bread core (code "*Cd*"); blue—raw bread shell (code "*S*"); red—pre-dried bread shell (code "*Sd*"). Meaning of color lines in (**d**): Green numbers—relevant eigenvalues.

4. Discussion

Water determination in bread samples is a challenge. The real water content of such complex food matrices is difficult to evaluate. There are many factors influencing the results, starting from the selection of the method to the composition and sample preparation. The water content ranges in the common values [20,26,48,49], but with differences between bread samples. However, the water content of bread was close to the upper limit of the common range of water content values determined with standard thermal methods or even higher, if the two-component volumetric KFT technique was applied [48,49]. The explanation is that the KFT is sensitive to water, while other methods such as oven drying determines the mass loss. This includes all volatiles that evaporate at the selected temperature, especially water but also aroma compounds. On the other hand, drying at elevated temperature can induces changes in the sample, sometimes with generation of volatiles [27,31–33,50]. KFT analysis allows selecting the polarity of the solvent according to sample type and a temperature program for a complete extraction and titration of water [3,37–39]. Generally, water content of non-packed bread or fresh from the manufacturer is higher, most probably due the water absorption from the environment during storage, transportation and commercialization. Ayub et al. (2003) evaluated the moisture content by oven drying and water activity of bread samples during storage. The moisture content increased by a few percent (sometimes by 12%) after four days, while the water activity also increased [20]. The water content of the bread crumb during the baking process can increase up to 2.5%, followed by a slow decrease up to the final baking step [51]. The maximum moisture content of the core can reach 51%, depending on the composition and baking conditions [26].

The differences between KFT water content of core (crumb) and shell (crust) of breads are obvious, the crust having consistently lower values (about half in comparison with the corresponding cores, Table 1). On the other hand, pre-drying of core and shell samples reveals completely different comparative values. Pre-drying lowered the water content of core samples only by a third, while the shell samples by 75–95%. It is clear that most of the water in the core samples are strongly retained, in comparison with that from shell samples, where a significant water fraction is "surface" water. This behavior can be explained by the presence in the bread core of less degraded starch and gluten, as well as other proteins, which absorb higher quantity of water. The absorbed water molecules are strongly retained by such flour and bread components. On the other hand, the bread crust (shell) has higher fraction of consistently degraded components and the absorbed water is mainly "surface" water. These water molecules are easily evaporated [3,52–54]. Ureta and co-workers (2018) performed baking tests and oven drying moisture determinations on various regions on the bread. The top crust had a moisture content at the final of the baking process of 3.7–9.5%, depending on the baking temperature. The upper, lower and core crumb regions had approximately constant moisture content during baking, with a slight increase at the end of the baking process (up to 42–43%). On the contrary, the moisture content of various crust regions consistently decreases during baking process [26,51].

There is a correlation between the water content determined by oven drying technique and water activity of food products. This is due to the fact that both are related through the sorption isotherm and the water content can be indirectly determined if the sorption isotherm and water activity are known [31]. Such correlation can be observed for the bran-based bread crumb. An increasing of the oven drying water content of bread crumb after the addition of bran fractions from 41% to 43% is correlated with an increasing of the water activity from 0.96 to 0.97 [13,55]. This is particularly due to the influence of bran on the starch–gluten–water interactions. Similar observations have been made for bread obtained with potato fiber in order to improve the product characteristics during storage [56]. On the other hand, the crust and under-crust of various traditional sourdough bread types had a water content of 19% and 23%, respectively (determined by oven drying). The corresponding crumb samples had a water content of ~46.4%. They are well correlated with the water activity. The crust had a water activity of 0.76–0.77, while this parameter was consistently higher for the crumb (~0.97). However, no significant changes on the water activity during eight days of storage have been observed [57]. Such correlations, including the influence on the crispness, fragility and plasticity, have also been obtained for flat extruded wheat and rye breads or bread spread samples containing eggshell powders [58,59]. Regarding the bread crust and crumb, there are many studies related to the influence of various ingredients on the water activity, water content and other physical chemical characteristics [60–65]. No studies on the correlation of the water content determined by KFT and water activity of bread have been made. However, such correlation cannot be simply made due to very different conditions of measurements. Water activity is the ratio between the partial vapor pressure of water in the food sample and the standard state partial vapor pressure of water, which is supposed to be analyzed in gaseous state. On the contrary, the KFT water content is determined in solution or suspension (the case of bread samples), which means water extraction in the liquid working medium.

To our knowledge, no significant studies on the application of KFT for evaluating the types of water molecules in bread have been performed. The strength of physically bonding of water molecules into a less soluble samples can be evaluated from the KFT curve. An efficient and simple way is based on KFT kinetics, especially the mean water reaction rates on specific time ranges. Only few studies on the KFT kinetics for estimating the water content in solids (i.e., magnesium gluconate dihydrate or sucrose) or solutions have been published [66,67]. Our previous studies demonstrate the possibility to evaluate the "surface" and "strongly retained" water molecules in solid samples by KFT kinetics [3,35–37,68–70]. The "surface" water molecules are weak bonded on the surface of the solid particle, especially by hydrogen bonding. They appear by absorption during storage, transportation and even prior analysis of a less soluble solid or semi-solid sample. They also exist in the raw samples. "Surface" water molecules are easily disposable to the KFT working medium and react quickly. This is observed on the first pseudolinear range of the KFT curve and are similar to the KFT water reaction in completely soluble samples [31,33]. On the other hand, "strongly retained" water molecules (physically bonded in the interior of the less soluble solid or semi-solid particles) react at a much slower rate, which principally depends on the diffusion of these water molecules to the surface of sample particles. Thus, this KFT water reaction rate consistently depends on the particle surface or dimensions. It is the case of bread samples, which are heterogeneous particles. The reaction rate of the "strongly retained" water molecules in the bread core samples is very different for white bread and whole meal or brown bread samples due to the heterogeneity and level of degradation in the last case. Moreover, this "strongly retained" water reaction rate can be easily determined for bread shell samples (Figure 2). Generally, water reaction rates are more than twenty times higher for "surface" water molecules in bread core samples, in comparison with the case of bread shell samples (v_1 is up to seven times higher than v_2, Table 2).

The absolute values for the KFT kinetic parameters are difficult to evaluate in order to identify the similarity/dissimilarity of the samples and the corresponding influence of these parameters. The multivariate analysis techniques are the most appropriate for such evaluation. One of these techniques is PCA, which is often used in food analysis. The bread samples are well classified if KFT parameters have been applied to PCA, especially for whole meal and brown bread samples. Moreover, core and shell samples are obviously grouped by means of KFT kinetic parameters. The KFT water reaction rates corresponding to "surface" and "strongly retained" water molecules (i.e., v_1 and v_2, respectively) are more important for discriminating along the second *PC*, while the parameters related to the final KFT water content are important for the first *PC* (Figures 3c and 4c). The latter clearly influence the discrimination between core and shell samples (the "*C*" (core) and "*S*" (shell) groups are separated along PC_1, Figure 4a), while the bread types are discriminated by KFT kinetic parameters, along PC_2 (Figure 3a). These two *PCs* retain the most useful information related for such discriminations and classifications (Figures 3d and 4d). The behavior of bread samples are similar to other less soluble samples such as cereal flour [3,36], meat products [35], cyclodextrins and their complexes [37,68–70].

5. Conclusions

This is the first study on the coupling of Karl Fischer water titration technique with a multivariate statistical analysis method in order to differentiate bread samples through water reaction kinetics. The first two principal components in PCA analysis explain more than 70% of the variance of the KFT data. The bread core and shell samples are completely dissimilar if the "surface" and "strongly retained" KFT water reaction rates are considered. Moreover, whole meal and white wheat bread samples are clearly differentiated through the same water reaction rates. These are correlated with the composition and the way of bonding of water molecules in the bread particles used in the KFT measurement, but the KFT–PCA technique does not need compositional information for discrimination. KFT is selective for water, while other methods such as oven drying determine the overall mass change ("moisture content", which are also comprises of other volatile compounds).

Consequently, the overall KFT titration parameters could be used for discriminating by sample region or parts, while KFT kinetics for classifications by sample types. However, such KFT–PCA coupling technique needs appropriate calibrations and model systems, possibly a database for every type of samples (e.g., bread products and specific solid ingredients for food such as flour, dairy products, etc.).

Supplementary Materials: The following are available online at http://www.mdpi.com/2076-3417/10/18/6518/s1, Figure S1: The *Volume/sample mass (V/m)* versus *Time* titration curves from KFT analysis of the bread samples, Figure S2: The *Volume/sample mass (V/m)* versus *Time* titration curves from KFT analysis of the pre-dried core, shell and pre-dried shell of bread samples, Table S1: The mean KFT water reaction rates on the third time range (pseudolinear, the "normal" drift) for the core, pre-dried core, shell and pre-dried shell of bread samples, Table S2: Significance levels (p-level) from the Tukey's HSD (honestly significant difference) test for the KFT volume/sample mass ratio (V/m) (mL/g) in the case of KFT analysis of all bread samples, Table S3: Significance levels (p-level) from the Tukey's HSD (honestly significant difference) test for the water content (W) (%) in the case of KFT analysis of all bread samples, Table S4: Significance levels (p-level) from the Tukey's HSD (honestly significant difference) test for the mean KFT reaction rate for the first time range (v_1) (mM/s) in the case of KFT analysis of all bread samples, Table S4: Significance levels (p-level) from the Tukey's HSD (honestly significant difference) test for the mean KFT reaction rate for the second time range (v_2) (mM/s) in the case of KFT analysis of all bread samples, Figure S3: PCA results for the KFT data of bread core samples, Figure S4: PCA results for the KFT data of all bread core and shell samples.

Author Contributions: Conceptualization, D.I.H. and N.G.H.; methodology, G.P., I.R., O.A.I., D.I.H., N.G.H. and M.R.; validation, D.I.H. and N.G.H.; formal analysis, D.I.H. and N.G.H.; investigation, L.R., D.I.H., C.A.L. and N.G.H.; resources, M.D.O., M.D., G.S.B. and I.D.; writing—original draft preparation, D.I.H. and N.G.H.; writing—review and editing, G.P., I.R., O.A.I., D.I.H., N.G.H. and M.R.; supervision, N.G.H.; project administration, N.G.H.; and funding acquisition, I.R. and N.G.H. All authors have read and agreed to the published version of the manuscript.

Funding: This research was funded by Ministry of Research and Innovation of Romania, Executive Unit for Financing Higher Education, Research, Development and Innovation (UEFISCDI), grant PNCDI III 2015–2020—ID 368, institutional development project "Ensuring excellence in R&D within USAMVBT", from the institutional performance subprogram 1.2, development of the R&D national system program 1.

Acknowledgments: The authors thank Heinz-Dieter Isengard (University of Hohenheim, Germany) for the help with the KFT equipment and analysis, as well as to Simona Funar-Timofei ("Coriolan Drăgulescu" Institute of Chemistry, Romanian Academy) for the help with Statistica 7.1 software.

Conflicts of Interest: The authors declare no conflict of interest. The funders had no role in the design of the study; in the collection, analyses, or interpretation of data; in the writing of the manuscript, or in the decision to publish the results.

References

1. Dewettinck, K.; Van Bockstaele, F.; Kühne, B.; Van de Walle, D.; Courtens, T.M.; Gellynck, X. Nutritional value of bread: Influence of processing, food interaction and consumer perception. *J. Cereal Sci.* **2008**, *48*, 243–257. [CrossRef]

2. Goesaert, H.; Brijs, K.; Veraverbeke, W.S.; Courtin, C.M.; Gebruers, K.; Delcour, J.A. Wheat flour constituents: How they impact bread quality, and how to impact their functionality. *Trends Food Sci. Technol.* **2005**, *16*, 12–30. [CrossRef]

3. Hădărugă, D.I.; Costescu, C.I.; Corpaş, L.; Hădărugă, N.G.; Isengard, H.-D. Differentiation of rye and wheat flour as well as mixtures by using the kinetics of Karl Fischer water titration. *Food Chem.* **2016**, *195*, 49–55. [CrossRef] [PubMed]

4. Mondal, A.; Datta, A.K. Bread baking—A review. *J. Food Eng.* **2008**, *86*, 465–474. [CrossRef]

5. Lopes Almeida, E.; Chang, Y.K.; Joy Steel, C. Dietary fibre sources in bread: Influence on technological quality. *LWT Food Sci. Technol.* **2013**, *50*, 545–553. [CrossRef]

6. Xu, J.; Wang, W.; Li, Y. Dough properties, bread quality, and associated interactions with added phenolic compounds: A review. *J. Funct. Foods* **2019**, *52*, 629–639. [CrossRef]

7. Birch, A.N.; Petersen, M.A.; Arneborg, N.; Hansen, A.S. Influence of commercial baker's yeasts on bread aroma profiles. *Food Res. Int.* **2013**, *52*, 160–166. [CrossRef]

8. Blandino, M.; Sovrani, V.; Marinaccio, F.; Reyneri, A.; Rolle, L.; Giacosa, S.; Locatelli, M.; Bordiga, M.; Travaglia, F.; Coïsson, J.D.; et al. Nutritional and technological quality of bread enriched with an intermediated pearled wheat fraction. *Food Chem.* **2013**, *141*, 2549–2557. [CrossRef]

9. Bosmans, G.M.; Lagrain, B.; Fierens, E.; Delcour, J.A. The impact of baking time and bread storage temperature on bread crumb properties. *Food Chem.* **2013**, *141*, 3301–3308. [CrossRef]

10. Caballero, P.A.; Gomez, M.; Rosell, C.M. Improvement of dough rheology, bread quality and bread shelf-life by enzymes combination. *J. Food Eng.* **2007**, *81*, 42–53. [CrossRef]

11. Cho, I.H.; Peterson, D.G. Chemistry of Bread Aroma: A Review. *Food Sci. Biotechnol.* **2010**, *19*, 575–582. [CrossRef]

12. Corpaş, L.; Hădărugă, N.G.; Codina, G.G.; Riviş, A.; Guran, E.; Baliţa, E.-N.; Hădărugă, D.I. Phospholipids in homemade bread. *J. Agroaliment. Process. Technol.* **2012**, *18*, 336–340.

13. Curti, E.; Carini, E.; Bonacini, G.; Tribuzio, G.; Vittadini, E. Effect of the addition of bran fractions on bread properties. *J. Cereal Sci.* **2013**, *57*, 325–332. [CrossRef]

14. de Lamo, B.; Gómez, M. Bread enrichment with oilseeds. A Review. *Foods* **2018**, *7*, 191. [CrossRef]

15. Goesaert, H.; Slade, L.; Levine, H.; Delcour, J.A. Amylases and bread firming—An integrated view. *J. Cereal Sci.* **2009**, *50*, 345–352. [CrossRef]

16. Jiang, D.; Peterson, D.G. Identification of bitter compounds in whole wheat bread. *Food Chem.* **2013**, *141*, 1345–1353. [CrossRef]

17. Moayedallaie, S.; Mirzaei, M.; Paterson, J. Bread improvers: Comparison of a range of lipases with a traditional emulsifier. *Food Chem.* **2010**, *122*, 495–499. [CrossRef]

18. Shin, D.-J.; Kim, W.; Kim, Y. Physicochemical and sensory properties of soy bread made with germinated, steamed, and roasted soy flour. *Food Chem.* **2013**, *141*, 517–523. [CrossRef]

19. Belz, M.C.E.; Ryan, L.A.M.; Arendt, E.K. The Impact of Salt Reduction in Bread: A Review. *Crit. Rev. Food Sci. Nutr.* **2012**, *52*, 514–524. [CrossRef]

20. Ayub, M.; Wahab, S.; Durrani, Y. Effect of Water Activity (Aw) Moisture Content and Total Microbial Count on the Overall Quality of Bread. *Int. J. Agric. Biol.* **2003**, *5*, 274–278.

21. Chen, G.; Öhgren, C.; Langton, M.; Lustrup, K.F.; Nydén, M.; Swenson, J. Impact of long-term frozen storage on the dynamics of water and ice in wheat bread. *J. Cereal Sci.* **2013**, *57*, 120–124. [CrossRef]

22. Miś, A.; Krekora, M.; Niewiadomski, Z.; Dziki, D.; Nawrocka, A. Water redistribution between model bread dough components during mixing. *J. Cereal Sci.* **2020**, *95*, 103035. [CrossRef]

23. Gil, M.J.; Callejo, M.J.; Rodriguez, G. Effect of water content and storage time on white pan bread quality: Instrumental evaluation. *Z. Lebensm. Unters. Forsch.* **1997**, *205*, 268–273. [CrossRef]

24. Cappelli, A.; Oliva, N.; Cini, E. Stone milling versus roller milling: A systematic review of the effects on wheat flour quality, dough rheology, and bread characteristics. *Trends Food Sci. Technol.* **2020**, *97*, 147–155. [CrossRef]

25. Ziobro, R.; Korus, J.; Juszczak, L.; Witczak, T. Influence of inulin on physical characteristics and staling rate of gluten-free bread. *J. Food Eng.* **2013**, *116*, 21–27. [CrossRef]

26. Wagner, M.J.; Lucas, T.; Le Ray, D.; Trystram, G. Water transport in bread during baking. *J. Food Eng.* **2007**, *78*, 1167–1173. [CrossRef]

27. Margolis, S.A.; Huang, P.H.; Hădărugă, N.G.; Hădărugă, D.I. Water determination. In *Encyclopedia of Analytical Science*, 3rd ed.; Elsevier: Oxford, UK, 2019; Volume 10, pp. 382–390.

28. Yazgan, S.; Bernreuther, A.; Ulberth, F.; Isengard, H.-D. Water—An important parameter for the preparation and proper use of certified reference materials. *Food Chem.* **2006**, *96*, 411–417. [CrossRef]

29. Isengard, H.-D.; Kling, R.; Reh, C.T. Proposal of a new reference method to determine the water content of dried dairy products. *Food Chem.* **2006**, *96*, 418–422. [CrossRef]

30. Isengard, H.-D.; Präger, H. Water determination in products with high sugar content by infrared drying. *Food Chem.* **2003**, *82*, 161–167. [CrossRef]

31. Isengard, H.-D. Water content, one of the most important properties of food. *Food Control.* **2001**, *12*, 395–400. [CrossRef]

32. Heinze, P.; Isengard, H.-D. Determination of the water content in different sugar syrups by halogen drying. *Food Control.* **2001**, *12*, 483–486. [CrossRef]

33. Isengard, H.-D. Rapid water determination in foodstuffs. *Trends Food Sci. Technol.* **1995**, *6*, 155–162. [CrossRef]

34. Sahin, A.W.; Wiertz, J.; Arendt, E.K. Evaluation of a new method to determine the water addition level in gluten-free bread systems. *J. Cereal Sci.* **2020**, *93*, 102971. [CrossRef]

35. Zippenfening, S.E.; Hădărugă, D.I.; Negrea, I.; Velciov, A.; Hădărugă, N.G. Karl Fischer titration of salami products: Variation of the water content during ripening and kinetics of the KFT process. *Chem. Bull. Politeh. Univ. Timişoara* **2015**, *64*, 18–24.

36. Corpaş, L.; Hădărugă, N.G.; David, I.; Pîrşan, P.; Hădărugă, D.I.; Isengard, H.-D. Karl Fischer water titration—Principal component analysis approach on wheat flour. *Food Anal. Methods* **2014**, *7*, 1353–1358. [CrossRef]

37. Hădărugă, N.G.; Hădărugă, D.I.; Isengard, H.-D. Water content of natural cyclodextrins and their essential oil complexes: A comparative study between Karl Fischer titration and thermal methods. *Food Chem.* **2012**, *132*, 1741–1748. [CrossRef]

38. Felgner, A.; Schlink, R.; Kirschenbuhler, P.; Faas, B.; Isengard, H.-D. Automated Karl Fischer titration for liquid samples—Water determination in edible oils. *Food Chem.* **2008**, *106*, 1379–1384. [CrossRef]

39. Schmitt, K.; Isengard, H.-D. Karl Fischer titration. A method for determining the true water content of cereals. *Fresenius' J. Anal. Chem.* **1998**, *360*, 465–469. [CrossRef]

40. Merkh, G.; Pfaff, R.; Isengard, H.-D. Capabilities of automated Karl Fischer titration combined with gas extraction for water determination in selected dairy products. *Food Chem.* **2012**, *132*, 1736–1740. [CrossRef]

41. Isengard, H.-D.; Kerwin, H. Proposal of a new reference method for determining water content in butter oil. *Food Chem.* **2003**, *82*, 117–119. [CrossRef]

42. Tomassetti, M.; Campanella, L.; Aureli, T. Thermogravimetric analysis of some spices and commercial food products. Comparison with other analytical methods for moisture content determination (Part 3). *Thermochim. Acta* **1989**, *143*, 15–26. [CrossRef]

43. Isengard, H.-D.; Haschka, E.; Merkh, G. Development of a method for water determination in lactose. *Food Chem.* **2012**, *132*, 1660–1663. [CrossRef]

44. Kestens, V.; Conneely, P.; Bernreuther, A. Vaporisation coulometric Karl Fischer titration: A perfect tool for water content determination of difficult matrix reference materials. *Food Chem.* **2008**, *106*, 1454–1459. [CrossRef]

45. Rückold, S.; Grobecker, K.H.; Isengard, H.-D. The effects of drying on biological matrices and the consequences for reference materials. *Food Control.* **2001**, *12*, 401–407. [CrossRef]

46. Kestens, V.; Charoud-Got, J.; Bau', A.; Bernreuther, A.; Emteborg, H. Online measurement of water content in candidate reference materials by acousto-optical tuneable filter near-infrared spectrometry (AOTF-NIR) using pork meat calibrants controlled by Karl Fischer titration. *Food Chem.* **2008**, *106*, 1359–1365. [CrossRef]

47. Isengard, H.-D.; Färber, J.-M. 'Hidden parameters' of infrared drying for determining low water contents in instant powders. *Talanta* **1999**, *50*, 239–246. [CrossRef]

48. Belitz, H.-D.; Grosch, W.; Schieberle, P. *Food Chemistry*; Springer: Berlin, Germany, 2009; pp. 670–745. [CrossRef]

49. Bhatt, C.M.; Nagaraju, J. Studies on electrical properties of wheat bread as a function of moisture content during storage. *Sens. Instrum. Food Qual. Saf.* **2010**, *4*, 61–66. [CrossRef]

50. Isengard, H.-D.; Heinze, P. Determination of total water and surface water in sugars. *Food Chem.* **2003**, *82*, 169–172. [CrossRef]

51. Ureta, M.M.; Diascorn, Y.; Cambert, M.; Flick, D.; Salvadori, V.O.; Lucas, T. Water transport during bread baking: Impact of the baking temperature and the baking time. *Food Sci. Technol. Int.* **2018**, *25*, 187–197. [CrossRef]

52. Hădărugă, D.I.; Birău (Mitroi), C.L.; Gruia, A.T.; Păunescu, V.; Bandur, G.N.; Hădărugă, N.G. Moisture evaluation of β-cyclodextrin/fish oils complexes by thermal analyses: A data review on common barbel (*Barbus barbus* L.), Pontic shad (*Alosa immaculata* Bennett), European wels catfish (*Silurus glanis* L.), and common bleak (*Alburnus alburnus* L.) living in Danube river. *Food Chem.* **2017**, *236*, 49–58. [CrossRef]

53. Hădărugă, N.G.; Bandur, G.N.; David, I.; Hădărugă, D.I. A review on thermal analyses of cyclodextrins and cyclodextrin complexes. *Environ. Chem. Lett.* **2019**, *17*, 349–373. [CrossRef]

54. Hădărugă, N.G.; Szakal, R.N.; Chirilă, C.A.; Lukinich-Gruia, A.T.; Păunescu, V.; Muntean, C.; Rusu, G.; Bujancă, G.; Hădărugă, D.I. Complexation of Danube common nase (*Chondrostoma nasus* L.) oil by β-cyclodextrin and 2-hydroxypropyl-β-cyclodextrin. *Food Chem.* **2020**, *303*, 125419. [CrossRef] [PubMed]

55. Curti, E.; Carini, E.; Tribuzio, G.; Vittadini, E. Effect of bran on bread staling: Physico-chemical characterization and molecular mobility. *J. Cereal Sci.* **2015**, *65*, 25–30. [CrossRef]

56. Curti, E.; Carini, E.; Diantom, A.; Vittadini, E. The use of potato fibre to improve bread physico-chemical properties during storage. *Food Chem.* **2016**, *195*, 64–70. [CrossRef]

57. Chiavaro, E.; Vittadini, E.; Musci, M.; Bianchi, F.; Curti, E. Shelf-life stability of artisanally and industrially produced durum wheat sourdough bread ("Altamura bread"). *LWT Food Sci. Technol.* **2008**, *41*, 58–70. [CrossRef]

58. Marzec, A.; Lewicki, P.P. Antiplasticization of cereal-based products by water. Part I. Extruded flat bread. *J. Food Eng.* **2006**, *73*, 1–8. [CrossRef]

59. Kobus-Cisowska, J.; Szymanowska-Powałowska, D.; Szymandera-Buszka, K.; Rezler, R.; Jarzębski, M.; Szczepaniak, O.; Marciniak, G.; Jędrusek-Golińska, A.; Kobus-Moryson, M. Effect of fortification with calcium from eggshells on bioavailability, quality, and rheological characteristics of traditional Polish bread spread. *J. Dairy Sci.* **2020**, *103*, 6918–6929. [CrossRef]

60. Primo-Martín, C.; de Beukelaer, H.; Hamer, R.J.; Van Vliet, T. Fracture behaviour of bread crust: Effect of ingredient modification. *J. Cereal Sci.* **2008**, *48*, 604–612. [CrossRef]

61. Purlis, E.; Salvadori, V.O. Modelling the browning of bread during baking. *Food Res. Int.* **2009**, *42*, 865–870. [CrossRef]

62. Primo-Martín, C.; van de Pijpekamp, A.; van Vliet, T.; de Jongh, H.H.J.; Plijter, J.J.; Hamer, R.J. The role of the gluten network in the crispness of bread crust. *J. Cereal Sci.* **2006**, *43*, 342–352. [CrossRef]

63. Pico, J.; Reguilón, M.P.; Bernal, J.; Gómez, M. Effect of rice, pea, egg white and whey proteins on crust quality of rice flour-corn starch based gluten-free breads. *J. Cereal Sci.* **2019**, *86*, 92–101. [CrossRef]

64. Lind, I.; Rask, C. Sorption isotherms of mixed minced meat, dough, and bread crust. *J. Food Eng.* **1991**, *14*, 303–315. [CrossRef]

65. Wolter, A.; Hager, A.-S.; Zannini, E.; Czerny, M.; Arendt, E.K. Influence of dextran-producing *Weissella cibaria* on baking properties and sensory profile of gluten-free and wheat breads. *Int. J. Food Microbiol.* **2014**, *172*, 83–91. [CrossRef] [PubMed]

66. Grünke, S.; Wünsch, G. Kinetics and stoichiometry in the Karl Fischer solution. *Fresenius' J. Anal. Chem.* **2000**, *368*, 139–147.

67. Yap, W.T.; Cummings, A.L.; Margolis, S.A.; Schaffer, R. Estimation of water content by kinetic method in Karl Fischer titration. *Anal. Chem.* **1979**, *51*, 1595–1596. [CrossRef]

68. Hădărugă, D.I.; Hădărugă, N.G.; Bandur, G.N.; Isengard, H.-D. Water content of flavonoid/cyclodextrin nanoparticles: Relationship with the structural descriptors of biologically active compounds. *Food Chem.* **2012**, *132*, 1651–1659. [CrossRef]

69. Hădărugă, N.G.; Hădărugă, D.I.; Isengard, H.-D. "Surface water" and "strong-bonded water" in cyclodextrins: A Karl Fischer titration approach. *J. Incl. Phenom. Macrocycl. Chem.* **2013**, *75*, 297–302. [CrossRef]

70. Ünlüsayin, M.; Hădărugă, N.G.; Rusu, G.; Gruia, A.T.; Păunescu, V.; Hădărugă, D.I. Nano-encapsulation competitiveness of omega-3 fatty acids and correlations of thermal analysis and Karl Fischer water titration for European anchovy (*Engraulis encrasicolus* L.) oil/β-cyclodextrin complexes. *LWT Food Sci. Technol.* **2016**, *68*, 135–144. [CrossRef]

applied
sciences

Review

Brewer's Spent Grains: Possibilities of Valorization, a Review

Ancuța Chetrariu and Adriana Dabija *

Faculty of Food Engineering, Stefan cel Mare University of Suceava, 720229 Suceava, Romania;
ancuta.chetrariu@fia.usv.ro
* Correspondence: adriana.dabija@fia.usv.ro; Tel.: +40-748-845-567

Received: 3 August 2020; Accepted: 12 August 2020; Published: 13 August 2020

Abstract: This review was based on updated research on how to use brewer's spent grains (BSG). The use of BSG was considered both in food, as an ingredient or using value-added components derived from brewer's spent grain, or in non-food products such as pharmaceuticals, cosmetics, construction, or food packaging. BSG is a valuable source of individual components due to its high nutritional value and low cost that is worth exploiting more to reduce food waste but also to improve human health and the environment. From the bioeconomy point of view, biological resources are transformed into bioenergetically viable and economically valuable products. The pretreatment stage of BSG biomass plays an important role in the efficiency of the extraction process and the yield obtained. The pretreatments presented in this review are both conventional and modern extraction methods, such as solvent extractions or microwave-assisted extractions, ultrasonic-assisted extractions, etc.

Keywords: brewer's spent grain; bioeconomy; valuable compounds

---

1. Introduction

While global hunger still plays an important role affecting millions of people, there is an overproduction of food in developed countries. High production of food includes a big amount of waste. Around 90 million tons of food waste per year are produced, which causes serious environmental problems [1], but the waste reduction has become of interest to researchers, thus alternative uses can be found [2]. Food waste can be an important source of carbohydrates, proteins, lipids, and complex nutraceuticals. The bioeconomy addresses the possibilities of transforming renewable biological resources into economically and bioenergetically viable products [3]. In industries around the world, there is a need to expand the basic resources for the production of fuel, chemicals, energy, and materials, but the pretreatment of biomass is the main impediment to the development of the bioeconomy [4]. Waste reduction has a positive impact on the environment (soil, water, atmosphere) and contributes to climate change mitigation [5]. Zero hunger, responsible food consumption, and production belong to the 17 global sustainability goals of the United Nations, which represents a global action by 2030 and defines sustainable development in the three dimensions: economic, social, and environmental [3]. Product Lifecycle Assessment (LCA) is an environmental management technique adopted by the main food industry companies regarding the environmental impact of the product and the flow of materials, energy, and waste [3]. In 2018, 1.94 billion hectolitres of beer were produced globally [6]. An hL of beer results in 20 kg of wet brewers spent grain (BSG) [7,8], which means that 2018 resulted in 38.8 million tons of wet BSG.

BSG consists of layers of peel, pericarp, and seeds with residual amounts of endosperm and aleurone from barley used as raw material (Figure 1). BSG has 80% moisture, sweet taste, malt smell, and can be considered lignocellulosic material [9–11]; it is characterized by large amounts of fiber (up to

70%), including cellulose, hemicellulose, and lignin, and protein content of 25–30% [3,12,13]. BSG is a little used by-product due to its high moisture content, which makes it difficult to transport and store and makes it an unstable product conducive to microbial growth [8]. Recycling and reuse of food waste and by-products benefits both the beer industry and the environment. EUROSTAT estimates approximately 90 million tons of food is wasted in the EU/year, which means 179 kg/person [14]. EU norms encourage the extraction of valuable components (functional foods, adjuvants, pharmaceutical preparations) from food industry by-products. Until recently, food waste was of no interest, being used only for animal feed or composting, but current trends—and because it represents a cheap and valuable raw material—have brought food waste to the attention of researchers. Waste is a valuable resource that we do not yet know how to use intelligently, and converting it from problem to resource needs to be more of an interest to us. By-products and waste can become a sustainable alternative source to reduce malnutrition and hunger in developing countries [15]. There is a growing interest in finding natural resources with antioxidant activity to effectively replace synthetic antioxidants with toxic and carcinogenic effects [16]. Oxidative stress occurs as an imbalance between the number of oxidants and the antioxidant defense mechanism, and regular consumption of cereals and by-products helps prevent many chronic diseases associated with oxidative stress. Since BSG is derived from food materials, it can be incorporated into food diets (bread, snacks, muffins, pasta, etc.) [17]. The purpose of this review is to update information on BSG, health roles, and methods of extraction of valuable components.

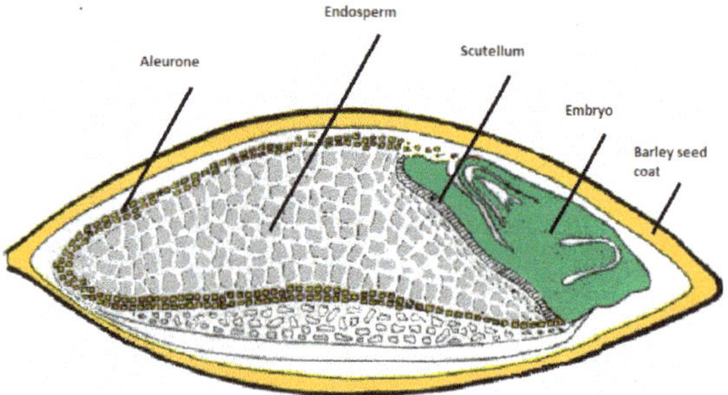

Figure 1. Barley structure.

2. BSG: A Valuable By-Product

BSG has a complex chemical composition, which varies according to variety of barley used, harvesting and malting time, and mixing time [8]. Due to the high moisture of the BSG (approximately 80%), shelf life is 7–10 days. Preservation can be done with acids (lactic acid, benzoic acid, formic acid, or acetic acid), which is contrary to the desire of consumers of the most natural food, or can be used in a mixture of benzoate-propionate-sorbate in concentrations 0.2–0.3% (v/v), which can extend its validity by 4–5 days [8]. Drying is considered the most effective method of preservation of BSG. A drying method can be used in two stages—pressing and drying—bringing moisture to 10% [9]. According to studies by Lynch et al. (2016), other preservation/drying methods that can be used are:

- drying in the oven is considered the most suitable but must be carried out at temperatures below 60 °C, with the disadvantage of a high energy consumption;
- drying by overheated steam in a thin layer brings the advantage of less consumption and improved drying efficiency;
- autoclaving at 121 °C for one hour has the disadvantage of solubilization of polysaccharides and phenolic compounds;
- drying by freezing with the disadvantage of the need for large storage spaces;

- pressing and filtration through the membrane followed by vacuum drying and drying 2 days in the air to bring moisture to 10% with the advantage that the products in which it was used no longer show microbial growth for 6 months [18].

Wet preservation can be achieved by adding salt, mixing with pomace, membrane filtration, or vacuum drying (moisture 20–30%) [13]. BSG has high proliferation post-production and for later use must be stabilized and properly stored. For long storage, a moisture of 10% is recommended.

The chemical composition of the brewer's spent grain is shown in the Table 1.

Table 1. Chemical composition of dry brewers spent grain *.

Proteins	Lipids	Fibers	Carbohydrates	Lignin	Arabinoxilan (AX)	Ashes	Lyzine	Study
234 mg/g	-	-	459 mg/g	-	-	-	-	[19]
24.69%	-	-	-	-	-	4.18%	-	[16]
15–28%	5–8%	-	-	-	-	4.5–6%	-	[13]
20%	-	50%	-	10–28%	40%	-	14.30%	[8]
18–35.4%	-	-	-	-	-	-	14.30%	[14]
14.2–31%	3–13%	59.1–74.1%	-	-	-	-	-	[20]
15–26%	3.9–10%	70%	-	-	-	-	-	[21]
19.20%	-	-	-	22.30%	-	4.54%	-	[22]
15.4–30%	10%	-	-	11.9–27.8%	-	2–5%	-	[3]
31%	9%	-	-	16%	-	4%	-	[18]
15.3–24.6%	-	-	-	11.9–27.8%	-	1.2–4.6%	-	[23]
22.44%	5.3%	-	46.52%	19.57%	-	3.54%	-	[24]
31.81%	-		3.07%	12.72%				[25]

* values are expressed of dry matter.

Fibers are the components with the greatest interest in health intake (arabinoxylans and β-glucans) but also phenolic components (hydroxycinnamic acid) [8]. Hemicelluloses consisting mainly of arabinoxylan (AX) can be present in up to 40% of the dry weight of BSG (AX is the main non-cellulosic polysaccharide present in cereals and herbs). Another important polysaccharide in BSG is cellulose, and among monosaccharides are found xylose, glucose, arabinose, and traces of rhamnose and galactose [8].

Proteins are found to be around 20% of the dry mass, and the most abundant are hordeins, glutelines, globulins, and albumins. Of the protein content, 30% are essential amino acids, the most present being lysine [8].

Lipids are found in a low percentage, of which triglycerides have a higher percentage (approximately 67% of total extract), followed by 18% fatty acids, according to del Rio et al. (2013) [26].

BSG also contains phosphorus (6000 mg/kg), calcium (3600 mg/kg), and magnesium (1900 mg/kg) [8,9], iron, copper, potassium, and manganese [12], and the vitamins present are biotin (0.1 ppm), niacin (44 ppm), choline (1800 ppm), folic acid (0.2 ppm), pantothenic acid (8.5 ppm), riboflavin (1.5 ppm), thiamine (0.7 ppm), and pyridoxine (0.7 ppm) [9,11,13]. In addition to all these components are extractives made up of waxes, gums, resins, tannins, essential oils, and other cytoplasmatic components [13].

It is of interest to separate BSG into individual components due to its high nutritional value and low cost, having applications in both food and non-food industries. When BSG was introduced into the rat diet, it had a beneficial effect on both constipation and diarrhea due to the high content of non-cellulosic polysaccharides, the small quantity of β-glucans, and the content of proteins rich in glutamine [27]. Due to high annual quantity and low cost, it is a source of interest worth exploiting [11].

3. Possible Uses of BSG

Due to the complex composition of BSG, there could be diverse uses, some of which are very well known while others are still developing.

3.1. BSG as Animal Feed

BSG is mainly used as animal feed. When used together with nitrogen sources (e.g., urea), it provides all the essential amino acids for ruminants [11,12]. BSG is suitable as feed for cattle, poultry, pigs, and fish. Used to feed the cattle, it leads to a milk production increase but decreases its lipid content [12]. BSG can be used as a source of protein in feeding *Pangasiusanodon hypophthalmus* fish, and the 50% substitution of soy flour proved most advantageous, with fish gaining weight at the end of the experiment; additionally, protein content was significantly higher, and the cost of feed decreased by 27.56%, as evidenced by Jayant et al. (2018) [28].

3.2. BSG in Food

Ingestion of BSG benefits human health, in such ways as: it accelerates intestinal transit and alleviates both diarrhea and constipation, decreases the incidence of gallstones, and reduces cholesterol and postprandial glucose level, all due to the protein-rich content in glutamine, non-cellulosic polysaccharides, and soluble dietary fibers [12]. BSG can be added to food to increase fiber and protein contents [29]. Due to biologically active compounds with physiological roles in the body, food fortified with BSG is considered functional food that offers health benefits and is used as an adjunct in a balanced diet [11].

The addition of functional ingredients to the bakery has been in the sights of scientists due to its ability to reduce the risk of chronic diseases beyond basic nutritional function [20]. A percentage of up to a maximum of 15% BSG can be used, as more would negatively influence sensory properties. Bread obtained from wheat flour and treated with four enzymes in which BSG (0–30%) was added had a higher shelf life as well as improved texture and volume, according to McCarthy et al. (2013) [11]. In the study on the rheological properties of the dough supplemented with BSG by Ktenoudaki et al. (2013), it was concluded that it negatively influences the texture and decreases the volume of the dough; the structure becomes dense and is negatively influenced by both the extensional biaxial viscosity and the uniaxial extensibility [30]. With the increase of the amount of BSG addition, dough strength, peak viscosity, final viscosity, and setback decrease.

When ready-to-eat snacks rich in fiber are desired, BSG [31] can be used in products with a double fiber content compared to the control samples. Studies have revealed that 10% BSG leads to baked snacks with high crispiness index (Ci), low crispiness work (Wc), and a large number of peaks during texture analysis, which means it does not adversely affect the product. BSG has a characteristic smell that contributes to the flavor of the products, but after the test acceptance, it follows that 10% BSG offers the possibility of incorporating this product into ripe snacks [32].

BSG can be added to frankfurters sausages to obtain low-fat products [11]. A maximum of 15% BSG can be added to sausages. In the case of extrusion cooking (reactive extrusion), percentages greater than 10% BSG change volume, color, texture, and structure of the products. It results in darker products with low volume, low hardness, and denser structure [8,11]. The addition of BSG to fruit juices and smoothies has a beneficial effect on increasing phenolic content and antioxidant activity [8].

In the study conducted by Spinelli et al. (2016) fish burgers prepared with various bioactive powders of BSG were obtained [33]. Polyphenols and flavonoids were extracted by supercritical CO_2 at 40 °C and 35 MPa. The extract was microencapsulated and dried by drying spray to prevent it from tasting unpleasant or bitter but also to protect it from the high temperatures during cooking. The 5% BSG powder (spray drying powders) gave the best ratio between the chemical properties of active powders and the sensory evaluation of samples. In this respect, antioxidant properties were also compared; the samples containing BSG had a 30% higher amount of polyphenols and 50% more flavonoids than the control sample. Attempts were made to replace feed with BSG for finishing steers fed and as a source of polyphenols, but it had no particular effects on the validity of meat [34].

BSG strengthens foods such as pasta, infant formula, meat, and meat products [14]. Cappa and Alamprese (2017) enriched fresh pasta with eggs with BSG in a quantity of 6.2 g/100 g, obtaining fiber-rich products, even if the BSG percentage was low [35]. BSG was successfully used to obtain a fermented beverage rich in phenolic compounds, the beverage that has a shelf life regarding the bioactive components for 15 days [36].

3.3. BSG Used as a Substrate

Lignocellulosic materials oppose hydrolysis, which is why lingnocellulosic biomass is subject to pre-treatments that make it accessible [5], then cellulose and hemicellulose undergo hydrolysis with their transformation into sugars that can be used by microorganisms [37].

Pretreatments are expensive and have a significant impact on the environment; thus far, no pretreatment with conversion factor 100 has been found, the most used being acidic solutions (cheap and effective) using low concentrations of acids (<5%), high temperatures (120–210 °C), pressure < 10 atm, or concentrated acids (<30%), temperatures < 100 °C, and atmospheric pressure [38]. To obtain fermentation media capable of converting polysaccharides into bioactive compounds, BSG is used as the raw material in the biorefinery process, because "green chemistry" requires finding alternatives instead of thermal or chemical hydrolytic processes. One option in this respect may be pretreatment with ion fluids, followed by enzymatic hydrolysis. Delignification with cholinium-based ionic liquids was achieved with a yield of 75.89% compared to traditional delignification with imidazole with a yield of 40.18% [39].

3.3.1. Substrate for the Cultivation of Microorganisms and the Production of Enzymes

BSG is mainly used for the growth of fungi for the production of enzymes (alpha-amylases, cellulases, hemicellulases), and amino acids, vitamins, and inorganic compounds are added to improve enzyme yield [8]. The microorganisms successfully grown on the medium of BSG are *Peurotus*, *Agrocybe*, *Lentinus*, and *Trichoderma*, and *Streptomyces* bacteria, according to reports of Mussatto (2009) [23]. BSG is suitable for the isolation and the maintenance of known and highly suitable strains for screening and production of new biologically active substances [17].

According to studies by Mussatto (2009), the substrate value is determined by the conversion yield of biomass [23]. An essential step in obtaining enzymes is the pretreatment of lignocelluloses, followed by enzymatic hydrolysis. The production of commercial enzymes is an expensive process [1]. Varieties of arabinoolignoxilans (AX) can be obtained with different structures through hydrolysis processes, products of interest to the food industry [1].

3.3.2. A Substrate in Fermentation Processes

Production of Xylitol

Xylitol is a polyol with sweetening power similar to sucrose but with lower caloric value (2.4 versus 4.0 cal/g), thus it can also be consumed by insulin-deficient people. The downside is the higher price of sucrose that can be solved by an affordable production technology [40]. Xylitol can be produced by microbial fermentation, the BSG proving to be a more economical alternative to other lignocellulosic materials [17]. Mussatto et al. (2008) and Mussatto and Roberto (2006) optimized the process for the hydrolysis of hemicelluloses from BSG by obtaining xylitol as the final product [41,42]. The first step in obtaining xylitol is the fermentation of the BSG substrate with the help of *Candida guilliermondii* yeast [43]. From the fermentation process results CO_2, which is separated from the bioreactor. The broth is followed by evaporation at a temperature of 40 °C and an ethanol crystallization process to increase xylitol insolubility to 5 °C. To separate the crystallized xylitol from molasses, centrifugation is used, resulting in efficiency between 87–92%, according to studies conducted by Mussatto et al. (2013) [44]. From rice straws was obtained a yield of xylitol (percentage of the theoretical yield of 0.917 g/g) of 78.5% in agitated flasks or 57.8% in a stirred-tank bioreactor with detoxified hydrolysate [45]. From BSG was obtained 0.107 g xylitol per gram dry matter [46], and from sugar cane biogases was obtained 0.633 g/g with an efficiency of 68.97% [47]. Xylitol can be given to diabetics due to low caloric value and has beneficial properties in otitis media, osteoporosis, and lung infections [46]. Xylitol is of great commercial importance due to its anti-cariogenic properties, preventing the formation of acids that attack tooth enamel and at the same time inducing the remineralization of enamel [47].

Production of Lactic Acid

BSG has been evaluated as the raw material for the production of lactic acid [17]. Polylactic acid (PLA) has bioplastic applications, and BSG is a precursor in obtaining it with the help of *Lactobacillus delbrueckii* from lignocellulosic material. Cellulose is subjected to a chemical treatment to make it more accessible to enzymes, enzymatic hydrolysis, which obtains a sugar solution with glucose as the main component, followed by the fermentation of hydrolysate with the help of *Lactobacillus delbrueckii* [21,48]. A similar way of producing lactic acid was also used by Mussatto et al. (2013). The process begins with enzymatic saccharification of cellulose; after, the cellulosic material is washed with water until neutralized and dried up to 10% [44]. Unconverted components are separated by vacuum filtration, and highly hydrolyzed glucose is fermented with *Lactobacillus delbrueckii*, resulting in a yield of 0.96 g lactic acid per gram glucose. In another study, Mussatto et al. (2008) obtained a lactic acid yield of 0.98 g/g glucose, similar to the theoretical maximum value of 1 g/g glucose [41]. Other microorganisms used to obtain lactic acid are *Lactobacillus pentosus* and *Lactobacillus rhamnosus* [23].

Obtaining lactic acid offers great opportunities in obtaining biodegradable polymers as plant growth regulators or in obtaining green chemicals/solvents [21]. Liang and Wan (2015) used BSG substrates to obtain carboxylic acids with mixed-culture fermentation, a study in which lactic acid was the dominant one in both acidic pH (9.2 g/L) and alkaline pH (6.7 g/L) [49].

Ethanol Production

Lignocellulosic hydrolysates can be used as a fermentation medium for obtaining ethanol, xylitol, and other products, which must be subjected to detoxification before fermentation to improve the efficiency of the process [37]. Applied treatments can be treatments with acids, microwave digestion, ultrasound, or enzymatic hydrolysis to make glucose extraction from cellulose easier [12]. Celluloses break down into glucose, and hemicelluloses break down into xylose, arabinosis, manosis, glucose, acetic acid, and galactose [37], which can be converted with microorganisms into ethanol. Two methods are used: separate fermentation hydrolysis (SHF) and hydrolysis and simultaneous fermentation (SSF) [1]. Mussatto and Roberto (2006) used an acid hydrolysis process resulting in a maximum concentration of 13.21 g/dm^3 xylose, 8.21 g/dm^3 arabinosis, and 0.31 g/dm^3 glucose under the conditions of a maximum sugar yield of 19.6 g xylose/100 g dry matter and 8.3 g arabinosis/100 g dry matter [42]. According to studies reported by Mussatto (2014), the production of combustible ethanol takes place in five steps: BSG pretreatment (with acid heat solutions), hydrolysis of BSG with enzymes to convert starch and cellulose into simple sugars, fermentation of sugars with ethanol result (with the help of *Zymomonas mobilis*, *Saccharomyces cerevisiae*, *Escherichia coli*, *Bacillus subtilis*, or *Pichia pastoris*), followed by ethanol distillation and its dehydration to remove its water [12]. Dragone et al. (2007) achieved maximum productivity of 2.09 g.L^{-1}.h^{-1} ethanol for wort with 19.6 °P by continuous fermentation versus a maximum of 0.47 g.L^{-1}.h^{-1} ethanol for wort with 20 °P by discontinuous fermentation [50].

Lignocellulosic Yeast Carrier (LCYC)

The continuous fermentation of beer is based on a population of high-density yeasts, most commonly obtained by attachment to solid substrates with the formation of biofilm. The important properties of the solid substrates are increased adhesion capacity, physical endurance, high availability, and low-cost of production [51]. Continuous fermentation systems represent an innovative process for obtaining products of uniform quality, and beer production is faster. Continuous fermentation systems with immobilized cells reduce the time of primary fermentation to 1 day, but the cost of immobilization support is high and is the main impediment in industrial application [50,52]. BSG is a good source for lignocellulosic yeast (LCYC) [17], because it is a residue of the beer industry, thus it can be easily integrated into the continuous fermentation of beer. BSG is a low-cost material with good stability and high retention capacity of yeast cells; it can be easily prepared without modifying the chemical content, then sterilized and regenerated by washing in a caustic solution. No investments are necessary, it being a byproduct of the beer industry [52,53], and it represents an alternative worth exploring in support of immobilized cells [52]. Studies say that protein and fat extraction is recommended both for obtaining valuable fractions of BSG and for improving the LCYC production yield [51].

3.3.3. Prebiotics

Xylan is the main component of hemicellulose present in the plant cell walls. Xilooligosaccharides (XOS) are mainly produced by hydrolysis of xylan, and the prebiotic effects have been demonstrated. According to Amorim et al. (2019), XOS shows stability to heat and pH, and its use in food gives acceptable organoleptic properties [54]. It has beneficial effects on health, namely in the prevention of diabetes, neurotoxicity, inflammation of the colon, detoxification, weight loss, treatment of constipation, and microencapsulation. The BSG lignocellulosic biomass is composed mainly of cellulose, hemicellulose, and lignin. Cellulose is composed of a chain of glucose molecules with hydrogen bonds between different layers of polysaccharides, giving a crystalline conformation. Hemicelluloses, consisting mainly of xylan, are the target component for XOS production. Lignin supports the integrity and the rigidity of cell walls [9], and it is composed of phenolic components (*p*-coumaryl alcohol, coniferyl alcohol, and sinapyl alcohol) [54]. BSG contains hemicelluloses in a percentage of 16.5%, composed mainly of xylan (10.3%), arabinan (5.1%), and a small group of acetyl groups (1.1%) [55]. Using a substrate of BSG, Amorim et al. (2019) obtained, with commercial xylanoses with *Trichoderma longibrachiatum*, 444.3 mg XOS/g xylan in 12 h, and with *Trichoderma reesei* by direct fermentation, 326.2 mg XOS/g xylan in 72 h, without prior treatment [54].

A novelty for the food industry is food containing symbiotic products, both probiotic bifidobacteria as well as prebiotic oligosaccharides, products enriched both in terms of physico-chemical properties and in terms of health benefits. Non-digestible oligosaccharides have potential as food ingredients to improve the quality of many foods in terms of flavor and physico-chemical characteristics [56].

3.4. Obtaining Building Materials (Bricks)

The high amount of fibrous material combined with the reduced amount of ash makes BSG suitable for obtaining bricks [13,17]. The sawdust is commonly used in obtaining bricks and can be replaced with BSG due to the increase in the porosity of the bricks without changing their color or quality [13].

3.5. Adsorbent

BSG has been tested as an adsorbent for the removal of volatile compounds from waste gases but also for heavy metals from aqueous solutions (cadmium, lead, or chromium adsorption at adsorption capacities of 17.3 mg/g, 35.5 mg/g, and 18.94 mg/g) [12,17]. Pyrolysed BSG can absorb volatile organic compounds similar to coal from coconut shells [12].

BSG has also been successfully used as an adsorbent of orange acid dye 7 in wastewater, with an adsorption capacity of 30.5 mg of orange acid 7 per gram BSG at 30 °C [12].

Activated carbon can be obtained using BSG as raw material, a material used to purify water and gas. It is obtained by an alkaline treatment applied to BSG for the recovery of lignin, followed by precipitation with sulphuric acid and impregnation/activation of recovered lignin with 3 g per gram phosphoric acid at 600 °C and has the ability to adsorption of phenolic compounds and metal ions in liquid media [12].

3.6. Source of Phenolic Compounds

Phenolic compounds in plants are made up of several components: phenolic acids, phenolic alcohols, flavonoids, tannins, stilbene, and lignans [57,58]. BSG is a generous source of hydroxycinnamic acids (HCA) [9], which accumulate in cell walls are considered the most important source of antioxidants in cereals, both in free form and the bound form [17]. Phenolic compounds are considered natural antioxidants associated with some chronic diseases, such as cardiovascular disease, neurodegenerative diabetes, and cancer [9].

Phenolic acids, including caffeic acid, vanillic acid, quercetin, and epigallocatechin-3-gallate, inhibit cyclooxygenase isoforma, reducing the risk of cancer according to the method evaluated by DNA fragmentation and the Hoechst staining test [11]. Phenolic acids are grouped into hydroxycinnamic acids (HCA) and hydroxybenzoic acids (HBA), the latter in low amounts in BSG. Ferulic acid (FA) and p-cumaric acid (*p*-CA) are the most abundant HCAs in BSG, with values between 35–490 mg/100 g dry matter for FA and 6.7–180 mg/100 g dry matter for *p*-CA [17]. Ferulic acid is the target product of BSG; it is an antioxidant found in cell walls with a key role in the development and the protection of the plant. Ferulic acid has similar properties in the human body, protecting the skin and keeping it young, with effects similar to vitamin C [1]. It is a stable and gentle compound with the skin that reduces the negative effects of free radicals, considerably slowing the aging of the skin. It has also been approved as a preservative in many countries to prevent food oxidation [9,11]. For these reasons, extraction and recovery of these compounds is of niche interest for research for both the food industry and the pharmaceutical/cosmetic industry [17].

The important steps in phenolic compounds extraction include pretreatment, extraction, isolation, and purification. There is an upward trend of "green chemistry" or "green extraction" that implies the use of solvents and substances obtained from renewable materials but which could keep the quality of the gained extraction [57]. The extraction methods and the used treatments are microwave-assisted extraction (MAE) and saponification with NaOH 1–4 M (alkaline hydrolysis is the most used). Contents of 27.31 ± 0.69 µg/mL ferulic acid and 0.732 ± 0.020 mg galic acid equivalent (GAE)/mL phenolic acid were obtained at a concentration of 1 M. Other methods imply acid hydrolysis, enzymatic hydrolysis, ultrasound extraction, liquid–liquid extraction or solid–liquid extraction, water bath extraction with solvents, supercritical CO_2 extraction, and cosolvent ethanol (extraction temperature 40 °C, 240 min, 35 MPa CO_2 pressure, and 60% ethanol [9,17].

Enzymatic hydrolysis proved to be less efficient and with a higher cost of the analyses due to the high cost of pure enzymes (cellulase, α-amylase, pectinase) [17]. Microwave-assisted extraction (MAE) is an efficient, promising, and fast technique. Microwave hydrodiffusion and gravity (MHG) is a new bioactive compounds plant extraction technique that mixes the microwave heating with the gravity, the whole process deploying at atmospheric pressure without the water implication or another solvent [57]. Solid-state fermentation (SSF) is when the microbial growth and formation of products take place almost in the absence of water; the substrate contains moisture to allow growth and metabolism of microorganisms. Submerged fermentation (SmF) assumes that microorganisms are grown in a liquid medium containing the necessary nutrients [59]. Studies have shown that SSF has a higher yield than SmF in the production of bioactive compounds in addition to lower costs and secondary compounds that occur in a shorter time than SmF without the need for aseptic conditions. Very important roles in the case of these fermentations lie with humidity and water activity, which show us the water available for microbial growth, which affects the development of biomass, metabolic reactions, and mass transfer processes [59]. According to studies carried out by Moreira et al. (2013), up to 20 mg GAE/g BSG can be extracted [51]. The most important factor in the extraction of phenolic compounds is the solvent used. Meneses et al. (2013) reported a total phenolic compound content of 9.90 ± 0.41 mg GAE/g dry matter, using 60% acetone (*v*/*v*) and 2.14 mg GAE/g BSG when ethyl acetate was used, with an increase in extraction yield of 4.6 times [16]. When acetone was used as a solvent, the flavonoid content was between 0.51–2.12 mg QE/g BSG. Guido and Moreira (2017) reported results of 1.26 ± 0.10 and 4.53 ± 0.16 mg GAE/g dry matter for pale BSG and dark BSG (obtained by drying at very high temperatures), using acidified methanol as a solvent (HCl/methanol/water 1:80:10 *v*/*v*/*v*) [17], while Moreira et al. (2013) reported contents of 19.5 ± 0.6 mg GAE/g dry matter for BSG light and 16.2 ± 0.6 mg GAE/g dry matter for BSG dark [51]. The type of malt used and the malting process (burning regime and roasting temperatures) influence the content of phenolic compounds, BSG light having a higher content than BSG dark [51]. According to studies by Vellingri et al. (2014), the total content of polyphenols in BSG was 167.07 mg GAE/l following a first extraction with 80% ethanol and 66.75 mg GAE/l following a second extraction with the same solvent [60]. When the extract is used in food, the solvents used are very important and are regulated from the perspective of food safety (according to good manufacturing practices are the mixture acetones:water or ethanol:water). Methanol is a very effective extraction solvent, but its toxic nature limits the use of the extract in the food industry or in the pharmaceutical industry [16]. An alternative and ecological method to conventional extractions with organic solvents is the method of supercritical CO_2 extract (SCE-CO_2), which uses CO_2 in a supercritical condition when both temperature and pressure are equal to or exceed the critical point of 31 °C and 73 atm, providing the ideal conditions for extracting compounds with a high degree of recovery in a short time [61]. BSG has a lower content of phenolic compounds than berries, which have a content between 28–51 mg GAE/g, but its phenolic content is much higher than many vegetables (e.g., onions 2.5 mg GAE/g, potato peel 4.3 mg GAE/g, tomatoes 2.0 mg GAE/g), making it an interesting by-product to evaluate [16]. An interesting application is an increase in the content of bioactive phenolic compounds in food, which have a big impact lately due to the desire of consumers to improve their health through food [59].

3.7. Biogas Production

The whole world is going through an energy crisis, a crisis that can be reduced by producing energy from the BSG due to its high availability and low cost [12]. Biogas refers to a mixture of gases produced by anaerobic decomposition of organic matter through a complex process that occurs naturally in an oxygen-free environment and that is considered an effective method of converting biomass into methane [62]. Biogas is composed of methane (40–75%), water (0–10%), carbon dioxide (25–55%), hydrogen sulfide (1–3%), ammonia 0–1%, nitrogen 0–5%, oxygen 0–1, and hydrogen 0–1%. Biogas has a thermal value of approximately 22 MJ/m^3 [63], is considered a clean, CO_2-free fuel, and can be used in fuel cells for electricity generation. Although it is a recyclable, efficient, and clean product, 96% of its production is made with fossil combustibles, resulting in environmental pollution and energy crises due to high energy consumption [64]. Obtaining biogas from BSG involves largely two steps: a hydrolytic stage that allows a complete degradation of the material, a very important step to obtain high yields, and a methanogenic stage, where, with the help of macromolecule microorganisms, they convert into volatile fatty acids, acetates, butyrate, propionate, and methane [12]. Pretreatments play an important role in the degradation of the crystalline structure of cellulose molecules and decrease the degree of polymerization, easing enzymatic hydrolysis into simple sugars. Alkaline pretreatment provides a pH-friendly environment for further fermentation that is more effective [64]. After 15 days of digestion in a batch of anaerobic fermentation, the yield of biogas was 3476 cm^3 per 100 g BSG, according to Mussatto studies (2014) [12]. A decisive factor in hydrolysis and the use of lignocellulosic biomass in the production of biofuels is crystallinity (the crystalline index indicates the amount of crystalline cellulose present in biomass). Zhang and Zang (2017) used BSG pretreated with calcined red mud (resulting from bauxite), which reduced the BSG's crystalline index, with positive effects on BSG solubility and biohydrogen production [64].

3.8. Food/Composite Packaging

Incorporation of chitosan into BSG proteins can give rise to microfilm with antimicrobial and antioxidant properties, according to food packaging [14]. The development of biological and biodegradable materials is influenced by the ability of BSG proteins to interact between polypeptide chains. It is necessary to add plasticizers to protein-based films, which play a role in reducing the fragility of film and giving it certain plastic properties, which helps to increase flexibility and film handling, for example, sorbitol, polyethylene glycol, and glycerol. Film formation is influenced by protein–protein interactions, which are more intense at pH close to the isoelectric point. According to studies conducted by Proaño et al. (2020), the films formed at pH = 2 were the most homogeneous and could easily detach from the support [65]. Films obtained with polyethylene glycol (PEG) as plasticizer show homogeneity and are fine. Water solubility increases with increasing the amount of PE. The films also show good properties of water barrier, color is influenced by protein concentration, and opacity increases with the increase of PEG. Opacity is very important when packing fatty foods, as the oxidative degradation given by light is alleviated. The obtained films can be used as a barrier against UV, as they do not present transmittance between 200 and 400 nm. The films obtained present antioxidant activity to be taken into account in the case of active packaging of food. Not many studies have been conducted on this segment of obtaining biodegradable films from BSG, thus it is a subject to consider.

Cellulose is used for food packaging due to its fine network, biodegradability, and high water resistance. It can be added as a source of fiber or as a thickening and stabilizing agent in functional foods and drinks. Cellulose can also be used to produce nanocomposite materials [4]. Natural cellulose-rich fibers, including BSG, can be used as an alternative to produce organic polyurethane, which has proven to be the most promising filler. Different BSG polyurethane rations and ground tire rubber slots have been studied by Formela et al. (2017), with improvements in physicomechanical properties (apparent density increases by 37%, compression resistance by 50%) and thermal stability [66].

3.9. Proteins, Protein Hydrolysates, Bioactive Peptides

Because of the upward trend of vegetarianism and veganism, it is necessary to find sources of protein and natural protein derivatives from plants. There are several possibilities of using proteins derived from plants due to the functional effects on the human body (related to physiological and nutritional properties) and technical positive properties (related to physicochemical properties: appearance, texture, stability) as well as for nutrients for fortified foods and dietary supplements, technical-functional ingredients in terms of emulsification and gelification properties, or as materials for the development of biopolymers [67]. The application of proteins from BSG include antioxidant and antimicrobial packaging materials production, edible films (obtained from proteins and polysaccharides/chitosan), complexing proteins, fermented beverages, cookies, enzymes, and edible proteins. Proteins and protein hydrolysates derived from BSG, as resulted from the study of Ikram et al. (2017) [9], present immunomodulatory effects that can be useful to anti-inflammatory diseases control, hypertension, and diabetes treatment [8]. The most abundant protein in BSG is hordein, with a content of 43% of the total content of proteins, followed by glutelins with 21.5% [9].

Amino acids play an important role in human health, and BSG is a good source of these components. Essential amino acids from BSG are methionine, phenylalanine, tryptophan, histidine, and lysine, and the non-essential are serine, alanine, glycine, and proline [9]. A study conducted by Wen et al. (2019) revealed the fact that BSG has a similar composition in amino acids as in germinated barley—rich in glutamine/glutamic acid, vanillin, and leucine but with a smaller content of cysteine and methionine [68].

According to the studies of Pojicet et al. (2018), protein products can be classified in:

- protein containing flour up to 65% protein;
- protein hydrolysates containing concentrates between 65–90% protein;
- protein isolates containing more than 90% protein [67].

For an increased extraction yield, pretreatment BSG is applied (with diluted alkaline, enzymatic, or hydrothermal acids, or combinations thereof). Pretreatment with diluted acids can extract 90% of the total protein; it is not a selective method, extracting carbohydrates and lignin along with proteins. Hydrothermal pretreatment offers this selectivity, although the extraction yield is 64–66%, but it is a more environmentally friendly method, thus it is carried out at low temperatures and does not require the addition of chemicals [69]. The extraction methods used are largely synthesized in the following categories:

- dry extraction techniques: fractions with high impurity and agglomerated particles. This category includes two-step electrostatic separation, which involves loading particles and separating them in an electric field [67];
- wet extraction techniques: acidic extractions are less effective because they do not degrade the cell wall, resulting in less protein in the extraction environment. Alkaline extractions are more effective but at high alkaline concentrations after the Maillard reaction, which affects the nutritional properties of proteins. Extracts with organic solvents are also used [68]. Combinations of water with enzymes, water in subcritical conditions, or protein extraction through reverse smalls attract more and more of the attention of researchers [67]. Alkalis are used as extraction solvents at high temperatures, followed by precipitation with alcohol or isoelectric precipitation, treatment with NaOH/KOH (0.1 M, 0.5 M, 4 M) for 24 h at room temperature, acidification with citric acid up to pH = 3, then precipitation with ethanol, an enzymatic hydrolysis-effective technique for extractions [9];
- other extraction methods are microwave-assisted extraction (MAE), ultrasonic-assisted extraction (EAU), electrically pulsed energy extraction, or extraction using high hydrostatic pressure [67].

Treatment of BSG with sodium dodecyl sulfate (SDS) and disodium phosphate at different temperatures, followed by the addition of ethanol and refrigeration of extracts, resulted in a yield of 49% protein [70]. Bioactive peptides (BAP) are specific amino acid sequences that bring health benefits of interest to the pharmaceutical industry in terms of the discovery of new nutraceuticals or the food industry as ingredients for functional foods [15]. Peptides derived from BSG have potential against cardiovascular diseases and type 2 diabetes [71]. The specific bioactivity of food peptides is given by the length of the amino acid chain and their hydrophobicity and molecular weight. There is a growing interest in the use of bioactive peptides derived from food proteins against chronic diseases and for maintaining health status [72], inhibitory activity on angiotensin conversion enzyme (ACE), and dipeptide peptidase IV (DPP IV), which is being studied in vitro by Cermeno et al. (2019) [71]. BAPs are encrypted in the primary structure of proteins in the form of inactive amino acids (inactive when part of the source protein) but are activated by fermentation and food processing with the help of enzymes or in the digestive tract after human consumption [72,73]. According to the BIOPEP-UWM Database of Bioactive Peptides [74], the biological functions of bioactive peptides are ACE inhibitors (angiotensin conversion enzyme I), influence blood pressure regulation, antioxidant activity, antimicrobial activity, opioid activity, and antithrombotic and immunomodulatory activities. Strong bioactivity and high yield are challenges in BAP research [72]. The introduction of bioactive peptides into functional food brings the benefits of the amino acids contained. The techniques used for the synthesis of bioactive peptides are chemical synthesis, enzyme synthesis (as the main technique, microbial fermentation of proteins or protein isolates) [75], and synthesis by recombinant DNA technology [15]. Protein hydrolysis is not only a way to improve nutritional values but also to release bioactive peptides [75]. Obtaining pure peptides involves a laborious process of splitting and isolating individual peptides by chromatographic methods and membrane technologies [75].

Studies in which hydrolyzed proteins in cereals have increased the shelf life of meat have been conducted as the incorporation of antioxidant peptide fractions decreased lipid oxidation from 19% to 15% after one week [75]. Protein hydrolysates can be used in the food industry as texture improvement agents and as food additives or in the pharmaceutical industry. Hydrolysis in BSG has had good rheological results; protein changes by enzymatic or chemical means bring improvements to certain functional properties (water/oil retention capacity, emulsion properties and foam expansion, turbidity) [76]. Hydrolyzed protein isolates with a number of enzymes (Alcalase, Corolase PP, Flavourzyme, and Promod 144 MG) had improvements in heat stability at pH = 6.0, even after 300 min of temperature maintenance at pH between 2.0 and 12.0, which increased solubility, emulsification, and foaming capacity.

Nitrogen solubility decreased when pH approaches isoelectric point (pH = 3.8), but the highest solubility was achieved at pH between 6.0 and 12.0. Compared to non-hydrolyzed soy protein, which is more stable at pH = 9.0 than at pH = 3.8, the protein hydrolysates in BSG showed no coagulation at pH = 8.0, but HCT (heat coagulation time) decreased at pH below 6.0 [77]. Protein isolates and protein hydrolysates associated with them can be used as techno-functional ingredients but carefully, as the pH at which they are used influences these properties according to the above-mentioned study.

3.10. Source of Fiber

BSG is a rich source of dietary fiber, especially viscous fiber, with a significant contribution in speeding up intestinal transit and improving the symptoms of ulcerative colitis [9]. Dietary fibers are classified according to solubility: soluble dietary fibers include β-glucans pectic polysaccharides, arabinogalactans, and xyloglucans, and insoluble dietary fibers include lignin, cellulose, and galactomannans [9].

Lignin can be partially degraded by intestinal microbiota, and the resulting compounds can be metabolized, according to studies by Lynch et al. (2016) [8]. Berglund et al. (2016) conducted studies to obtain cellulosic nanofibers using a method of bleaching and mechanical separation but obtained low yields (22%) compared to carrot residues from juice processing (32%) [78]. Energy consumptions were high at 21 kWh/kg and 0.9 kWh/kg, respectively.

The method of separating proteins and fibers proposed by He et al. (2019) involves grinding with a disk mill, then mixing the BSG with deionized water [79]. Then, one adds the reagent (NaOH, sodium bisulfite, or alcalase) to the suspension on the water bath for 4 h at 60 °C, then transfers the samples to a sieve shaker where it is shaken for 15 min to separate the small solubilized proteins from the large insoluble fibers. During the sifting, the samples are washed with deionized water, the proteins pass through the sieve, and the fibers remain on the sieve.

3.11. Polymers

Food waste may be a good source for the production of polyhydroxyalcanates (PHA) and poly-3-hydroxybutyrate (PHB) due to abundance and low cost. PHA is a material similar to plastics and can replace plastics derived from oil. The main disadvantage of producing PHA is the high operational cost of production, reduced by the use of raw materials resulting from food waste, such as BSG [1]. The production of volatile fatty acids without pretreatment by anaerobic digestion helps to obtain bioplastic materials (e.g., PHA) by using BSG as a substrate [80].

3.12. Other Application

BSG can also be used for obtaining paper-serves-cards, production of coal (inferior in terms of burning properties), production of resins, production of an antifoaming agent in beer [12,13], and production of bacteriocin using *Lactococcus lactis* and *Enterococcus mundtii* with antimicrobial activity against *Lysteria monocytogenes* as a bioindicator. Lignocellulosic material is subjected to treatment, usually hydrolysis, in order to solubilize in the constituent monomers. The sugars formed are subject to microbial fermentation as precursors of added-value compounds (e.g., bacteriocins) or enzymes [25]. Ndayishimiye et al. (2020) conducted a study on the encapsulation of oils recovered from BSG by an eco-friendly technique of saturated gas solutions (CO_2 supercritical), aiming to obtain a product with improved physical properties and with oxidative stability [81].

The growing demand for products obtained with stable ingredients obtained from food by-products stimulates the finding of innovative alternatives to obtain those [81]. The concept of biorefinery says that all components of raw material are converted into products of commercial importance (e.g., biofuel, enzymes, oils, nutraceuticals) [82] and is adopted in many sectors for integrated food production. Mussatto and coworkers (2013) developed an integrated system for the production of lactic acid, xylitol, activated carbon, and phenolic acids [44]. Bio-based economy is an economy based on the use of feedstock biomass for food, energy, chemicals, and other materials. The use of biomass as a raw material for the production of the above mentioned brings social benefits and economic potential and ensures a reduction in carbon emissions [5].

Pyrolyzed BSG is used to obtain biochar, which is a porous material with stable physical and chemical properties and a high content of C. It is used to increase the amount of C in the soil and to reduce nutrient leakage; it is also used for soil amendment and as a nutrient supplier for plant growth and improving soil characteristics. Biochar has a large amount of ash, stable aromatic C structures, low bulk density, and moderate cation exchange capacity, which makes it suitable for this use [83]. The submerged fermentation of BSG using *Aspergillus niger* and *Saccharomyces cerevisiae* can obtain citric acid in amounts of 0.512% and 0.312%, respectively [84–86].

4. Conclusions

The use of food waste brings benefits to both reducing environmental pollution and to industry. Turning by-products into value-added components reduces food production costs and quantifies their nutritional value. It is necessary to find low-cost sources of valuable compounds of plant origin given the upward trend of vegetarian and vegan diets. The researchers seek new alternatives for food fortification, and the conversion of vegetable by-products in higher value products for obtaining functional compounds has their attention, with researchers showing more and more interest.

Author Contributions: A.C. and A.D. contributed equally to the collection of data and preparation of the paper. All authors have read and agreed to the published version of the manuscript.

Funding: This research received no external funding.

Conflicts of Interest: The authors declare no conflict of interest.

References

1. Ravindran, R.; Jaiswal, A.K. Exploitation of Food Industry Waste for High-Value Products. *Trends Biotechnol.* **2016**, *34*, 58–69. [CrossRef]
2. Orzua, M.C.; Mussatto, S.I.; Contreras-Esquivel, J.C.; Rodriguez, R.; de la Garza, H.; Teixeira, J.A.; Aguilar, C.N. Exploitation of agro industrial wastes as immobilization carrier for solid-state fermentation. *Ind. Crops Prod.* **2009**, *30*, 24–27. [CrossRef]
3. Roth, M.; Jekle, M.; Becker, T. Opportunities for upcycling cereal byproducts with special focus on Distiller's grains. *Trends Food Sci. Technol.* **2019**, *91*, 282–293. [CrossRef]
4. Mussatto, S.I.; Loosdrecht, M. Cellulose: A key polymer for a greener, healthier, and bio-based future. *Biofuel Res. J.* **2016**, *3*, 482. [CrossRef]
5. Yamakawa, C.K.; Qin, F.; Mussatto, S.I. Advances and opportunities in biomass conversion technologies and biorefineries for the development of a bio-based economy. *Biomass Bioenerg.* **2018**, *119*, 54–60. [CrossRef]
6. Statista. Beer Production Worldwide from 1998 to 2018. Available online: https://www.statista.com/statistics/270275/worldwide-beer-production/ (accessed on 20 May 2020).
7. Fărcaş, A.C.; Socaci, S.A.; Dulf, F.V.; Tofană, M.; Mudura, E.; Diaconeasa, Z. Volatile profile, fatty acids composition and total phenolics content of brewers' spent grain by-product with potential use in the development of new functional foods. *J. Cereal Sci.* **2015**, *64*, 34–42. [CrossRef]
8. Lynch, K.M.; Steffen, E.J.; Arendt, E.K. Brewers' spent grain: A review with an emphasis on food and health. *J. Inst. Brew.* **2016**, *122*, 553–568.
9. Ikram, S.; Huang, L.Y.; Zhang, H.; Wang, J.; Yin, M. Composition and Nutrient Value Proposition of Brewers Spent Grain. *J. Food Sci.* **2017**, *82*, 2232–2242. [CrossRef]
10. Mussatto, S.I.; Roberto, I.C. Acid hydrolysis and fermentation of brewer's spent grain to produce xylitol. *J. Sci. Food Agric.* **2005**, *85*, 2453–2460. [CrossRef]
11. McCarthy, A.L.; O'Callaghan, Y.C.; Piggott, C.O.; FitzGerald, R.J.; O'Brien, N.M. Brewers' spent grain; Bioactivity of phenolic component, its role in animal nutrition and potential for incorporation in functional foods: A review. *Proc. Nutr. Soc.* **2013**, *72*, 117–125. [CrossRef]
12. Mussatto, S.I. Brewer's spent grain: A valuable feedstock for industrial applications. *J. Sci. Food Agric.* **2014**, *94*, 1264–1275. [CrossRef]
13. Mussatto, S.I.; Dragone, G.; Roberto, I.C. Brewers' spent grain: Generation, characteristics and potential applications. *J. Cereal Sci.* **2006**, *43*, 1–14. [CrossRef]

14. Nazzaro, F.; Fratianni, F.; Ombra, M.N.; D'Acierno, A.; Coppola, R. Recovery of biomolecules of high benefit from food waste. *Curr. Opin. Food Sci.* **2018**, *22*, 43–54. [CrossRef]

15. Montesano, D.; Gallo, M.; Blasi, F.; Cossignani, L. Biopeptides from vegetable proteins: New scientific evidences. *Curr. Opin. Food Sci.* **2020**, *31*, 31–37. [CrossRef]

16. Meneses, N.G.T.; Martins, S.; Teixeira, J.A.; Mussatto, S.I. Influence of extraction solvents on the recovery of antioxidant phenolic compounds from brewer's spent grains. *Sep. Purif. Technol.* **2013**, *108*, 152–158. [CrossRef]

17. Guido, L.F.; Moreira, M.M. Techniques for Extraction of Brewer's Spent Grain Polyphenols: A Review. *Food Bioprocess Technol.* **2017**, *10*, 1192–1209. [CrossRef]

18. El-Shafey, E.I.; Gameiro, M.L.F.; Correia, P.F.M.; De Carvalho, J.M.R. Dewatering of brewer's spent grain using a membrane filter press: A pilot plant study. *Sep. Sci. Technol.* **2004**, *39*, 3237–3261. [CrossRef]

19. Treimo, J.; Westereng, B.; Horn, S.J.; Forssell, P.; Robertson, J.A.; Faulds, G.B.; Waldron, K.W.; Buchert, J.; Eijsink, V.G.H. Enzymatic solubilization of brewers' spent grain by combined action of carbohydrases and peptidases. *J. Agric. Food Chem.* **2009**, *57*, 3316–3324. [CrossRef]

20. Martins, Z.E.; Pinho, O.; Ferreira, I.M.P.L.V.O. Food industry by-products used as functional ingredients of bakery products. *Trends Food Sci. Technol.* **2017**, *67*, 106–128. [CrossRef]

21. Nigam, P.S. An overview: Recycling of solid barley waste generated as a by-product in distillery and brewery. *Waste Manag.* **2017**, *62*, 255–261. [CrossRef]

22. Amorim, C.; Silvério, S.C.; Silva, S.P.; Coelho, E.; Coimbra, M.A.; Prather, K.L.J.; Rodrigues, L.R. Single-step production of arabino-xylooligosaccharides by recombinant *Bacillus subtilis 3610* cultivated in brewers' spent grain. *Carbohydr. Polym.* **2018**, *199*, 546–554. [CrossRef]

23. Mussatto, S.I. Biotechnological Potential of Brewing Industry By-Products. In *Biotechnology for Agro-Industrial Residues Utilisation: Utilisation of Agro-Residues*; Nigam, S.N., Pandey, A., Eds.; Springer: Berlin/Heidelberg, Germany, 2009; pp. 313–326.

24. Qin, F.; Johansen, A.Z.; Mussatto, S.I. Evaluation of different pretreatment strategies for protein extraction from brewer's spent grains. *Ind. Crops Prod.* **2018**, *125*, 443–453. [CrossRef]

25. Paz, A.; da Silva Sabo, S.; Vallejo, M.; Marguet, E.; Pinheiro de Souza Oliveira, R.; Domínguez, J.M. Using brewer's spent grain to formulate culture media for the production of bacteriocins using Patagonian strains. *LWT Food Sci. Technol.* **2018**, *96*, 166–174. [CrossRef]

26. Del Río, J.C.; Prinsen, P.; Gutiérrez, A. Chemical composition of lipids in brewer's spent grain: A promising source of valuable phytochemicals. *J. Cereal Sci.* **2013**, *58*, 248–254. [CrossRef]

27. Tang, D.S.; Tian, Y.J.; He, Y.Z.; Li, L.; Hu, S.Q.; Li, B. Optimisation of ultrasonic-assisted protein extraction from brewer's spent grain. *Czech J. Food Sci.* **2010**, *28*, 9–17. [CrossRef]

28. Jayant, M.; Hassan, M.A.; Srivastava, P.P.; Meena, D.K.; Kumar, P.; Kumar, A.; Wagde, M.S. Brewer's spent grains (BSGs) as feedstuff for striped catfish, *Pangasianodon hypophthalmus* fingerlings: An approach to transform waste into wealth. *J. Clean. Prod.* **2018**, *199*, 716–722. [CrossRef]

29. Stojceska, V.; Ainsworth, P. The effect of different enzymes on the quality of high-fibre enriched brewer's spent grain breads. *Food Chem.* **2008**, *110*, 865–872. [CrossRef]

30. Ktenioudaki, A.; O'Shea, N.; Gallagher, E. Rheological properties of wheat dough supplemented with functional by-products of food processing: Brewer's spent grain and apple pomace. *J. Food Eng.* **2013**, *116*, 362–368. [CrossRef]

31. Stojceska, V.; Ainsworth, P.; Plunkett, A.; Ibanoğlu, S. The recycling of brewer's processing by-product into ready-to-eat snacks using extrusion technology. *J. Cereal Sci.* **2008**, *47*, 469–479. [CrossRef]

32. Ktenioudaki, A.; Crofton, E.; Scannell, A.G.M.; Hannon, J.A.; Kilcawley, K.N.; Gallagher, E. Sensory properties and aromatic composition of baked snacks containing brewer's spent grain. *J. Cereal Sci.* **2013**, *57*, 384–390. [CrossRef]

33. Spinelli, S.; Conte, A.; Del Nobile, M.A. Microencapsulation of extracted bioactive compounds from brewer's spent grain to enrich fish-burgers. *Food Bioprod. Process.* **2016**, *100*, 450–456. [CrossRef]

34. Stefanello, F.S.; Fruet, A.P.B.; Trombetta, F.; Franco da Fonseca, P.A.; dos Santos da Silva, M.; Stefanello, S.; Nörnberg, J.L. Stability of vacuum-packed meat from finishing steers fed different inclusion levels of brewer's spent grain. *Meat Sci.* **2019**, *147*, 155–161. [CrossRef]

35. Cappa, C.; Alamprese, C. Brewer's spent grain valorization in fiber-enriched fresh egg pasta production: Modelling and optimization study. *LWT Food Sci. Technol.* **2017**, *82*, 464–470. [CrossRef]

36. Gupta, S.; Jaiswal, A.K.; Abu-Ghannam, N. Optimization of fermentation conditions for the utilization of brewing waste to develop a nutraceutical rich liquid product. *Ind. Crops Prod.* **2013**, *44*, 272–282. [CrossRef]

37. Mussatto, S.I.; Roberto, I.C. Alternatives for detoxification of diluted-acid lignocellulosic hydrolyzates for use in fermentative processes: A review. *Bioresour. Technol.* **2004**, *93*, 1–10. [CrossRef]

38. Mussatto, S.I. *Biomass Fractionation Technologies for a Lignocellulosic Feedstock Based Biorefinery*; Elsevier: Amsterdam, The Netherlands, 2016; pp. 169–185.

39. Outeiriño, D.; Costa-Trigo, I.; Paz, A.; Deive, F.J.; Rodríguez, A.; Domínguez, J.M. Biorefining brewery spent grain polysaccharides through biotuning of ionic liquids. *Carbohydr. Polym.* **2019**, *203*, 265–274. [CrossRef]

40. Da Silva, S.S.; Chandel, A.K. *D-Xylitol: Fermentative Production, Application and Commercialization*; Springer: Berlin/Heidelberg, Germany, 2012; Volume 1, p. 345.

41. Mussatto, S.I.; Dragone, G.; Teixeira, J.A.; Roberto, I.C. Total reuse of brewer's spent grain in chemical and biotechnological processes for the production of added—Value compounds. In Proceedings of the International Conference and Exibition on Bioenergy, Guimarães, Portugal, 6–9 April 2008; pp. 4–5.

42. Mussatto, S.I.; Roberto, I.C. Chemical characterization and liberation of pentose sugars from brewer's spent grain. *J. Chem. Technol. Biotechnol.* **2006**, *81*, 268–274. [CrossRef]

43. Silva, C.J.S.M.; Mussatto, S.I.; Roberto, I.C. Study of xylitol production by *Candida guilliermondii* on a bench bioreactor. *J. Food Eng.* **2006**, *75*, 115–119. [CrossRef]

44. Mussatto, S.I.; Moncada, J.; Roberto, I.C.; Cardona, C.A. Techno-economic analysis for brewer's spent grains use on a biorefinery concept: The Brazilian case. *Bioresour. Technol.* **2013**, *148*, 302–310. [CrossRef]

45. Mussatto, S.I.; Santos, J.C.; Roberto, I.C. Effect of pH and activated charcoal adsorption on hemicellulosic hydrolysate detoxification for xylitol production. *J. Chem. Technol. Biotechnol.* **2004**, *79*, 590–596. [CrossRef]

46. Mussatto, S.I.; Dragone, G.; Roberto, I.C. Kinetic behavior of *Candida guilliermondii* yeast during xylitol production from brewer's spent grain hemicellulosic hydrolysate. *Biotechnol. Prog.* **2005**, *21*, 1352–1356. [CrossRef] [PubMed]

47. Martinez, E.A.; Villarreal, M.L.M.; Almeida e Silva, J.B.; Solenzal, A.I.N.; Canilha, L.; Mussatto, S.I. Uso De Diferentes Materias Primas Para La Producción Biotecnológica De Xilitol. *Cienc. Tecnol. Aliment.* **2002**, *3*, 295–301. [CrossRef]

48. Mussatto, S.I.; Fernandes, M.; Dragone, G.; Mancilha, I.M.; Roberto, I.C. Brewer's spent grain as raw material for lactic acid production by *Lactobacillus delbrueckii*. *Biotechnol. Lett.* **2007**, *29*, 1973–1976. [CrossRef] [PubMed]

49. Liang, S.; Wan, C. Carboxylic acid production from Brewer's spent grain via mixed culture fermentation. *Bioresour. Technol.* **2015**, *182*, 179–183. [CrossRef]

50. Dragone, G.; Mussatto, S.I.; De Almeida, E.; Silva, J.B. Use of concentrated worts for high gravity brewing by continuous process: New tendencies for the productivity increase. *Cienc. Tecnol. Aliment.* **2007**, *27*, 37–40. [CrossRef]

51. Moreira, M.M.; Morais, S.; Carvalho, D.O.; Barros, A.A.; Delerue-Matos, C.; Guido, L.F. Brewer's spent grain from different types of malt: Evaluation of the antioxidant activity and identification of the major phenolic compounds. *Food Res. Int.* **2013**, *54*, 382–388. [CrossRef]

52. Dragone, G.; Mussatto, S.I.; Almeida, E.; Silva, J.B. Influence of temperature on continuous high gravity brewing with yeasts immobilized on spent grains. *Eur. Food Res. Technol.* **2008**, *228*, 257–264. [CrossRef]

53. Brányik, T.; Vicente, A.A.; Machado Cruz, J.M.; Teixeira, J.A. Spent grains—A new support for brewing yeast immobilisation. *Biotechnol. Lett.* **2001**, *23*, 1073–1078. [CrossRef]

54. Amorim, C.; Silvério, S.C.; Prather, K.L.J.; Rodrigues, L.R. From lignocellulosic residues to market: Production and commercial potential of xylooligosaccharides. *Biotechnol. Adv.* **2019**, *37*, 107397. [CrossRef]

55. Amorim, C.; Silvério, S.C.; Rodrigues, L.R. One-step process for producing prebiotic arabino-xylooligosaccharides from brewer's spent grain employing *Trichoderma* species. *Food Chem.* **2019**, *270*, 86–94. [CrossRef]

56. Mussatto, S.I.; Mancilha, I.M. Non-digestible oligosaccharides: A review. *Carbohydr. Polym.* **2007**, *68*, 587–597. [CrossRef]

57. Asofiei, I. Nonconventional Techniques for the Separation of Valuable Compounds from Plants. Ph.D. Thesis, Politehnica University, Bucharest, Romania, 2017.

58. Oarcea, A.I. Effect of Plant Polyphenolson Cancer Cells. Ph.D. Thesis, West University Vasile Goldis of Arad, Arad, Romania, 2016.

59. Martins, S.; Mussatto, S.I.; Martínez-Avila, G.; Montañez-Saenz, J.; Aguilar, C.N.; Teixeira, J.A. Bioactive phenolic compounds: Production and extraction by solid-state fermentation. A review. *Biotechnol. Adv.* **2011**, *29*, 365–373. [CrossRef]

60. Vellingiri, V.; Amendola, D.; Spigno, G. Screening of four different agro-food by-products for the recovery of antioxidants and cellulose. *Chem. Eng. Trans.* **2014**, *37*, 757–762.

61. Kitryte, V.; Šaduikis, A.; Venskutonis, P.R. Assessment of antioxidant capacity of brewer's spent grain and its supercritical carbon dioxide extract as sources of valuable dietary ingredients. *J. Food Eng.* **2015**, *167*, 18–24. [CrossRef]

62. Ghasemi Ghodrat, A.; Tabatabaei, M.; Aghbashlo, M.; Mussatto, S.I. Waste Management Strategies; the State of the Art. In *Biogas: Fundamentals, Process, and Operation*; Tabatabaei, M., Ghanavati, H., Eds.; Biofuel and Biorefinery Technologies; Springer International Publishing AG: Berlin/Heidelberg, Germany, 2018; Volume 6, pp. 1–33.

63. Nassary, E.K.; Nasolwa, E.R. Unravelling disposal benefits derived from underutilized brewing spent products in Tanzania. *J. Environ. Manag.* **2019**, *242*, 430–439. [CrossRef] [PubMed]

64. Zhang, J.; Zang, L. Enhancement of biohydrogen production from brewers' spent grain by calcined-red mud pretreatment. *Bioresour. Technol.* **2016**, *209*, 73–79. [CrossRef]

65. Proaño, J.L.; Salgado, P.R.; Cian, R.E.; Mauri, A.N.; Drago, S.R. Physical, structural and antioxidant properties of brewer's spent grain protein films. *J. Sci. Food Agric.* **2020**. [CrossRef]

66. Formela, K.; Hejna, A.; Zedler, Ł.; Przybysz, M.; Ryl, J.; Saeb, M.R.; Piszczyk, L. Structural, thermal and physico-mechanical properties of polyurethane/brewers' spent grain composite foams modified with ground tire rubber. *Ind. Crops Prod.* **2017**, *108*, 844–852. [CrossRef]

67. Pojić, M.; Mišan, A.; Tiwari, B. Eco-innovative technologies for extraction of proteins for human consumption from renewable protein sources of plant origin. *Trends Food Sci. Technol.* **2018**, *75*, 93–104. [CrossRef]

68. Wen, C.; Zhang, J.; Duan, Y. A mini-review on brewer's spent grain protein: Isolation, physicochemical properties, application of protein, and functional properties of hydrolysates. *J. Food Sci.* **2019**, *84*, 3330–3340. [CrossRef]

69. Qin, F.; Mussatto, S.I. Protein and amino acids recovery from brewer's spent grains by different pretreatment technologies. In Proceedings of the Brazilian Bioenergy Science and Technology Conference 2017 (BBEST 2017): Designing a Sustainable Bioeconomy, Campos do Jordão, Brazil, 17–19 October 2017; p. 193.

70. Ervin, V.; Alli, I.; Smith, J.P.; Li, Z. Extraction and Precipitation of Proteins From Brewer's Spent Grain. *Can. Inst. Food Sci. Technol. J.* **1989**, *22*, 216–221. [CrossRef]

71. Cermeño, M.; Connolly, A.; O'Keeffe, M.B.; Flynn, C.; Alashi, A.M.; Aluko, R.E.; FitzGerald, R.J. Identification of bioactive peptides from brewers' spent grain and contribution of Leu/Ile to bioactive potency. *J. Funct. Foods* **2019**, *60*, 103455. [CrossRef]

72. Udenigwe, C.C.; Aluko, R.E. Food protein-derived bioactive peptides: Production, processing, and potential health benefits. *J. Food Sci.* **2012**, *77*, R11–R24. [CrossRef] [PubMed]

73. Toldrá, F.; Reig, M.; Aristoy, M.C.; Mora, L. Generation of bioactive peptides during food processing. *Food Chem.* **2018**, *267*, 395–404. [CrossRef] [PubMed]

74. Minkiewicz, P.; Iwaniak, A.; Darewicz, M. BIOPEP-UWM Database of Bioactive Peptides: Current Opportunities. *Int. J. Mol. Sci.* **2019**, *20*, 5978. [CrossRef] [PubMed]

75. Esfandi, R.; Walters, M.E.; Tsopmo, A. Antioxidant properties and potential mechanisms of hydrolyzed proteins and peptides from cereals. *Heliyon* **2019**, *5*, e01538. [CrossRef]

76. Kotlar, C.E.; Ponce, A.G.; Roura, S.I. Improvement of functional and antimicrobial properties of brewery byproduct hydrolysed enzymatically. *LWT Food Sci. Technol.* **2013**, *50*, 378–385. [CrossRef]

77. Connolly, A.; Piggott, C.O.; FitzGerald, R.J. Technofunctional properties of a brewers' spent grain protein-enriched isolate and its associated enzymatic hydrolysates. *LWT Food Sci. Technol.* **2014**, *59*, 1061–1067. [CrossRef]

78. Berglund, L.; Noël, M.; Aitomäki, Y.; Öman, T.; Oksman, K. Production potential of cellulose nanofibers from industrial residues: Efficiency and nanofiber characteristics. *Ind. Crops Prod.* **2016**, *92*, 84–92. [CrossRef]

79. He, Y.; Kuhn, D.D.; Ogejo, J.A.; O'Keefe, S.F.; Fernández Fraguas, C.; Wiersema, B.D.; Jin, Q.; Yu, D.; Huang, H. Wet fractionation process to produce high protein and high fiber products from brewer's spent grain. *Food Bioprod. Process.* **2019**, *117*, 266–274. [CrossRef]

80. Ribau, M.; Guarda, E.C.; Freitas, E.B.; Galinha, C.F.; Duque, A.F.; Reis, M.A.M. Valorization of raw brewers' spent grain through the production of volatile fatty acids. *New Biotechnol.* **2020**, *57*, 4–10. [CrossRef] [PubMed]

81. Ndayishimiye, J.; Ferrentino, G.; Nabil, H.; Scampicchio, M. Encapsulation of Oils Recovered from brewer's Spent Grain by Particles from Gas Saturated Solutions Technique. *Food Bioprocess Technol.* **2020**, *13*, 256–264. [CrossRef]

82. Abasiekong, S.F. Effects of Fermentation on Crude Protein Content of Brewers Dried Grains and Spent Sorghum Grains. *Bioresour. Technol.* **1991**, *35*, 99–102. [CrossRef]

83. Amoriello, T.; Vecchiarelli, V.; Pagano, M. Evaluation of Spent Grain Biochar Impact on Hop (*Humulus lupulus* L.) Growth by Multivariate Image Analysis. *Appl. Sci.* **2020**, *10*, 533. [CrossRef]

84. Rachwał, K.; Waśko, A.; Gustaw, K.; Polak-Berecka, M. Utilization of brewery wastes in food industry. *PeerJ* **2020**, 1–28. [CrossRef]

85. Atere, V.A. Citric acid production from brewers spent grain by *Aspergillus niger* and *Saccharomyces cerevisiae*. *Int. J. Res. Biosci.* **2013**, *2*, 30–36.

86. Dabija, A.; Tulbure, M. *Capitalization of By-Products in the Beer Industry*; PIM Publishing House: Iasi, Romania, 2010.

Review

Oil Press-Cakes and Meals Valorization through Circular Economy Approaches: A Review

Petraru Ancuța and Amariei Sonia *

Faculty of Food Engineering, Ștefan cel Mare University of Suceava, 720229 Suceava, Romania; ancuta.petraru@fia.usv.ro
* Correspondence: sonia@usm.ro; Tel.: +40-740-311-291

Received: 23 September 2020; Accepted: 20 October 2020; Published: 22 October 2020

Abstract: The food industry generates a large amount of waste every year, which opens up a research field aimed at minimizing and efficiently managing this issue to support the concept of zero waste. From the extraction process of oilseeds results oil cakes. These residues are a source of bioactive compounds (protein, dietary fiber, antioxidants) with beneficial properties for health, that can be used in foods, cosmetics, textile, and pharmaceutical industries. They can also serve as substrates for the production of enzymes, antibiotics, biosurfactants, and mushrooms. Other applications are in animal feedstuff and for composites, bio-fuel, and films production. This review discusses the importance of oilseed and possible valorization methods for the residues obtained in the oil industry.

Keywords: residues; sustainability; oil cake; bioactive compounds; edible films

1. Introduction

In recent decades, due to the current planet issues (e.g., over-exploitation and mismanagement of resources, the unsustainable consumption behaviors from consumers, the degradation of the environment and equilibrium of the ecosystems, climate change), it is necessary to transition to a circular economy model based on the development of new strategies for making the best use of our resources and for the elimination of the concept of wastes along the supply chain. In this model, materials are recycled (a process in which wastes are transformed into value-added products by making them input elements for other products) and re-circulated during processing to create the concept "waste = food" [1–3].

Any substance or object already disposed or intended to be discarded by holders is called waste. Food wastes are edible materials (lost, discarded, or consumed by pest) collected from the food industry in all phases, from primary agriculture up to production, processing, and direct consumption in the households. The Food Agriculture Organization declared that annually one third of the total food production became lost, about 1,3 billion tons (20% in oilseeds, meat and dairies, 30% in cereals, 35% in fish and, 40–50% in fruit and vegetables). During the supply chain, a quantity of 54% of the total waste results from cultivation and post-harvest and 46% from processing, distribution, and utilization. Administration of these residues/wastes/by-products could be problematic due to the high disposal costs in landfills and the creation of serious environment problems [4,5]. A solution for reducing food wastes is extracting the maximum value from wastes and by-products. The last ones contain high amounts of bioactive compounds (carbohydrates, lipids, organic acids, proteins, vitamins, minerals, and antioxidants) with numerous health benefits (anti-tumoral, viral, bacterial, and mutagenic abilities) that can be isolated and employed in foods, pharmaceutics, cosmetics, and textiles. Achieving this can be a challenge, but at the same time could add more value to food, reducing disposal costs and risks caused by residues [6]. Other alternatives, after the extraction are: conversion in energy, introduction in human food, livestock and fish feedstuff, production of fertilizers and compost (generated from the aerobic decomposition of organic matter by microorganisms and insects) [1].

This paper provides a critical review ofthe existing literature about the possible ways to re-utilized the by-products obtained from the oil industry.

2. Oilseed, General Aspects

Oilseeds are seeds used in most countries as a source of vegetable oils. Soybean and walnut are cultivated in USA, Brazil and Argentina [7,8]. The largest rapeseed producers are the European countries (Germany, France), Russia, Pakistan, Canada, Australia, China, and India [9,10]. The world leading producers of sunflower seeds are Russia, Ukraine, Argentina, USA, China, India, and Turkey [10]. Sesame is grown primarily in Asia, Africa, India, China, and South America [11]. Pumpkin is largely cultivated in America, Europe (such as Austria, Slovenia, Croatia, Hungary, Greece, Bulgaria, Romania, and Turkey), Africa, and Asia [12–14]. In India, China, United States, and Ethiopia linseed is mainly cultivated flaxseed [15].

Sunflower, rapeseed, coconut, mustard, soybean, cotton, corn, oat, peanut, olive, and palm are oil-producing crops. Oil can be extracted also from nuts as almond, hazelnut, walnut, and pumpkin. Depending on the level of production oilseeds are classified as major (soybean, rapeseed, sunflower, sesame, niger, castor) and minor (coconut, palm seed). Globally the most produced oilseed is soybean, followed by rapeseed/canola, sunflower, cottonseed, groundnut, coconut, safflower, and flaxseed. Oilseeds cultivated mainly for oil extraction are rapeseed and sunflower, for proteins is soybean and for fibers is cotton. Linseed can be used for edible and industrial oil production but peanuts non solely for the oil but also for direct consumption [16]. The chemical composition and thus the nutritive values of oilseeds depend on the following factors: oilseed genotype, soil type, climatic condition, growing area, agricultural practices, and processing conditions [8]. They are rich in phytochemicals (phenolic compounds, carotenoids, polyphenols, flavonoids, lignans, phytosterols, tocopherols, and tocotrienols), protein, fat, ash, fiber, carbohydrate, vitamins (A, K, E, C, B_1, B_2, B_3, B_6), and minerals (P, Cu, Ca, K, Mg, Mn, Fe, Zn, Se). The vitamins are important for the functioning of the skin, nerves, and digestive system, maintenance of cells and have an antioxidant role, the minerals are involved in the metabolic and enzymatic processes and the phytochemical possess antioxidant activity, reduces tumors, inflammations, and improve immunity. For these reasons, oilseeds have an important role in human nutrition [17].

3. Oilseed Production Process and by-Products Physiochemical Properties

Oilseeds are cleaned, dehulled, grinded, cooked (destruction of microorganism and oil cell, enzymes inactivation, coagulation of proteins), and then are subjected to oil extraction. Each phase of the process generates waste with residual bioactive compounds that can be recovered: steams, leaves, pods, and broken grain in the cleaning steps and in the dehulling phase, hulls [16,18].

The two traditional methods to extract oil involve either the use of a mechanical press (hot and cold pressing) or a solvent. When the process is mechanical the oilseeds are heated (100 degrees) and the oil is extracted with a screw-press. Alternately, in cold pressing, the temperature of heating is kept at 50–60 degrees. The use of a solvent helps to maximize the oil recovery. The differences in chemical compositions between the cold-pressed and expeller cakes were in crude protein, fat, and crude fiber content. The protein content of the cold-extracted cakes was lower in comparison to that extracted at high temperatures. Regarding the fat content, the mechanical expeller cakes have 6–7% and the solvent extracted ones have <1%. The product resulting directly from the expeller is called a cake, and when it has undergone an additional solvent de-oiling process the term meal is used. This conventional oil extraction method can be replaced by modern/green techniques (pressurized liquid, pulsed electric field, high hydrostatic pressure, and high voltage electrical discharges) because they are more rapid and less solvent consuming [19–23].

After the extraction of oil from seeds, the principal by-product obtained is oil cake/meal. Oilcakes/meals are classified into two categories—edible and non-edible. Edible oil cake (soybean, groundnut, rapeseed, sunflower, coconut, cottonseed, safflower, flaxseed) have a high nutritional value and can be used in animal and human consumption as processed ingredients (protein concentrate, isolate, hydrolysate), a substrate (in the production of bioactive compounds, surfactants, enzymes, antibiotics, vitamins, pigments, flavors, and amino acids) and source of antioxidants. The defatted edible cakes can be used in the diet of undernourished people by incorporating them in bakery, infant products, and multipurpose supplements. Non-edible oilseed cake/meal (castor, neem, mauha, karanja and linseed) are used as manure due to the presence of toxic compounds [22].

According to the U.S. Department of Agriculture (USDA), the world production of oilseeds in 2018/2019 was 600.47 million metric tons, hence a large amount of press cakes and residues are available. The use of these by-products and residues from the oil industries is a sustainable alternative to reduce waste disposal and also contributes to the development at low cost, of new products rich in nutrients [17].

Oilcakes/meals composition depends on variety, extraction method, and growing condition. Taste and smell are characteristics to the initial raw materials without musty, mold, pro-rancid, and foreign smells. Walnut oilcakes are yellow to light brown, and sweetish. The color of pumpkin seeds oilcake is brown to brownish green with an insipid and a sweetish flavor. Sesame oilcake color varies from cream to light brown and has sweetish, insipid taste. Oilcake from flaxseeds have different shades of brown with a neutral and insipid taste [24].

As shown in Table 1, the highest content in proteins was found in groundnut cake then soybean, almond, chia, rapeseed, sunflower, cottonseed, pumpkin, hemp, safflower, sesame, coconut, flaxseed, and olive oilcakes. The oil content in sunflower and rapeseed is over 40%, 15–25% in soybean and cotton, another major source is peanut (56%) [18].

Despite the fact that oilcakes are low-processed materials they are safe. Improper storage and handling can cause rancidity (unpleasant taste and smell). During four months of storage, Tarek-Tilistyak et al. [25] observed that the water activity remains stable inhibiting bacterial and mold growth. The macronutrients decrease after one month and the lowest level of contamination was detected in walnut and the highest in linseed. Further, study investigated the influence of different packaging. More precisely, if flaxseed oilcake is packed either in paper or polypropylene and stored for six months at room temperature it can maintain the healthy properties, nutrients, and nutraceuticals contents typical of the raw material. Cakes stored either in paper or plastic show similar behavior—a modest decrease of antioxidant activity, optimal omega-6/omega-3 ratio, increase in α-tocopherol (possible from the conversion of the γ-form), decrease in crude fat (48% in paper and 51% in plastic) and proteins (12% and 6% respectively). It was concluded that this method of preservation was efficient and it is preferable the use of paper because is more eco-friendly [26]. It was demonstrated that peanut oilcake, stored twelve months in vacuum packaging with low permeability, regardless of the temperature retained its quality (good water/oil absorption and high content of unsaturated fatty acids), antioxidant capacity, and low microorganism yeast (due to the oxygen restriction that delays their growth and lipid oxidation) [27].

Table 1. Chemical composition of oilcake/meals obtained from different methods of extraction.

Type	Moisture %	Dry Matter %	Crude Protein %	Crude Lipid %	Ash %	Crude Fiber %	Carbohydrates %	Reference
				RAPESEED				
SEC [1]	3.96–10.59	88–96.04	16–45	1.1–10	6.1–15.8	8.2–17.5	21.38–47.72	[11,28–36]
CPC [2]	6–10.8	86.3–94.3	14.03–40.1	5.14–23.1	5–19.7	5.5–15.46	25.1–48	[17,20,24,32,34,37–47]
HPC [3]	4.7–10.1	89.9–95.3	36.1–39.1	9.2–12.2	6.9–7.1	13.1	31.5	[20,48]
M [4]	5.8–12	88–94.5	33.9–40.6	1.7–23	7–13.9	11–14	35.5–44	[17,40–42,49]
				CANOLA				
SEC	7.3–12.09	87.91–92.7	34.5–40	2.46–19.34	4.19–6.34	12.05–20.11	8.35–17	[50,51]
CPC	9.07–10	90–90.93	33.9–37.75	17.96	4.50–6.2	9.7–11.63	19.08	[52,53]
M	2.86	97.14	55.47	3.08	5.56	12.05	20.98	[53]
				CHIA				
SEC	10.47	89.53	41.36	0.21	7.24	27.57	23.62	[54]
CPC	6.8–10.84	89.16–93.2	28.2–35	6.52–11.39	4.58–6.27	23.81–30.46	23.53–30.24	[54,55]
				SESAME				
SEC	4–10	90–96	22.65–48.5	1.07–29.94	5.27–13	4–10.36	25.5–34	[21,28,56–60]
CPC	1.17–16.8	83.2–98.83	16.96–45.9	5.10–48	3.80–12.4	3.28–22,7	22.5–46.96	[24,58,60–70]
HPC	5.05–9.3	90.7–94.95	39.1–47.1	3–13.31	3.41–10.35	3.48–9.97	20.08–42.85	[58,71]
M	0.25–10.38	89.62–99.75	30.3–59	0.40–5	6.9–11.87	3–29.37	39.9–55.4	[12,72–75]
				COCONUT				
SEC	5.75–9.9	90.10–94.25	17.8–19.95	3.22–9.4	2.024.9	10.3	47.7	[21,76]
CPC	6.52–11.2	88.8–93.48	17.40–25.2	15.02–15.3	4.13–8.34	10.8	53.94	[52,77–79]
				OLIVE				
SEC	N.a [5]	N.a	0.3–10.6	4	3.4–9.1	N.a	N.a	[80]
CPC	14.8	85.2	0.4–4.77	8.72–11.1	4.1–6.6	40–60.1	10.1–20.6	[80–82]
				WALNUT				
SEC	10.59	89.41	13–38.87	2,45–10	7.48	5–33	12.65	[24,25,55,83–88]
CPC	3.6–10,5	91.8–96.4	10.30–50.4	7.95–36.80	2.79–10	6.79–18.5	17.4–49.75	[84,89]
				ALMOND				
CPC	5.80	94.2	51.3	19	5.30	5.6	18.80	[55]

Table 1. *Cont.*

Type	Moisture %	Dry Matter %	Crude Protein %	Crude Lipid %	Ash %	Crude Fiber %	Carbohydrates %	Reference
SAFFLOWER								
CPC	6.8	93.2	23.25	9.14	5.10	34.41	55.71	[55]
SOYBEAN								
SEC	88.7	11.3	46–52.4	0.55–8	2–6.96	4.4–75.44	42.8–47.05	[59,90,91]
CPC	8.4–9.66	90.34–91.6	43.3–45.5	9.3–15.55	5.71–5.91	4.95–11.28	14.98–21.76	[44,92,93]
HPC	89.9	10.1	39.1	9.2	6.9	N.a.	N.a	[48]
M	12	88–95.79	44–49	0.5–0.65	4.21–7.5	3.5–7	35.85–47.79	[70,94]
GROUNDNUT								
CPC	8.22–4.57	91.78–95.43	41.73–60	7–22.59	1.5–5.76	4.33–8.55	14.1–36.6	[55,67,95,96]
SUNFLOWER								
SEC	2.56–10	90–97.44	31.9–43.38	1–23.6	6.4–7.83	13.07–28.85	25.99	[25,90,97,98]
CPC	2.5–11	89–97.5	19.93–44.9	7–16.6	4.69–8	17.4–33.4	15–28.2	[44,77,95,96,99–101]
FLAXSEED								
CPC	6.89–9.27	90.73–93.1	14.4–41.97	6.11–21.4	4.7–6.27	6.29–12.9	16.26–52.45	[24–26,44,47,52,68,76, 102–105]
HPC	2.65–6.14	93.86–97.35	28.13–59.62	1.85–2.10	5.26–9.40	9.35–34.77	21.27–25.6	[53,105]
M	5.75–9.37	90.63–94.25	19.95–34.13	1.87–3.22	2.02–5.72	9.38	48.9	[103]
PUMPKIN								
CPC	5–8.2	91.8–96.5	29.39–53.98	5.92–36.22	4.20–8.7	3.89–7.1	15.88–19.73	[24,47,55,77,92,106]
M	8.01	91.99	60.94	0.94	9.93	N.a.	28.19	[107]
HEMP								
CPC	6.35–13.61	86.39–93.65	23.25–33.45	0.51–14.02	3.30–9.78	17.41–60.38	2.80–48.54	[47,53,55,99,106,108]
HPC	2.15	97.85	52.48	1.54	6.52	18	19.31	[52]
M	9.37	90.63	32.07	0.97	9.65	N.a.	57.31	[107]
COTTONSEED								
CPC	N.a	N.a	24.79	8.91	6.2	7	32.2	[96]

[1] SEC-solvent extracted cake; [2] CPC- cold-pressed cake; [3] HPC- expeller cake, [4] M-meal; [5] N.a- not analyzed

4. Anti-Nutritional Factors

Anti-nutritional factors are substates of natural or synthetic origin, found in the human diet or animal feed, which can affect the health, and growth performance of livestock. These factors can have different effects on animals depending on their digestive process, for example, trypsin inhibitors have a negative effect on monogastric animals, but not in ruminants because it is degraded [109]. The direct utilization of oilseed cakes in human or animal nutrition is limited by the presence of antinutrients, which influences the organoleptic properties, protein digestibility, and macro-/micro-elements bioavailability [22]. The major antinutrient in oil cakes are presented in Table 2.

Canola meal contains phenolics, phytic acid, and glucosinolates, that can cause problems in food. The main antinutritional factor (30 times higher compared to soybean) are the phenolics acid esters, especially sinapine. This binds to protein and creates a complex that confers dark color, bitter taste, and through oxidation, poor digestibility. Regarding glucosinolates, it was demonstrated to have a positive effect on health in low proportion [110]. When seeds are crushed, enzymes such as myrosinase are liberated, which are responsible for the hydrolyzation of glucosinolates in glucose and various toxic chemicals (isothiocyanate or thiocyanate ions) [10].

Cyanogenic glycosides are the major antinutrient in flaxseed cake. In the gastrointestinal tract, during digestion they form hydrogen cyanide, causing below 50 mg/Kg acute toxicity in adults. Linatine, another antinutritional factor can cause pyridoxine (B_6) deficiency [111].

The majority of oilcakes (rapeseed/canola, sesame, sunflower, soybean, groundnut, hempseed) contain antinutrients such as phytic acid and trypsin inhibitors, which can diminish protein and vitamin utilization. More precisely the second one (trypsin inhibitors) connects to the digestive enzyme trypsin, thus reducing proteins adsorption and digestion. [112].

Saponins are a group of steroidal glycosides (natural foam producers) that cause hemolysis and interference in bile acids, lipid-soluble vitamins, cholesterol, and dietary lipids [111].

Phytic acid can bind with minerals, proteins and amino acids forming phytates and insoluble complexes, thus reducing essential nutrient bioavailability, and digestibility. They can also lower the minerals availability and amylase activity [67].

Table 2. Main antinutritional factors.

Types	Oilcakes	Effect	Reference
Sinapine	Rapeseed/Canola	dark color; undesirable taste, indigestibility; lower nutritional value	[51,94,110]
Phytic acid	Rapeseed/Canola, Sesame, Sunflower, Soybean, Groundnut, Hempseed	decrease mineral availability and absorption and digestion of proteins and starch	[90,94,113]
Tannis	Soybean, Sesame, Sunflower, Groundnut, Rapeseed, Hempseed	astringency, inhibit protein adsorption, produce amino acid imbalance	[90,94,113]
Saponins	Soybean, Sunflower, Hempseed	hemolysis, interference in bile acids, lipid-soluble vitamins, cholesterol and dietary lipids	[90,94]
Glucosinolates	Rapeseed/Canola, Hempseed	reduce feed intake, impaired thyroid function, liver enlargement	[10,94,113]
Cyanogenic glycosides	Hempseed, Flaxseed	pyridoxine (B_6) deficiency	[94,111,114]
Chlorogenic acid	Sunflower	protein indigestibility and alter their organoleptic properties, storage life and stability in food systems	[94,101]
Trypsin inhibitors	Soybean, Sunflower, Sesame, Groundnut, Hempseed, Flaxseed	reduce protein digestibility	[90,94,112,113]

The elimination or inactivation of these toxic substances can be achieved by various methods: physical (dehulling, heat-cooking, autoclaving, toasting), chemical (ammoniation, the addition of choline, methionine, ferrous sulphate, sodium carbonate), enzymatic and fermentative.

Heat treatment can remove or reduce effectively cyanogenic glycosides, proteases, trypsin inhibitors, glucosinolates, and phytic acid (the last one was reduced by 43%). The extraction of protein isolates is another method useful to eliminating the anti-nutritive factors, due to the high pH used [109,111].

Extrusion, solvent extraction and biological, thermal, or microwave treatment are various methods used to reduce or remove cyanogenic glycosides. The heat hydrolyzed the latter in hydrogen cyanide, which is then evaporated [114].

Phytates and tannin levels in sesame oil cake can be reduced by fermentation with *Lactobacillus acidophilus*, after which it can be used as feed in the diets of *Labeo rohita* [67]. Tannins from groundnut oil cake can be removed through solid-state fermentation (SSF) with the help of *Pichia kudriavzevii*, a process that also enhances its chemical properties [115]. The same process with *Rhizopus oligosporus* DSM 1964 and ATCC 64063 was performed for flaxseed press-cake to reduce phytate by 48% and 33% respectively, and which enhances the bioavailability of calcium, magnesium, and phosphor by 14%, 3.3% and 4% [116].

Methods for glucosinolates removal in canola are: extraction of protein based on protein micellar mass formation with minimal loss of proteins (ultrafiltration because the toxic compounds have low molecular weight than proteins), heat treatment (reduction of 94% and improved flavor and palatability), use of enzyme (pectinase, protease, hemicellulase), application of organic solvent (ethanol, methanol, and acetone). During the traditional extraction of proteins, the amount can be reduced due to the dilution process during washing [22,110].

5. Possible Ways to Capitalize the by-Products Obtained from the Oil Industry

The most conventional practices of disposal/capitalizing oil cakes are: animal feed, landfilling, and biofuel conversion. They can be used as animal feed due to the rich content in protein (suitable for omnivore), cellulose, and hemicellulose (suitable for ruminants). Unfortunately, the presence of toxic compounds (with antinutritive effects and unbalanced nutrients) can affect both animals and humans. In landfilling, the decomposition of wastes leads to the production of methane and to water pollution. For these reasons, governments are trying to reduce landfilling through regulations (that impose the following measures: landfill prohibition, taxes, development of alternative solutions, and development of infrastructures), and public awareness. Food wastes contain organic compounds that can be converted first in energy through incineration. Increasing concerns about the high cost and negative impact on the environment due to the emissions [4,5].

The creation of a sustainable food chain and the increasing environmental issues lead to new valorization methods that imply the recovering/recapturing of valuable components, production of functional ingredients, development of new products and biopolymer films [4].

5.1. Extraction of Bioactive Compounds

Industrial residues contain valuable compounds (pigments, fibers, minerals, antioxidants) that can be reintegrated in the food industry and other fields (agriculture, cosmetics, pharmaceutics) [117]. Their extraction is done in three stages. The first step is pretreatment, which helps remove microbes from materials without affecting biological activities, using the following methods: foam mat, electro-osmotic, de-watering, and micro-filtration. The second step is extraction that can be realized through classical (solvents alone or mixed together, maceration, steam distillation) or advanced/novel/greener methods: extraction with water or ethanol, pulsed electric field, super/subcritical fluid extraction, enhanced solvent extraction, use of mixture water/organic solvent with dioxide carbon, accelerated solvent extraction, microwave/ultrasound-assisted methods and high voltage electric discharge, enzyme-assisted extraction. Moreover, these green, sustainable and innovative methods present some advantages and drawbacks. Advantages in using modern techniques are: high-quality extraction, selectivity in recovery, use of a small volume of solvent, a small amount of waste remained, environment friendly and less extraction time. Further, they provide a way to use wastes and by-products in industrial applications, based on the sustainable circular economy concept.

The main drawbacks are high-energy consumption when using microwaves, separation issues in ultrasonic extraction, and lack of user-friendliness in the pulsed electric field. Traditional methods require long time extraction and large quantities of expensive, toxic, and hazardous solvents but respect of the novel ones have better efficiency reproducibility, efficiency, and extract manipulation. The final step implies purification with alcohol precipitation, ultrafiltration, or chromatographic technique [118–120].

From oil press-cakes and meals can be extracted high-value components such as proteins, antioxidants, phytochemicals, and dietary fiber [118].

5.1.1. Protein Isolate, Concentrate and Hydrolysate

Every year, the world population increases creating an increasing demand for food supply and proteins. Oil cakes/meals are a suitable alternative because they are a valuable source of proteins [111]. Plant proteins, recently, became an alternative for replacing those from animal sources because they are versatile, easily digestible, non-toxic, and nutritional sufficient. Moreover, the isolation process is less costly [121]. However, usually they lack sulfur amino acids and can negatively influence the flavor, color, and texture of foods [17].

Of all plants, proteins from rapeseed contain a large amount of S-amino acids (in an amount that exceeds the requirement for adults and children) and in terms of nutritional value can be compared with those from soy [122]. Napin and cruciferin are the dominating proteins groups (together 85–90%) in rapeseed. Proteins recovered from cold-pressed rape oil cake have higher recovery yield and emulsion properties compared with those extracted with hot-press and solvents. In the recovery process, when the heat was applied the recovery-yield improved slightly ($p < 0.05$, however only for solvent extracted rape oil cake. Conversely, the ability to stabilize emulsion decreased [22]. Soybean and groundnut contain proteins that can be easily absorbed, digested and are nutritionally equivalent to animal protein, the amount is 20–26% and covers the recommended daily dose for adults and children. Regarding the amino acid profile, soybean lacks methionine, and groundnut is rich in arginine. The proteins from linseeds are comparable with those from soya and are also rich in glutamic and aspartic acid and arginine. Furthermore, those from flaxseed have antifungal properties [22,53,123]. In hemp oil cake they are present two main proteins, histidine and albumin (a high amount of essential amino acids, especially arginine) comparable with those from egg and soy [17].

Press-cakes/meals are used to prepare protein isolates (with a content of proteins > 90%), concentrates (with 30–80%) and hydrolysates (resulted from the hydrolysis of protein isolates) [22]. By first solubilizing protein with alkali at a high pH, then isoelectric precipitation with acid, washing, and drying the isolates are prepared. The isolates present high water-holding capacity and emulsifying activity and stability. They find application an as emulsifier and functional food (food and texture retainer and food system size expander) [53]. Proteins isolated from canola by the alkali method are not suitable as food ingredients because they have poor technological properties and solubility. Furthermore, during the process an irreversible denaturation takes place. For this reason, they need chemical or enzymatic improvement [110]. Those obtained from flaxseed and sesame are highly soluble in acidic and alkaline pH with good functional properties such as water holding and fat absorption capacities, emulsifying and foaming properties, bulk density, and poor gelling capacity [68]. Salgado et al. used sunflower oil cake and different procedures to obtain protein concentrates and isolates with high solubility (75%). These were then subjected to extraction with various solvents (water, ethanol 70%, methanol 80%, acidic butanol, and sodium thiosulfate 0,1%) to remove phenolic compounds and/or isoelectric precipitation. The residual phenolic compounds (remaining due to the association with proteins) confer to the final products antioxidant properties (without affecting water solubility) and a strong dark coloration (that could limit their potential application) [101].

Hydrolysates resulted from the hydrolysis of protein isolates: the protein structure is modified resulting in improvements of their functionality, solubility, surface activity, hydration, and gelling properties. Through this process, a fragment of protein is created known as a bioactive peptide with biological activity and a positive impact on human health (antioxidant, antithrombotic, hypercholesterolemic, bile acid binding and immunomodulatory activities). These changes depend on the type of enzyme used and the degree of hydrolysis. Peptides with antihypertensive and antioxidant activity and high digestibility are developed by subjecting the sesame oilcake/meal hydrolysate to an additional process of hydrolysis with alcalase, papain, and pepsin [22,124]. Protein hydrolysates from canola can be used in enhancing cooking yield, water holding capacity, and organoleptic attributes in meat formulation. Furthermore, enzymatic hydrolysates can be used for the formulation of a flavor with meat characteristics [125]. Rapeseed bioactive peptides in concentrations between 30 and 50 mg/L inhibit thrombosis activity up to 90%. They also have antioxidant properties (malondialdehyde inhibition by 50% in blood serum, inhibition of lipid peroxidation), blood pressure regulation capacity (a peptide called rapakinin induced vasorelaxation), and bile acid-binding capacity. Peanut peptide instead has high antithrombotic activities [125–127].

Kavitha. et al. [128] use sesame seed cake for the production of vegetable peptones. On a media containing protein hydrolysates, yeast extract, sodium chloride, and agar were carried out morphological and growth analysis of different organisms (*Candida albicans*, *Bacillus subtilis*, *Escherichia coli*, and *Saccharomyces cerevisiae*). On the vegetal peptone, both prokaryotes and eukaryotes present higher microbial growth and thus can act as a replacement for those of animal origin. Some advantages were observed: vegetal peptone can be used for the production of amylase and the yeast produced has a short fermentation time.

Protein concentrates from rapeseed have poor gelation ability than soybean that can be improved by glutaminase enzyme [129]. Results also showed that presents darker color, lower purity, and yield (due to the wide range of isoelectric points) [130]. Sensory evaluation of sausages, formulated with canola protein concentrate in place of casein demonstrated that improves taste, texture, and aroma [125]. Sesame protein concentrate introduced in the proportion of 10% in extruded snacks lower the carbohydrate content and enhance organoleptic properties, color, and protein content [131].

5.1.2. Antioxidants

Oil cakes/meals contains free, esterified or condensed form of phenolic acids, flavonoids and lignans that help in reducing oxidative stress and thus preventing various types of cancer. They can be extracted with solvent (it can be used organic solvents, alone or mixed together or non-toxic solvents such as water, but these need to be combined with other mild extraction techniques), high pressure, microwave, and supercritical fluid. Antioxidants extracted in this way can be used in the preparation of foods (beverages, energy bars, bakery, and extruded products) [22,118]. The major antioxidants in oil cakes are presented in Table 3. The antioxidative activity in soybean cake/meal is due to isoflavones and cinnamic acid derivative and for groundnut cake/meal, phenolics, caffeic, and chlorogenic acids. Phenolic compounds are the principal antioxidants in rapeseed, coconut, sunflower, mustard, sesame and cotton cakes/meals [22].

Fermentation helps to improve the nutrient levels of oil cake, because during the process a microbial synthesis of biomolecules takes place. Following this idea, Stodolak et al. [132] observed that in flaxseed oil cake fermented for 48 h with *Rhizopus oligosporus* DSM 1964 and/or ATTC 6403 there was an increase of 13–85% in phenolic levels and a decrease in scavenging activity by 20–30%. The same results were obtained when peanut oil cake was fermented for 120 h with *Aspergillus awamori* [133]. Another treatment useful in enhancing antioxidant capacity involves the use of ultrasound. Maximum polyphenol content was obtained when the treatment time was 20 min and with the temperature at 70 °C. The high heat is able to break the bonding between the phenolic compounds and matrix, modifying the structure of plant membranes and causing lipoproteins coagulation. Furthermore, prolongation of exposure means that the ultrasonic wave divided the phenolic extract in longer periods [134].

Cold-pressed cake from walnut is a source of natural antioxidants. Bakkalbaşi [135] extracted phenolic compound from walnut pressed cake. This extract was added in different concentrations

(50, 100, and 200 mg/kg) to increase the oxidative stability of walnut oil. The highest result was obtained with the addition of 200 mg/Kg. When oilcake is added directly to oil (1:100) through ultrasonic maceration the oxidative stability decreases although the amount of total phenolic compounds increased. This may be caused by free fatty acids (created in sonification) with a pro-oxidative effect.

The cake obtained from sesame oil contained bioactive compounds such as phytochemicals with numerous benefits (phenolic compounds, flavonoids, tocopherols, vitamins, pigments, steroids, carotenoids, and lignans) for health. Lignans such sesamin, sesamolin, sesamol, and sesaminol glucosides have the following effect: anticarcinogenic, antiproliferative, antioxidant, antimicrobial, anti-inflammatory, antidepressant, neuroprotective, and hypocholesterolemic activities [136]. Sesame meal introduced in concentrations ranging from 5 to 200 ppm acts as a stabilizer for sunflower and soybean oil, by inhibiting double bond conjugation, thermal deterioration, and losses of polyunsaturated fatty acids. Furthermore, in high concentrations, the efficiency is equivalent to butylated hydroxy anisole and butylated hydroxy toluene. Products were introduced to an oven temperature of 70 °C for 72 h and subjected to oxidation in the dark, and a protective effect was observed during the initial and final steps. This activity is attributed to the phenolic compounds and their redox properties (extinguish singlet and triplet oxygen, decompose peroxides, adsorb, and neutralize free radicals) [61,137]. Reshma [74] analyzed the potential of using defatted sesame cake to inhibit two digestive enzymes (α-amylase, α-glucosidase). The result exhibits a strong inhibitory potential for α-amylase and mild for α-glucosidase, due to the presence of phenolic compounds and flavonoids. These modulate the enzymatic decomposition of carbohydrates by inhibiting glucosidases and amylases.

Natural phenolic antioxidants (flavonoids, phenolic acids) from sesame and coconut oilcake and synthetic butylated hydroxytoluene (BHT) were introduced in vanilla cake composition to compare their potential in improve the oxidative stability and both microbial and chemical shelf-life, without changing consumer acceptability. Both oilcakes improved chemical and microbiological stability up to 13 days, and only up to 11 days for BHT. The sensory quality in cakes with by-products from the oil industry was retained from the first day to day 12. Moreover, the activity of natural antioxidants is stable thermally [137].

Sesame oilseed cake extract (lignans) at a lower amount (150 ppm) can be used as a food additive with a role in oil stability improvement. In butter, it can reduce the lipid oxidation time much better than BHT without changing the sensory properties [57,136].

Table 3. Antioxidants in oil cakes.

Antioxidants	Oilcake	Reference
Gallic acid	Canola	[111,136,138]
Tannic acid	Linseed	[111]
p-coumaric	Flax, Peanut, Sunflower, Linseed, Sesame, Mustard, Rapeseed, Palm, Olive	[111,136,138]
Catechin	Canola/ Rapeseed, Sunflower	[111,136,138]
Caffeic acid	Canola/Rapeseed, Hemp, Peanut, Mustard, Palm	[111,136,138]
Epicatechin	Canola/Rapeseed, Sunflower	[111,136,138,139]
Ferulic acid	Canola/Rapeseed, Flax, Linseed, Sesame, Cottonseed, Mustard, Palm	[111,136,138]
Quercetin	Hemp, Canola/Rapeseed, Cottonseed, Olive	[111,136,138]
Luteolin	Hemp, Canola/Rapeseed, Olive	[111,136,138]
Lignans	Sesame, Olive, Linseed	[136–141]
Chlorogenic acid	Sunflower	[136,139]
Sinapic acid	Rapeseed, Mustard	[125]
p-hydroxybenzoic acid	Linseed, Olive, Palm	[136]

5.1.3. Dietary Fiber

Dietary fiber (DF) is defined as a group of carbohydrate polymers (oligosaccharides and polysaccharides) that cannot be hydrolyzed (digested and absorbed) in the small intestine by human digestive enzymes [142]. In this category are included insoluble components (cellulose, hemicelluloses, lignin), soluble components (gums, pectin substances, mucilages), inulin, and resistant starch [143,144]. DF has different functions and effects, depending on the physicochemical properties. For example, DF with high water-holding and/or swelling capacities reduce defecation time and produce satiety, on the other hand, those with high binding activity hinder DNA and epithelial cell damage [76].

A diet rich in dietary fiber brings many health benefits, such as reducing the risk of coronary heart disease, cancer, obesity, and diabetes. In the food sector DF is used to improve nutritive, organoleptic (color, flavor, taste), and textural properties (increasing water/oil holding, emulsion, or/and gel formation capacities and avoiding syneresis). They can also prolong the shelf life and improve oxidative stability [76]. Recovery and utilization of fiber from wastes represent key factors in implementing a sustainable production (they have health benefits, are available in large quantities, have low cost and are environment friendly), zero wastes, and circular economy [119,145].

DF can be extracted through different methods: traditional (dry/wet processing, chemical, enzymatic gravimetric, microbial) and green/innovative extraction (with ethanol, water, steam, ultrasonic, hydrostatic pressure, and pulsed electric field). The method used can influence the composition, trait, and behavior (in food and human body) of DF. From the first category, the most rentable, effective (purity of final product 50–90%) is wet processing. The utilization of alkali/acid can damage the structure and through enzymatic path the extraction can be incomplete. The innovative methods present advantages such as high recovery rate and low environmental impact [119,143].

Dietary fiber extracted from by-products can be used in foods and pharmaceutics as a functional ingredient, supplements, or additive [144]. Zheng and Li [76] studied how particles size, extraction with cellulase, and acid treatment affect the structure and properties of the dietary fiber from coconut oil cake. Reduction of the first factor (particle size) led to an increase in entrapment, water holding, and swelling capacities. The cellulase hydrolysis had an undesirable effect on color and oil holding capacity but increased carbohydrate content, porosity, water holding, swelling and alpha-amylase inhibition capacities. Compared to the previous (cellulase hydrolysis), the last parameter (acid treatment) led to inverse results.

5.2. Substrate for Functional Ingredient Production

Wastes from the oil industry can be used as substrate for the production of enzymes, antibiotics, biopesticides, biofertilizers, aroma compounds, pigments, vitamins, biosurfactants, and mushrooms due to their low cost, accessibility, and nutrient composition that require small or no supplementation. These are called functional ingredients and have important benefits for consumers [22].

Their production can be realized with a submerged fermentative medium (SFM) or solid-state fermentation (SSF). The first method involves the use of a liquid substrate rich in various nutrients for microbial growth. Unfortunately, the substrate is used rapidly and needs to be supplemented constantly [79]. The second one is a process in which an organism (fungi, yeast, and bacteria as single or mixed cultures) is grown on a moist solid substrate, in an environment with low or nil water content. The microorganisms, in the presence of an abundant substrate (glycerol, glucose, or carbon source) are opposed to the catalytic processes (inhibition of enzyme synthesis). It presents various advantages such high-stability and productivity, reduction of cost process, energy, and pollution [146,147].

5.2.1. Enzymes

Various enzymes have an important role in industrial application. They can be produced by solid-state fermentation using oil cakes/ meals as substrate, because they are cheap materials (reduce the cost production) rich in nutrients, carbon and nitrogen and are compatible for fungal species. Some examples are tannase, inulinase, α-amylase, glucoamylase, protease, phytase, mannase, lipase and L-glutaminase (Table 4). Their production depends on incubation time and pH, moisture and particle sizes of substrate. Large particles and too much content in water prevent the digestion of raw material and microbial respiration which leads to poor growth and enzyme production [148].

Proteases are a group of enzymes with a role in the protein hydrolysis. Bacterial and fungal species such as *Bacillus horikoshii, Bacillus clausi I52, Penicillium sp., Candida utilis, Bacillus sp. 1-312, Aspergillus* oryzae *(NCIM No.649, NRRL 1808), Streptomyces termovulgaris* were used for protease production with soybean (overheated), coconut, olive, palm, sesame and rapeseed oilcakes/meals as substrates [81,148]. Gupta et al. [149] studied the ability of non-edible Mahua and Jastropha cake to act as substrates for *Aspergillus niger* and *Paecilomyces variotii* growth and produce enzyme. The study concluded that despite the low protease activities, solid-state fermentation is suitable in antinutrient removal.

Lipases are hydrolytic enzymes with numerous applications in the pharmaceutical and cosmetic industries and in dairy and bakery foods. Parihar et al. [147], Oliveira et al. [50] optimized the production of lipase through SSF with *Pseudomonas aeruginosa* and *Aspergillus ibericus*. The highest production was obtained when the substrate used resulted by combining two oilcakes: linseed/ olive and sesame/ palm kernel. The obtained enzyme can be used in esterification (butyl decanoate formation using 5% biocatalyst) and hydrolysis reactions (increase production in shorth-chain triacylglycerols). Lipase extracted from *Rhizomucos pusillus, Rhizopus rhizopodiformis* and *Pseudomonas sp. S1* cultivated on olive oil cake, are suitable for oily waste water treatment and biodiesel production [150]. The enzyme could be extracted also from *Penicillium simplicissimum, Candida rugosa, Penicillium Chrysogenum* S_1, and *Candida utilis* inoculated on soybean, coconut, sesame and olive oilcakes [81,147,151].

L-asparaginase is an enzyme that can be used as a drug, due to its therapeutic effect against leukemia. A content of 90.025 IU, 310 U/gds and 5.75 U/gds was obtained in solid SSF from *Aspergillus wentii* NCIM 941, *Aspergillus niger* C4 and *Serratia marcescens*, cultivated respectively on palm, sesame oil and coconut cake [63,152,153].

α-amylase used in starch liquefaction was produced from *Aspergillus oryzae, Aspergillus terreus* UF39 cultivated on two oilcakes: coconut and groundnut [79,147].

Phytase, and glucoamylase have been produced using bacterial and fungal species such as *Bacillus licheniformis, Aspergillus ficuum*, and *Thermomucor indicae-seudaticae* respectively growing on sunflower, rape/canola, and cotton oilcakes as substrates in solid state fermentation [10].

Xylanase, xylosidase, acetyl-xylan esterase and cellulase were produced in a SSF process with rapeseed/canola oilcake as the substrate, on which microorganisms such as *Trichoderma reesei* and *Streptomyces termovulgaris* were grown. Alternately, *Aspergillus oryzae* produced protease, phytase and phosphatase [10,89].

L-glutaminase and phytase were produced using sesame oilcake as substrate and microorganisms such as *Zygosaccharomyces rouxii, Mucor racemosus* and *Rhizopus oryzae* [151].

Use of coconut oilcake as a nutrient source grew a bacterial strain of *Staphlococcus* sp. and fungal species such as *Aspergillus ficuum, Mucor racemosus, Rhizopus oryzae, Rhizopus oligosporus* and *Aspergillus niger* to produce inulase, phytase, and glucoamylase [22,82,147].

Lomascolo, et al. [10] reported the production (through submerged culture medium and SSF) of alpha-amylase and endoxylanase when using sunflower meal. The first one is obtained from *Bacillus licheniformis* and the second one from *Humicola lanuginose*.

Table 4. Production of enzymes through solid state fermentation using as substrates oil cakes.

Enzyme	Oilcakes	Microorganism	Reference
Lipase	Coconut	*Penicillium simplicissimum, Candida Rugosa*	[147,151]
	Sesame	*Penicillium Chrysogenum*	[151]
	Olive	*Rhizomucos pusillus, Pseudomonas* sp1, *Rhizopus rhizopodiformis, Candida Utilis*	[81,150]
	Soybean	*Penicillium simplicissimum*	[147]
	Cotton	*Penicillium strain X*	[151]
	Flaxseed–Olive	*Pseudomonas aeruginosa*	[147]
	Sesame–Palm kernel	*Aspergillus ibericus*	[50]
Protease	Soybean	*Bacillus horikoshi*	[81,148]
	Coconut	*Bacillus clausi* I52	[81,148]
	Palm kernel	*Bacillus* sp. I312	[81,148]
	Sesame	*Aspergillus oryzae* NRRL 1808, *Aspergillus oryzae* No. 649	[81,148]
	Olive	*Penicillium* sp., *Candida utilis*	[81]
	Mahua	*Aspergillus niger*	[149]
	Jastropha	*Paecilomyces varioti*	[149]
Phytase	Rapeseed	*Aspergillus ficuum, Streptomyces thermovulgaris, Aspegillus oryzae*	[151]
	Coconut	*Aspergillus ficuum, Rhizopus oligosporus* NRRL 5905, *Mucor racemosus, Rhizopus oryzae*	[151]
	Sesame	*Mucor racemosus, Rhizopus oryzae*	[22,82,151]
	Canola/Rapeseed	*Aspergillus ficuum, Aspergillus oryzae*	[10,22,82]
α-Amylase	Coconut	*Aspergillus oryzae, Aspergillus terreus* UF39	[79,147]
	Sunflower	*Bacillus licheniformis*	[10]
	Groundnut	*Aspergillus oryzae*	[79,147]
Glucoamylase	Coconut	*Aspergillus niger*	[22,82]
	Cotton	*Thermomucor indicae-seudaticae*	[10]
L-Asparaginase	Sesame	*Aspergillus niger*	[152]
	Palm	*Aspergillus wentii*	[63]
	Coconut	*Serratia marcescens*	[153]
L-Glutaminase	Sesame	*Zygosaccharomyces rouxii*	[151]
Inulase	Coconut	*Staphylococcus* sp.	[22,82,147]
Xylanase	Rapeseed	*Trichoderma reesi*	[10,89]
Endoxylanase	Sunflower	*Humicola lanuginosa*	[10]
Phosphatase	Rapeseed	*Aspergillus oryzae*	[151]

5.2.2. Mushrooms

A possible valorization of agro-industrial by-products is in mushroom cultivation. Mushrooms with good antioxidant properties were produced by SSF, using oil cakes as substrate. Moreover, some nutrients may transfer during the growth making them, food with therapeutic properties and high nutritional value. For example, *Lentinula edodes, Ganoderma lucidum,* and *Grifola frondosa* are hepatoprotective, antidiabetic, anticholesterolemic, and immunologic [154,155].

Krupodorova, et al. [156] investigate the possibility of using alternative and cheap substrates (wastes from the oil industry) for mycelial growth of 29 mushroom species. Oilcakes studied were those obtained from walnut and soybean, rape, sunflower, linseed, pumpkin, and mustard seeds. Soybean and walnut oilcakes were found respectively most and least suitable for mushroom growth (24 and 2 species respectively out of 29). Low results were found also for mustard and pumpkin (6 and 8 species). These may be due to the composition (small amount of proteins and high amount in lipids; deficiencies in essential nutrients) and pH of the substrate. The replacement of peptone with sunflower cake led to the growth of 12 out of 29 species of mushroom, more precisely to *Ganoderma lucidum* and *Laetiporus sulphureus* species. Linseed/flaxseed cake supported 13 species of mushroom growth including the *Pleurotus* species, results that was observed also by Jape, et al. [157].

Gregori and Pohleven [154] tested how different proportions of olive oil cake influence the production of *Ganoderma lucidum, Lentinula edodes,* and *Grifola frondosa*. The mushroom production decreased with the increase in proportion, they also change in color and deformed (content >40%). The same negative effect was observed with *Pleurotus ostreatus, Pleurotus eryngii, Pleurotus pulmonarius, Pleurotus cystidiosus* and *Agrocybe cylindracea* species (content >60%). This was probably due to the presence of polyphenolic compounds and low porosity, substrate aeration, and water retention capacity [158,159].

5.2.3. Antibiotics

Antibiotics are defined as substances produced by microorganisms with the ability at low concentration to inhibit/kill other microorganisms. Oilseeds cakes can be used as a substrate for various antimicrobial agents and antibiotics production, doing so also reduces the cost [22,160].

Sunflower, soybean, and sesame meal were used for the production of clavulanic acid and cephamycin C and by SSF with *Bacillus licheniformis* for endotoxin and bacitracin. The maximum synthesis of the latter was found in soybean meal [22,160–162].

Neomycin is a low toxic antibiotic used in tuberculosis, wound injury, and skin diseases. It was extracted by SSF process from *Streptomyces fradiae* NCIM 2418, grown on coconut cake [163].

Rifamycin is a broad-spectrum antibiotic with the ability to inhibit RNA- dependent DNA polymerase, commercial fermentation process with industrial strain generates only rifamycin B (modest activity) that can be chemically converted to rifamycin SV (a direct derivate product of rifamycin, it is also more efficient). The latter was isolated in SSF from *Amycolatopsis mediterranei* OVA5-E7. The maximum production was observed on the de-oiled cotton cake, followed by mustard, sunflower, groundnut, coconut, and sesame de-oiled cake [164]. Vastrad, et al. [163,165] also extracted rifamycin B (it is very powerful, easily biodegradable, and less toxic) from *Nocardia mediterranei* MTCC14 growth on sunflower oil cake and *Amycolatopsis mediterranei* MTCC14 growth on coconut and groundnut cake.

Sunflower meal can act as a substrate in the production of two antibiotics. Clavulanic acid with broad-spectrum antibacterial activity and potent β-lactamases inhibitor could be obtained in a solid-state fermentation process by *Streptomyces clavuligerus*. Cephamycin C was successfully achieved in both submerged culture medium and solid-state fermentation (SSF), the meal served also as pH regulator and foam controller [10]. From *Bacillus subtilis* grown by SSF with rapeseed meal, 5.3 g/Kg of iturin A, and 51.3 g/Kg of poly-γ-glutamic acid were obtained in about 90 h of fermentation [166].

5.2.4. Biosurfactants

Surfactants are amphiphatic molecules (derivate from petroleum compounds) composed of a hydrophilic and hydrophobic group that decrease interfacial and surface tension with water/hydrophobic systems. When synthesized by different microorganisms they are called biosurfactants with numerous advantages over the synthetic ones: low toxicity and good biodegradability, biocompatibility, chemical diversity, and stability. Based on chemical composition they are divided into glycolipids, lipopeptides, lipopolysaccharides, and oligosaccharides. Examples include rhamnolipids, trehalolipids, sophorolipids, mannosylerythritol lipids. Due to these properties they can be used in the environmental sector (bioremediation of pesticides, oil, and heavy metals, enhanced oil recovery, antimicrobial agents and biopesticides), petrochemicals, cosmetics, food, agriculture, and pharmaceutics [167–169].

Sophorolipids, microbial biosurfactants used in detergents and cosmetics formulation, were synthesized by *Starmerella bombicola* growth on residual sunflower cake through SSF. Their effectiveness is equivalent to Triton X-100, that is a considerate environmental pollutant [170].

Thavasi et al. [171] cultivated *Lactobacillus delbrueckii* and *Pseudomonas aeruginosa* on peanut oil cake for the production of biosurfactants (glycolipids and lipopeptides) with potential use in biodegradation of crude oil and in bioremediation of hydrocarbon pollution environments. Another example of biosurfactant (rhamnolipid) utilized for this purpose was synthesized from *Ochrobactrum anthropi* cultivated on palm oil decanter cake [168,172].

In the pharmaceutical and food industries surfactants from *Enterobacter* sp. MS16 cultivated on sunflower oil cake can be utilized as fungal inhibitor [95]. Antibacterial and antifungal activities were shown by lipopeptide from *Brevibacterium aureum MSA13* cultivated on olive/safflower oilcake.

Glycolipids obtained from *Lactobacillus delbrueckii* on peanut cake also has good foaming and emulsion capacity and for this reason, can be used in emulsion food [22].

Lipopeptides are biosurfactants isolated from *Bacillus pseudomycoides* OR1 cultivated on groundnut oilcake and can be used for their antibacterial activity against foodborne, pathogens. At 50 µg/mL they inhibit *Escherichia coli*, *Staphylococcus aureus* and *Klebsiella pneumoniae*. They bind to the bacterial surface, modifying the lipid organization by changing fatty acids and thus preventing cellular processes [173].

5.3. Animal Feedstuff

A possible valorization of the residues resulted from oil extraction process is in animal feedstuff. Their utilization can be limited by: seasonal production (that can lower spatial and temporal availability), low levels of nutritive components, presence of toxic compounds (with possible negative effect on animal performances), and composition variability (diet formulation becomes difficult). This valorization strategy improves economy, and environment sustainability. Oil cakes rich in fat and proteins can be suitable for feeding omnivores, while those with high cellulose content for ruminants [4,99]. Soybean oilcake may be used primely in the diet of monogastric species (chickens and pigs), palm oilcake in ruminants, rapeseed, sunflower, and pumpkin oilcakes in pigs [174].

Ensuring durable and economic fish farming involves the use of fishmeal substitutes. Different studies investigate the possibility to use oilcakes as feed in aquaculture. Pellets were produced by replacing 25% of the reference fishmeal with rapeseed, sunflower, soybean, and linseed press-cakes. The products obtained had similar nutritional values to the reference with high sedimentation velocity and less water stability, abrasion resistance, and expansion. This can be improved by using finer milled particles and altering the screw configuration of the extruder [21]. Nang et al. [175] observed that the replacement of fish meal with sesame press-cake (without amino acid supplementation) up to 52% did not reduce rainbow trout growth performances. The feed cost can also be reduced and made more sustainable. The performance of mustard and soybean press-cake in fish production was analyzed. The first cake led to poorer growth and food conversion ratio, but the second one is advantageous for crap breeding [93]. Fish meal can be replaced also with solid state fermented oilcakes, this process increases the levels of amino acids and decreases crude fiber content and antinutritional factors. Fermented soybean meal (200–400 g/Kg), sunflower cake (0–100 g/Kg) and groundnut oilcake (20–50%) were used in different levels as a replacement to fish meal in the diets of shrimps and carps. The inclusion up to 325.1 g/Kg, 32.6 g/Kg and 40% respectively had no adverse effect on growth, carcass composition and feed utilization efficiency in comparison to the reference diets [90,115].

In ruminants feeding is necessary to cover the nutritional needs (energy, protein and mineral intake) for animal production. Serrapica et al. [99] investigate the potential use of fifteen oilseeds cake (from hemp, sunflower, tobacco, cardoon and pomegranate) as protein supplements in their feeding. Of all this, hemp and tobacco oilcakes are the most suitable due to the high content in carbohydrate and slow degradable fractions of crude protein. Olive cake contains a low amount of proteins and considerable percentage of tannins, which makes it unpalatable and poorly digestible. In order to remove these antinutritional components and increase the protein content, olive cake was subjected to a biological process (SSF). The cake was inoculated with four fungi (*Beauveria bassiana*, *Rhizodiscina* cf. *lignyota*, *Aspergillus niger* and *Fusarium flocciferum*) for 15 days. Results showed a significant ($p < 0.05$) increase in proteins (94%) and decrease in phenolic compounds (43%), flavonoids (70%) and condensed tannins (42%) due to the production of enzymes (oxidases, peroxidases) by fungi [176].

In dairy sheep feed concentrate, two levels of oil cake (rapeseed and sunflower) inclusion (50% in winter and 30% in spring) did not affect the cheesemaking (crude fiber/crude protein ratio) or milk production parameters. Analysis on the milk showed that in winter, when the pasture is unavailable, the amount of healthy fatty acid increased [37]. Satisfactory carcass yield and feedlot performance (body weight gain, feed conversion efficiency) also were observed with the addition up to 20% (300g) of groundnut and sesame oilcakes [177,178]. Sesame meal also improved intake, digestibility and rumen condition in lambs without changing performance and carcass composition [179].

In adult male goats, the incorporation in a concentrate mixture of walnut oilcake up to 10% did not have negative effects on the calcium and phosphorus balance, however the high amount in aminotransferase may cause hepatotoxic problems [180]. Similar results were concluded by Mir [86], furthermore the level of oilcake (<10%) did not affect total gas production, efficiency biomass production, in vitro digestibility of dry matter, microbial biomass production and true degradable organic matter.

Chipa et al. [181] evaluated the possibility of incorporation soybean press-cake, in proportions of 0–20%, as a source of proteins in the diets of beef cattle. The experiment was conducted on 20 heifers and 20 steers for a period of 98 days. At the end of the treatment the steers were heavier that the heifers, due to the high feed intake and conversion ratio. An increasing in growth was observed in the diet with 6% and 13%.

5.4. Applications in Food Products

New food products can be developed with two categories of ingredients: firstly, it includes new, alternative, and never used before on a regular basis in food products and secondly, they are obtained by the valorization of by-products. Their successful implementation can be achieved by creating a marketing strategy. This involves new regulations, customer education and clear transparency (better communication, adequate labeling) [182]. Introduction to the food of ingredients with important benefits for consumers (functional ingredient) is called functional [22]. Residues from the oil industry could be used as co-products for high value-added products, food additive, or supplements [71]. Bochkarev, et al. [24] studied the physicochemical composition of four oilcakes, and they divided them into three groups. In sesame oilcake, which is used for enrichment, proteins and lipids dominate. When proteins and carbohydrates dominate in oilcakes (in the case of pumpkin and walnut), it can be used as fillers in dairies and meat. A possible use of oilcakes rich in proteins and fibers (flax) can be found in bakery and confectionery.

Four different cold-pressed cakes (coconut, flax, sunflower, pumpkin) were mixed and then compacted to create tablets with functional properties, that can be used as supplements. When coated with starch and enriched with honey, they have better hardness and resistance to cutting. These tablets can be directly consumed if coated with various flavors as chocolate or caramel [77].

Application of treatments on oilcakes made them more appealing, palatable, edible, and suitable for human consumption as supplements. In this regard, coconut and sesame cakes were mixed with water and cooked so that the solid part can be separated from the liquid, and then subjected to drying. After that, the nutritional composition of the raw and cooked residue (solid and extract) was analyzed. Overall, the cooked residues retain more nutrients than the extract [21]. Starting from this premise, Sunil, et al. [183] used sesame and coconut solid cake for the formulation of four healthy foods. The first and third formulation contained coconut cake in proportions of 10% and 20%. The other two, instead, contained 15% coconut and sesame cake. All foods were considerate acceptable in sensory evaluation.

5.4.1. Bread

The most consumed product is bread and can act as a vector for bioactive ingredients. For this purpose, Pycia K., et al. [184] incorporated walnut oil and oilcake (1%, 3%, 5%) as a replacement for wheat flour. Increasing the level of oilcake enrichment led to harder, smaller (the replacement led to a decrease and weakening of gluten), darker and chewy bread, but with a high antioxidant potential compared to the control. Rheological properties of the dough with different ratios (7.5%, 10% and, 12.5%) of flaxseed oil cakes were studied in the Mixolab. As the levels increased the water absorption capacity increased, but the viscosity, amylolytic activity, and retrogradation decreased. All of these lead to a rapid dough maturation, thus reducing by 1.5-times the total fermentation duration [185,186].

5.4.2. Biscuits

Sesame oilcake flour was incorporated in five biscuits formulation (0%, 10%, 25%, 50%, and 75%) as a replacement for wheat flour. Products with white sesame oil cake (up to 50%) did not change taste and texture, but at 75% it becomes frail. Texture and taste of the first three biscuits formulation with black sesame oil cake were good. The last two formulations were produced after taste, but only the last one became brittle. With the previous incorporation, the diameter and weight loss (so the microbial contamination is reduced) were reduced, but the thickness, hardness, and fracturability increased. [71]. The same results were obtained when soybean and groundnut were used, due to the fact that high protein flours present high water retention and the biscuits cannot develop well. Unlike the previously analyzed products, these biscuits are considered acceptable at low-level incorporation. Regarding the rheological properties, the first oilcake increased the water absorption more, and the mixing characteristics of doughs. With the increasing proportion of the second oilcake (groundnut) the farinograph bandwidth increased but became slightly irregular. This led to an increase development of gluten protein and therefore of the dough consistency. The color of the final products become darker with the addition of the two oilcakes, due to the brownish and yellowish color of the raw materials used [187].

5.4.3. Snacks

On corn extrudate the addition of cold-pressed cake from sesame (0–20%) caused an imbalance in the amino acid profile, reduced carbohydrates content and sectional expansion, and increased the protein, fat and ash contents. Products became more darker but were organoleptically acceptable, especially those with 20% [62].

Healthy, non-caloric, and high nutritional snacks can be achieved with the incorporation of food industry by-products rich in antioxidants, essential fatty acids, fiber, and minerals. Cottonseed meal was incorporated, by Jáquez, et al. [188], in corn extruded snacks at different levels (5–98%) with a 10% optimum. With the increase in oil meal the consumer acceptance decreases, the surface becomes rougher, lumpier, and disrupted. The same conclusions and optimum levels were found for rice/corn snacks when defatted flaxseed meal was incorporated [96]. In accordance with the first two studies related previously, a proportion of 5% defatted hemp oilcake can be successfully introduced in corn snacks. The level of acceptance is lower because of the high fiber content [108]. Radočaj, et al. [106] developed gluten-free snacks/crackers with hemp flour (from cold oilcake) and decaffeinated green tea leaves. All formulations were qualified as healthy (high minerals, fibers, and a desirable omega-6/omega-3 ratio) and are suitable for people suffering from celiac disease. The highest appreciation was achieved with the highest addition level.

5.4.4. Desserts

Macaron is a luxury and expensive dessert, and by adding walnut oilcake (0–50%) their quality can be improved, and at the same time the production cost can be reduced. The improvements refer to the decrease in carbohydrates, omega-6/omega-3 ratio, and energy which is favorable in the problems associated with obesity, normal, and under nutrition. The addition strongly influences the amount of linoleic and caproic acid but not the aroma of macarons. However, the sensory analysis demonstrated that macarons with an addition of more than 10% were not appreciated, due to the increase in phenolic and volatile compounds that affect the taste [88]. Walnut oilcake can be introduced, at different levels 0–20%, as a supplement in cake formulations. Increasing the addition, it led to a decrease in firmness, cakes become softer due to the high-fat content of press-cake. The substitution levels increase the total phenolic compounds and antioxidant ability but were not enough for increasing the proteins. Sensory evaluations demonstrated that cakes supplemented with 15% were the most appreciated. The crust color became lighter due to the addition of the skin portion of walnut [83].

5.4.5. Dairy Products

Increasing interest for new functional food, ideal different typologies of consumers such as vegetarian/vegan and with intolerance/allergy to dairies leads to the production of kefir-like fermented beverages using as substrate different levels of flaxseed cake (5, 10, 15%). After the inoculation of kefir grains on the substrate, the mixtures were incubated for 24 h at 25 °C and then stored at 6 °C for 21 days. Acid lactic and yeast grew well on flaxseed cake without any supplementation and their viability exceeded the recommended level for kefir products. With the increase in cake percent, the beverages had high viscosity (due to the presence of mucilage and protein), firmness (due to the production of polysaccharide kefiran), and antioxidant activity (due to the production of phenolic compounds and bioactive peptides). Another advantage is in the economic field; oilcakes are cheap, safe material available throughout the year [114].

5.5. Biopolymer Packaging

In order to improve the quality and shelf life of foods but also to minimize the petroleum derivates packaging residues, there were synthesized biodegradable polymeric materials [189]. Films and coatings are biopolymer packaging materials with the same chemical composition, but different appearance: the first ones are separate materials while the second ones are created on the surface of a product. The two processes used for films development are casting (or wet process) and compression/extrusion (or dry process). The most used is the first one and involves the dispersion of the forming macromolecules in an appropriate medium followed by solvent evaporation [190,191].

Their applicability in packaging is influenced by the following properties: optical, thermal, mechanical (must preserve the integrity of packaged products, films must have high tensile strength and elasticity), structural, microbiological, sensory (should be tasteless, clean, transparent, odorless) and barrier (against aroma compounds, water, gas, and light permeability). The films have numerous advantages such biodegradability, biocompatibility, protection against oxidation, and microbiological failures [98,192].

Biodegradable films/coatings are made with biopolymers, such lipids, proteins, polysaccharides, used alone, or in combination. Among these, superior films (good gas barrier and mechanical properties) were obtained with proteins. A single natural polymer produces films with various good characteristics but poor aspects quality. Improvements can be achieved when combining two or more polymers from different origins [8].

5.5.1. Sunflower Oil Cake (SuOC)

Suput et al. [98] investigated the possibility to use SuOC for the production of biopolymer films and the effects of pH and temperature on them. The final products were firm, smooth, flexible, dark brownish-green, shiny, with a sunflower flagrance; with the increase in temperature and pH tensile strength increased; water vapor permeability, swelling, and solubility were uniform but decreased at high temperatures. Optimal films were obtained at pH12 and 90 °C. Addition of 0.25–1% parsley and rosemary essential oils decreased tensile strength and increased elongation at break and the antioxidant activity. Water permeability increased with rosemary and decreased parsley [193].

Films prepared with protein isolates and concentrates from SuOC have good barrier properties, high adhesive characteristics, high water solubility oil, and organic solvents resistance [192]. Comparing them films prepared with soy protein isolates and different amounts of phenolic compounds, they have low elongation, deformation, and elasticity, but higher antioxidant capacity [194,195].

5.5.2. Pumpkin Oil Cake (PuOC)

Cakes resulted from cold pressing of pumpkin seeds are a source of natural macromolecules with film-forming ability. By substituting gelatin with an increasing percentage of pumpkin oilcake, Popovic, et al. [196] created films with adequate characteristics. The best tensile strength was observed with a 40% addition. Solubility, antioxidant activity, and swelling capacity did not change when PuOC was added, but elongation at break increased by 20% at 95% addition. Subsequent studies investigated the possibility of using only pumpkin oilcake in film-forming and the possible impact of temperature and pH on them. Films with the highest antioxidant activity were obtained at pH 10 and 60 °C. At 90 °C and pH12 were created films with the highest tensile strength, elongation at break, and lowest gas permeability. Films produced at the same pH and temperature 50 °C have total soluble protein and matter content very high. When films are prepared at high temperature (90 °C) the surface is compact with nanoholes. At low temperature (50 °C) the holes are elder and this explains the low gas barrier ability [197].

Films obtained with pumpkin cake are not heat-sealable, so they were laminated with zein. Bulut et al. [198] analyzed the changes in barrier, mechanical and physicochemical properties in bilayer films made with pumpkin oil cakes and zein and their pouches during four weeks of storage. During this time, the film loses in moisture, total soluble matter, and swelling ability. It was observed also a good barrier for oxygen but moderate for carbon dioxide. The pouches present a high percentage of oxygen and low in dioxide carbon due to the decrease in heat deal strength and gas transmission rate. The addition of stabilizers as 30% glycerol and 0.5% xanthan, produced films with good mechanical properties, poor barrier for water vapor, and very good to oxygen. With the increase in concentrations, the transmission rate of dioxide carbon also increased. This can be advantageous for maintaining the quality and shelf-life of foods such as cheese, vegetables, and fruits [199].

The incorporation of different essential oils (caraway, wintery savory, and basil essential oils) in the biopolymer matrix (chitosan, Tween 20) increased the edible packaging antimicrobial and antioxidant activity. Chitosan films with caraway are elastic, oil resistance, present low permeability to gases, effective against *Escherichia coli* and *Staphylococcus aureus*, but only with the lowest amount of caraway the antioxidant activity increases [194]. When increasing the concentration of Tween 20 in films with winter savory the antioxidant activity also increases, reverse was observed in films with basil [98,200].

The ability of protein isolates to create biopolymer films with different amounts of plasticizer and pH values was analyzed. At pH 4-8, they could not be synthesized and only moderate amount of plasticizer produce suitable final products. Besides this, they present excellent oxygen, nitrogen, dioxide carbon, and gas barrier properties (150–250 better than plastic) [201].

5.5.3. Soybean Oil Cake

In comparison to other plant protein sources, soy proteins are superior materials for the production of films, they are much cleaner, smoother, and flexible. Their adhesive ability is characterized by low water resistance and gluing strength [8]. However, during long time storage, they suffer aging (aggregation, thiol oxidation, molecular rearrangement, and spontaneous relaxation). Improvement of their properties can be done by: application of cellulose and starch, enzyme-mediated and chemical crosslinking (with rutin, epicatechin, genipin, metal methacrylate, and dialdehyde carboxymethyl cellulose). Moreover, the film's submission to high temperature and/or pressure can increase their tensile strength and decrease their permeability and water sensitivity [192].

5.5.4. Rapeseed Oil Cake

The formation of films only with rapeseed protein cannot be done, because they present bad mechanical properties and no antimicrobial activity, thus plasticizer and emulsifier can be used. For example, gelatin and agarose increase tensile strength, and decrease elongation at break [98]. The use of protein hydrolysate with chitosan gives anti-microbial films with high activity against *Staphylococcus aureus*, *Bacillus subtilis*, and *Escherichia coli* [127].

5.5.5. Peanut Oil Cake

The utilization of only peanut protein isolate form brittle films, glycerol was found to be the most suitable plasticizer, producing optimal mechanical properties without influencing the barrier properties. Moreover, at high pH and temperature, they have good appearance and color, low water sensitivity, and, vapor, oxygen permeability [98].

Improvement of the characteristics can be realized, also with crosslinking with citric acid and a mixture of pea starch. The first one led to the improvement of tensile strength and a decrease in elongation at break, and water permeability. The second one improves coloration and elongation at break, and decrease tensile strength, water permeability, and total soluble matter [202,203].

Films have antimicrobial properties when are activated with different proportion of thymol. The latter, inhibit *l. plantarum, E. Coli, S. aureus, and P. aeruginosa,* at minimum concentration [127].

Comparing to the protein-based films, those create with defatted peanut flour create films, that are more transparent and excellent water barrier [204].

Edible fiber films are difficult to make because they have poor film-forming properties. To improve this, Wan [205] performed a dynamic high-pressure micro-fluidization treatment. The final films have excellent solubility and mechanical properties and can be used for packaging noodles, spices, biscuits, and candies.

5.6. Landfill and Bio-Fuel Production

The most common and cheap waste valorization is landfilling, defined as the embankment, compression, and disposal in appropriate sites. The decomposition of residues leads to methane production from heavy metals and organic compounds [206].

Due to the high content in an organic compound, energy can be obtained from food wastes/residues. There are two methods for biofuel conversion: thermochemical, more suitable for residues with low moisture and anaerobic digestion, for residues with high organic and water content. The first one converts biomass rich in energy into intermediate products (liquid or gaseous) and includes pyrolysis, combustion, and gasification. Incineration is another viable option but involves high cost, and negative environmental impact [4].

Economics of the bio-diesel production improved by using residual seedcakes as the main co-product. Because of their toxic compounds some oilcakes are unusable for consumption but have a potential energy production. The high protein content of oilseed cakes can cause some disadvantages on the pyrolysis process (decrease of the rate of degradation) [206].

6. Challenges and Future Perspectives

Oilseed by-products are an important source of dietary fiber, proteins and compounds with antioxidants properties. Further studies, need to be done on the extraction, utilization, and incorporation of dietary fiber and antioxidants in food products.

The researches accomplished until now highlights the possibility of adding appropriate amounts of by-products from oil-industry as functional ingredients in bakery, and dairy products without influencing negatively the quality of the final products. More investigations are required at the molecular structure. Additionally, there are limited studies about the rheological properties of dough with different oilcakes levels and particle sizes.

7. Conclusions

This review highlights the potential application of the ample amounts of wastes and by-products generated during the supply chain, primarily used as energy sources, landfills, and animal feedstuffs. Moreover, we have described new green methods for their recycling and further reuse in various food and pharmaceutical industries. This green approach also allows the protection of both the environment and people.

Oil cake/meal are the by-products obtained after the extraction of oil from seeds. It can be used in both animal and human diet because contain high amounts of bioactive compounds that can be isolated and used in foods. These residues can also act as a substrate in the production of low-cost value-added products such as fuel, surfactants, enzymes, antibiotics, vitamins, natural pigments, flavors components, and health promoting components, such as dietary fibers, amino acids, flavonoids, phytochemicals and proteins. All bioactive components recovered from residues can be used as essential ingredients to produce functional ingredient, achieving thus effective waste utilization and the successful realization of the zero waste and circular economy concepts

Waste valorization, in the perception of the circular economy, represents a challenge but at the same time could allow the reuse of materials into the supply chain as they add more value to foods, reduce costs, allow economic growth and reduce risks caused by their disposal in the environment.

Author Contributions: P.A. and A.S. contributed equally to the collection of data and preparation of the paper. All authors have read and agreed to the published version of the manuscript

Funding: This work was supported by a grant of Romanian Ministry of Education and Research, CCCDI-UEFISCDI, project number PN-III-P2-2.1-PED-2019-3863, within PNCDI III

Conflicts of Interest: The authors declare no conflict of interest

References

1. Borrello, M.; Caracciolo, F.; Lombardi, A.; Pascucci, S.; Cembalo, L. Consumers' Perspective on Circular Economy Strategy for Reducing Food Waste. *Sustainability* **2017**, *9*, 141. [CrossRef]
2. Pesce, M.; Tamai, I.; Guo, D.; Critto, A.; Brombal, D.; Wang, X.; Cheng, H.; Marcomini, A. Circular Economy in China: Translating Principles into Practice. *Sustainability* **2020**, *12*, 832. [CrossRef]
3. Esposito, B.; Sessa, M.R.; Sica, D.; Malandrino, O. Towards Circular Economy in the Agri-Food Sector. A Systematic Literature Review. *Sustainability* **2020**, *12*, 7401. [CrossRef]
4. Otles, S.; Despoudi, S.; Bucatariu, C.; Kartal, C. Food waste management, valorization, and sustainability in the food industry. In *Food Waste Recovery*, 1st ed.; Galanakis, C.M., Ed.; Elsevier Inc.: London, UK, 2015; pp. 3–23.
5. Mateos-Aparicio, I.; Matias, A. Food industry processing by-products in foods. In *The Role of Alternative and Innovative Food Ingredients and Products in Consumer Wellness*, 1st ed.; Galanakis, C., Ed.; Elsevier Inc.: London, UK, 2019; pp. 239–281.
6. Kumar, S.; Kushwaha, R.; Verma, M.L. Recovery and utilization of bioactives from food processing waste. In *Biotechnological Production of Bioactive Compounds*, 1st ed.; Verma, M.L., Chandel, A.K., Eds.; Elsevier: Amsterdam, The Netherlands, 2020; pp. 37–68.
7. García-Rebollar, P.; Cámara, L.; Lázaro, R.; Dapoza, C.; Pérez-Maldonado, R.; Mateos, G. Influence of the origin of the beans on the chemical composition and nutritive value of commercial soybean meals. *Anim. Feed. Sci. Technol.* **2016**, *221*, 245–261. [CrossRef]
8. Arrutia, F.; Binner, E.; Williams, P.; Waldron, K.W. Oilseeds beyond oil: Press cakes and meals supplying global protein requirements. *Trends Food Sci. Technol.* **2020**, *100*, 88–102. [CrossRef]
9. Wanasundara, J.; Tan, S.; Alashi, A.; Pudel, F.; Blanchard, C.L. Proteins from Canola/Rapeseed: Current Status. In *Sustainable Protein Sources*; Nadathur, S.R., Wanasundara, J.P.D., Scalin, L., Eds.; Elsevier Inc.: London Wall, UK, 2017; pp. 285–304.
10. Lomascolo, A.; Uzan-Boukhris, E.; Sigoillot, J.-C.; Fine, F. Rapeseed and sunflower meal: A review on biotechnology status and challenges. *Appl. Microbiol. Biotechnol.* **2012**, *95*, 1105–1114. [CrossRef]
11. Ranganayaki, S.; Vidhya, R.; Jaganmohan, R. Isolation and proximate determination of protein using defatted sesame seed oil cake. *Int. J. Nutr. Metab.* **2012**, *4*, 141–145.
12. Krimer-Malešević, V. Pumpkin Seeds: Phenolic Acids in Pumpkin Seed (*Curcubita pepo* L.). In *Nuts and Seeds in Health and Disease Prevention*, 2nd ed.; Preedy, V., Watson, R., Eds.; Elsevier Inc.: London, UK, 2020; pp. 533–542.
13. Özbek, Z.A.; Ergönül, P.G. Cold pressed pumpkin seed oil. In *Cold Pressed Oils*, 1st ed.; Ramadan, F.M., Ed.; Elsevier Inc.: London, UK, 2020; pp. 219–229.
14. Apostol, L.; Berca, L.; Mosoiu, C.; Badea, M.; Bungau, S.; Oprea, O.B.; Cioca, G. Partially Defatted Pumpkin (*Cucurbita maxima*) Seeds—A Rich Source of Nutrients for Use in Food Products. *Rev. Chim.* **2018**, *69*, 1398–1402. [CrossRef]
15. Eyres, L.; Eyres, M. Flaxseed (linseed) fibre-nutritional and culinary uses—A review. *Food N. Z.* **2014**, *14*, 26.
16. Mullen, A.M.; Alvarez, C.; Pojić, M.; Hadnadev, T.D.; Papageorgiou, M. Classification and target compounds. In *Food Waste Recovery: Processing Technologies and Industrial Techniques*, 1st ed.; Galanakis, C., Ed.; Elsevier Inc.: London, UK, 2015; pp. 25–57.
17. Sarwar, F.; Qadri, N.A.; Moghal, S. The role of oilseeds nutrition in human health: A critical review. *J. Cereals Oilseeds* **2013**, *4*, 97–100. [CrossRef]

18. Savoire, R.; Lanoisellé, J.-L.; Vorobiev, E. Mechanical Continuous Oil Expression from Oilseeds: A Review. *Food Bioprocess Technol.* **2012**, *6*, 1–16. [CrossRef]

19. Ramadan, M.F. Introduction to cold pressed oils: Green technology, bioactive compounds, functionality, and applications. In *Cold Pressed Oils*, 1st ed.; Ramadan, F.M., Ed.; Elsevier Inc.: London, UK, 2020; pp. 1–5.

20. Leming, R.; Lember, A. Chemical composition of expeller-extracted and cold-pressed rapeseed cake. *Agraarteadu* **2005**, *16*, 103–109.

21. Sunil, L.; Appaiah, P.; Kumar, P.K.P.; Krishna, A.G.G. Preparation of food supplements from oilseed cakes. *J. Food Sci. Technol.* **2014**, *52*, 2998–3005. [CrossRef] [PubMed]

22. Gupta, A.; Sharma, R.; Sharma, S.; Singh, B. Oilseed as potential functional food Ingredient. In *Trends & Prospects in Food Technology, Processing and Preservation*, 1st ed.; Prodyut Kumar, P., Mahawar, M.K., Abobatta, W., Panja, P., Eds.; Today and Tomorrow's Printers and Publishers: New Delhi, India, 2018; pp. 25–58.

23. Östbring, K.; Malmqvist, E.; Nilsson, K.; Rosenlind, I.; Rayner, M. The Effects of Oil Extraction Methods on Recovery Yield and Emulsifying Properties of Proteins from Rapeseed Meal and Press Cake. *Foods* **2019**, *9*, 19. [CrossRef] [PubMed]

24. Bochkarev, M.S.; Egorova, E.Y.; Reznichenko, I.Y.; Poznyakovskiy, V.M. Reasons for the ways of using oilcakes in food industry. *Foods Raw Mater.* **2016**, *4*, 4–12. [CrossRef]

25. Tarek-Tilistyák, J.; Juhász-Román, M.; Jeko, J.; Mathe, E. Short-term storability of oil seed and walnut cake—Microbiological aspect. *Acta Aliment.* **2014**, *43*, 632–639. [CrossRef]

26. Mannucci, A.; Castagna, A.; Santin, M.; Serra, A.; Mele, M.; Ranieri, A. Quality of flaxseed oil cake under different storage conditions. *LWT* **2019**, *104*, 84–90. [CrossRef]

27. Marchetti, L.; Romero, L.; Andrés, S.C.; Califano, A. Characterization of pecan nut expeller cake and effect of storage on its microbiological and oxidative quality. *Grasas Aceites* **2018**, *68*, 226. [CrossRef]

28. Volli, V.; Singh, R. Production of bio-oil from de-oiled cakes by thermal pyrolysis. *Fuel* **2012**, *96*, 579–585. [CrossRef]

29. Rezvani, M.; Kluth, H.; Bulang, M.; Rodehutscord, M. Variation in amino acid digestibility of rapeseed meal studied in caecectomised laying hens and relationship with chemical constituents. *Br. Poult. Sci.* **2012**, *53*, 665–674. [CrossRef]

30. Thanaseelaan, V. Proximate Analysis, Mineral and Amino Acid Profiles of Deoiled Rapeseed Meal. *Int. J. Food Agric. Vet. Sci.* **2013**, *3*, 66–69.

31. Sari, Y.W.; Bruins, M.E.; Sanders, J.P. Enzyme assisted protein extraction from rapeseed, soybean, and microalgae meals. *Ind. Crop. Prod.* **2013**, *43*, 78–83. [CrossRef]

32. Maison, T. Evaluation of the Nutritional Value of Canola Meal, 00-Rapeseed Meal, and 00-Rapeseed Expellers Fed To Pigs. Ph.D. Thesis, University of Illinois at Urbana-Champaign, Urbana-Champaign, IL, USA, 2013.

33. Thiel, A.; Muffler, K.; Tippkötter, N.; Suck, K.; Sohling, U.; Hruschka, S.M.; Ulber, R. A novel integrated downstream processing approach to recover sinapic acid, phytic acid and proteins from rapeseed meal. *J. Chem. Technol. Biotechnol.* **2015**, *90*, 1999–2006. [CrossRef]

34. Egorova, T.A.; Lenkova, T.H. Rapseed (*Brassica napus* L.) and its prospective useage (review). *Agric. Biol.* **2015**, *50*, 172–182.

35. Choi, H.B.; Jeong, J.H.; Kim, D.H.; Lee, Y.; Kwon, H.; Kim, Y.Y. Influence of Rapeseed Meal on Growth Performance, Blood Profiles, Nutrient Digestibility and Economic Benefit of Growing-finishing Pigs. *Asian-Australasian J. Anim. Sci.* **2015**, *28*, 1345–1353. [CrossRef] [PubMed]

36. Tofanica, B.M. Rapeseed—A Valuable Renewable Bioresource. *Cellul. Chem. Technol.* **2019**, *53*, 837–849. [CrossRef]

37. Amores, G.; Virto, M.; Nájera, A.I.; Mandaluniz, N.; Arranz, J.; Bustamante, M.Á.; Valdivielso, I.; De Gordoa, J.R.; Garcia-Rodriguez, A.; Barron, L.J.; et al. Rapeseed and sunflower oilcake as supplements for dairy sheep: Animal performance and milk fatty acid concentrations. *J. Dairy Res.* **2014**, *81*, 410–416. [CrossRef]

38. Jabłoński, S.J.; Biernacki, P.; Steinigeweg, S.; Łukaszewicz, M. Continuous mesophilic anaerobic digestion of manure and rape oilcake–Experimental and modelling study. *Waste Manag.* **2015**, *35*, 105–110. [CrossRef]

39. Shi, C.; He, J.; Yu, J.; Yu, B.; Huang, Z.; Mao, X.; Zheng, P.; Chen, D. Solid state fermentation of rapeseed cake with Aspergillus niger for degrading glucosinolates and upgrading nutritional value. *J. Anim. Sci. Biotechnol.* **2015**, *6*, 1–7. [CrossRef]

40. Kaczmarek, P.; Korniewicz, D.; Lipinski, K.; Mazur, M. Chemical composition of rapeseed products and their use in pig nutrition. *Pol. J. Nat. Sci.* **2016**, *31*, 545–562.

41. Salazar-Villanea, S.; Bruininx, E.M.; Gruppen, H.; Hendriks, W.H.; Carré, P.; Quinsac, A.; Van Der Poel, A.F. Physical and chemical changes of rapeseed meal proteins during toasting and their effects on in vitro digestibility. *J. Anim. Sci. Biotechnol.* **2016**, *7*, 1–11. [CrossRef] [PubMed]

42. Mosenthin, R.; Messerschmidt, U.; Sauer, N.; Carré, P.; Quinsac, A.; Schöne, F. Effect of the desolventizing/toasting process on chemical composition and protein quality of rapeseed meal. *J. Anim. Sci. Biotechnol.* **2016**, *7*, 36. [CrossRef]

43. Rommi, K.; Hakala, T.K.; Holopainen, U.; Nordlund, E.; Poutanen, K.; Lantto, R. Effect of Enzyme-Aided Cell Wall Disintegration on Protein Extractability from Intact and Dehulled Rapeseed (Brassica rapaL. andBrassica napusL.) Press Cakes. *J. Agric. Food Chem.* **2014**, *62*, 7989–7997. [CrossRef]

44. Tyapkova, O.; Osen, R.; Wagenstaller, M.; Baier, B.; Specht, F.; Zacherl, C. Replacing fishmeal with oilseed cakes in fish feed—A study on the influence of processing parameters on the extrusion behavior and quality properties of the feed pellets. *J. Food Eng.* **2016**, *191*, 28–36. [CrossRef]

45. Nega, T. Review on Nutritional Limitations and Opportunities of using Rapeseed Meal and other Rape Seed by—Products in Animal Feeding. *J. Nutr. Health Food Eng.* **2018**, *8*, 1. [CrossRef]

46. Martin, A.; Osen, R.; Greiling, A.; Karbstein, H.P.; Emin, A. Effect of rapeseed press cake and peel on the extruder response and physical pellet quality in extruded fish feed. *Aquaculture* **2019**, *512*, 734316. [CrossRef]

47. Cozea, A.; Ionescu, N.; Popescu, M.; Neagu, M.; Gruia, R. Comparative study concerning the composition of certain oil cakes with phytotherapeutical potential. *Rev. Chim.* **2016**, *67*, 422–425.

48. Halmemies-Beauchet-Filleau, A.; Rinne, M.; Lamminen, M.; Mapato, C.; Ampapon, T.; Wanapat, M.; Vanhatalo, A. Alternative and novel feeds for ruminants: Nutritive value, product quality and environmental aspects. *Animal* **2018**, *12*, s295–s309. [CrossRef]

49. Fetzer, A.; Herfellner, T.; Stäbler, A.; Menner, M.; Eisner, P. Influence of process conditions during aqueous protein extraction upon yield from pre-pressed and cold-pressed rapeseed press cake. *Ind. Crop. Prod.* **2018**, *112*, 236–246. [CrossRef]

50. Oliveira, F.; Souza, C.E.; Peclat, V.R.; Salgado, J.M.; Ribeiro, B.D.; Coelho, M.A.; Venâncio, A.; Belo, I. Optimization of lipase production by Aspergillus ibericus from oil cakes and its application in esterification reactions. *Food Bioprod. Process.* **2017**, *102*, 268–277. [CrossRef]

51. Tan, S.H.; Mailer, R.J.; Blanchard, C.L.; Agboola, S.O. Extraction and characterization of protein fractions from Australian canola meals. *Food Res. Int.* **2011**, *44*, 1075–1082. [CrossRef]

52. Sadh, P.K.; Duhan, S.; Duhan, J.S. Agro-industrial wastes and their utilization using solid state fermentation: A review. *Bioresour. Bioprocess.* **2018**, *5*, 1. [CrossRef]

53. Teh, S.S.; Bekhit, A.E.; Carne, A.; Birch, J. Effect of the defatting process, acid and alkali extraction on the physicochemical and functional properties of hemp, flax and canola seed cake protein isolates. *J. Food Meas. Charact.* **2013**, *8*, 92–104. [CrossRef]

54. Capitani, M.; Spotorno, V.; Nolasco, S.; Tomás, M. Physicochemical and functional characterization of by-products from chia (Salvia hispanica L.) seeds of Argentina. *LWT* **2012**, *45*, 94–102. [CrossRef]

55. Montrimaitė, K.; Moščenkova, E. Food Science and Applied Biotechnology. *Food Sci. Appl. Biotechnol.* **2018**, *1*, 154–164. [CrossRef]

56. Kenari, R.E.; Mohsenzadeh, F.; Amiri, Z.R. Antioxidant activity and total phenolic compounds of Dezful sesame cake extracts obtained by classical and ultrasound-assisted extraction methods. *Food Sci. Nutr.* **2014**, *2*, 426–435. [CrossRef]

57. Nadeem, M.; Situ, C.; Mahmud, A.; Khalique, A.; Imran, M.; Rahman, F.; Khan, S. Antioxidant Activity of Sesame (Sesamum indicum L.) Cake Extract for the Stabilization of Olein Based Butter. *J. Am. Oil Chem. Soc.* **2014**, *91*, 967–977. [CrossRef]

58. Yasothai, R. Chemical composition of sesame oil cake—Rieview. *Int. J. Sci. Environ. Technol.* **2014**, *3*, 827–835.

59. Grasso, S. Extruded snacks from industrial by-products: A review. *Trends Food Sci. Technol.* **2020**, *99*, 284–294. [CrossRef]

60. Benítez, R.B.; Bonilla, R.A.O.; Franco, J.M. Comparison of two sesame oil extraction methods: Percolation and pressed. *Biotecnoloía Sect. Agropecu. Agroind.* **2016**, *14*, 10. [CrossRef]

61. Mohdaly, A.A.; Smetanska, I.; Ramadan, M.F.; Sarhan, M.A.; Mahmoud, A.; Ramadan, M.F. Antioxidant potential of sesame (Sesamum indicum) cake extract in stabilization of sunflower and soybean oils. *Ind. Crop. Prod.* **2011**, *34*, 952–959. [CrossRef]

62. Carvalho, C.W.; Takeiti, C.Y.; Freitas, D.D.; Ascheri, J.L. Use of sesame oil cake (*Sesamum indicum* L.) on corn expanded extrudates. *Food Res. Int.* **2012**, *45*, 434–443. [CrossRef]
63. Uppuluri, K.B.; Dasari, R.K.; Sajja, V.; Jacob, A.S.; Reddy, D.S. Optimization of l-Asparaginase Production by Isolated Aspergillus niger C4 from Sesame (black) Oil Cake under SSF using Box–Behnken Design in Column Bioreactor. *Int. J. Chem. React. Eng.* **2013**, *11*, 103–109. [CrossRef]
64. Saleh, M.; Ghazzawi, H.; Al-Ismail, K.; Akash, M.; Al-Dabbas, M. Sesame-oil-cake (SOC) impacted consumer liking of a traditional Jordanian dessert; a mixture response surface model approach. *Int. Food Res. J.* **2016**, *23*, 2096–2102.
65. Capellini, M.C.; Chiavoloni, L.; Giacomini, V.; Rodrigues, C.E. Alcoholic extraction of sesame seed cake oil: Influence of the process conditions on the physicochemical characteristics of the oil and defatted meal proteins. *J. Food Eng.* **2019**, *240*, 145–152. [CrossRef]
66. Nagendra Prasad, M.N.; Sanjay, K.R.; Prasad, D.S.; Vijay, N.; Kothari, R.; Nanjunda Swamy, S. A Review on Nutritional and Nutraceutical Properties of Sesame. *J. Nutr. Food Sci.* **2012**, *2*, 1–6. [CrossRef]
67. Das, P.; Ghosh, K. Improvement of nutritive value of sesame oil cake in formulated diets for rohu, Labeo rohita (Hamilton) after bio-processing through solid state fermentation by a phytase-producing fish gut bacterium. *Int. J. Aquat. Biol.* **2015**, *3*, 89–101.
68. Elsorady, M.E. Characterization and functional properties of proteins isolated from flaxseed cake and sesame cake. *Croat. J. Food Sci. Technol.* **2020**, *12*, 77–83. [CrossRef]
69. Hejazi, A.; Omar, J.M.A. Effect of Feeding Sesame Oil Cake on Performance, Milk and Cheese Quality of Anglo-Nubian Goats. *Hebron Univ. Res. J.* **2009**, *4*, 81–91.
70. Omer, H.A.A.; Ahmed, S.M.; Abdel-Magid, S.S.; Bakry, B.A.; El-Karamany, M.F.; El-Sabaawy, E.H. Nutritional impact of partial or complete replacement of soybean meal by sesame (*Sesamum indicum*) meal in lambs rations. *Bull. Natl. Res. Cent.* **2019**, *43*, 98. [CrossRef]
71. Prakash, K.; Naik, S.; Vadivel, D.; Hariprasad, P.; Gandhi, D.; SaravanaDevi, S. Utilization of defatted sesame cake in enhancing the nutritional and functional characteristics of biscuits. *J. Food Process. Preserv.* **2018**, *42*, e13751. [CrossRef]
72. Bigoniya, P.; Nishad, R.; Singh, C.S. Preventive effect of sesame seed cake on hyperglycemia and obesity against high fructose-diet induced Type 2 diabetes in rats. *Food Chem.* **2012**, *133*, 1355–1361. [CrossRef]
73. Shu, Z.; Liu, L.; Geng, P.; Liu, J.; Shen, W.; Tu, M. Sesame cake hydrolysates improved spatial learning and memory of mice. *Food Biosci.* **2019**, *31*, 100440. [CrossRef]
74. Reshma, M.; Namitha, L.; Sundaresan, A.; Kiran, C.R. Total Phenol Content, Antioxidant Activities and α-Glucosidase Inhibition of Sesame Cake Extracts. *J. Food Biochem.* **2012**, *37*, 723–731. [CrossRef]
75. Achouri, A.; Nail, V.; Boye, J.I. Sesame protein isolate: Fractionation, secondary structure and functional properties. *Food Res. Int.* **2012**, *46*, 360–369. [CrossRef]
76. Zheng, Y.; Li, Y. Physicochemical and functional properties of coconut (*Cocos nucifera* L) cake dietary fibres: Effects of cellulase hydrolysis, acid treatment and particle size distribution. *Food Chem.* **2018**, *257*, 135–142. [CrossRef] [PubMed]
77. Paweł, S.; Kazimierz, Z.; Starek, A.; Żukiewicz-Sobczak, W.; Sagan, A.; Zdybel, B.; Andrejko, D. Compaction Process as a Concept of Press-Cake Production from Organic Waste. *Sustainability* **2020**, *12*, 1567. [CrossRef]
78. Te, K.G.; Go, A.W.; Wang, H.J.; Guevarra, R.G.; Cabatingan, L.K.; Tabañag, I.D.; Angkawijaya, A.E.; Ju, Y.-H. Extraction of lipids from post-hydrolysis copra cake with hexane as solvent: Kinetic and equilibrium data. *Renew. Energy* **2020**, *158*, 311–323. [CrossRef]
79. Akhter, P.; Mishra, A.; Mishra, V.; Raghav, A. Alpha amylase production from Aspergillus terreus UF39 using oil cakes. *Eurpean J. Biotechnol. Bioci.* **2017**, *5*, 15–21.
80. Contreras, M.D.M.; Romero, I.; Moya, M.; Castro, E. Olive-derived biomass as a renewable source of value-added products. *Process Biochem.* **2020**, *97*, 43–56. [CrossRef]
81. Moftah, O.A.S.; Grbavcic, S.; Zuza, M.; Luković, N.; Bezbradica, D.; Knežević-Jugović, Z.D. Adding Value to the Oil Cake as a Waste from Oil Processing Industry: Production of Lipase and Protease by Candida utilis in Solid State Fermentation. *Appl. Biochem. Biotechnol.* **2011**, *166*, 348–364. [CrossRef]
82. Ramachandran, S.; Singh, S.K.; Larroche, C.; Soccol, C.R.; Pandey, A. Oil cakes and their biotechnological applications—A review. *Bioresour. Technol.* **2007**, *98*, 2000–2009. [CrossRef] [PubMed]

83. Bakkalbaşi, E.; Meral, R.; Dogan, I.S. Bioactive Compounds, Physical and Sensory Properties of Cake Made with Walnut Press-Cake. *J. Food Qual.* **2015**, *38*, 422–430. [CrossRef]

84. Grosu, C.; Boaghi, E.; Paladi, D.; Deseatnicova, O.; Vladislav, R. Prosects of using walnut oil cake in food industry. In Proceedings of the International Conference Modern Technologies in the Food Industry, Chișinău, Moldova, 1–3 November 2012; pp. 362–365.

85. Jokić, S.; Moslavac, T.; Bošnjak, A.; Aladić, K.; Rajić, M.; Bilić, M. Optimization of walnut oil production. *Croat. J. Food Sci. Technol.* **2014**, *6*, 27–35.

86. Mir, M.A.; Sharma, R.K.; Rastogi, A.; Barman, K. Effect of incorporation of walnut cake (Juglans regia) in concentrate mixture on degradation of dry matter, organic matter and production of microbial biomass in vitro in goat. *Veter-World* **2015**, *8*, 1172–1176. [CrossRef]

87. Bardeau, T.; Savoire, R.; Cansell, M.; Subra-Paternault, P. Recovery of oils from press cakes by CO_2-based technology. *OCL* **2015**, *22*, D403. [CrossRef]

88. Pop, A.; Paucean, A.; Socaci, S.A.; Alexa, E.; Man, S.M.; Muresan, V.; Chis, M.S.; Salanta, L.; Popescu, I.; Berbecea, A.; et al. Quality characteristics and volatile profile of macarons modified withwalnut oilcake by-product. *Molecules* **2020**, *25*, 2214. [CrossRef]

89. Wang, R.; Shaarani, S.M.; Casas-Godoy, L.; Melikoğlu, M.; Vergara, C.S.; Koutinas, A.; Webb, C. Bioconversion of rapeseed meal for the production of a generic microbial feedstock. *Enzym. Microb. Technol.* **2010**, *47*, 77–83. [CrossRef]

90. Jannathulla, R.; Dayal, J.S.; Ambasankar, K.; Muralidhar, M. Effect of Aspergillus niger fermented soybean meal and sunflower oil cake on growth, carcass composition and haemolymph indices in Penaeus vannamei Boone, 1931. *Aquaculture* **2018**, *486*, 1–8. [CrossRef]

91. Zamindar, N.; Bashash, M.; Khorshidi, F.; Serjouie, A.; Shirvani, M.A.; Abbasi, H.; Sedaghatdoost, A. Antioxidant efficacy of soybean cake extracts in soy oil protection. *J. Food Sci. Technol.* **2017**, *54*, 2077–2084. [CrossRef] [PubMed]

92. Gerliani, N.; Hammami, R.; Aïder, M. A comparative study of the functional properties and antioxidant activity of soybean meal extracts obtained by conventional extraction and electro-activated solutions. *Food Chem.* **2020**, *307*, 125547. [CrossRef]

93. Jahan, D.; Hussain, L.; Islam, M.; Khan, M. Comparative Study of Mustard Oil Cake and Soybean Meal Based Artificial Diet for Rohu, Labeo rohita (Ham.) Fingerlings. *Agriculturists* **2013**, *11*, 61–66. [CrossRef]

94. Tangendjaja, B. Quality control of feed ingredients for aquaculture. In *Feed and Feeding Practices in Aquaculture*, 1st ed.; Davis, D.A., Ed.; Woodhead Publishing: Cambridge, UK, 2015; pp. 141–169.

95. Jadhav, M.; Kagalkar, A.; Jadhav, S.; Govindwar, S.P.; Chakankar, M. Isolation, characterization, and antifungal application of a biosurfactant produced by Enterobacter sp. MS16. *Eur. J. Lipid Sci. Technol.* **2011**, *113*, 1347–1356. [CrossRef]

96. Grosu, C. Valorificarea Șrotului de nuci și Obținerea Produselor de Cofetărie. Ph.D. Thesis, Technical University of Moldova, Chişinău, Moldova, 2016.

97. Bhise, S.; Kaur, A.; Manikantan, M.; Singh, B. Development of textured defatted sunflower meal by extrusion using response surface methodology. *Acta Aliment.* **2015**, *44*, 251–258. [CrossRef]

98. Šuput, D.; Lazić, V.; Mađarev-Popović, S.; Hromiš, N.; Bulut, S.; Pezo, L.; Banićević, J. Effect of process parameters on biopolymer films based on sunflower oil cake. *J. Process. Energy Agric.* **2018**, *22*, 125–128. [CrossRef]

99. Serrapica, F.; Masucci, F.; Raffrenato, E.; Sannino, M.; Vastolo, A.; Barone, C.M.A.; Di Francia, A. High Fiber Cakes from Mediterranean Multipurpose Oilseeds as Protein Sources for Ruminants. *Animal* **2019**, *9*, 918. [CrossRef] [PubMed]

100. Lazaro, E.; Benjamin, Y.; Robert, M. The Effects of Dehulling on Physicochemical Properties of Seed Oil and Cake Quality of Sunflower. *Tanzan. J. Agric. Sci.* **2014**, *13*, 41–47.

101. Salgado, P.R.; Ortiz, S.E.M.; Petruccelli, S.; Mauri, A.N. Sunflower Protein Concentrates and Isolates Prepared from Oil Cakes Have High Water Solubility and Antioxidant Capacity. *J. Am. Oil Chem. Soc.* **2010**, *88*, 351–360. [CrossRef]

102. Budžaki, S.; Strelec, I.; Krnić, M.; Alilović, K.; Tišma, M.; Zelić, B. Proximate analysis of cold-press oil cakes after biological treatment with Trametes versicolor and Humicola grisea. *Eng. Life Sci.* **2018**, *18*, 924–931. [CrossRef]

103. Tirgar, M.; Silcock, P.; Carne, A.; Birch, E.J. Effect of extraction method on functional properties of flaxseed protein concentrates. *Food Chem.* **2017**, *215*, 417–424. [CrossRef]

104. Stodolak, B.; Starzyńska-Janiszewska, A.; Bączkowicz, M. Aspergillus oryzae (Koji Mold) and Neurospora intermedia (Oncom Mold) application for flaxseed oil cake processing. *LWT* **2020**, *131*, 109651. [CrossRef]

105. Ganorkar, P.; Patel, J.M.; Shah, V.; Rangrej, V.V. Defatted flaxseed meal incorporated corn-rice flour blend based extruded product by response surface methodology. *J. Food Sci. Technol.* **2015**, *53*, 1867–1877. [CrossRef]

106. Radočaj, O.; Dimic, E.; Tsao, R. Effects of Hemp (Cannabis sativa L.) Seed Oil Press-Cake and Decaffeinated Green Tea Leaves (Camellia sinensis) on Functional Characteristics of Gluten-Free Crackers. *J. Food Sci.* **2014**, *79*, C318–C325. [CrossRef] [PubMed]

107. Panak Balentić, J.; Jozinović, A.; Ačkar, Đ.; Babić, J.; Miličević, B.; Benšić, M.; Jokić, S.; Šarić, A.; Šubarić, D. Nutritionally improved third generation snacks produced by supercritical CO 2 extrusion I. Physical and sensory properties. *J. Food Process. Eng.* **2018**, *42*, e12961. [CrossRef]

108. Jozinović, A.; AčkAr, Đ.; Jokić, S.; BABić, J.; BAlentić, J.P.; BAnožić, M.; ŠuBArić, D. Optimisation of extrusion variables for the production of corn snack products enriched with defatted hemp cake. *Czech J. Food Sci.* **2017**, *35*, 507–516. [CrossRef]

109. Bello, F.; Salami, I.; Sani, I.; Abdulhamid, A.; Musa, I. Evaluation of Some Antinutritional Factors in Oil-Free White *Sesamum indicum* L. Seed Cake. *Int. J. Food Nutr. Saf.* **2013**, *4*, 27–33.

110. Tan, S.H.; Mailer, R.J.; Blanchard, C.L.; Agboola, S.O. Canola Proteins for Human Consumption: Extraction, Profile, and Functional Properties. *J. Food Sci.* **2010**, *76*, R16–R28. [CrossRef] [PubMed]

111. Teh, S.S.; Bekhit, A.E. Utilization of Oilseed Cakes for Human Nutrition and Health Benefits. In *Agricultural Biomass Based Potential Materials*; Hakeem, K.R., Jawaid, M., Alothman, O.Y., Eds.; Springer International Publishing: Cham, Switzerland, 2015; pp. 191–229.

112. Dong, F.M.; Hardy, R.W.; Higgs, D.A. Antinutritional factors. In *The Encyclopedia of Aquaculture*; Stickney, R.R., Ed.; John Wiley and Sons: New York, NY, USA, 2000; pp. 45–51.

113. Pojić, M.; Hadnađev, T.R.D.; Hadnađev, M.; Rakita, S.; Brlek, T. Bread Supplementation with Hemp Seed Cake: A By-Product of Hemp Oil Processing. *J. Food Qual.* **2015**, *38*, 431–440. [CrossRef]

114. Łopusiewicz, Ł.; Drozłowska, E.; Siedlecka, P.; Mężyńska, M.; Bartkowiak, A.; Sienkiewicz, M.; Zielińska-Bliźniewska, H.; Kwiatkowski, P. Development, Characterization, and Bioactivity of Non-Dairy Kefir-Like Fermented Beverage Based on Flaxseed Oil Cake. *Foods* **2019**, *8*, 544. [CrossRef]

115. Ghosh, K.; Mandal, S. Nutritional evaluation of groundnut oil cake in formulated diets for rohu, Labeo rohita (Hamilton) fingerlings after solid state fermentation with a tannase producing yeast, Pichia kudriavzevii (GU939629) isolated from fish gut. *Aquac. Rep.* **2015**, *2*, 82–90. [CrossRef]

116. Duliński, R.; Stodolak, B.; Byczyński, Ł.; Poreda, A.; Starzyńska-Janiszewska, A.; Żyła, K. Solid-State Fermentation Reduces Phytic Acid Level, Improves the Profile of Myo-Inositol Phosphates and Enhances the Availability of Selected Minerals in Flaxseed Oil Cake. *Food Technol. Biotechnol.* **2017**, *55*, 413–419. [CrossRef] [PubMed]

117. Fărcaş, A.C.; Socaci, S.A.; Diaconeasa, Z.M. Introductory Chapter: From Waste to New Resources. In *Food Preservation and Waste Exploitation.*, 1st ed.; Socaci, S.A., Fărcaș, A.C., Aussenac, T., Eds.; IntechOpen: London, UK, 2019; pp. 1–11.

118. Baiano, A. Recovery of Biomolecules from Food Wastes—A Review. *Molecules* **2014**, *19*, 14821–14842. [CrossRef] [PubMed]

119. Hussain, S.; Jõudu, I.; Bhat, R. Dietary Fiber from Underutilized Plant Resources—A Positive Approach for Valorization of Fruit and Vegetable Wastes. *Sustainability* **2020**, *12*, 5401. [CrossRef]

120. Fierascu, R.C.; Fierascu, I.; Avramescu, S.M.; Sieniawska, E. Recovery of Natural Antioxidants from Agro-Industrial Side Streams through Advanced Extraction Techniques. *Molecules* **2019**, *24*, 4212. [CrossRef] [PubMed]

121. Sá, A.G.A.; Moreno, Y.M.F.; Carciofi, B.A.M. Plant proteins as high-quality nutritional source for human diet. *Trends Food Sci. Technol.* **2020**, *97*, 170–184. [CrossRef]

122. Östbring, K.; Tullberg, C.; Burri, S.C.; Malmqvist, E.; Rayner, M. Protein Recovery from Rapeseed Press Cake: Varietal and Processing Condition Effects on Yield, Emulsifying Capacity and Antioxidant Activity of the Protein Rich Extract. *Foods* **2019**, *8*, 627. [CrossRef] [PubMed]

123. Kajla, P.; Sharma, A.; Sood, D.R. Flaxseed—A potential functional food source. *J. Food Sci. Technol.* **2014**, *52*, 1857–1871. [CrossRef] [PubMed]

124. Chatterjee, R.; Dey, T.K.; Ghosh, M.; Dhar, P. Enzymatic modification of sesame seed protein, sourced from waste resource for nutraceutical application. *Food Bioprod. Process.* **2015**, *94*, 70–81. [CrossRef]

125. Aider, M.; Barbana, C. Canola proteins: Composition, extraction, functional properties, bioactivity, applications as a food ingredient and allergenicity—A practical and critical review. *Trends Food Sci. Technol.* **2011**, *22*, 21–39. [CrossRef]

126. Yamada, Y.; Iwasaki, M.; Usui, H.; Ohinata, K.; Marczak, E.D.; Lipkowski, A.W.; Yoshikawa, M. Rapakinin, an anti-hypertensive peptide derived from rapeseed protein, dilates mesenteric artery of spontaneously hypertensive rats via the prostaglandin IP receptor followed by CCK1 receptor. *Peptides* **2010**, *31*, 909–914. [CrossRef]

127. Zhang, S.B. In vitro antithrombotic activities of peanut protein hydrolysates. *Food Chem.* **2016**, *202*, 1–8. [CrossRef] [PubMed]

128. Kavitha, N.; Selvakumar, K.; Srinivasan, G.; Madhan, R. Production and analysis of microbial growth in enzyme hydrolysate of sesame oil seed cake. *J. Biol. Inf. Sci.* **2012**, *1*, 2–5.

129. Alashi, A.M.; Blanchard, C.L.; Mailer, R.J.; Agboola, S.O. Technological and Bioactive Functionalities of Canola Meal Proteins and Hydrolysates. *Food Rev. Int.* **2013**, *29*, 231–260. [CrossRef]

130. Campbell, L.; Rempel, C.; Wanasundara, J.P. Canola/Rapeseed Protein: Future Opportunities and Directions—Workshop Proceedings of IRC 2015. *Plants* **2016**, *5*, 17. [CrossRef] [PubMed]

131. Sisay, M.T.; Emire, S.A.; Ramaswamy, H.S.; Workneh, T.S. Effect of feed components on quality parameters of wheat–tef–sesame–tomato based extruded products. *J. Food Sci. Technol.* **2018**, *55*, 2649–2660. [CrossRef] [PubMed]

132. Stodolak, B.; Starzyńska-Janiszewska, A.; Wywrocka-Gurgul, A.; Wikiera, A. Solid-State Fermented Flaxseed Oil Cake of Improved Antioxidant Capacity as Potential Food Additive. *J. Food Process. Preserv.* **2016**, *41*, e12855. [CrossRef]

133. Sadh, P.K.; Chawla, P.; Duhan, J.S. Fermentation approach on phenolic, antioxidants and functional properties of peanut press cake. *Food Biosci.* **2018**, *22*, 113–120. [CrossRef]

134. Teh, S.-S.; Birch, E.J. Effect of ultrasonic treatment on the polyphenol content and antioxidant capacity of extract from defatted hemp, flax and canola seed cakes. *Ultrason. Sonochemistry* **2014**, *21*, 346–353. [CrossRef]

135. Bakkalbaşi, E. Oxidative stability of enriched walnut oil with phenolic extracts from walnut press-cake under accelerated oxidation conditions and the effect of ultrasound treatment. *J. Food Meas. Charact.* **2018**, *13*, 43–50. [CrossRef]

136. Şahin, S.; Elhussein, E.A.A. Assessment of sesame (*Sesamum indicum* L.) cake as a source of high-added value substances: From waste to health. *Phytochem. Rev.* **2018**, *17*, 691–700. [CrossRef]

137. Mohdaly, A.A.A.; Hassanien, M.F.R.; Mahmoud, A.; Sarhan, M.A.; Smetanska, I. Phenolics Extracted from Potato, Sugar Beet, and Sesame Processing By-Products. *Int. J. Food Prop.* **2013**, *16*, 1148–1168. [CrossRef]

138. Senanayake, C.M.; Algama, C.H.; Wimalasekara, R.L.; Weerakoon, W.N.; Jayathilaka, N.; Seneviratne, K.N. Improvement of Oxidative Stability and Microbial Shelf Life of Vanilla Cake by Coconut Oil Meal and Sesame Oil Meal Phenolic Extracts. *J. Food Qual.* **2019**, *2019*, 1–8. [CrossRef]

139. Teh, S.-S.; Bekhit, A.E.-D.; Birch, J. Antioxidative Polyphenols from Defatted Oilseed Cakes: Effect of Solvents. *Antioxidants* **2014**, *3*, 67–80. [CrossRef]

140. Pickardt, C.; Eisner, P.; Kammerer, D.R.; Carle, R. Pilot plant preparation of light-coloured protein isolates from de-oiled sunflower (*Helianthus annuus* L.) press cake by mild-acidic protein extraction and polyphenol adsorption. *Food Hydrocoll.* **2015**, *44*, 208–219. [CrossRef]

141. Sarkis, J.R.; Michel, I.; Tessaro, I.C.; Marczak, L.D.F. Optimization of phenolics extraction from sesame seed cake. *Sep. Purif. Technol.* **2014**, *122*, 506–514. [CrossRef]

142. Elleuch, M.; Bedigian, D.; Roiseux, O.; Besbes, S.; Blecker, C.; Attia, H. Dietary fibre and fibre-rich by-products of food processing: Characterisation, technological functionality and commercial applications: A review. *Food Chem.* **2011**, *124*, 411–421. [CrossRef]

143. Maphosa, Y.; Jideani, V.A. Dietary fiber extraction for human nutrition—A review. *Food Rev. Int.* **2015**, *32*, 98–115. [CrossRef]

144. Dhingra, D.; Michael, M.; Rajput, H.; Patil, R.T. Dietary fibre in foods: A review. *J. Food Sci. Technol.* **2011**, *49*, 255–266. [CrossRef]

145. Dungani, R.; Karina, M.; Sulaeman, A.; Hermawan, D.; Hadiyane, A. Agricultural Waste Fibers towards Sustainability and Advanced Utilization: A Review. *Asian J. Plant Sci.* **2016**, *15*, 42–55. [CrossRef]

146. Lizardi-Jiménez, M.A.; Hernández-Martínez, R. Solid state fermentation (SSF): Diversity of applications to valorize waste and biomass. *3 Biotech* **2017**, *7*, 44. [CrossRef]

147. Parihar, D.K. Production of Lipase Utilizing linseed oilcake as fermentation substrate. *Int. J. Sci. Environ. Technol.* **2012**, *1*, 135–143.

148. Thakur, S.A.; Nemade, S.N.; Sharanappa, A. Solid State Fermentation of Overheated Soybean Meal (Waste) For Production of Protease Using Aspergillus Oryzae. *Int. J. Innov. Res. Sci. Eng. Technol.* **2015**, *4*, 18456–18461. [CrossRef]

149. Gupta, A.; Sharma, A.; Pathak, R.; Kumar, A.; Sharma, S. Solid state fermentation of non-edible oil seed cakes for production of proteases and cellulases and degradation of anti-nutritional factors. *J. Food Biotechnol. Res.* **2018**, *1*, 3–8.

150. Sahoo, R.; Subudhi, E.; Kumar, M. Quantitative approach to track lipase producingPseudomonassp. S1 in nonsterilized solid state fermentation. *Lett. Appl. Microbiol.* **2014**, *58*, 610–616. [CrossRef]

151. Medhe, S.; Anand, M.; Anal, A.K. Dietary Fibers, Dietary Peptides and Dietary Essential Fatty Acids from Food Processing By-Products. In *Food Processing By-Products and Their Utilization*, 1st ed.; Anil, A.K., Ed.; John Wiley & Sons Ltd.: Hoboken, NJ, USA, 2018; pp. 111–136.

152. Satya, C.V.; Anuradha, C.; Reddy, D.S. Palm Oil Cake: A Potential Substrate for L-Asparaginase production. *Int. J. Innov. Res. Sci. Eng. Technol.* **2014**, *3*, 14627–14632.

153. Ghosh, S.; Murthy, S.; Govindasamy, S.; Chandrasekaran, M. Optimization of L-asparaginase production by Serratia marcescens (NCIM 2919) under solid state fermentation using coconut oil cake. *Sustain. Chem. Process.* **2013**, *1*, 9. [CrossRef]

154. Gregori, A.; Pohleven, F. Cultivation of three medicinal mushroom species on olive oil press cakes containing substrates. *Acta Agric. Slov.* **2014**, *103*, 49–54. [CrossRef]

155. Song, B.; Ye, J.; Sossah, F.L.; Li, C.; Li, D.; Meng, L.; Xu, S.; Fu, Y.; Li, Y. Assessing the effects of different agro-residue as substrates on growth cycle and yield of Grifola frondosa and statistical optimization of substrate components using simplex-lattice design. *AMB Express* **2018**, *8*, 46. [CrossRef]

156. Krupodorova, T.A.; Barshteyn, V.Y. Alternative substrates for higher mushrooms mycelia cultivation. *J. BioSci. Biotechnol.* **2015**, *1*, 339–347.

157. Jape, A.; Salunke, D.; Dighe, S.; Harsulkar, A. The Improvement of Pleuorotus Species Cultivated On Soybean Straw Bed Supplemented with Flax Seed Meal. *Int. J. Sci. Res.* **2014**, *3*, 2012–2015.

158. Atila, F. Cultivation of Pleurotus spp., as an Alternative Solution to Dispose Olive Waste. *J. Agric. Ecol. Res. Int.* **2017**, *12*, 1–10. [CrossRef]

159. Zervakis, G.I.; Koutrotsios, G.; Katsaris, P. Composted versus Raw Olive Mill Waste as Substrates for the Production of Medicinal Mushrooms: An Assessment of Selected Cultivation and Quality Parameters. *BioMed Res. Int.* **2013**, *2013*, 1–13. [CrossRef] [PubMed]

160. Farzana, K.; Shah, S.N.H.; Butt, F.B.; Awan, S.B. Biosynthesis of bacitracin in solid-state fermentation by Bacillus licheniformis using defatted oil seed cakes as substrate. *Pak. J. Pharm. Sci.* **2005**, *18*, 55–57.

161. Zarei, I. Biosynthesis of Bacitracin in Stirred Fermenter by Bacillus Licheniformis Using Defatted Oil Seed Cakes as Substrate. *Mod. Appl. Sci.* **2012**, *6*, 30–36. [CrossRef]

162. Ali, S.; Nelofer, R.; Andleeb, S.; Baig, S.; Shakir, H.A.; Qazi, J.I. Biosynthesis and optimization of bacitracin by mutant Bacillus licheniformis using submerged fermentation. *Indian J. Biotechnol.* **2018**, *17*, 251–260.

163. Vastrad, B.M.; Neelagund, S.E. Optimization of Medium Composition for the Production of Neomycin byStreptomyces fradiaeNCIM 2418 in Solid State Fermentation. *Biotechnol. Res. Int.* **2014**, *2014*, 1–11. [CrossRef]

164. Nagavalli, M.; Ponamgi, S.P.; Girijashankar, V.; Venkateswar Rao, L. Solid State Fermentation and production of Rifamycin SV usingAmycolatopsis mediterranei. *Lett. Appl. Microbiol.* **2014**, *60*, 44–51. [CrossRef]

165. Vastrad, B.M.; Neelagund, S.E. Optimization of Process Parameters for Rifamycin B Production under Solid State Fermentation from Amycolatopsis Mediterranean Mtcc 14. *Int. J. Curr. Pharm. Res.* **2012**, *4*, 101–108.

166. Yao, D.; Ji, Z.; Wang, C.; Qi, G.; Zhang, L.; Ma, X.; Chen, S. Co-producing iturin A and poly-γ-glutamic acid from rapeseed meal under solid state fermentation by the newly isolated Bacillus subtilis strain 3-10. *World J. Microbiol. Biotechnol.* **2011**, *28*, 985–991. [CrossRef]

167. Banat, I.M.; Franzetti, A.; Gandolfi, I.; Bestetti, G.; Martinotti, M.G.; Fracchia, L.; Smyth, T.J.; Marchant, R. Microbial biosurfactants production, applications and future potential. *Appl. Microbiol. Biotechnol.* **2010**, *87*, 427–444. [CrossRef]

168. Thavasi, R. Microbial Biosurfactants: From an Environmental Application Point of View. *J. Bioremediation Biodegrad.* **2011**, *2*, 1000104. [CrossRef]

169. Fracchia, L.; Cavallo, M.; Martinotti, G.M.; Banat, I.M. Biosurfactants and Bioemulsifiers Biomedical and Related Applications—Present Status and Future Potentials. In *Biomedical Science, Engineering and Technology*; Ghista, D.N., Ed.; InTecch Europe: Rijeka, Croatia, 2012; pp. 325–370.

170. Jiménez-Peñalver, P.; Koh, A.; Gross, R.; Gea, T.; Font, X. Biosurfactants from Waste: Structures and Interfacial Properties of Sophorolipids Produced from a Residual Oil Cake. *J. Surfactants Deterg.* **2019**, *23*, 481–486. [CrossRef]

171. Thavasi, R.; Nambaru, V.S.; Jayalakshmi, S.; Balasubramanian, T.; Banat, I.M. Biosurfactant Production by Pseudomonas aeruginosa from Renewable Resources. *Indian J. Microbiol.* **2011**, *51*, 30–36. [CrossRef]

172. Noparat, P.; Maneerat, S.; Saimmai, A. Utilization of palm oil decanter cake as a novel substrate for biosurfactant production from a new and promising strain of Ochrobactrum anthropi 2/3. *World J. Microbiol. Biotechnol.* **2014**, *30*, 865–877. [CrossRef] [PubMed]

173. Chittepu, O.R.; Reddy, C.O. Isolation and characterization of biosurfactant producing bacteria from groundnut oil cake dumping site for the control of foodborne pathogens. *Grain Oil Sci. Technol.* **2019**, *2*, 15–20. [CrossRef]

174. Zentek, J.; Knorr, F.; Mader, A. Reducing waste in fresh produce processing and households through use of waste as animal feed. In *Global Safety of Fresh Produce: A Handbook of Best Practice, Innovative Commercial Solutions and Case Studies*, 1st ed.; Hoorfar, J., Ed.; Woodhead Publishing Limited: Cambridge, UK, 2014; pp. 140–152.

175. Thu, T.N.; Bodin, N.; De Saeger, S.; Larondelle, Y.; Rollin, X. Substitution of fish meal by sesame oil cake (*Sesamum indicum* L.) in the diet of rainbow trout (*Oncorhynchus mykiss* W.). *Aquac. Nutr.* **2010**, *17*, 80–89. [CrossRef]

176. Chebaibi, S.; Grandchamp, M.L.; Burgé, G.; Clément, T.; Allais, F.; Laziri, F. Improvement of protein content and decrease of anti-nutritional factors in olive cake by solid-state fermentation: A way to valorize this industrial by-product in animal feed. *J. Biosci. Bioeng.* **2019**, *128*, 384–390. [CrossRef]

177. Hassan, H.E.; Elhashmi, Y.H.A.; Eldar, A.A.T.; Elamin, K.M.; Elbushra, M.E. Effects of Feeding Different Levels of Sesame Oil Cake (*Sesamum indicum* L.) On Performance and Carcass Characteristics of Sudan Desert Sheep. *J. Anim. Sci. Adv.* **2013**, *3*, 91. [CrossRef]

178. Fitwi, M.; Tadesse, G. Effect of sesame cake supplementation on feed intake, body weight gain, feed conversion efficiency and carcass parameters in the ration of sheep fed on wheat bran and teff (*Eragrostis teff*) straw. *Momona Ethiop. J. Sci.* **2013**, *5*, 89. [CrossRef]

179. Ghorbani, B.; Yansari, A.T.; Sayyadi, A.J. Effects of sesame meal on intake, digestibility, rumen characteristics, chewing activity and growth of lambs. *S. Afr. J. Anim. Sci.* **2018**, *48*, 151. [CrossRef]

180. Mir, M.A.; Sharma, R.K.; Rastogi, A.; Haq, Z.; Ganai, I.A. Effect of dietary incorporation of walnut cake (*Juglans regia*) on calcium—Phosphorus balance and blood biochemical parameters in goats. *Vet. Arhiv* **2018**, *88*, 763–771. [CrossRef]

181. Chipa, M.J.; Siebrits, F.K.; Ratsaka, M.M.; Leeuw, K.J.; Nkosi, B.D. Growth performance of feedlot weaners cattle fed diet containing different levels of cold press soya bean oilcake. *S. Afr. J. Anim. Sci.* **2010**, *40*, 499–501.

182. Stübler, A.-S.; Heinz, V.; Aganovic, K. Development of food products. *Curr. Opin. Green Sustain. Chem.* **2020**, *25*, 100356. [CrossRef]

183. Sunil, L.; Prakruthi, A.; Prasanth Kumar, P.K.; Gopala Krishna, A.G. Development of Health Foods from Oilseed Cakes. *J. Food Process. Technol.* **2016**, *7*, 1–6. [CrossRef]

184. Pycia, K.; Kapusta, I.; Jaworska, G. Walnut oil and oilcake affect selected the physicochemical and antioxidant properties of wheat bread enriched with them. *J. Food Process. Preserv.* **2020**, *44*, e14573. [CrossRef]

185. Koneva, S.I.; Egorova, E.Y.; Kozubaeva, L.A.; Kuzmina, S.S.; Zakharova, A.S. Influence of Flaxseed Flour on Dough Rheology from Wheat-Flaxseed Meal. In *International Scientific and Practical Conference "AgroSMART—Smart Solutions for Agriculture" (AgroSMART 2018)*; Atlantis Press: Paris, France, 2018; pp. 370–377.

186. Behera, S.; Indumathi, K.; Mahadevamma, S.; Sudha, M.L. Oil cakes—A by-product of agriculture industry as a fortificant in bakery products. *Int. J. Food Sci. Nutr.* **2013**, *64*, 806–814. [CrossRef]

187. Jáquez, D.R.; Casillas, F.; Flores, N.; Cooke, P.; Licon, E.D.; Soto, S.; González, I.A.; Carreón, F.O.C.; Roldán, H.M. Effect of Glandless Cottonseed Meal Content on the Microstructure of Extruded Corn-Based Snacks. *Adv. Food Sci.* **2014**, *36*, 125–130.

188. Bulut, S.; Lazić, V.; Mađarev-Popović, S.; Hromiš, N.; Šuput, D. Mono- and bilayer biopolymer films: Synthesis and characterisation. *J. Process. Energy Agric.* **2017**, *21*, 214–218. [CrossRef]

189. Garrido, T.; Leceta, I.; Cabezudo, S.; Guerrero, P.; De La Caba, K. Tailoring soy protein film properties by selecting casting or compression as processing methods. *Eur. Polym. J.* **2016**, *85*, 499–507. [CrossRef]

Appl. Sci. **2020**, *10*, 7432

190. Garrido, T.; Etxabide, A.; Peñalba, M.; De La Caba, K.; Guerrero, P. Preparation and characterization of soy protein thin films: Processing–properties correlation. *Mater. Lett.* **2013**, *105*, 110–112. [CrossRef]

191. Puscaselu, R.G.; Gutt, G.; Amariei, S. Rethinking the Future of Food Packaging: Biobased Edible Films for Powdered Food and Drinks. *Molecules* **2019**, *24*, 3136. [CrossRef] [PubMed]

192. Popović, S.; Hromiš, N.; Šuput, D.; Bulut, S.; Romanić, R.; Lazić, V. Valorization of by-products from the production of pressed edible oils to produce biopolymer films. In *Cold Pressed Oils*, 1st ed.; Ramadan, F.M., Ed.; Elsevier Inc.: London, UK, 2020; pp. 15–30.

193. Šuput, D.; Mađarev-Popović, S.; Hromiš, N.; Bulut, S.; Lazić, V. Biopolymer films properties change affected by essential oils addition. *J. Process. Energy Agric.* **2019**, *23*, 61–65. [CrossRef]

194. Salgado, P.R.; Ortiz, S.E.M.; Petruccelli, S.; Mauri, A.N. Biodegradable sunflower protein films naturally activated with antioxidant compounds. *Food Hydrocoll.* **2010**, *24*, 525–533. [CrossRef]

195. Mađarev-Popović, S.; Lazic, V.; Popovic, L.; Vaštag, Ž.; Pericin, D. Effect of the addition of pumpkin oil cake to gelatin to produce biodegradable composite films. *Int. J. Food Sci. Technol.* **2010**, *45*, 1184–1190. [CrossRef]

196. Mađarev-Popović, S.; Pericin, D.; Vaštag, Ž.; Popovic, L.; Lazic, V. Evaluation of edible film-forming ability of pumpkin oil cake; effect of pH and temperature. *Food Hydrocoll.* **2011**, *25*, 470–476. [CrossRef]

197. Bulut, S.; Lazic, V.; Popovic, S.; Hromiš, N.; Šuput, D. Influence of storage period on properties of biopolymer packaging materials and pouches. *Acta Period. Technol.* **2017**, *48*, 53–62. [CrossRef]

198. Bulut, S.; Lazic, V.; Popovic, S.; Hromis, N.; Suput, D. Influence of different concentrations of glycerol and guar xanthan on properties of pumpkin oil cake-zein bi-layer film. *Ratar. Povrt.* **2017**, *54*, 19–24. [CrossRef]

199. Bulut, S.; Mađarev-Popović, S.; Hromiš, N.; Šuput, D.; Lazić, V.; Malbaša, R.; Vitas, J.S. Incorporation of essential oils into biopolymer films based on pumpkin oil cake in order to improve their antioxidant activity. *J. Process. Energy Agric.* **2019**, *23*, 162–166. [CrossRef]

200. Mađarev-Popović, S.; Peričin, D.; Vaštag, Ž.; Lazić, V.; Popović, L.M. Pumpkin oil cake protein isolate films as potential gas barrier coating. *J. Food Eng.* **2012**, *110*, 374–379. [CrossRef]

201. Reddy, N.; Jiang, Q.; Yang, Y. Preparation and properties of peanut protein films crosslinked with citric acid. *Ind. Crop. Prod.* **2012**, *39*, 26–30. [CrossRef]

202. Sun, Q.; Sun, C.; Xiong, L. Mechanical, barrier and morphological properties of pea starch and peanut protein isolate blend films. *Carbohydr. Polym.* **2013**, *98*, 630–637. [CrossRef]

203. Riveros, C.G.; Martin, M.P.; Aguirre, A.; Grosso, N.R. Film preparation with high protein defatted peanut flour: Characterisation and potential use as food packaging. *Int. J. Food Sci. Technol.* **2017**, *53*, 969–975. [CrossRef]

204. Wan, J.; Liu, C.; Liu, W.; Tu, Z.; Wu, W.; Tan, H. Optimization of instant edible films based on dietary fiber processed with dynamic high pressure microfluidization for barrier properties and water solubility. *LWT* **2015**, *60*, 603–608. [CrossRef]

205. Ghodrat, A.G.; Tabatabaei, M.; Aghbashlo, M.; Mussatto, S.I. Waste Management Strategies the State of the Art. In *Biogas Fundamentals, Process and Operation*; Tabatabaei, M., Ghanavati, H., Eds.; Springer International Publishing: Cham, Switzerland, 2018; Volume 6, pp. 1–33.

206. Cheng, F.; Bayat, H.; Jena, U.; Brewer, C.E. Impact of feedstock composition on pyrolysis of low-cost, protein- and lignin-rich biomass: A review. *J. Anal. Appl. Pyrolysis* **2020**, *147*, 104780. [CrossRef]

Publisher's Note: MDPI stays neutral with regard to jurisdictional claims in published maps and institutional affiliations.

Article

Production of Cellulosic Ethanol from Enzymatically Hydrolysed Wheat Straws

Vasile-Florin Ursachi and Gheorghe Gutt *

Faculty of Food Engineering, Stefan cel Mare University of Suceava, 720229 Suceava, Romania;
florin.ursachi@fia.usv.ro
* Correspondence: g.gutt@fia.usv.ro

Received: 18 September 2020; Accepted: 26 October 2020; Published: 29 October 2020

Abstract: The aim of this study is to find the optimal pretreatment conditions and hydrolysis in order to obtain a high yield of bioethanol from wheat straw. The pretreatments were performed with different concentrations of sulphuric acid 1, 2 and 3% (v/v), and were followed by an enzymatic hydrolysis that was performed by varying the solid-to-liquid ratio (1/20, 1/25 and 1/30 g/mL) and the enzyme dose (30/30 μL/g, 60/60 μL/g and 90/90 μL/g Viscozyme® L/Celluclast® 1.5 L). This mix of enzymes was used for the first time in the hydrolysis process of wheat straws which was previously pretreated with dilute sulfuric acid. Scanning electron microscopy indicated significant differences in the structural composition of the samples because of the pretreatment with H_2SO_4 at different concentrations, and ATR-FTIR analysis highlighted the changes in the chemical composition in the pretreated wheat straw as compared to the untreated one. HPLC-RID was used to identify and quantify the carbohydrates content resulted from enzymatic hydrolysis to evaluate the potential of using wheat straws as a raw material for production of cellulosic ethanol in Romania. The highest degradation of lignocellulosic material was obtained in the case of pretreatment with 3% H_2SO_4 (v/v), a solid-to-liquid ratio of 1/30 and an enzyme dose of 90/90 μL/g. Simultaneous saccharification and fermentation were performed using *Saccharomyces cerevisiae* yeast, and for monitoring the fermentation process a BlueSens equipment was used provided with ethanol, O_2 and CO_2 cap sensors mounted on the fermentation flasks. The highest concentration of bioethanol was obtained after 48 h of fermentation and it reached 1.20% (v/v).

Keywords: wheat straws; pretreatment; hydrolysis; fermentation; bioethanol

---

1. Introduction

In the past decade, due to climate changes there has been an increasing attention on reducing greenhouse gas (GHG) emissions. In accordance with the Paris Agreement (Council Decision (EU) 2016/1841 of the 5 October 2016) from 2023, every 5 years a comprehensive assessment of the progress of the parties will be made on the basis of scientific data and the situation regarding the reduction of emissions, the adjustments made and the support provided will be analysed. Compared to 1990, the mandatory target set for 2030 is at least a 40% domestic reduction economy-wide greenhouse gas (GHG) emissions [1].

Agricultural biomass is considered one of the most important renewable energy resources and contributes to the development of bioenergy generation. It consists of annual and perennial energy crops (green biomass for animals feed), residues from agricultural production (straw, corn stalks, corn cobs, sugar cane, etc.) and the food industry (residues from dairy industry, sugar industry, etc.). In recent years, there has been an increasing trend in obtaining bioethanol from renewable resources such as straws resulted from cereals harvesting. However, huge quantities of wheat straws (WS) are generated annually which could be used for the production of cellulosic bioethanol [2,3].

Liquid biofuels can partially or completely replace conventional fuels and can be an alternative source in the transport sector (aviation, shipping and heavy freight trucks). Therefore, liquid biofuels are an important solution because they do not require major changes in distribution infrastructure or the transport fleet. International Renewable Energy Agency (IRENA) argues that a reduction in carbon emissions by 2050 is only possible if there is a five-fold increase in biofuel consumption, from 130 billion litres in 2016 to almost 650 billion litres in 2050 [4].

Wheat (Triticum sp.) is the third most cultivated cereal in the world. Based on the data provided by Food and Agriculture Organization of the United Nations (FAOSTAT) in 2020, the worldwide cereal production is 2789.8 million tons. The worldwide cereal production in 2020 (2789.8 million tons) is higher by 3% compared to 2019 (2708.5 million tons) [5]. In 2019/2020 the estimated global wheat production is 762.2–764.1 million tons in comparison with the production of 2018/2019 which was 732.1 million tons (Figure 1) [6,7]. Therefore, if we take into account a coefficient of 1.3 [8,9], in 2019/2020 an amount of 990.86–993.33 million tons of wheat residues were produced.

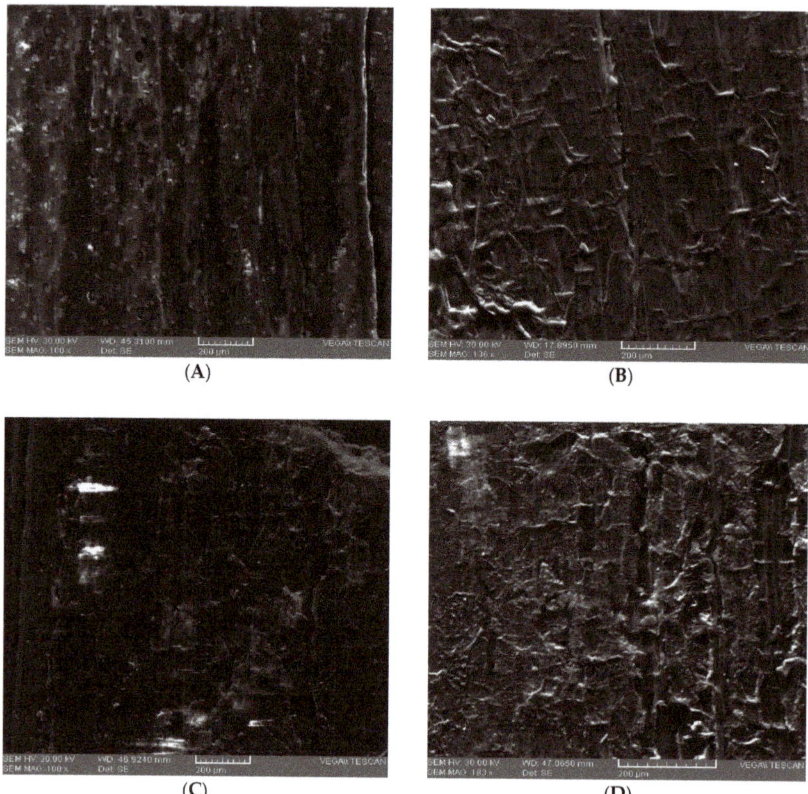

Figure 1. SEM images for untreated WS ((**A**)—E detector-exterior view; and for pretreated WS ((**B**) —SE detector-1% H_2SO_4 (*v/v*); (**C**)—SE detector-2% H_2SO_4 (*v/v*); (**D**)—SE detector–3%; H_2SO_4 (*v/v*)).

The wheat straw is consisted of internodes (57 ± 10%), nodes (10 ± 2%), leaves (18 ± 3%), straw (9 ± 4%) and the central axis (6 ± 2 %). The chemical composition of the wheat straw contains cellulose (34–40%), hemicellulose (20–25%) and lignin (20%) [10–13].

Cellulose is a linear polymer composed of β-*D*-glucopyranosyl units linked at 1,4 positions [14]. The utilisation of cellulose and hemicellulose sugars present in the hydrolysate of lignocellulosic biomass is essential for the production of bioethanol [15]. Hemicellulose is a complex polysaccharide, found in the cell walls of plants, composed of neutral monosaccharide units: *D*-xylose, *D*-manose, *D*-galactose, *D*-glucose, *L*-arabinose, *D*-glucuronic acid etc. Hemicellulose fractions consist of β-*D*-xylopyranose backbone residues linked at 1,4 positions to the units of *L*-arabinose, *D*-galactose and/or *D*-glucuronic acid [16].

The mechanical pretreatment of lignocellulosic materials is an important stage in the process of obtaining bioethanol which reduces the particle size and crystallinity of cellulose, increases the contact surface, dissociates tissues and disintegration of the cell wall [17–19]. All these parameters contribute significantly to the conversion of saccharides during hydrolysis. Talebnia et al. (2010) reported in their study that mechanical pretreatment (chopping and grinding) contributes significantly to the improvement of enzymatic hydrolysis of WS. It was established that, after 24 h of hydrolysis, the content of glucose and xylose was increased by 39% and 20% respectively when the particle size was reduced from 2–4 cm to 53–149 μm [8]. However, the very small particles is a disadvantage because it leads to a high energy consumption in the milling phase. On the other hand, the additional elimination of lignin contributes significantly to the decrease of energy consumption [20].

The acid pretreatment leads to the formation of inhibitory compounds at a low value of pH, high temperature and pressure. The degradation of lignin and hemicellulose produces free organic acids (acetic, formic and levulinic acid) or phenolic derivatives (4-hydroxynenzoic acid or vanillin), and 2-furaldehyde (furfural–FF) with 5-hydroxymethyl-2-furaldehyde (hydroxymethylfurfural–HMF) which affect negatively the chemical composition, enzymatic hydrolysis and fermentation. The FF and HMF are obtained from the transformation of pentoses (xylose and arabinose) and hexoses (glucose, mannose and galactose) [21]. Acid pretreatment is a significant phase for obtaining bioethanol which destroys the crystalline structures of cellulose and increases the accessibility of the enzyme during enzymatic hydrolysis [22,23]. Zheng et al. (2017) studied the impact of acid pretreatment on WS, using 2% and 4% of H_2SO_4. This pretreatment led to the elimination of high amount of hemicellulose [24]. The acid pretreatment is most suitable for WS, but also has some disadvantages, such as acid pollution of the environment and feed, and production of secondary compounds [25–27]. Tian et al. (2018) established that pretreatment with 2% diluted H_2SO_4 improved the rate of lignin removal from WS [27]. Mardetko et al. (2018) reported that pretreatment with 0.5% H_2SO_4 was more effective in obtaining a higher amount of glucose (4.84 g/L) for 10 min at 200 °C compared to 4.09 g/L glucose obtained by using 2% H_3PO_4 [28].

During pretreatment, HMF can be converted to formic and levulinic acid, while FF to formic acid [19]. The structure of hemicellulose is composed of acetylated sugars which can turn into acetic acid during pretreatment. Also, during lignin pretreatment, significant amount of phenolic and aromatic compounds was achieved [29]. These compounds limit the transformation of sugars into bioethanol which reduce the final yield of this alcohol.

Enzymatic hydrolysis is the most efficient method of releasing carbohydrates from lignocellulosic materials. Thus, the hydrolysis of cellulose is catalysed by a class of enzymes, called cellulases. This type of hydrolysis is influenced by the lignocellulosic substrate, enzyme activity and process conditions [8]. By hydrolysis, high concentrations of acetic and formic acids were detected, with values between 4.9–9.4 g/L and 1.6–10.3 g/L, respectively, and levulinic acid of 0.3–0.6 g/L. Acetic acid, furfural and HMF are the products that result from hydrolysis and has inhibitory effects on the fermentation process [30].

The aim of this paper is to obtain bioethanol as a result of the superior recovery of wheat straw. A number of objectives have also been set for monitoring the whole process, such as the pretreatment, enzymatic hydrolysis and fermentation steps. To find out which is the most indicated option for pretreatment of wheat straws and its effect on the hydrolysis step, a diluted concentration of 1, 2 and 3% (*v/v*) H_2SO_4 was used.

2. Materials and Methods

2.1. Materials

Wheat straws (WS) used in this study were collected on 21st of July 2020, from Mitocu Dragomirnei (Suceava county, Romania, 47°45′14.0″ N 26°13′44.0″ E).

The analytical reagents, standard materials and enzymes used in this study were purchased from Sigma-Aldrich. The utilised enzymes Viscozyme® L and Celluclast® 1.5 L (Novozyme Corp, Bagsvaerd, Denmark) were used for enzymatic hydrolysis. The yeast, fermentation activator and diammonium phosphate (DAP) were purchased from Enzymes & Derivates, Neamț. Milli-Q water (Direct-Q® 3 UV, Milipore SAS 67120, Molsheim, France) was used in the preparation of reagents, standards and samples.

Viscozyme® L contains a multienzymatic complex of carbohydrates such as arabanase, cellulase, β-glucanase, hemicellulase and xylanase. This is a clear liquid enzyme produced by *Aspergillus aculeatus* which has brown color and has a density of approx. 1.2 g/mL, enzymatic activity ≥ 100 FBG/g (β-glucanase fungal units) with activity under optimal conditions at pH between 3.3 and 5.5 and a temperature of 25–55 °C. The enzyme must be stored at a temperature of 2–8 °C.

Celluclast® 1.5 L is an enzyme (endoglucanase) that hydrolyses the (1,4)-β-D-glucoside bonds in cellulose and other β-glucans. This is a brown liquid enzyme and is produced by *Trichoderma reesei*, has a density of approx. 1.22 g/mL, enzymatic activity ≥ 700 EGU/g (β-glucanase fungal units) with activity in optimal conditions at pH between 4 and 6 and a temperature of 25–55 °C. At lower temperatures (5–10 °C) the shelf life is considerably increased.

Hydra PC is a yeast activator that helps strengthen its plasma membrane and gives it increased resistance in unfavourable environments. This product contains a significant amount of magnesium that contributes to cell division, increases the speed of yeast development.

DistillaMax SR is a special yeast of the species *Saccharomyces cerevisiae* that produces low levels of higher alcohols and has good resistance to osmotic pressure, organic acids and high temperatures. The recommended amount is between 10 and 50 g/hL, and the fermentation temperature is 30–35 °C.

2.2. Methods

2.2.1. Chemical Composition of WS

Total of 5 g of ground WS was accurately weighed on analytical balance (Partner Corporation, Bracka 28, Poland) and heated at 105 °C for 4 h to a constant mass in an oven [31]. Then, the ash content was determined by weighing 1 g of ground WS and calcining at 575 ± 25 °C for 3 h [32] in a furnace (Thermo Scientific Thermolyne, Kerper blvd Dubuque, Iowa, USA). The content of cellulose [31] and lignin [32] was determined using the methods described by Ishtiaq et al. (2010) and Sluiter et al. (2012) respectively.

2.2.2. Pretreatments of WS

A. Mechanical pretreatment. The WS were cut into small pieces and then heated at 40 °C for 24 h in an oven (Memmert, Schwabach, Germany). Then, WS were milled and sieved in a shaker (Sieve shaker Retsch, Haan, Germany). In this study, the particle size used of wheat straws grounded (WSG) was <1 mm.

B. Physico-chemical pretreatment. Total of 2 g of WSG were weighed with accuracy and precision of 0.001. The samples were added in borosilicate glass bottles with polypropylene cap and pouring ring and were boiled for 1 h at 100 °C in a water bath (Precisdig JP Selecta, Abrera, Barcelona, Spain) with different concentration of H_2SO_4 (1, 2 and 3% (*v/v*)) and solid/liquid (S/L) ratio (1/20, 1/25 and 1/30 *w/v*). Then, the samples were rapidly cooled using cold water.

2.2.3. Enzymatic Hydrolysis of Pretreated WS

The preatreated samples were filtered under vacuum using Whatman qualitative filter paper, grade 5 and washed with 100 mL of Milli-Q water to remove enzymatic inhibitory compounds. Then, the solid fraction was transferred into borosilicate glass bottles with polypropylene cap and pouring ring, and Milli-Q water was added over the solid fraction, maintaining the S/L ratio of 1/20, 1/25 and 1/30 respectively. The pH of obtained samples was corrected to 4.5 with 3 M NaOH using a pH meter (Mettler-Toledo, model SevenCompact S210). The enzyme activities were described by Ghose (1987) and used by Vintilă et al. (2019) [33–35]. Afterwards, 30/30 µL/g or 20 FPU/g, 60/60 µL/g or 40 FPU/g and 90/90 µL/g or 60 FPU/g (Viscozyme® L and Celluclast® 1.5 L) were dosed in the pretreated WS. The order of addition of the enzymes was Viscozyme® L and Celluclast® 1.5 L, respectively, which were allowed to act for 24 h at 52 °C.

2.2.4. Scanning Electron Microscope (SEM) Analysis

The acid pretreated and enzyme-hydrolysed samples were dried and analysed with SEM Tescan Vega II LMU (Tescan Orsay Holding, Brno, Czech Republic), operated at 30 kV.

2.2.5. ATR-FTIR Analysis

FT-IR analysis was performed to detect the modifications in the functional groups in dried raw material, pretreated with acid and enzyme-hydrolysed WS. The FT-IR spectra of samples were achieved with a FT-IR spectrometer (Thermo Scientific, Karlsruhe, Dieselstraße, Germany) with ATR IX option. The results were obtained within a range of 400–4000 cm^{-1} with a detector at 4 cm^{-1}.

2.2.6. HPLC Instrumentation and Separation Conditions

The individual phenolic compounds, organic acids and individual carbohydrates were analysed using a high-performance liquid chromatography (HPLC) (Shimadzu, Kyoto, Japan) equipped with a LC-20 AD liquid chromatograph, SIL-20A autosampler, CTO-20AC coupled with a SPD-M20A diode array detector (DAD) and RID-10A refractive index detector (RID) respectively. The separation of phenolic compounds, organic acids and individual carbohydrates was performed in a column specific to each constituent analysed. The standards of phenolic compounds, organic acids and individual carbohydrates were determined based on the retention times and quantified based on their calibration curves (all the curves had R^2 higher than 0.98). For samples, the limits of detection (LOD) and limits of quantification (LOQ) were calculated according to Kuppusamy et al. (2018) [36–38].

Determination of Individual Phenolic Compounds

For the analysis of the individual phenolic compounds resulting from the acid pretreatment and from the enzymatic hydrolysates, the following was performed: 1 mL of solution from each sample pretreated with acid and respectively 1 mL of solution from the enzyme-hydrolysed samples for 24 h were filtered through PTFE membrane with 0.45-µm dimension of pores and were stored at −20 °C until analysis. The obtained samples were analysed using HPLC-DAD. The separation of individual polyhenols was performed into a Phenomenex Kinetex® 2.6 µm Biphenyl 100 Å column, LC Column 150 × 4.6 mm and thermostated at 25 °C (Column oven). The utilised method of analysis was described by Palacios et al. (2011) and used by Pauliuc et al. (2020) with some modifications [39–42]. The identification of the 12 phenolic compounds from pretreated and enzyme-hydrolysed samples of WS was performed at 280 nm for gallic, protocatechuic, vanillic and p-hydroxybenzoic acid; at 320 nm for chlorogenic, caffeic, p-coumaric and rosmarinic acid, quercitin, luteolin and kaempferol. The 12 phenolic compounds were injected individually to identify their retention time, then a mix of them was made. The final concentration of each identified phenolic compound was expressed in mg/L.

Determination of Organic Acids

In order to identify the organic acids resulting from the acid pretreatment and from the enzymatic hydrolysates, the following was performed: 1 mL of solution pretreated with acid samples and 1 mL of solution enzyme-hydrolysed samples for 24 h were filtered through PTFE membrane with 0.45 μm dimension of pores and were stored at −20 °C until analysis. The obtained mixtures were analysed using HPLC-DAD equipment. The separation of organic acids was performed into a Phenomenex Kinetex® 5 μm C18 100 Å HPLC Column 250 × 4.6 mm. The utilised method of analysis was described by Özcelik et al. [42,43]. The standards of individual organic acid (gluconic, acetic, formic, succinic, propionic, lactic and butyric acid) and mixed solution of them were prepared to determine the concentration of organic acids, expressed in mg/L.

Determination of Individual Carbohydrates After Enzymatic Hydrolysis

After 24 h of enzymatic hydrolysis, 1 mL of each sample was filtered through PTFE 0.45 μm dimension of pores and analysed for determination of carbohydrates using HPLC-RID. The separation of carbohydrates was performed in a Phenomenex Luna® Omega 3 μm de SUGAR 100 Å column, 150 × 4.6 mm. The utilised method of analysis was described by Bogdanov et al. (2002) [44]. Before analysis, the inactivation of enzymes was achieved by exposing the samples for 5 min at 121 °C [45]. The identification of carbohydrates was performed based on the standards (Carbohydrates Kit (CAR10-KIT) de D-(−)-Arabinose, ≥98% (A3131-5G), D-(−)-Ribose, ≥99%, (R7500-5G), D-(+)-Xylose, ≥99% (X1500-5G), D-(−)-Fructose, ≥99% (F0127-5G), D-(+)-Glucose, ≥99.5%, (G8270-5G-KC), D-(+)-Galactose ≥99.5% (G0750-5G), α-Lactose monohydrate, ≥99% total lactose basis, (L3625-5G), Sucrose ≥99.5% (S9378-5G-KC), D-(+)-Mannose, wood, ≥99% (M2069-5G), D-(+)-Maltose monohydrate, from potato, ≥99%, (M5885-5G) which were injected to identify the retention times, then a mixed solution of them was prepared to determinate the concentration of carbohydrates, expressed in mg/L.

2.2.7. Monitoring of Bioethanol Concentration

About 20 ± 0.001 g of WSG with <1 mm particle size was pretreated with 3% H_2SO_4 (*v/v*) following the procedure in Section 2.2.1, then the resulting solid fraction was hydrolysed using enzymes as described in Section 2.2.2. After 24 h of hydrolysis, the mash was cooled to 35 °C (optimal condition for yeast *Saccharomyces cerevisiae*). The pH of the mash was adjusted to 4.5 using a 3 M NaOH solution. The 2.5 g of Hydra PC activator was dissolved in 50 mL Milli-Q water (42 °C, conc. 5%), then the obtained mix was cooled to 38 °C and pH was adjusted to 4.5 using a solution of 3 M NaOH. Afterward, the cooled mix was used for the activation of 40 mg of dry yeast *Saccharomyces cerevisiae*. After 20 min, mash and 50 mL of solution with activated yeast were transferred to the fermentation flask. The temperature of the mash was maintained at 35 °C throughout the fermentation process.

The fermentation process was conducted using BlueSens gas sensors GmbH, Germany which monitors the content of carbon dioxide (CO_2), oxygen (O_2) and ethanol (C_2H_5OH) of the mash from lignocellulosic materials. The CO_2 (H31953 series), O_2 (H32132 series) and C_2H_5OH (H32132) sensors were connected via the BACCom 12 multiplexer data which allows a connection of 12 sensors to the software. The processing and transmission data were performed in real time via BacVis software. The equipment used to monitor the fermentation process processes the information by means of three spectral sensors in the IR range, which are mounted on each hole of the fermentation vessel. BlueSens sensors allow continuous monitoring of the content of CO_2, oxygen and ethanol in fermenters [46]. The use of this equipment facilitates the conditions of controlled study by the simultaneous analysis of metabolic processes. The monitoring of CO_2 and O_2 concentrations is performed in the fermenter continuously and directly, where the fermentation processes take place. The parallel measurement of CO_2, O_2 and ethanol allows the analysis of metabolic processes without interruption during the entire fermentation process.

2.2.8. Experimental Design and Statistical Analysis

The experiment was conducted into three factor full factorial experiment for each type of carbohydrate (glucose, fructose and xylose) and total of carbohydrates. Each independent variable (concentration of sulphuric acid, ratio S/L and enzymes dosage) had 3 levels, as follows: concentration of sulphuric acid (1%, 2% and 3% respectively), ratio S/L (1/20 *w/v*, 1/25 *w/v* and 1/30 *w/v* respectively) and enzymes dosage (30/30 μL, 60/60 μL and 90/90 μL). The response of design were considered total of carbohydrates, glucose, fructose and xylose. The full factorial designed was made using Design-Expert 10.0 (Stat-Ease, Inc., Minneapolis, MN, USA).

The model used to predict the evolution of carbohydrates was a quadratic polynomial response surface model which can be applied to fit the experimental data obtained by Box-Behnken design. The quadratic polynomial response surface model which describes the relationship between the experimental data is [47,48]:

$$Y = a_0 + \sum_{i=1}^{n}\left(a_i X_i\right) + \sum_{i=1}^{n}\left(a_{ii} X_{ii}^2\right) + \sum_{i=1}^{n}\left(a_{ii} X_i X_j\right) \tag{1}$$

where: Y—predicted response, X_i stands for the coded levels of the design variable (concentration of sulphuric acid, ratio S/L and enzymes dosage—Table 1), a_0 is a constant, a_i—linear effects, a_{ii}—quadratic effects and a_{ij}—interaction effects.

Table 1. Levels in full factorial experiments for carbohydrates, glucose, xylose and fructose.

Factors	Level		
	−1	**0**	**1**
Sulphuric acid (%), X_1	1	2	3
Ratio S/L (*w/v*), X_2	1/20	1/25	1/30
Enzymes dosage (μL/g of sample), X_3	30/30	60/60	90/90

The results of acid organics and individual phenolic compounds were presented to analysis of variance (ANOVA) using Statgraphics Centurion XVIII software (Manugistics Corp., Rockville, MD, USA trial version).

3. Results and Discussions

3.1. Chemichal Composition of WS

Table 2 shows the chemical composition of WS used in the experiments compared to the composition of WS reported in different studies (Table 3).

Table 2. The chemical composition of wheat straws (WS) used in the experiments.

Cellulose (%)	Acid Insoluble Lignin (%)	Acid Soluble Lignin (%)	Dry Substance (%)	Humidity (%)	Ash (%)
37.53 ± 1.15	14.35 ± 0.53	1.42 ± 0.18	92.32 ± 0.23	7.68 ± 0.54	3.87 ± 0.07

Table 3. The chemical composition of WS reported in different studies.

Raw Material	Cellulose (%)	Lignin (%)	Ash (%)	Reference
	44.8 ± 0.7	8.46 ± 0.31	5.68	[49]
	44.2 ± 1.8	22.4 ± 1.7	2.8 ± 0.6	[50]
Wheat straws	39.8	22.6	4.2	[51]
	39	17	1.8	[52]
	38–40.8	8.9–10.5	1.4	[53]

3.2. Scanning Electron Microscope (SEM) Analysis

The scanning electron microscope (SEM) analysis was used to investigate the changes in the structure of WS samples after acid pretreatment. From the images (Figure 1) a significant difference of the structural composition due to pretreatment with H_2SO_4 at different concentrations can be observed. The highest degradation of lignocellulosic material was obtained in the case of pretreatment with 3% H_2SO_4 (v/v) compared to other concentrations of the same acid. Figure 1D shows that initial uniform and rigid structure of the WS was changed after pretreatment, obtaining a porous structure which can positively influence the enzymatic action. This modification has also been reported by Zheng et al. (2018) and Momayez et al. (2019) [24,54].

3.3. ATR-FTIR Analysis

The changes in the functional groups as a result of the pretreatment of the straw biomass were analysed by means of attenuated total reflectance-Fourier transform infrared spectroscopy (ATR-FTIR). Figure 2 shows the spectra recorded for the untreated sample, the liquid fraction of the sample following sulfuric acid pretreatment 1%, 2% and 3% (v/v) (Figure 2A) and for the solid fractions of the samples after enzymatic hydrolysis (B, C, D). As seen in Figure 2A, the chemical shifts in the liquid fraction of the sample resulted from the sulfuric acid treatment were similar irrespective of the volume of acid used and the S/L ratio and included a high absorbance peak around 3400 cm^{-1} corresponding to O–H stretching, a sharp peak at 1640 cm^{-1} assigned to the C–O bonds in the alkyl groups of lignin side chains [55], and some small peaks around 1200 and 1050 cm^{-1}, which were due to C–O stretching vibrations in cellulose and hemicellulose structure [56].

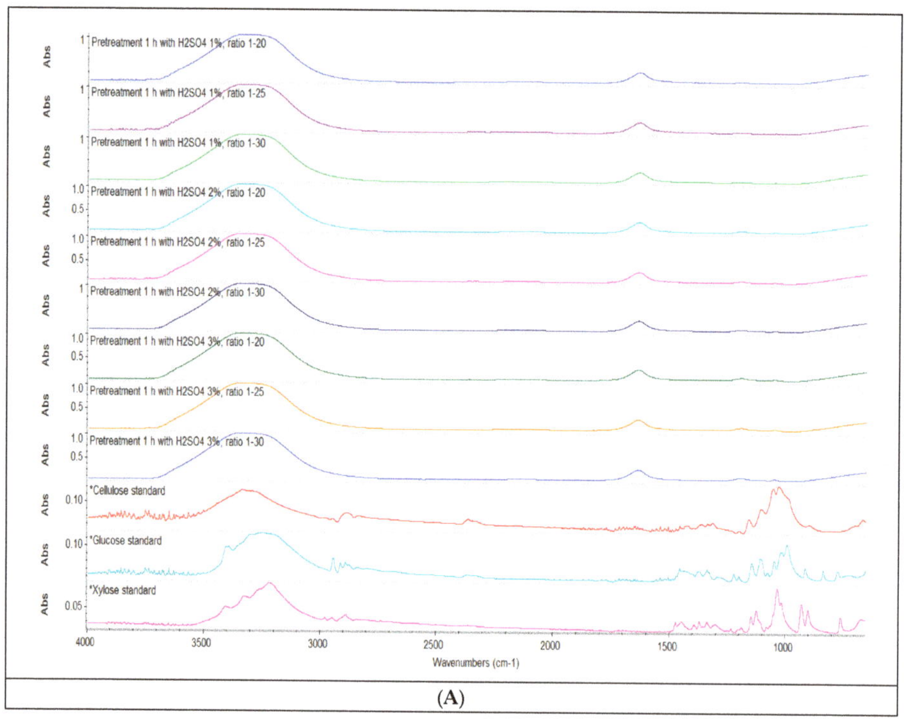

Figure 2. *Cont.*

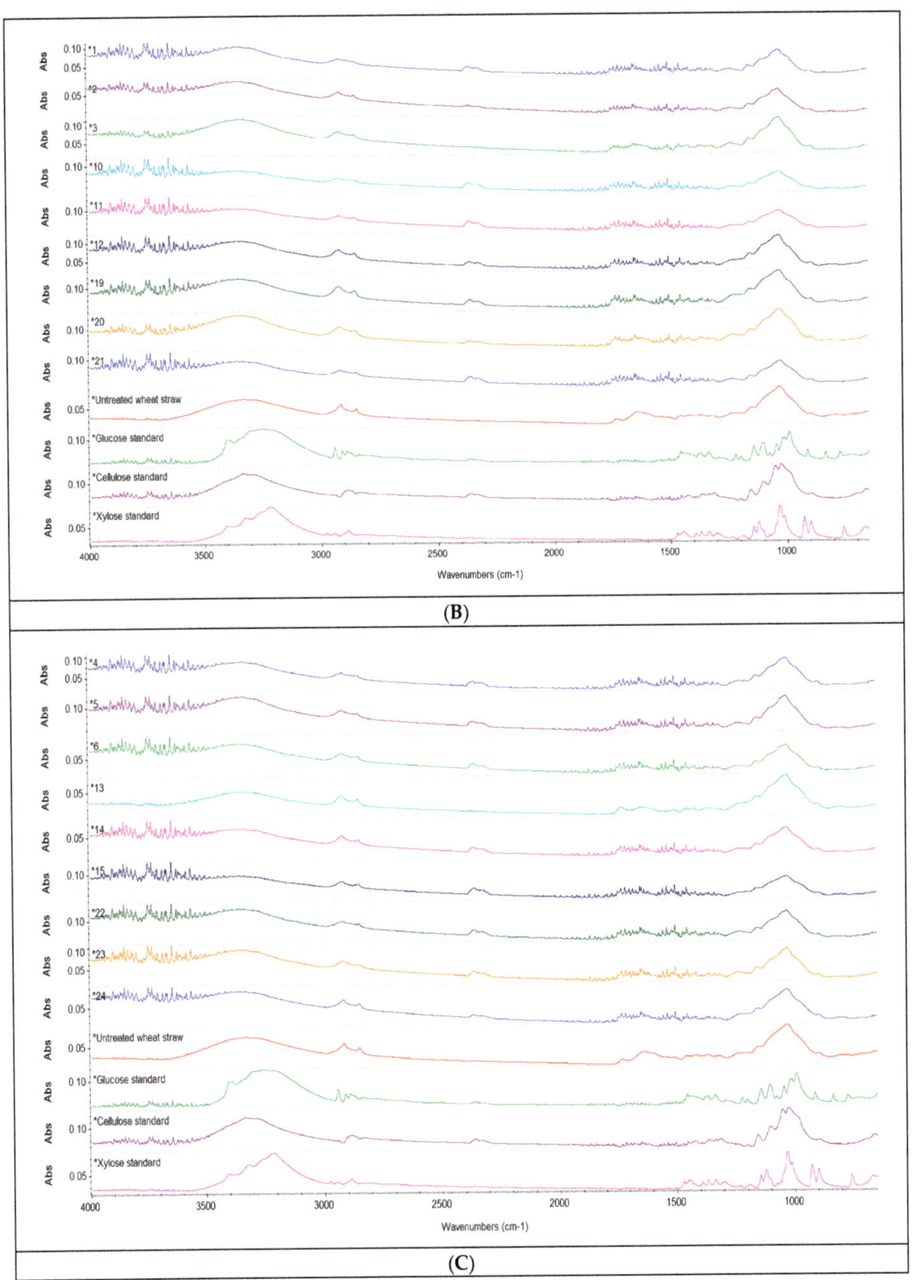

Figure 2. *Cont.*

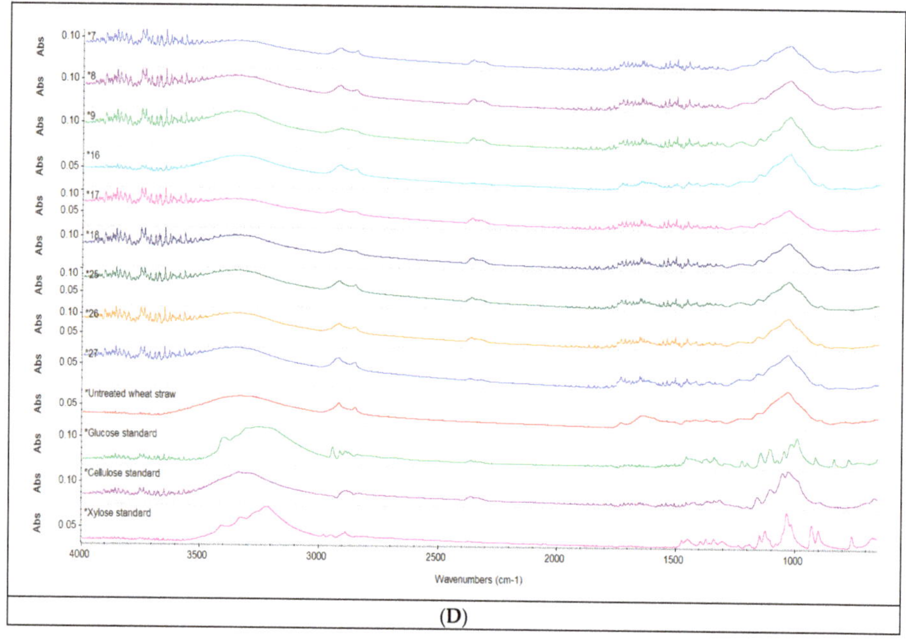

Figure 2. Spectroscopic analysis. ATR-FTIR spectra of the liquid fraction after acid pretreatment with 1% H_2SO_4 (*v/v*), 2% H_2SO_4 (*v/v*), 3% H_2SO_4 (*v/v*) (**A**) and of the solid fraction of enzymatically hydrolysed straw (**B–D**), compared to the untreated sample and standards of cellulose, xylose and glucose respectively. Depending on the concentration of H_2SO_4 used, the spectra were stacked to make their interpretation easier.

By comparison to the spectra of the untreated sample, the solid fraction of pretreated and enzymatically hydrolysed straw presented notable changes in the wide absorption band at 3400 cm^{-1} in function of the volume of sulfuric acid, the enzyme dose and the S/L ratio. A decrease of the intensity of the absorption band in this region indicated a reduction of the cellulose content, as previously reported by Zheng et al. (2018) [24]. When considering the conditions of the pretreatment and enzymatic hydrolysis applied, the reduction of cellulose content seems to be more pronounced when an enzyme dose of 90 µL was used. The peaks at 2910 cm^{-1}, corresponding to bending vibration of C–H hemicelluloses and cellulose [57], and the peaks around 2300 cm^{-1}, also denoting the presence of cellulose in the sample, showed similar changes of intensity determined by the pretreatment and enzymatic hydrolysis. The prominent sharp peak around 1060–1000 cm^{-1} was attributed to C=C, C–O and C–C–O groups stretching in lignin, cellulose and hemicellulose [58]; the highest decrease in the intensity of this peak was also determined by the use of an enzyme dose of 90 µL associated with a higher S/L ratio in the pretreatment and enzymatic hydrolysis.

3.4. Determination of Individual Phenolic Compounds

The WS contain a wide range of individual phenolic compounds (caffeic, p-coumaric, 4-hydroxybenzoic, protocatechuic, chlorogenic, vanillic acid etc.,). The phenols are inhibitory compounds for fermentative microorganisms and cellulases in the process of obtaining bioethanol [59]. The polyphenols must be removed before the fermentation to improve the yield of bioethanol in WS and decrease the amount of acid lactic, thus inhibiting the activity of lactic bacteria [60]. During pretreatment, phenolic compounds are removed from the lignocellulosic material because of the action of acid which leads to the breaking of bonds between polysaccharides and polyphenols.

Chen et al. (2018) studied the chemical compounds of rice straws after pretreatment with 2% H_2SO_4 (*v/v*) and obtained that p-coumaric acid has the highest concentration (0.40 mg/g) compared to vanillic (0.09 mg/g), chlorogenic (0.02 mg/g) and caffeic acid (0.04 mg/g) [61].

In the studied samples it can be observed that increasing the concentration of sulphuric acid results in an enhancement of the phenolic compounds. Thus, the highest value for the protocatechuic (34.72 ± 0.06 µg/g), vanillic (18.92 ± 0.02 µg/g), 4-hydroxybenzoic (18.74 ± 0.01 µg/g) were obtained at the 3% concentration of sulphuric acid and 1/30 S/L ratio, but p-coumaric (3.21 ± 0.01 µg/g), chlorogenic (3.36 ± 0.04 µg/g) and caffeic acid (3.96 ± 0.03 µg/g) were obtained at the 2% concentration of sulphuric acid and 1/30 S/L ratio. The values for individual phenolic compounds are shown in Table 4 (the chromatograms obtained for standards and one of the samples can be found in Supplementary Materials).

3.5. Determination of Organic Acids

The presence and amount of organic acids depend on the nature of the material and the conditions of the solvent pretreatment. The significant parameter that improves the formation of organic acids are the pH with high values and acid pretreatment which imply the risk of production of furaldehydes (FF and HMF) and aliphatic acids (formic and acetic acid) [62]. Lu et al. (2010) founded 3300 mg/L acetic acid, which are 10–15 times higher concentrations in comparison with other acids [63]. Also, Erdei et al. (2010) identified a concentration of 1.7 g/L acetic acid in liquid (prehydrolysate) fractions in steam-pretreated (temp. 190 °C for 10 min) WS slurry [64] compared to that of Linde et al.'s (2008) who obtained 0.04–1.01 g/L acetic acid (temp. 190–210 °C, residence time 2–10 min, sulfuric acid 0.2%) [65]. In the study conducted by Djioleu (2015), the highest concentration for acetic acid and formic acid was 11.04 g/L and 6.08 g/L, respectively [66]. Rajan et al. (2014) pretreated WS at 140 °C with 10 dm^3/m^3 H_2SO_4 concentration for 30 min and have obtained concentrations of formic acid and acetic acid of 32.37 ± 4.91, 7.98 ± 1.02 g/kg, respectively [67]. In another study Rajan et al. (2014) pretreated rice straws at 220 °C for 52 min, pH 7.0 and have obtained concentrations of formic acid and acetic acid of 6.32 ± 1.46 and 8.45 ± 0.59 g/L, respectively [68].

In this study, organic acids from WS samples pretreated with H_2SO_4 of different concentrations (1, 2 and 3% (*v/v*)) were analysed and observed that the highest content of acetic acid (0.94 ± 0.01 mg/g) was founded at the correlation between 3% H_2SO_4 (*v/v*) and 1/30 ratio, but the lowest values of acetic acid (0.40 ± 0.02 mg/g) was identified at the pretreatment with 1% H_2SO_4 (*v/v*) and 1/30 ratio. The content of organic acids are shown in Table 5 (the chromatograms obtained for standards and one of the samples can be found in Supplementary Materials).

The lowest value of gluconic acid was obtained at 1% (*v/v*) H_2SO_4, s/l ratio 1/20—13.77 ± 0.05 mg/g, but highest value at 3% (*v/v*) H_2SO_4, s/l ratio 1/30—22.59 ± 0.05 mg/g. Gluconic acid is present in various plants, fruits, wine and honey. The obtained concentration of gluconic acid can be explained by the oxidation of aldehyde group (C1) of D-glucose to a carboxyl group [69,70].

3.6. Determination of Individual Carbohydrates After Enzymatic Hydrolysis

Saha and Cotta (2010) analysed the content of neutral monosaccharides in WS after pretreatment with 0.75% H_2SO_4 (*v/v*) at 121 °C for 1 h and hydrolysis (using three commercial enzyme preparations Celluclast 1.5, Novozym 188 and ViscoStar 150 L) obtained the following concentration for glucose—282 mg/g, xylose—180 mg/g and total content of carbohydrates—504 mg/g [71]. Saha et al. (2005) were established from barley straw a concentration of 214 mg/g for glucose, 208 mg/g for xylose+galactose and 452 mg/g for total carbohydrates after pretreatment with 1% (*v/v*) H_2SO_4 at 121 °C for 1 h and hydrolysis using the same three commercial enzyme preparations [15]. Analysing different varieties of rice straws, Park et al. (2011) showed that these contain free fructose with values between 7.4 ± 0.1–22.9 ± 0.2 g/Kg [72]. Also, Park et al. (2009) obtained 0.62–2.32% fructose (per dry weight of rice straw (*w/w*)) [73].

Table 4. Content of phenolic compounds in pretreated WS with different acid concentrations and variations of the S/L ratio (µg/g sample).

Phenolic Compounds	Pretreatment H_2SO_4 1%, s/l ratio 1/20	Pretreatment H_2SO_4 1%, s/l ratio 1/25	Pretreatment H_2SO_4 1%, s/l ratio 1/30	Pretreatment H_2SO_4 2%, s/l ratio 1/20	Pretreatment H_2SO_4 2%, s/l ratio 1/25	Pretreatment H_2SO_4 2%, s/l ratio 1/30	Pretreatment H_2SO_4 3%, s/l ratio 1/20	Pretreatment H_2SO_4 3%, s/l ratio 1/25	Pretreatment H_2SO_4 3%, s/l ratio 1/30	F Value
Protocatecuic acid	24.73 ± 0.07 (0.35) [h]	25.94 ± 0.03 (0.37) [g]	28.94 ± 0.04 (0.41) [e]	27.045 ± 0.05 (0.39) [f]	30.00 ± 0.08 (0.43) [d]	32.59 ± 0.09 (0.47) [b]	28.23 ± 0.05 (0.40) [e]	31.53 ± 0.02 (0.45) [c]	34.72 ± 0.06 (0.49) [a]	118.71 ***
4-hidroxibenzoic acid	14.51 ± 0.04 (0.21) [f]	15.11 ± 0.02 (0.22) [e]	16.91 ± 0.03 (0.24) [c]	14.9 ± 0.06 (0.21) [e,f]	15.93 ± 0.01 (0.23) [d]	16.52 ± 0.04 (0.23) [c]	15.86 ± 0.02 (0.23) [d]	18.1 ± 0.01 (0.26) [b]	18.74 ± 0.01 (0.27) [a]	75.77 ***
Vanillic acid	9.38 ± 0.05 (0.13) [e]	9.79 ± 0.05 (0.14) [e]	9.81 ± 0.04 (0.14) [e]	11.93 ± 0.05 (0.17) [d]	12.36 ± 0.04 (0.18) [c]	14 ± 0.03 (0.20) [b]	12.71 ± 0.02 (0.18) [c]	18.53 ± 0.02 (0.26) [a]	18.92 ± 0.02 (0.27) [a]	690.71 ***
Cafeic Acid	3.12 ± 0.02 (0.04) [d]	3.37 ± 0.05 (0.05) [c]	3.86 ± 0.04 (0.06) [a]	3.22 ± 0.04 (0.04) [d]	3.50 ± 0.03 (0.05) [b]	3.96 ± 0.03 (0.06) [a]	2.84 ± 0.02 (0.04) [e]	3.35 ± 0.01 (0.05) [c]	2.9 ± 0.03 (0.04) [e]	127.72 ***
Clorogenic acid	2.74 ± 0.01 (0.04) [e]	3.01 ± 0.03 (0.04) [c]	3.24 ± 0.02 (0.05) [b]	2.8 ± 0.02 (0.04) [d,e]	3.06 ± 0.02 (0.04) [c]	3.36 ± 0.04 (0.05) [a]	2.86 ± 0.02 (0.04) [d]	3.1 ± 0.03 (0.04) [c]	3.21 ± 0.01 (0.04) [b]	46.46 ***
p-cumaric acid	2.74 ± 0.01 (0.04) [g]	3.01 ± 0.01 (0.04) [d,e]	3.24 ± 0.02 (0.04) [c]	2.8 ± 0.03 (0.04) [f]	3.06 ± 0.02 (0.04) [d]	3.36 ± 0.04 (0.04) [b]	2.86 ± 0.03 (0.04) [e,f]	3.1 ± 0.01 (0.04) [a,b]	3.21 ± 0.01 (0.04) [a]	105.81 ***

ns–not significant ($p > 0.05$), * $p < 0.05$, ** $p < 0.01$, *** $p < 0.001$, a–h–different letters in the same row indicate significant differences between samples ($p < 0.001$).

Table 5. Content of organic acids in pretreated straws with different acid concentrations and variations of the S/L ratio (mg/g sample).

Organic Acids	Pretreatment H_2SO_4 1%, s/l ratio 1/20	Pretreatment H_2SO_4 1%, s/l ratio 1/25	Pretreatment H2SO4 1%, s/l ratio 1/30	Pretreatment H_2SO_4 2%, s/l ratio 1/20	Pretreatment H_2SO_4 2%, s/l ratio 1/25	Pretreatment H_2SO_4 2%, s/l ratio 1/30	Pretreatment H_2SO_4 3%, s/l ratio 1/20	Pretreatment H_2SO_4 3%, s/l ratio 1/25	Pretreatment H_2SO_4 3%, s/l ratio 1/30	F Value
Gluconic acid	13.77 ± 0.05 (196.84) [g]	15.80 ± 0.07 (190.13) [e]	18.49 ± 0.05 (178.45) [c]	14.80 ± 0.1 (240.11) [f]	16.35 ± 0.8 (233.64) [d]	20.23 ± 0.06 (203.34) [b]	15.71 ± 0.02 (127.66) [e]	20.50 ± 0.06 (292.95) [b]	22.59 ± 0.05 (279.97) [a]	366.72 ***
Formic acid	1.03 ± 0.02 (14.71) [h]	1.27 ± 0.03 (18.18) [g]	1.49 ± 0.02 (21.36) [e]	1.37 ± 0.05 (19.59) [f]	1.67 ± 0.03 (23.94) [d]	1.93 ± 0.05 (27.60) [c]	1.95 ± 0.06 (27.90) [c]	2.38 ± 0.05 (34.08) [b]	2.69 ± 0.02 (38.43) [a]	851.02 ***
Acetic acid	0.41 ± 0.03 (5.94) [g]	0.41 ± 0.01 (5.88) [g]	0.40 ± 0.02 (5.79) [g]	0.48 ± 0.07 (14.52) [f]	0.71 ± 0.03 (4.80) [d]	0.81 ± 0.04 (11.65) [c]	0.65 ± 0.02 (18.38) [e]	0.87 ± 0.07 (19.71) [b]	0.94 ± 0.01 (53.57) [a]	205.44 ***

ns–not significant ($p > 0.05$), * $p < 0.05$, ** $p < 0.01$, *** $p < 0.001$, a–h–different letters in the same row indicate significant differences between samples ($p < 0.001$). The F-value represents that the analysis of variance (ANOVA) is statistically significant. The obtained values of this parameter for content of phenolic compounds and content of organic acids represent the significant values of the test.

In this study, the Box Behnken design based on the response surface methodology with the representation of the experimental values of the independent variables was used to identify the optimal condition in order to obtain the highest carbohydrates content.

Figure 3 shows that the highest glucose content was obtained at the correlation of the following parameters—3% H_2SO_4 (v/v), 1/30 ratio and 90/90 µL/g of added enzyme content, and the lowest—1% H_2SO_4 (v/v), 1/20 ratio and 30/30 µL/g of added enzyme content. This tendency is also manifested for the other analysed parameters (fructose content–Figure 4, xylose–Figure 5 and the total carbohydrates content—Figure 6).

The equation for glucose content (mg/g sample) is presented in the Equation (2):

$$\begin{aligned}
Glucose = {} & 59.5633 + 1.7338 \cdot X_1 + 24.3427 \cdot X_2 + 2.2761 \cdot X_3 - 0.1350 \cdot X_1 \cdot X_2 \\
& + 0.1666 \cdot X_1 \cdot X_3 + 0.2800 \cdot X_2 \cdot X_3 + 0.2050 \cdot X_1^2 \\
& - 2.1916 \cdot X_2^2 - 0.7816 \cdot X_3^2
\end{aligned} \tag{2}$$

where: Y_1—the value of the glucose content parameter, mg/g sample; X_1, X_2, X_3—coded values for sulfuric acid content (%), S/L ratio (w/v) and added enzyme content (µL/g sample).

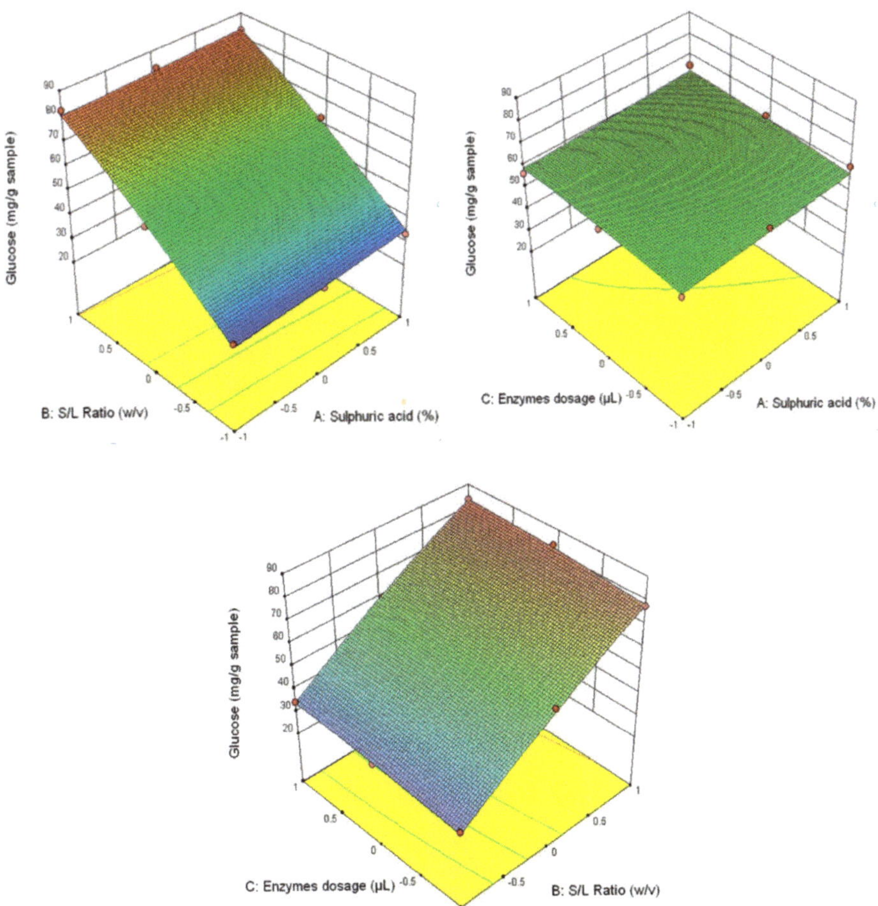

Figure 3. 3D diagrams of glucose content.

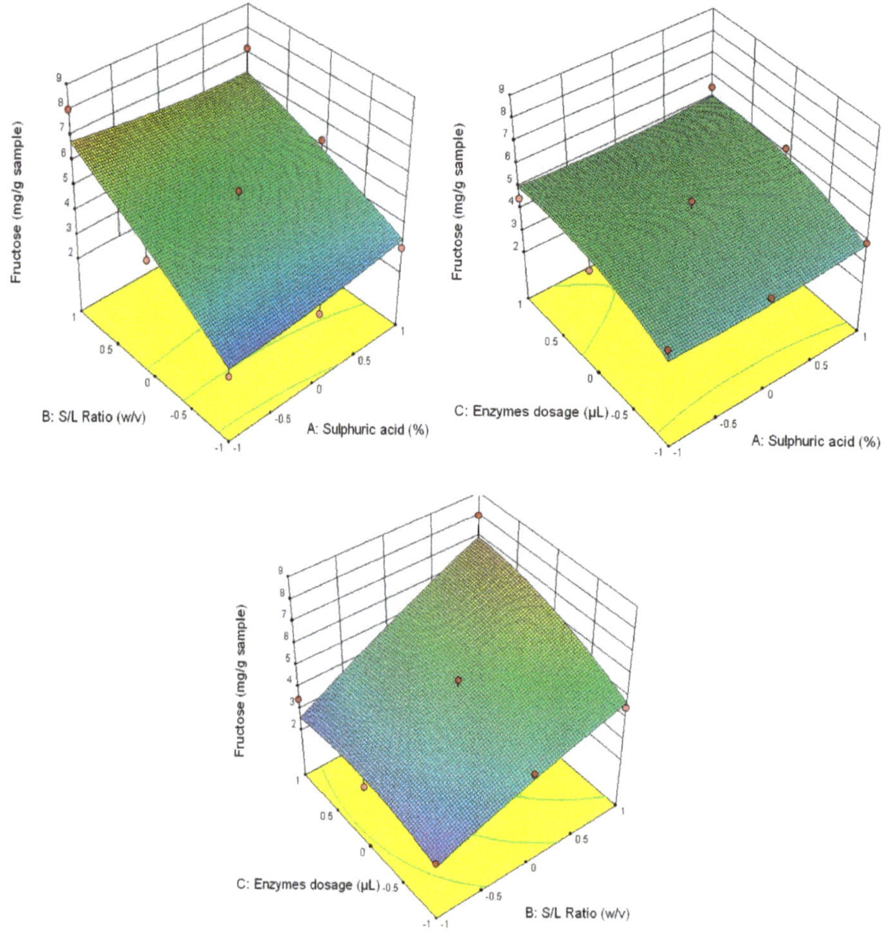

Figure 4. 3D diagrams of fructose content.

The equation for fructose content (mg/g sample) is presented in the Equation (3):

$$Fructose = 4.7733 + 0.0277{\cdot}X_1 + 1.7327{\cdot}X_2 + 0.6527{\cdot}X_3 - 0.3141{\cdot}X_1{\cdot}X_2 + \\ 8.3{\cdot}10^4{\cdot}X_1{\cdot}X_3 + 0.4766{\cdot}X_2{\cdot}X_3 + 0.1533{\cdot}X_1^2 - 0.1516{\cdot}X_2^2 - 0.4916{\cdot}X_3^2 \tag{3}$$

where: Y_1—the value of the fructose content parameter, mg/g sample; X_1, X_2, X_3—coded values for sulfuric acid content (%), S/L ratio (w/v) and added enzyme content (μL/g sample).

The equation for xylose content (mg/g sample) is presented in the Equation (4):

$$Xylose = 6.3244 + 2.0000{\cdot}X_1 + 2.2583{\cdot}X_2 + 0.2638{\cdot}X_3 + 0.3458{\cdot}X_1{\cdot}X_2 \\ + 0.3175{\cdot}X_1{\cdot}X_3 + 0.1358{\cdot}X_2{\cdot}X_3 + 1.3333{\cdot}X_1^2 \\ + 0.8450{\cdot}X_2^2 - 0.1116{\cdot}X_3^2 \tag{4}$$

where: Y_1—the value of the xylose content parameter, mg/g sample; X_1, X_2, X_3—coded values for sulfuric acid content (%), S/L ratio (w/v) and added enzyme content (μL/g sample).

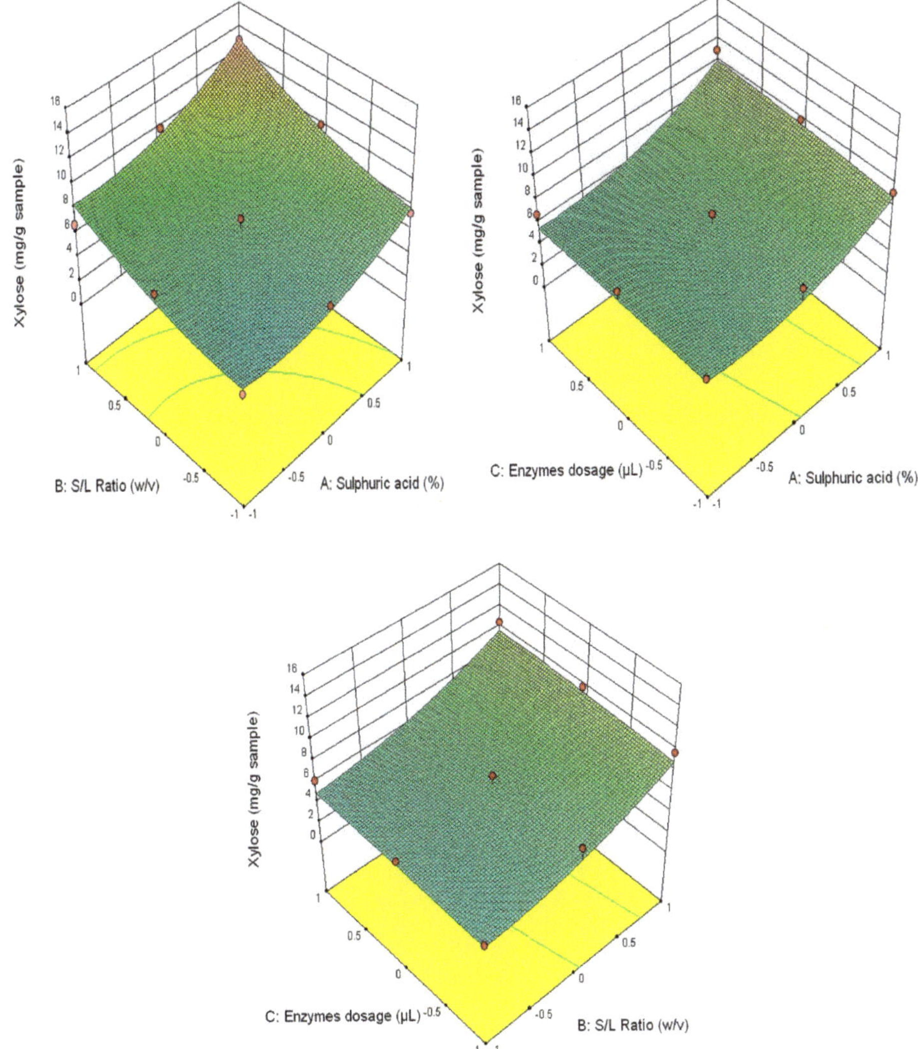

Figure 5. 3D diagrams of xylose content.

The equation for carbohydrates content (mg/g sample) is presented in the Equation (5)

$$\begin{aligned}
Total\ of\ carbohydrates &= 70.6592 + 3.7616{\cdot}X_1 + 28.3344{\cdot}X_2 + 3.1911{\cdot}X_3 \\
&\quad -0.1016{\cdot}X_1{\cdot}X_2 + 0.4850{\cdot}X_1{\cdot}X_3 + 0.8916{\cdot}X_2{\cdot}X_3 + 1.6905{\cdot}X_1^2 \\
&\quad -1.4977{\cdot}X_2^2 - 1.3844{\cdot}X_3^2
\end{aligned} \tag{5}$$

where: Y_1—the value of the carbohydrates content parameter, mg/g sample; X_1, X_2, X_3—coded values for sulfuric acid content (%), S/L ratio (*w/v*) and added enzyme content (µL/g sample).

The statistical parameters of each model are represented in Table 6. All the models proposed are significant ($p < 0.0001$). The coefficents of regression obtained for above quadratic equation indicated that the variation of glucose, fructose, xylose and total carbohydrates content can be explained by the correlation between independent variables.

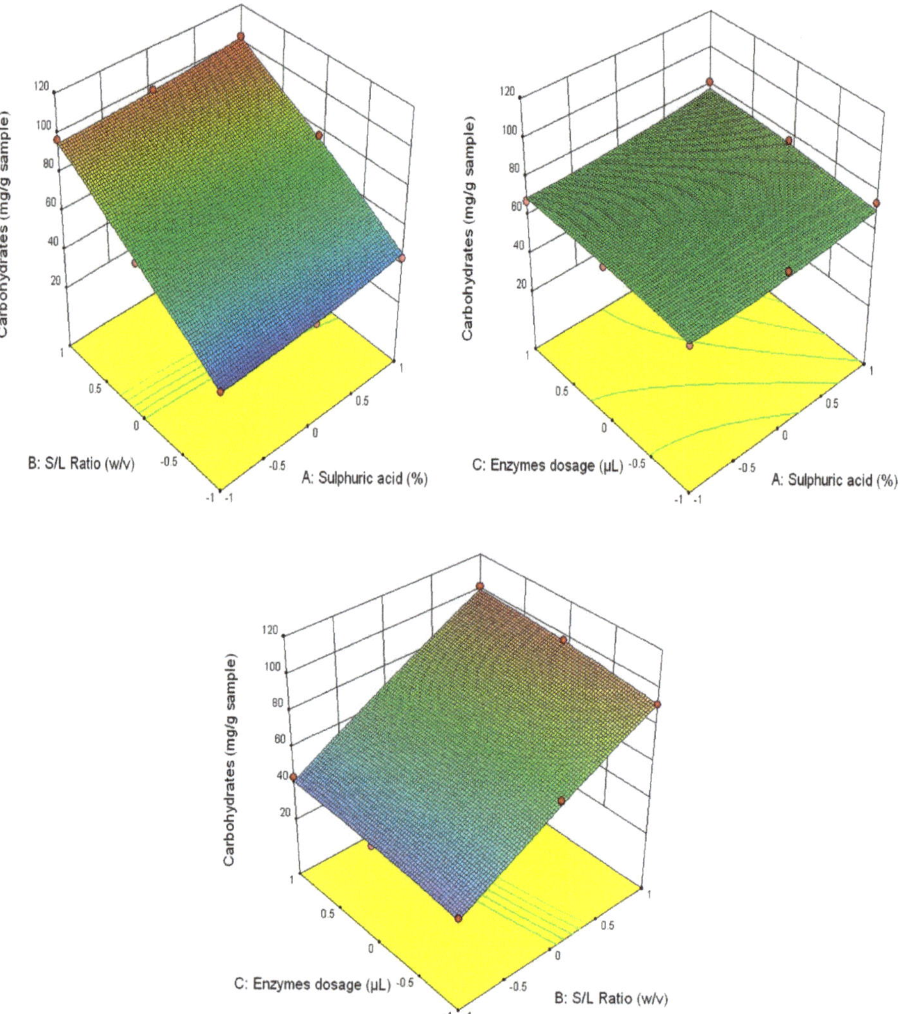

Figure 6. 3D diagrams of total carbohydrates content.

Table 6. Sum of square, mean square, *F*-value and R^2.

Parameter	Sum of Square	Mean Square	*F*-Value	R^2
Carbohydrates	14,943.73	1660.41	193.62	0.9903
Glucose	10,847.87	1205.32	392.16	0.9952
Xylose	182.95	20.33	7.20	0.7922
Fructose	67.37	7.49	13.70	0.8788

$p < 0.0001$.

3.7. Monitoring of Bioethanol Concentration

Following the enzymatic hydrolysis, organic acids and phenolic compounds from the total of 7 organic acids and 12 individual phenolic compounds studied were not detected, this is explained by the efficient washing of hydrolysed straw. Bellido et al. (2011) reported that regarding the individual effect of acetic acid, it was observed that the inhibition increased with the concentration of acetic acid and the media containing 3.5 g/L of acetic acid completely inhibited both the yeasts growth and production of ethanol. The theoretical yield of ethanol, defined as the percentage of the total amount of ethanol that could be produced from all available carbohydrates, decreased when acetic acid was present in mash in high concentration [74].

Fermentation Process

Vintilă et al. (2010) using a similar equipment of BlueSens sensors reported that after pretreatment with 2% NaOH of the raw material and applying the SSF process, after 48 h a maximum concentration of 1.5% ethanol was obtained in the case of corn ostriches, while for wheat straw the highest concentration of bioethanol was 1.33%. The enzymes used in the study were Trichoderma cellulases Onozuka (15 units/g cellulose), Aspergillus cellulases (15 units/g cellulose), Aspergillus cellobiase Novo (90 units/g cellulose) and the yeast strain was *Saccharomyces cerevisiae CMIT2*. [75]. Saha and Cotta (2006) obtained 15.1 mg/g ethanol following the fermentation process after enzymatic hydrolysis and alkaline pretreatment combined with H_2O_2 [76].

Patel et al. (2019) used a combined treatment (microwave-assisted and 0.5% NaOH), 5 FPU/g commercial cellulase (SIGMA) + 5 FPU/g of in-house enzyme (*Aspergillus niger ADH 11*) and followed by SSF with *Sacchromyces cerevisiae 3570*. After 34.42 h a maximum ethanol concentration and productivity of 32.44 g/L (0.95 g/L/h) was obtained from WS with a yield of 0.30 g ethanol/g reducing sugar consumed [77].

Novy et al. (2015) pretreated the wheat straw by steam explosion at 200 °C, 15 bar for 10 min, with a water to wheat straw ratio of 1, added 30 FPU/g (T. reesei SVG17) and followed by SHCF at 30 °C, pH 4.5 with S. cerevisiae IBB10B05. The ethanol obtained was 71.2 g/kg WS [78].

Singhania et al. (2014) applied a pretreatment with dilute acid (2.5% (w/w) H_2SO_4) on WS, 20 FPU/g Sacchari-SEB-C6 (advanced enzyme) + 20 FPU/g of in-house cellulase (*Penicillium janthinellum*) and followed by SSF with *K. marxianus* MTCC 4136 at 40 °C. After 48 h of fermentation resulted 12 g/L ethanol [79].

Xu et.al. (2011) pretreated with the ratio of WS to liquid at 80 g/kg, the NaOH concentration of 10 kg/m^3, the microwave power of 1000 W for 15 min, prehydrolysis at 50 °C for 24 h with Cellubrix® L, pH = 4.8, followed by SSF with *Sacchromyces cerevisiae* at 32 °C. After 120 h, the ethanol yield was 148.93 g/kg WS compared to the untreated material which was only 26.78 g/kg [80].

In this study, the activity of yeast *Saccharomyces cerevisiae* (DistillaMax SR) was not inhibited by the presence of acetic acid, as it was not identified after enzymatic hydrolysis in the analysed samples.

After inoculation of the slurry with the yeast *Saccharomyces cerevisiae* (DistillaMax SR) it can be seen based on the graph generated by the BACVis software that the fermentation process has started. The maximum fermentation was recorded approximately 12–13 h after the inoculation of the yeast, where the highest consumption of O_2 (from 22% vol. initially to approx. 15.1% vol.) and carbohydrates and the releasing of CO_2 (approx. 26.3% vol.), and the bioethanol content was approx. 0.853% vol. After 12–13 h from the maximum of the fermentation process, it can be observed that toward the end of the fermentation, the concentration of ethyl alcohol increased by approximately 0.4% vol. The maximum concentration of bioethanol obtained in gaseous phase from the fermentation process under the conditions presented was 1.20% (*v*/*v*) (Figure 7), which means the yield of bioethanol was 47.61 ± 2.3 g/Kg WS.

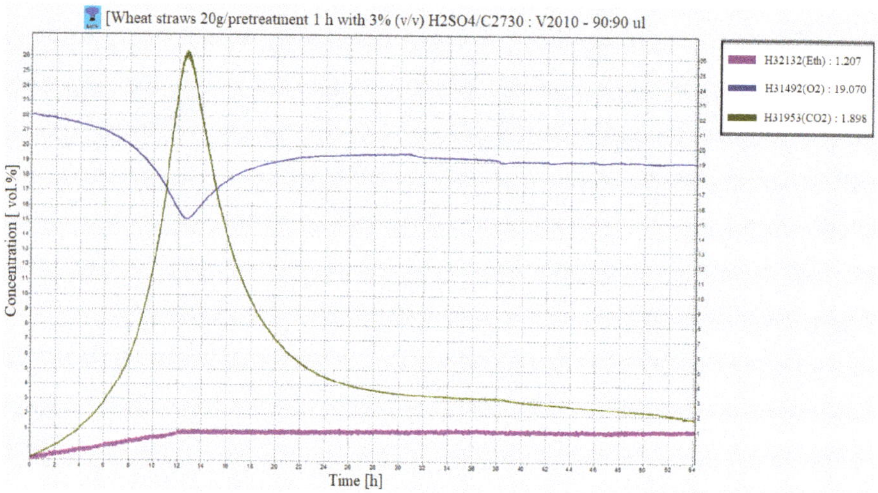

Figure 7. Evolution of CO_2, O_2 and ethanol concentrations during the saccharification and fermentation (SSF) process of WS mash.

4. Conclusions

Our results highlighted a highest degradation of lignocellulosic material in the case of pretreatment with 3% (*v/v*) H_2SO_4 compared to other concentrations of the same acid. This pretreatment led to an increasing of phenolics and organic acids content, especially of protocatechuic (34.72 ± 0.06 µg/g), vanillic (18.92 ± 0.02 µg/g) and 4-hydroxybenzoic (18.74 ± 0.01 µg/g), while the value of acid gluconic was 22.59 ± 0.05 mg/g, formic acid was 2.69 ± 0.02 mg/g and acetic acid was 0.94 ± 0.01 mg/g. The SEM microstructure of the pretreated samples revealed the changes which occurred on the surface of the straws undergoing pretreatment. Significant results were obtained in the case of treatment with 3% (*v/v*) H_2SO_4. ATR-FTIR analysis showed reduction in the intensity of the peaks characteristic to cellulose, hemicellulose and lignin at increased enzyme doses and S/L ratios, thus confirming the efficiency of the pretreatment and the enzymatic hydrolysis. This study showed that the pretreatment with 3% (*v/v*) H_2SO_4, in a S/L ratio of 1/30 (*w/v*) and with a dose of enzyme of 90/90 µL/g wheat straw led to the highest concentrations of carbohydrates (xylose 14.31 ± 0.11 mg/g, fructose 5.94 ± 0.13 mg/g, glucose 84.75 ± 0.23 mg/g and total carbohydrates 105.01 mg/g) and as a result the highest ethanol yield after the fermentation process.

The yield of ethanol (47.61 ± 2.3 g/Kg WS) could be improved if the concentration of H_2SO_4 (*v/v*) is increased in the pretreatment step, which would probably lead to a more significant disruption of the complex structure of the WS, hence facilitating the action of enzyme mixtures.

Supplementary Materials: The following are available online at http://www.mdpi.com/2076-3417/10/21/7638/s1, Figure S1. HPLC—DAD chromatogram at 280 nm (A) and 320 nm (B) for standard (100 mg/l) for gallic acid—peak 1 (8.69 min), protocatechuic acid—peak 2 (15.953 min), 4-hydroxybenzoic acid—peak 3 (20.886 min), vanillic acid—peak 5 (25.631 min), caffeic acid—peak 4 (23.277 min), chlorogenic acid—peak 6 (25.831 min), p-coumaric acid—peak 7 (32.011 min), rosmarinic acid—peak 8 (39.69 min), myricetin—peak 9 (43.216 min), luteolin—peak 10 (49.737 min), quercetin—peak 11 (50.128 min) and kaempferol—peak 12 (56.52 min) and (C—D) content of individual phenolic compounds in acid-pretreated wheat straw (liquid fraction)., Figure S2. HPLC-RID chromatogram for standard carbohydrates mix (CAR10-KIT) (A) and hydrolysed WS, D – (–) – Ribose—peak 1 (3.738 min), D – (+) – Xylose—peak 2 (4.235 min), D – (–) – Arabinose—peak 3 (4.771 min), D – (–) – Fructose—peak 4 (5.298 min), D – (+) – Mannose—peak 5 (5.921 min), D – (+) – Glucose—peak 6 (6.273 min), D – (+) – Galactose—peak 7 (6.77 min), Sucrose—peak 8 (9.886 min), D – (+) – Maltose—peak 9 (12.155 min), α-Lactose monohydrate—peak 10 (14.302 min) and (B) the carbohydrates content of enzymatically hydrolysed WS, Table S1. Box-Behnken design with coded values.

Author Contributions: Conceptualization, V.-F.U.; formal analysis, V.-F.U.; funding acquisition, G.G.; investigation, V.-F.U.; methodology, V.-F.U. and G.G.; project administration, G.G.; software, V.-F.U.; supervision, G.G.; validation, G.G.; writing—original draft, V.-F.U.; writing—review and editing, V.-F.U. and G.G. All authors have read and agreed to the published version of the manuscript.

Funding: This work was supported by contract no. 18PFE/16.10.2018 funded by Ministry of Research and Innovation within Program 1: Development of national research and development system, Subprogram 1.2: Institutional Performance—RDI excellence funding projects.

Conflicts of Interest: The authors declare no conflict of interest.

References

1. International Agreements Council Decision (EU) 2016/1841 of 5 October 2016. Available online: https://eur-lex.europa.eu/eli/dec/2016/1841/oj (accessed on 29 October 2020).

2. Perea-Moreno, M.A.; Samerón-Manzano, E.; Perea-Moreno, A.J. Biomass as renewable energy: Worldwide research trends. *Sustainability* **2019**, *11*, 863. [CrossRef]

3. Marks-Bielska, R.; Bielski, S.; Novikova, A.; Romaneckas, K. Straw stocks as a source of renewable energy. A case study of a district in Poland. *Sustainability* **2019**, *11*, 4714. [CrossRef]

4. IRENA—International Renewable Energy Agency. *Advanced Biofuels. What Holds Them Back?* International Renewable Energy Agency: Abu Dhabi, UAE, 2019; ISBN 978-92-9260-158-4. Available online: https://www.irena.org/-/media/Files/IRENA/Agency/Publication/2019/Nov/IRENA_Advanced-biofuels_2019.pdf (accessed on 29 October 2020).

5. FAOSTAT—Food and Agriculture Organization of the United Nations. *Crop Prospects and Food Situation—Quarterly Global Report No. 2*; FAOSTAT: Rome, Italy, July 2020.

6. European Commision. *Cereals Market Situation*; Committee for the Common Organisation of Agricultural Markets: Brussels, Belgium, 27 August 2020.

7. FAOSTAT—Food and Agriculture Organization of the United Nations. *2020 Food Outlook—Biannual Report on Global Food Markets*; FAOSTAT: Rome, Italy, June 2020; Food Outlook, 1.

8. Talebnia, F.; Karakashev, D.; Angelidaki, I. Production of bioethanol from wheat straw: An overview on pretreatment, hydrolysis and fermentation. *Bioresour. Technol.* **2010**, *101*, 4744–4753. [CrossRef] [PubMed]

9. Tishler, Y.; Samach, A.; Rogachev, I.; Elbaum, R.; Levy, A.A. Analysis of Wheat Straw Biodiversity for Use as a Feedstock for Biofuel Production. *BioEnergy Res.* **2015**, *8*, 1831–1839. [CrossRef]

10. Harper, S.H.T.; Lynch, J.M. The chemical components and decomposition of wheat straw leaves, internodes and nodes. *J. Sci. Food Agric.* **1981**, *32*, 1057–1062. [CrossRef]

11. Motte, J.-C.; Escudié, R.; Beaufils, N.; Steyer, J.-P.; Bernet, N.; Delgenès, J.-P.; Dumas, C. Morphological structures of wheat straw strongly impacts its anaerobic digestion. *Ind. Crops Prod.* **2014**, *52*, 695–701. [CrossRef]

12. Ghaffar, S.H. Aggregated understanding of characteristics of wheat straw node and internode with their interfacial bonding mechanisms. In *Design and Physical Sciences*; Department of Mechanical, Aerospace and Civil Engineering College of Engineering: Manchester, UK, 2016.

13. Chougan, M.; Ghaffara, S.H.; Al-Kheetan, M.J.; Gecevicius, M. Wheat straw pre-treatments using eco-friendly strategies for enhancing the tensile properties of bio-based polylactic acid composites. *Ind. Crops Prod.* **2020**, *155*, 112836. [CrossRef]

14. Alemdar, A.; Sain, M. Isolation and characterization of nanofibers from agricultural residues–Wheat straw and soy hulls. *Bioresour. Technol.* **2008**, *99*, 1664–1671. [CrossRef]

15. Saha, B.C.; Iten, L.B.; Cotta, M.A.; Wu, Y.V. Dilute acid pretreatment, enzymatic saccharification and fermentation of wheat straw to ethanol. *Process Biochem.* **2005**, *40*, 3693–3700. [CrossRef]

16. Lawther, J.M.; Sun, R.; Banks, W.B. Extraction, fractionation, and characterization of structural polysaccharides from wheat straw. *J. Agric. Food Chem.* **1995**, *43*, 667–675. [CrossRef]

17. Amin, F.R.; Khalid, H.; Zhang, H.; Rahman, S.; Zhang, R.; Liu, G.; Chen, C. Pretreatment methods of lignocellulosic biomass for anaerobic digestion. *AMB Express* **2017**, *7*, 72. [CrossRef] [PubMed]

18. Barakat, A.; Mayer, C.; Solhy, A.; Arancon, R.A.D.; De Vries, H.; Luque, R. Mechanical pretreatments of lignocellulosic biomass: Towards facile and environmentally sound technologies for biofuels production. *RSC Adv.* **2014**, *4*, 48109–48127. [CrossRef]

19. Galbe, M.; Zacchi, G. Pretreatment of lignocellulosic materials for efficient bioethanol production. In *Biofuels*; Olsson, L., Ed.; Springer: Berlin/Heidelberg, Germany, 2007; Volume 108, pp. 41–65.

20. Ibrahim, H.A.H. Pretreatment of straw for bioethanol production. *Energy Proc.* **2012**, *14*, 542–551. [CrossRef]

21. Jönsson, L.J.; Alriksson, B.; Nilvebrant, N.-O. Bioconversion of lignocellulose: Inhibitors and detoxification. *Biotechnol. Biofuels* **2013**, *6*, 1–6. [CrossRef]

22. Albu, E.; Apostol, L.-C. Study regarding the optimal conditions for reducing sugars production from wheat straws by enzymatic hydrolysis. *Environ. Eng. Manag. J.* **2014**, *16*, 2347–2352. [CrossRef]

23. Sasmal, S.; Mohanty, K. Pretreatment of Lignocellulosic Biomass Toward Biofuel Production. In *Biorefining of Biomass to Biofuels*; Springer: Berlin/Heidelberg, Germany, 2017; pp. 203–221.

24. Zheng, Q.; Zhou, T.; Wang, Y.; Cao, X.; Wu, S.; Zhao, M.; Guan, X. Pretreatment of wheat straw leads to structural changes and improved enzymatic hydrolysis. *Sci. Rep.* **2018**, *8*, 1–9. [CrossRef] [PubMed]

25. Chen, W.H.; Pen, B.L.; Yu, C.T.; Hwang, W.S. Pretreatment efficiency and structural characterization of rice straw by an integrated process of dilute-acid and steam explosion for bioethanol production. *Bioresour. Technol.* **2011**, *102*, 2916–2924. [CrossRef]

26. Zhang, J.; Shao, S.; Bao, J. Long term storage of dilute acid pretreated corn stover feedstock and ethanol fermentability evaluation. *Bioresour. Technol.* **2016**, *201*, 355–359. [CrossRef]

27. Tian, S.Q.; Zhao, R.Y.; Chen, Z.C. Review of the pretreatment and bioconversion of lignocellulosic biomass from wheat straw materials. *Renew. Sustain. Energy Rev.* **2018**, *91*, 483–489. [CrossRef]

28. Marđetko, N.; Novak, M.; Trontel, A.; Grubišić, M.; Galić, M.; Šantek, B. Bioethanol production from dilute-acid pre-treated wheat straw liquor hydrolysate by genetically engineered Saccharomyces cerevisiae. *Chem. Biochem. Eng. Q.* **2018**, *32*, 483–499. [CrossRef]

29. Taherzadeh, M.J.; Karimi, K. Acid-based hydrolysis processes for ethanol from lignocellulosic materials: A review. *BioResources* **2007**, *2*, 472–499.

30. Guo, B.; Zhang, Y.; Yu, G.; Lee, W.H.; Jin, Y.S.; Morgenroth, E. Two-stage acidic–alkaline hydrothermal pretreatment of lignocellulose for the high recovery of cellulose and hemicellulose sugars. *Appl. Biochem. Biotechnol.* **2013**, *169*, 1069–1087. [CrossRef] [PubMed]

31. Ishtiaq, A.; Muhammad, A.Z.; Hafiz, M.; Nasir, I. Bioprocessing of Proximally Analyzed Wheat Straw for Enhanced Cellulase Production through Process Optimization with Trichoderma viride under SSF. *Int. J. Biol. Life Sci.* **2010**, *6*, 164–170.

32. Sluiter, A.; Hames, B.; Ruiz, R.; Scarlata, C.; Sluiter, J.; Templeton, D.; Crocker, D. *Determination of Structural Carbohydrates and Lignin in Biomass*; Laboratory Analytical Procedure (LAP); NREL—National Laboratory of the U.S. Department of Energy, Office of Energy Efficiency & Renewable Energy: Golden, CO, USA, 2012; Issue Date: April 2008, Revision Date: August 2012 (Version 8 March 2012); Available online: https://www.nrel.gov/docs/gen/fy13/42618.pdf.

33. Ghose, T.K. Measurement of cellulase activities. *Pure Appl. Chem.* **1987**, *59*, 257–268. [CrossRef]

34. Vintilă, T.; Ionel, I.; Tiegam, R.F.T.; Wächter, A.R.; Julean, C.; Gabche, A.S. Residual Biomass from Food Processing Industry in Cameroon as Feedstock for Second-generation Biofuels. *BioResources* **2019**, *14*, 3731–3745.

35. Gama, R.; Van Dyk, J.S.; Pletschke, B.I. Optimisation of enzymatic hydrolysis of apple pomace for production of biofuel and biorefinery chemicals using commercial enzymes. *3 Biotech* **2015**, *5*, 1075–1087. [CrossRef]

36. Kuppusamy, P.; Lee, K.D.; Song, C.E.; Ilavenil, S.; Srigopalram, S.; Arasu, M.V.; Choi, K.C. Quantification of major phenolic and flavonoid markers in forage crop Lolium multiflorum using HPLC-DAD. *Rev. Bras. Farmacogn.* **2018**, *28*, 282–288. [CrossRef]

37. Wenzl, T.; Haedrich, J.; Schaechtele, A.; Piotr, R.; Stroka, J.; Eppe, G.; Scholl, G. *Guidance Document on the Estimation of LOD and LOQ for Measurements in the Field of Contaminants in Food and Feed*; Institute for Reference Materials and Measurements (IRMM): Geel, Belgium, 2016.

38. Yaman, N.; Durakli Velioglu, S. Use of Attenuated Total Reflectance—Fourier Transform Infrared (ATR-FTIR) Spectroscopy in Combination with Multivariate Methods for the Rapid Determination of the Adulteration of Grape, Carob and Mulberry Pekmez. *Foods* **2019**, *8*, 231. [CrossRef] [PubMed]

39. Palacios, I.; Lozano, M.; Moro, C.; D'Arrigo, M.; Rostagno, M.A.; Martinez, J.; García-Lafuente, A.; Guillamón, E.; Villares, A. Antioxidant properties of phenolic compounds occurring in edible mushrooms. *Food Chem.* **2011**, *128*, 674–678. [CrossRef]

40. Oroian, M.; Dranca, F.; Ursachi, F. Comparative evaluation of maceration, microwave and ultrasonic-assisted extraction of phenolic compounds from propolis. *J. Food Sci. Technol.* **2019**, *57*, 70–78. [CrossRef]

41. Oroian, M.; Ursachi, F.; Dranca, F. Influence of ultrasonic amplitude, temperature, time and solvent concentration on bioactive compounds extraction from propolis. *Ultrason. Sonochem.* **2020**, *64*, 105021. [CrossRef] [PubMed]

42. Pauliuc, D.; Oroian, M. Organic acids and physico-chemical parameters of romanian sunflower honey. *Food Environ. Safety* **2020**, *19*, 148–155.

43. Özcelik, S.; Kuley, E.; Özogul, F. Formation of lactic, acetic, succinic, propionic, formic and butyric acid by lactic acid bacteria. *LWT* **2016**, *73*, 536–542. [CrossRef]

44. Bogdanov, S.; Martin, P. Honey authenticity. *Mitteilungen Leb. Hyg.* **2002**, *93*, 232–254.

45. Dranca, F.; Oroian, M. Optimisation of Pectin Enzymatic Extraction from Malus domestica 'Fălticeni' Apple Pomace with Celluclast 1.5L. *Molecules* **2019**, *24*, 2158. [CrossRef]

46. Gâtlan, A.M.; Gutt, G.; Naghiu, A. Capitalization of sea buckthorn waste by fermentation: Optimization of industrial process of obtaining a novel refreshing drink. *J. Food Process. Preservat.* **2020**, *44*, e14565. [CrossRef]

47. Dranca, F.; Ursachi, F.; Oroian, M. Bee Bread: Physicochemical Characterization and Phenolic Content Extraction Optimization. *Foods* **2020**, *9*, 1358. [CrossRef]

48. Palamakula, A.; Nutan, M.T.; Khan, M.A. Response surface methodology for optimization and characterization of limonene-based coenzyme Q10 self-nanoemulsified capsule dosage form. *AAPS PharmSciTech* **2004**, *5*, 114–121. [CrossRef]

49. Liu, Q.; He, W.Q.; Aguedo, M.; Xia, X.; Bai, W.B.; Dong, Y.Y.; Goffin, D. Microwave-assisted alkali hydrolysis for cellulose isolation from wheat straw: Influence of reaction conditions and non-thermal effects of microwave. *Carbohydr. Polym.* **2020**, *253*, 117170. [CrossRef]

50. Chen, H.Z.; Liu, Z.H. Multilevel composition fractionation process for high-value utilization of wheat straw cellulose. *Biotechnol. Biofuels* **2014**, *7*, 137. [CrossRef]

51. Kristensen, J.B.; Thygesen, L.G.; Felby, C.; Jørgensen, H.; Elder, T. Cell-wall structural changes in wheat straw pretreated for bioethanol production. *Biotechnol. Biofuels* **2008**, *1*, 1–9. [CrossRef] [PubMed]

52. Sun, X.F.; Xu, F.; Sun, R.C.; Fowler, P.; Baird, M.S. Characteristics of degraded cellulose obtained from steam-exploded wheat straw. *Carbohydr. Res.* **2005**, *340*, 97–106. [CrossRef] [PubMed]

53. Bjerre, A.B.; Olesen, A.B.; Fernqvist, T.; Plöger, A.; Schmidt, A.S. Pretreatment of wheat straw using combined wet oxidation and alkaline hydrolysis resulting in convertible cellulose and hemicellulose. *Biotechnol. Bioeng.* **1996**, *49*, 568–577. [CrossRef]

54. Momayez, F.; Karimi, K.; Horváth, I.S. Sustainable and efficient sugar production from wheat straw by pretreatment with biogas digestate. *RSC Adv.* **2019**, *9*, 27692–27701. [CrossRef]

55. Liu, T.; Zhou, X.; Li, Z.; Wang, X.; Sun, J. Effects of liquid digestate pretreatment on biogas production for anaerobic digestion of wheat straw. *Bioresour. Technol.* **2019**, *280*, 345–351. [CrossRef] [PubMed]

56. Horikawa, Y.; Hirano, S.; Mihashi, A.; Kobayashi, Y.; Zhai, S.; Sugiyama, J. Prediction of lignin contents from infrared spectroscopy: Chemical digestion and lignin/biomass ratios of *Cryptomeria japonica*. *Appl. Biochem. Biotechnol.* **2019**, *188*, 1066–1076. [CrossRef]

57. Yang, Y.; Shen, H.; Qiu, J. Bio-inspired self-bonding nanofibrillated cellulose composite: A response surface methodology for optimization of processing variables in binderless biomass materials produced from wheat-straw-lignocelluloses. *Ind. Crops Prod.* **2020**, *149*, 112335. [CrossRef]

58. Singh, B.; Kumar, A. Process development for sodium carbonate pretreatment and enzymatic saccharification of rice straw for bioethanol production. *Biomass Bioenergy* **2020**, *138*, 105574.

59. Oliva-Taravilla, A.; Tomás-Pejó, E.; Demuez, M.; González-Fernández, C.; Ballesteros, M. Phenols and lignin: Key players in reducing enzymatic hydrolysis yields of steam-pretreated biomass in presence of laccase. *J. Biotechnol.* **2016**, *218*, 94–101. [CrossRef]

60. Xue, Y.; Wang, X.; Chen, X.; Hu, J.; Gao, M.T.; Li, J. Effects of different cellulases on the release of phenolic acids from rice straw during saccharification. *Bioresour. Technol.* **2017**, *234*, 208–216. [CrossRef]

61. Chen, X.; Wang, X.; Xue, Y.; Zhang, T.A.; Hu, J.; Tsang, Y.F.; Gao, M.T. Tapping the bioactivity potential of residual stream from its pretreatments may be a green strategy for low-cost bioconversion of rice straw. *Appl. Biochem. Biotechnol.* **2018**, *186*, 507–524. [CrossRef] [PubMed]

62. Toquero, C.; Bolado, S. Effect of four pretreatments on enzymatic hydrolysis and ethanol fermentation of wheat straw. Influence of inhibitors and washing. *Bioresour. Technol.* **2014**, *157*, 68–76. [CrossRef]

63. Lu, Y.; Wang, Y.; Xu, G.; Chu, J.; Zhuang, Y.; Zhang, S. Influence of high solid concentration on enzymatic hydrolysis and fermentation of steam-explodedcorn sto ver biomass. *Appl. Biochem. Biotechnol.* **2010**, *160*, 360–369. [CrossRef] [PubMed]

64. Erdei, B.; Barta, Z.; Sipos, B.; Réczey, K.; Galbe, M.; Zacchi, G. Ethanol production from mixtures of wheat straw and wheat meal. *Biotechnol. Biofuels* **2010**, *3*, 16. [CrossRef] [PubMed]

65. Linde, M.; Jakobsson, E.L.; Galbe, M.; Zacchi, G. Steam pretreatment of dilute H_2SO_4-impregnated wheat straw and SSF with low yeast and enzyme loadings for bioethanol production. *Biomass Bioenergy* **2008**, *32*, 326–332. [CrossRef]

66. Djioleu, A. *Compounds Released from Biomass Deconstruction: Understanding Their Effect on Cellulose Enzyme Hydrolysis and Their Biological Activity*; University of Arkansas: Fayetteville, AR, USA, 2015.

67. Rajan, K.; Carrier, D.J. Effect of dilute acid pretreatment conditions and washing on the production of inhibitors and on recovery of sugars during wheat straw enzymatic hydrolysis. *Biomass Bioenergy* **2014**, *62*, 222–227. [CrossRef]

68. Rajan, K.; Carrier, D.J. Characterization of rice straw prehydrolyzates and their effect on the hydrolysis of model substrates using a commercial endo-Cellulase, β-Glucosidase and cellulase cocktail. *ACS Sustain. Chem. Eng.* **2014**, *2*, 2124–2130. [CrossRef]

69. Karaffa, L.; Kubicek, C.P. *Production of Organic Acids by Fungi*; Springer: Berlin/Heidelberg, Germany, 2020.

70. Rodrigues, A.G. Secondary metabolism and antimicrobial metabolites of Aspergillus. In *New and Future Developments in Microbial Biotechnology and Bioengineering*; Elsevier: Amsterdam, The Netherlands, 2016; pp. 81–93.

71. Saha, B.C.; Cotta, M.A. Comparison of pretreatment strategies for enzymatic saccharification and fermentation of barley straw to ethanol. *N. Biotechnol.* **2010**, *27*, 10–16. [CrossRef]

72. Park, J.Y.; Kanda, E.; Fukushima, A.; Motobayashi, K.; Nagata, K.; Kondo, M.; Tokuyasu, K. Contents of various sources of glucose and fructose in rice straw, a potential feedstock for ethanol production in Japan. *Biomass Bioenergy* **2011**, *35*, 3733–3735. [CrossRef]

73. Park, J.Y.; Seyama, T.; Shiroma, R.; Ike, M.; Srichuwong, S.; Nagata, K.; Tokuyasu, K. Efficient recovery of glucose and fructose via enzymatic saccharification of rice straw with soft carbohydrates. *Biosci. Biotechnol. Biochem.* **2009**, *73*, 1072–1077. [CrossRef]

74. Bellido, C.; Bolado, S.; Coca, M.; Lucas, S.; González-Benito, G.; García-Cubero, M.T. Effect of inhibitors formed during wheat straw pretreatment on ethanol fermentation by *Pichia stipitis*. *Biores. Technol.* **2011**, *102*, 10868–10874. [CrossRef]

75. Vintilă, T.; Vintila, D.; Neo, S.; Tulcan, C.; Hadaruga, N. Simultaneous hydrolysis and fermentation of lignocellulose versus separated hydrolysis and fermentation for ethanol production. *Rom. Biotechnol. Lett.* **2011**, *16*, 106–112.

76. Saha, B.C.; Cotta, M.A. Ethanol production from alkaline peroxide pretreated enzymatically saccharified wheat straw. *Biotechnol. Prog.* **2006**, *22*, 449–453. [CrossRef] [PubMed]

77. Patel, A.; Patel, H.; Divecha, J.; Shah, A.R. Enhanced production of ethanol from enzymatic hydrolysate of microwave-treated wheat straw by statistical optimization and mass balance analysis of bioconversion process. *Biofuels* **2019**, 1–8. [CrossRef]

78. Novy, V.; Longus, K.; Nidetzky, B. From wheat straw to bioethanol: Integrative analysis of a separate hydrolysis and co-fermentation process with implemented enzyme production. *Biotechnol. Biofuels* **2015**, *8*, 46. [CrossRef] [PubMed]

79. Singhania, R.R.; Saini, J.K.; Saini, R.; Adsul, M.; Mathur, A.; Gupta, R.; Tuli, D.K. Bioethanol production from wheat straw via enzymatic route employing *Penicillium janthinellum cellulases*. *Bioresour. Technol.* **2014**, *169*, 490–495. [CrossRef]

80. Xu, J.; Chen, H.; Kádár, Z.; Thomsen, A.B.; Schmidt, J.E.; Peng, H. Optimization of microwave pretreatment on wheat straw for ethanol production. *Biomass Bioenergy* **2011**, *35*, 3859–3864. [CrossRef]

Publisher's Note: MDPI stays neutral with regard to jurisdictional claims in published maps and institutional affiliations.